biosocial genetics

You cannot teach a man anything; you can only help him find it within himself.—*Galileo*

Department of Natural Sciences, University of North Florida

uman heredity and social issues

acmillan publishing co., inc.

ew York

ollier macmillan publishers

ndon

Printed in the United States of America

Macmillan Publishing Co., Inc.
866 Third Avenue, New York, New York 10022

Collier Macmillan Canada, Ltd.

Library of Congress Cataloging in Publication Data

Stine, Gerald James.
  Biosocial genetics.

  Bibliography: p.
  Includes index.
  1. Human genetics—Social aspects. I. Title.
[DNLM: 1. Genetics, Human. QH431 S859b]
QH438.7.S73 573.2'1 76-1954
ISBN 0-02-416490-9

Printing: 1 2 3 4 5 6 7 8 Year: 7 8 9 0 1 2 3

7

I dedicate this book to my daughter Sherri and son Garrett. May you inherit the capacity to love as I so deeply love you.

Dad

This text assumes no scientific training in genetics and can be used in a class that is oriented to review the application of genetics in society and in a beginning course of human genetics.

The text was written for anyone who is interested in the historical and current events that influence developments in biology, genetics, and medicine and for those concerned about what is happening at the frontiers of the science of genetics and about the social, ethical, and legal implications of genetic research, for society.

The Author

Perhaps you have heard people say that "man" at a given point in time is the result of his history, meaning that one is the sum total of his experiences: love and laughter, depression and shock, joys and pleasure, sadness and remorse, trial and error, success and failure, achievement and disappointment—life and, finally, death. These many faces or moods of man are a part of one's image or makeup and as real as the clothes one wears. But what controls the behavior of man? Is it his genetic potential and/or the environment to which he is exposed? Why, for example, are some people

more susceptible to diseases than others?
heavy or thin?
tall or short?
white or black?

Obviously, the list of differences between people is endless. In fact, we are recognized to a great extent by our differences. Have you ever tried to distinguish between identical twins? How can we account for human differences? How do differences among people influence society? These questions are the basis for this text, and it presents the information necessary for at least partial answers to these and other questions concerning the impact of genetics on society.

My deep appreciation goes to B. J. Brown, Sybil Jones, and Janice Sconyers for the wear and tear on their anatomy in trying to read my writing and convert it into typescript; to Jack Funkhouser, Wellington Morton, and Kevin Inyang of the University of North Florida Instructional Communications Center for their advice in the creation of line drawings and photography; to C. A. Taylor,

Larry Freeman, Mike Shelby, Alice Hardigree, Cathy Cohen, Nancy Vermeulen, and Mary Wright who helped with library research; to Kathi McDown for the glossary and proofreading; and to the students of the University of North Florida, on whom this course was tried out over a period of years, for their opinions, which helped to fashion the final draft. A special thank you to Dr. Charles H. Carter, Director of Medicine, The Orlando Sunland Training Center, to Dr. Meyer Melicow, and to Dr. Nyhan for the use of their private photographs in this text and to Dr. Nicholas Sturm for the many hours of editorial work that he gave unselfishly. And finally, to my wife, Dolores, my daughter, Sherri Elizabeth, and my son, Garrett James, whose tolerance and understanding permitted the many hours of uninterrupted time required in writing of this text.

G. J. S.

Jacksonville, Florida

contents

Part 4
The Physical Basis: Transmission of Units of Hereditary Material Through the Genes / Some Social Implications

Part 5
Mutation / A Source of Change During the Continuation of Life

Part 6
Aspects of Human Genealogy

Part 7
Application of Counseling, Screening, and Medicine to Genetic Problems

Part 8
Politics, Social Outcasts, and Genetics in Law

Part 9
What We Can Anticipate from Continued Genetic Research

1

genetics and society / a brief overview

Men love to wonder and that is the seed of our science.

Ralph Waldo Emerson

Our species has come to dominate the earth through a series of pivotal mutations that, over many generations of trial-and-error existence, have produced a unique life form (Figure 1–1). We have the ability to determine, at least in part, our own destiny. We have developed, through events not yet clear, a complex nervous system that allows us to project our thoughts beyond our immediate environment, to entertain visions of history and insight into our future, to restyle our lives, and even to determine whether other life on our planet shall survive.

We humorously envision our ancestors, the cavemen of 25,000 years ago, as having been stoop-shouldered, dull-witted, filthy, and covered with matted hair (Figure 1–2). According to anthropologists, however, cavemen were very much like ourselves. Today the common belief is that we are, as a species, the most advanced form of life ever to have lived on this planet. Until there is conclusive scientific evidence that ancient astronauts visited our planet or that societies more advanced than ours do exist elsewhere in the universe, we can be rather smug about the accomplishments of modern man.

In our pursuit of a higher standard of living, we have transformed ourselves from lean and hungry scavengers to rather complacent men of leisure, our clothing from animal skins to synthetic furs, and our shelter from caves barely warmed by fire to skyscrapers with air conditioning. And during our headlong development of technology, we paused for just a moment to hear the explosions at Hiroshima and Nagasaki. Our technological power was finally realized at that moment of nuclear explosion, and since then our lives have been unquestionably altered.

While we were learning to live with the fear of nuclear holocaust, scientists found still another way to change the course of civilization when they

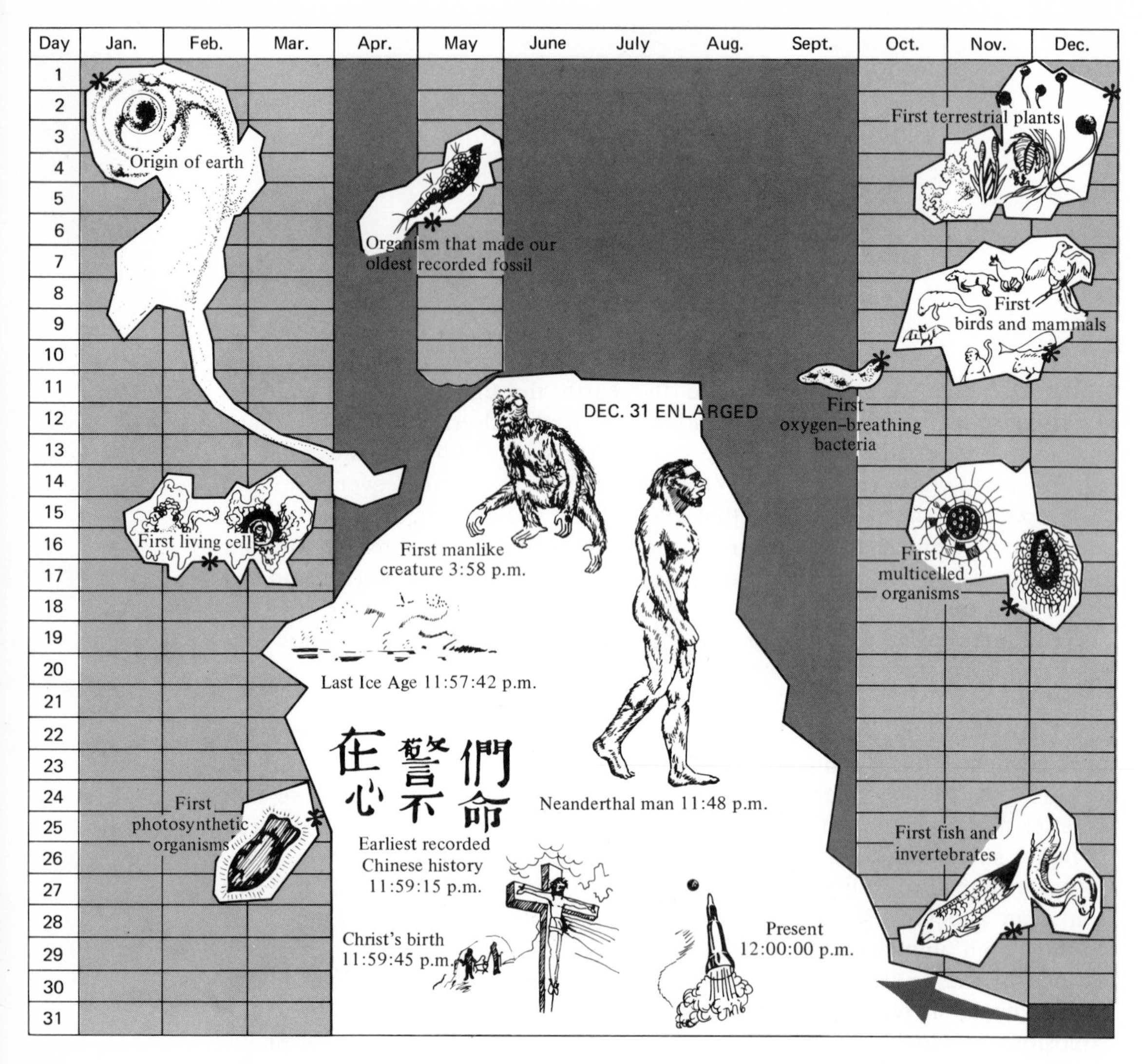

Figure 1–1. Evolution of the Earth, Organisms, and Man on a Common Time Scale. The span of the earth's history has been adjusted to fit within one year, from January 1 to midnight December 31. The month and day for each event are indicated by the position of the asterisk. Many species of life developed from February 17 to noon of December 31, but the entire history of man as a species occupies only about eight hours of December 31. The first multicellular organisms appeared on November 18; Neanderthal man appeared at 11:48 P.M. December 31; Jesus Christ was born at 11:59:45, only 15 seconds ago! (Courtesy of Dr. Ray Bowman, University of North Florida.)

deciphered the genetic code of life. We now know the universal genetic language of life at the molecular level. This knowledge will ultimately lead to the ability to regulate life. One day in our future, as Aldous Huxley implied in his book *Brave New World,* a minority may genetically manipulate man's nervous system and brain and thus control the behavior of the majority. Man's ability to change human life at its genetic source, that is, at the level of our molecules, will be the ultimate power; nothing up to that time will be of equal

Figure 1–2. Early Man. It used to be thought that early man was stoop-shouldered and covered with matted hair that was usually soiled. (Redrawn from Carolina Biological Supply Company picture.)

significance in the biological and social evolution of man. It is difficult to conceive of anything that could be of more value, except perhaps the finding of an inexpensive, inexhaustible supply of energy—and peace among all mankind. The years ahead promise to be exciting and medically revolutionary through the continuous revelations arising from the research in human genetics.

the gene

Slightly more than a century ago Gregor Mendel stated that one generation of a living species passes on its hereditary message for the continuation of the species through the transfer of "factors." Today we recognize these factors as genes. We know that each gene consists of a given number of nitrogen-containing molecules called organic bases. Genes contain four distinct nitrogenous organic bases: adenine (A), thymine (T), guanine (G), and cytosine (C). These bases are linked together into a chainlike polymer with each base attached to a specific molecule of deoxyribose (a sugar) and phosphate. The combination of base, sugar, and phosphate is referred to as a nucleotide. The collective molecule composed of connected nucleotides is referred to DNA, or deoxyribonucleic acid, and a given segment of DNA represents a gene. (Details of the composition, structure, and function of DNA are presented in Chapter 5.)

The double-stranded, spiral-shaped molecule of nucleic acid contains the intricate messages that define and distinguish all species and all individuals. The genetic code for every species is defined by the number, kind, and arrangement of bases in the species' DNA molecule. The DNA molecule directs the reproduction and maintenance of each living cell and organism on earth. It performs this astonishing feat through specifying the composition and conformation of the many thousands of structural and functional proteins that regulate life.

genetics in medicine

In 1909 an English physician, A. E. Garrod, published a book entitled *Inborn Errors of Metabolism,* a text that is a classic scientific contribution to the understanding of the biochemical genetics of man. Garrod suggested two important relationships, which later turned out to be correct.

1. Variation among humans is the result of our biochemistry.
2. Our biochemical activity depends on our supply of "factors," Mendel's terms for genes.

Thus Garrod implied that there is a relationship between a given gene and a specific biochemical reaction. He stated, for example, that alkaptonuria (the body's inability to degrade homogentisic acid) results from a biochemical defect in metabolism and that this defect is related to a defective gene. By 1930 we knew that a number of inherited human diseases resulted from the absence of particular proteins called enzymes. An enzyme is a catalytic agent that makes possible a chemical reaction under more favorable conditions than would be required without it, such as a shorter time or a lower temperature. The unique feature of a catalyst is that it is not used up in the reaction it controls. In 1941 George Beadle and Edward Tatum announced that a specific gene controls the synthesis of a single specific enzyme. They were saying what Garrod has suggested earlier, that one gene is responsible for the formation of one enzyme. Today we know that a gene codes for only one polypeptide, which is only a part of a total functioning enzyme.

For example, about 1 person in 12,000 has a defective gene such that he cannot produce phenylalanine hydroxylase, the enzyme required for the conversion of phenylalanine to tyrosine. Normally, when the enzyme is present in our cells, excess phenylalanine from our diets is converted to tyrosine; as conversion takes place, the phenylalanine hydroxylase enzyme is released to convert other molecules of phenylalanine to tyrosine. If the enzyme phenylalanine hydroxylase is *absent,* however, by-products of improper phenylalanine metabolism will increase in the person's blood to levels that cause brain damage. The result is the mentally retarded individual referred to as a phenylketonuric. That is, in the genetically mutant person who lacks the enzyme, phenylalanine accumulates in the bloodstream, is circulated through the brain, and in a manner not yet understood, causes mental retardation. Thus, this enzyme is a necessary part of either the cell's or the whole organism's health plan.

In short, each enzyme synthesized by a cell has a specific function. Certain enzymes, for example, break down the food we ingest into free amino acids essential for the synthesis of new protein. Other enzymes function in the synthesis of various amino acids, proteins, fatty acids, vitamins, and the purines and pyrimidines that make up DNA. Still other enzymes are responsible for controlling DNA replication and repairing cellular damage caused by radiation or chemicals. Many enzymes in human cells remain to be discovered, but one thing is certain: *many human diseases result from metabolic errors caused by the cell's failure to synthesize a specific enzyme.*

the organisms / genetics and man

Our earliest knowledge of the mechanisms of genetics was gained through studies of plants and later the common fruit (vinegar) fly, *Drosophila* (Figure 1–3). Much of what we now refer to as "classical genetics," or Mendelism, was learned from experiments using *Drosophila.* As biochemists became more sophisticated, they turned to even smaller organisms (there appears to be an inverse relationship between the sophistication of technical hardware and the size of the research organism used). In early biochemical genetics, Beadle and Tatum used a common bread mold, *Neurospora crassa,* to formulate the one gene–one enzyme hypothesis. Then as genetics became molecularly oriented, investigators turned to bacteria and viruses. Here the overwhelming choice of research organism was the bacterium *Escherichia coli,* the common inhabitant of the human colon, and the viruses that are specific to *E. coli* (Figure 1–4). With each new organism that was studied, man learned more about himself and his chromosomes, which carry genetic information, a fact that emphasizes the universality of genetics and chemical molecules. True, we cannot apply to man all the data derived from the many organisms utilized in genetic studies, but we can and do apply the *principles* established by these investigations.

In certain respects man is not a suitable subject for the study of genetics. Families are too small for establishing reliable ratios; the use of test matings raises ethical questions and in most cases is prevented by law; and study of more than a few generations for a particular purpose is generally impossible. However, the social implications are so great that human genetics *must* be investigated. Fortunately, some areas are open to study. We are learning a great deal about ourselves through studies of the biochemistry of hemoglobin and blood groups, by the examination of human tissue from abortions and

Figure 1–3. Adult *Drosophila melanogaster* (×20). The adult fruit fly is used extensively in the study of genetics. (Courtesy Carolina Biological Supply Company.)

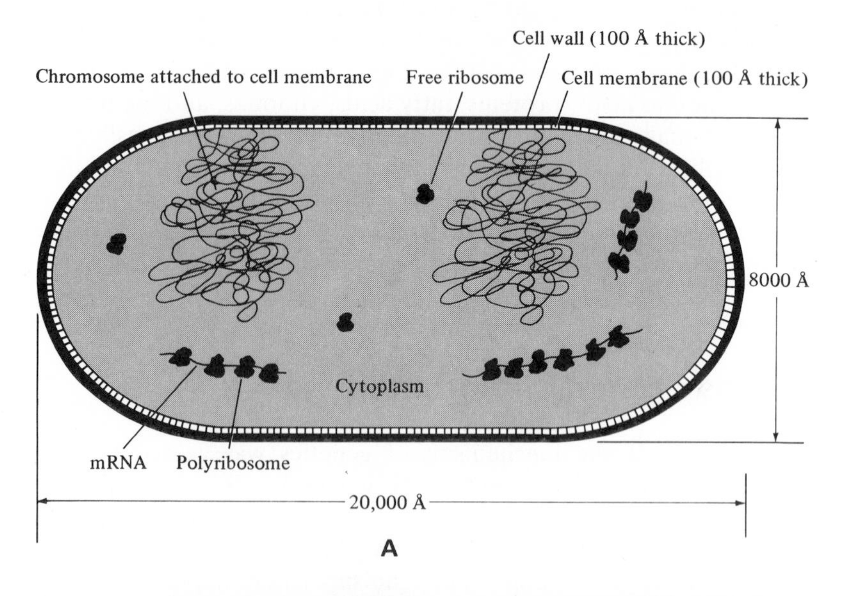

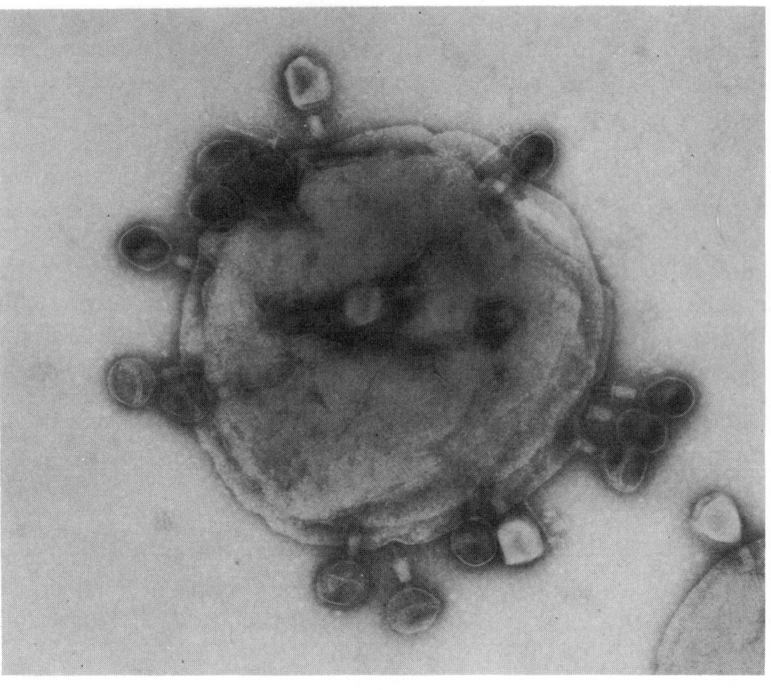

B

Figure 1–4. *Escherichia coli.* **A.** Diagram of an *E. coli* cell containing two identical chromosomes and polyribosomes attached to mRNA. The approximate weight of the cell is 10^{-12} gram, or one trillionth of a gram. **B.** Bacterial virus specific for *E. coli* attached to an *E. coli* cell (×65,000). (**A** from G. J. Stine, *Laboratory Exercises in Genetics,* Macmillan, 1973; **B** courtesy of David P. Allison.)

postmortems, and by extraction of placental enzymes from placental tissues to determine the association of certain enzymes and birth defects.

After 1900 a body of information accumulated concerning the Mendelian inheritance of a large series of aberrant conditions in man. In 1902 Garrod suggested that alkaptonuria is due to a single defective gene; Farabee reported the inheritance of brachydactyly (abnormally short fingers) in 1905; and by 1909 Garrod's evidence that the inheritance of alkaptonuria is caused by a recessive gene was conclusive.

The more obvious and familiar human characteristics, such as height and shape, eye, hair, and skin color, right- and left-handedness, and fingerprint patterns, although inherited, are difficult to analyze. In some mammals the inheritance of hair and eye color is well understood, but in man there are so many intermediates that analysis is often uncertain. Even more difficult to analyze is human mental behavior, which obviously is of interest and importance to society, although some Mendelian differences that must have indirect effects on our behavior—such as taste sensitivity, night blindness, color blindness, skin color, and physical malformations—are well established. In fact, a considerable number of more or less clear-cut Mendelian differences in man have been recognized in spite of the difficulties of proper identification of defects and their patterns of human inheritance. Even though these are largely concerned with relatively rare defects or with less obvious biochemical variations in human proteins, future knowledge of human genetics will undoubtedly help alleviate certain types of inherited birth defects.

guarding the genetic quality of man

Our human gene pool—our DNA, which makes us biologically distinct from all other life forms—is the culmination of more than 4 billion years of organic evolution controlled by the process of natural selection. We continue our species through the process of sexual reproduction, which ensures that our offspring belong in the human species, yet each is genetically unique, different from any other human being who ever lived. We are not created as biological equals, for each of us is born with biological individuality. Each of us who is capable of producing human offspring is a momentary keeper of our human heritage with the understanding that chance and nature, not man, will largely decide the biological similarities and differences among men. We do have a social and moral obligation, however, to protect our gene pool from obvious genetic damage.

Until recently the principal source of damage to our pool was exposure to the increasing radiation in our environment. Now we know that many mutagenic chemicals are introduced continuously into our environment in pesticides, food preservatives, and industrial pollution. Although the magnitude of these hazards is undefined, the effects on our gene pool are beginning to be seen in a noticeable rise in the rate of birth defects. Victor McKusick's text. *Mendelian Inheritance in Man* (1968), lists 1545 genetically determined variations that are a part of our human gene pool. Since 1968 the number has swelled to over 2000 genetic defects, many of which are severe and grossly debilitating.

who, then, are the genetically healthy?

Most genetically controlled or determined disorders await detection. If individually we do not display one of the various manifestations of a hereditary defect, we must not make the mistake of assuming that we are genetically healthy. In fact, if genetic health were defined as the condition of being free of every defective gene, no individual who ever lived would have known perfect genetic health. Today, it has been estimated, every human being carries an average of five to eight recessive lethal genes and perhaps many other recessive genes that are masked by normal genes. Thus, a search for utopian genetic health is pure fantasy. If, collectively, we are not genetically healthy, we must settle for something less. How well a person is depends on the lottery of genetic shuffling that occurs during the formation of his parents' gametes and on the response of his inherited genotype to the environment it encounters. Since, at the moment, we cannot control either, we really have no way of controlling genetic health. Diseases involving disorders of the hereditary material—the genes and chromosomes—are the oldest, most widespread, and probably the most burdensome of all human afflictions.

One authority, J. E. Seegmiller of the University of California, San Diego, said:

> The fact that many victims of hereditary diseases are completely unable to care for themselves and require full-time care by their relatives, or become lifetime wards of the state in mental institutions or nursing homes, makes the total cost to the nation far greater than that of the diseases that kill (more or less) outright, such as cancer, stroke or heart disease which usually affect individuals who have already lived out major portions of their lives as contributing members of society.*

and C. J. Epstein of the University of California, San Francisco, said:

> Carried to the *extreme*, virtually all of human disease and deterioration can be considered as birth defects, since genetic factors are undoubtedly of importance in their origin and pathogenesis.*

The common element in inherited diseases is that the individuals affected are born with the condition or the predisposition to develop the disease. If the disease causes sterility, those affected cannot transmit the defective gene to their offspring. Some genetic diseases are inherited in a complex manner involving several genes. Some involve genes plus certain environmental factors, such as diet, for the condition to express itself. Thus, it is difficult to assess the burden of inheritable diseases in the nation today. We lack a clear definition of how many diseases that show a tendency to appear in given families are transmitted. Thus discussions of genetic disease are usually limited to those defects in which genetic components are reasonably established. The list of genetic diseases is growing constantly, however, as we bring "nongenetic" diseases within definition. For example, certain cardiovascular problems have recently been found to result from specific genetic factors. The cardiovas-

* *What Are the Facts About Genetic Disease?* DHEW Publication No. NIH 74–370. Department of Health, Education and Welfare, Washington, D.C., 1974.

cular genetic factors account for approximately one fifth of all heart attacks in man under the age of 60. The three genetic abnormalities predisposing people to heart attack were estimated to occur in 1 in every 160 Americans, making these abnormalities "among the most common disease-producing genes in our population."

Even when we recognize a person with poor genetic health, it is difficult to judge the consequence of the defect for him or for his contribution to society. Certainly some severely defective individuals nevertheless made a major contribution to society; the famous French artist Toulouse-Lautrec, for instance, had pyknodysostosis, an inherited disorder of the connective tissue.

Recall, if you will, a scene from the movie *Deliverance.* Four adventurers stop for directions in the backcountry of the Appalachian Mountains. As they review the scenery, they notice a young boy with transfixed eyes and are so sickened by the sight that one of them exclaims, "My God, look at that genetic depravity!" To be sure, the youth showed manifestations of microphthalmia (an abnormally small eye size), a recessive hereditary disorder. This same boy might have been seen on any street. Whatever caused the problem, the adventurer formed his judgment immediately. But what of this boy's condition? Was he completely useless to society? Should he have been kept off the street? The answer was forthcoming in the next scene, as the boy bent over his banjo and, to the accompaniment of an adventurer's guitar, picked out what was to become a best-selling tune, "Dueling Banjos." The "genetic depravity" did not appear to stop his nimble fingers as they danced over the strings of that banjo. What, then, is genetic health?

A knowledge of the mechanics of heredity and awareness of our many inherited diseases induced Theodosius Dobzhansky, a well-known population geneticist, to say, "If we enable the weak and the deformed to live and to propagate their kind, we face the prospect of a genetic twilight. But if we let them die or suffer when we can save or help them, we face the certainty of moral twilight." Perhaps you would like to contrast this statement with that of Helmut Thielicke, a preeminent Protestant ethicist, who said that men must recognize that "the act of compassion to one generation can be an act of oppression to the next." One could interpret Thielicke's statement as meaning that (some?) men must be willing to make difficult decisions for the benefit of all. Two of the most difficult questions are: Who may live? and Who shall have the right to reproduce?

part 1

the beginning of life and the development of the genetic material

How did life begin? Have you ever asked this age-old question? If you have, you realize that your options as to how life began are limited. Two widely held views are (a) God created life and (b) life began billions of years ago without the aid of supernatural powers. One could suppose that such statements mean simply the opinion that a divine creator (God) was or was not involved in the creation of life. But this is not the case; various religious groups have very different points of view on what is meant by the "creation of life." For example, the teachings of the Catholic and the fundamentalist churches differ to a great extent on when the divine creator created life. The Catholic Church teaches that life may have originated, as scientists say it did, through "spontaneous generation" billions of years ago and proliferated by the process of natural selection during organic evolution. Man, however, has a God-given soul, and this separates humans from all other evolved life forms. The fundamentalist religions teach that man and all living things, as well as heaven and the earth, were created by God only about 4000 B.C. The creation took six "days," after which God rested (Genesis 2:1–3). The fundamentalists believe that the Bible *is* the unmistakable word of God. The Bible reads, "And the Lord God formed man of the dust of the ground, and breathed into his nostrils the breath of life; and man became a living soul" (Genesis 2:7). Since spontaneous generation, or abiogenesis, is defined as the production of living from nonliving matter, a belief in Genesis 2:7 (God turned dust into life) does not rule out acceptance of spontaneous generation. The difference between religion and science seems to be in the interpretation of *when* life began and the form in which life was "created" by spontaneous generation. Those who believe life arose billions of years ago also believe that man and other species evolved one from the other. The fundamentalist believes that man and all

Organic evolution has solved the problems that mortals could not envision until there was a solution.

primitive earth and chemical evolution / the origin of life

species were spontaneously generated (or created) by God in fixed and nonchanging forms a few thousand years ago.

A firm belief in the literal interpretation of the Bible led to the fundamentalist antievolution resolutions that challenged the right of educators to teach the theory of evolution in public schools. (The heated arguments between "fundamentalists" and "evolutionists," so dramatically presented at the trial of John Scopes, are described in Chapter 3.)

Many still crusade today for textbooks that teach divine creation as well as the theory of evolution. They have formed the Creation Research Society, which, although basically fundamentalist, includes hundreds of voting members with master's or doctoral degrees in various fields of science (Peter, 1970). The society's chief purpose is to publish scientific research in support of the thesis that the material universe, including plants, animals, and man, is the result of direct creative action by God.

At the moment, science has no method for measuring the accuracy of fundamentalist beliefs. It can, however, provide some factual materials, through the use of modern experimental procedures, supporting the ideas that the spontaneous generation of life could have occurred billions of years ago, that the various species of life evolved after the appearance of the first living forms, and that the spontaneous generation of life would not occur under present environmental conditions (probably it could not have occurred at any time after organisms became numerous and atmospheric conditions were altered from those of 4 to 5 billion years ago).

early ideas on spontaneous generation

Although most scientists believe that life now comes only from preexisting life, the idea that life can arise spontaneously has a long history. For example, the early Greeks believed that lifeless decaying materials could turn into living things if wetted and warmed in the sun. Such beliefs were championed by Aristotle, who felt that humans may have begun in this manner. The Roman poet-philosopher Lucretius wrote that "many animals spring forth from the earth, formed by rain and the heat of the sun." The writings of scholars of the Middle Ages abound in descriptions and pictures of the spontaneous generation of insects, worms, and fish from slime and moist earth, of frogs from May dew, and even of lions from the stones of the deserts.

Especially characteristic of the medieval perception of the origin of life were the legends of the goose tree and the lamb plant. In the eleventh century Cardinal Pietro Damiana wrote that geese and ducks were born on goose and duck trees (Figure 2–1). An early traveler, Odorico di Pordenona, told of succulent lambs that grew in gourds of tartary trees in the north of Scotland; when the gourds were ripe, they split open and released lambs of tasty meat and covered with white wool. Viewing ducks, geese, and lambs as the vegetable products of trees, the men and women of the Middle Ages ate these "vegetable meats" on days of meat fast. Since ducks and geese swam, they were also served as "fish."

Figure 2–1. Goose, Duck, and Lamb Trees of the Middle Ages. The animals, it was believed, were produced as "vegetables" within various kinds of gourds—apparently goose (**A**), duck (**B**), or lamb (**C**) gourds.

Paracelsus (1492–1541), an alchemist and physician, offered an original recipe for the generation of mice, frogs, toads, and turtles from water, air, straw, and rotting wood (Oparin, 1968). Early in the seventeenth century, Jan van Helmont (1577–1644) wrote out recipes for producing mice and scorpions, and William Harvey (1578–1657), who described the human blood circulatory system, maintained that worms and insects were produced spontaneously.

This theory of spontaneous generation—that life arose from nonliving material—was not without its opponents even in the seventeenth century. Francesco Redi (1626–1697) demonstrated that the maggots that develop in rotting meat come from eggs previously deposited on the meat by flies and do not arise spontaneously. His triumph was short-lived, however. The Dutch naturalist Anton van Leeuwenhoek (1632–1723), using the newly developed microscope, discovered the world of microorganisms (single-celled organisms too small to be seen with the naked eye). Now, to those who were equipped with a hand lens these tiny organisms seemed to be everywhere. John Needham (1713–1781), a Welsh priest, placed mutton broth in a flask, boiled it for two minutes, and stoppered the flask. Surely, he thought, if life does not arise spontaneously, the flask will remain empty. Within a few days, however, Needham's flask was rich in organisms. Others who repeated the experiment also were convinced that life had occurred spontaneously.

An Italian, Lazzaro Spallanzani (1729–1799), conducted an experiment similar to Needham's except that he boiled his broth for 45 minutes. His flasks remained sterile! This work convinced Appert, who heated airtight sealed foods and succeeded in preventing spoilage of food for Napoleon's troops in the field. Appertization, as the heating process came to be known, was successfully used for a number of years before the controversy over spontaneous generation was settled.

Spallanzani did not receive credit for disproving spontaneous generation, for his adversaries claimed that his sealed flasks destroyed the "vital force" (later identified as oxygen) necessary for life. Finally, the French Academy of

Sciences offered a prize for a final resolution of the argument. The winner was a French chemist, Louis Pasteur (1822–1895), who published his results in 1861. In retrospect, his innovative solution is deceptively simple. Pasteur used flasks similar to Spallanzani's, but to overcome the "destruction of the vital force" criticism, he drew out the unsealed necks of the flasks so that air could enter during or after the boiling process (Figure 2–2). The drawn-out curved neck of his flask was long enough for dust particles to settle out before the air reached the liquid. Thus the air directly above the liquid in the flask was dust-free and did not carry any airborne microorganisms that could grow in the broth. Pasteur, flushed with excitement at the force of his experiment, remarked "Never will the doctrine of spontaneous generation recover from the mortal blow of this simple experiment." He was right (Vallery-Radot, 1960).

Pasteur had demonstrated that life did not arise spontaneously and, further, that life comes only from preexisting life. That is, it takes life to make new life, or life begets life.

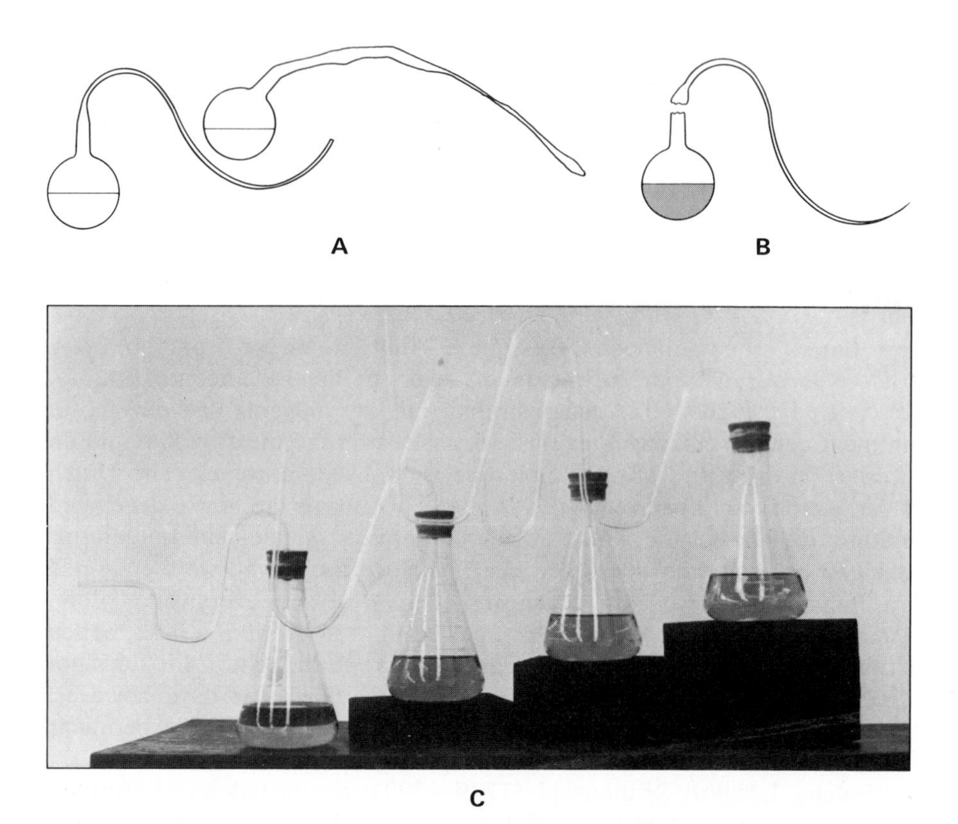

Figure 2–2. Pasteur's Experiment on Spontaneous Generation. **A.** With the neck of each flask drawn out into an odd shape, air entered the flask slowly and airborne organisms and dust particles settled out before the air reached the sterilized broth. Pasteur thus met Needham's objection to Spallanzani's sealed flasks, and still no growth occurred in the broth. **B.** As soon as the neck of the flask was opened directly to air, growth did appear. **C.** A repetition of Pasteur's experiment. Each flask contains nutrient broth, and the pieces of glass tubing, which allow for air exchange with the broth, are of increasing length and/or number of bends. These flasks were set up in 1971 and remain sterile.

Since Pasteur's work it has been almost universally accepted that all organisms now living are descended from preexisting organisms. No instance of spontaneous generation has ever been demonstrated. Moreover, it is doubtful that, once life forms became numerous, spontaneous generation could have been observed even if someone had been there to look. Once such microorganisms as bacteria, algae, protozoans, and fungi evolved, new spontaneously arising life forms would most likely have been used as nutrients by the many cellular species that already existed. Yet the question remains: How did the first living organism occur? The question of how life originated has not been answered, and perhaps it never can be. Could life have come about spontaneously billions of years ago? Pasteur himself, although he proved that spontaneous generation was not occurring in the nineteenth century, believed that it must have occurred in the very beginning—that initially life did arise spontaneously.

In fact, the assumption that life first appeared as a spontaneous event is essential to the theory of organic evolution via natural selection (see Chapter 3). A. I. Oparin and J. B. S. Haldane theorized that life could have evolved several billion years ago from inorganic (simple carbon-containing) chemicals in an early atmosphere that contained no free oxygen, and tests by a number of contemporary scientists, most notably Stanley Miller and Sidney Fox, appear to confirm their hypothesis.

the chemical origin of life

Since human life is thought to be the result of more than 4 billion years of organic evolution, we can only speculate about the composition of the earth's atmosphere at the time life began. Such speculation is based on our conception of how the earth was formed and our knowledge of the present makeup of life on earth, both of which offer clues about the first chemical molecules.

It is assumed that chemistry is the same whenever conditions are the same. If, in fact, chemistry then was similar to chemistry now, at the level of chemical molecules, we may view ourselves as a complex system of interacting chemical molecules that must at one time have been present in less complex organisms. A less complex system, then, would be older than a more complex one. Thus, looking backward in time we expect to find a succession of less and less complex systems and, ultimately, to find that all life has been derived from a common source of chemical molecules. If this is the case, we expect that all life shares the same chemical molecules arranged into units of varying complexity.

One way of looking backward through chemistry is to study the construction of a protein called myoglobin—a red-pigmented, iron-containing protein that carries oxygen in muscle tissue. Myoglobin protein is a chain of 153 amino acid units. Amino acids are the building blocks of protein. The thousands of different proteins that are found in plant and animal life are made up of various sequences of about 20 different amino acids (Figure 2–3). Any protein made by a living system has its specified number and sequence of these amino acids. Myoglobin is composed of the same number of amino acids regardless of what animal it appears in. The differences among the various animal myoglobins are

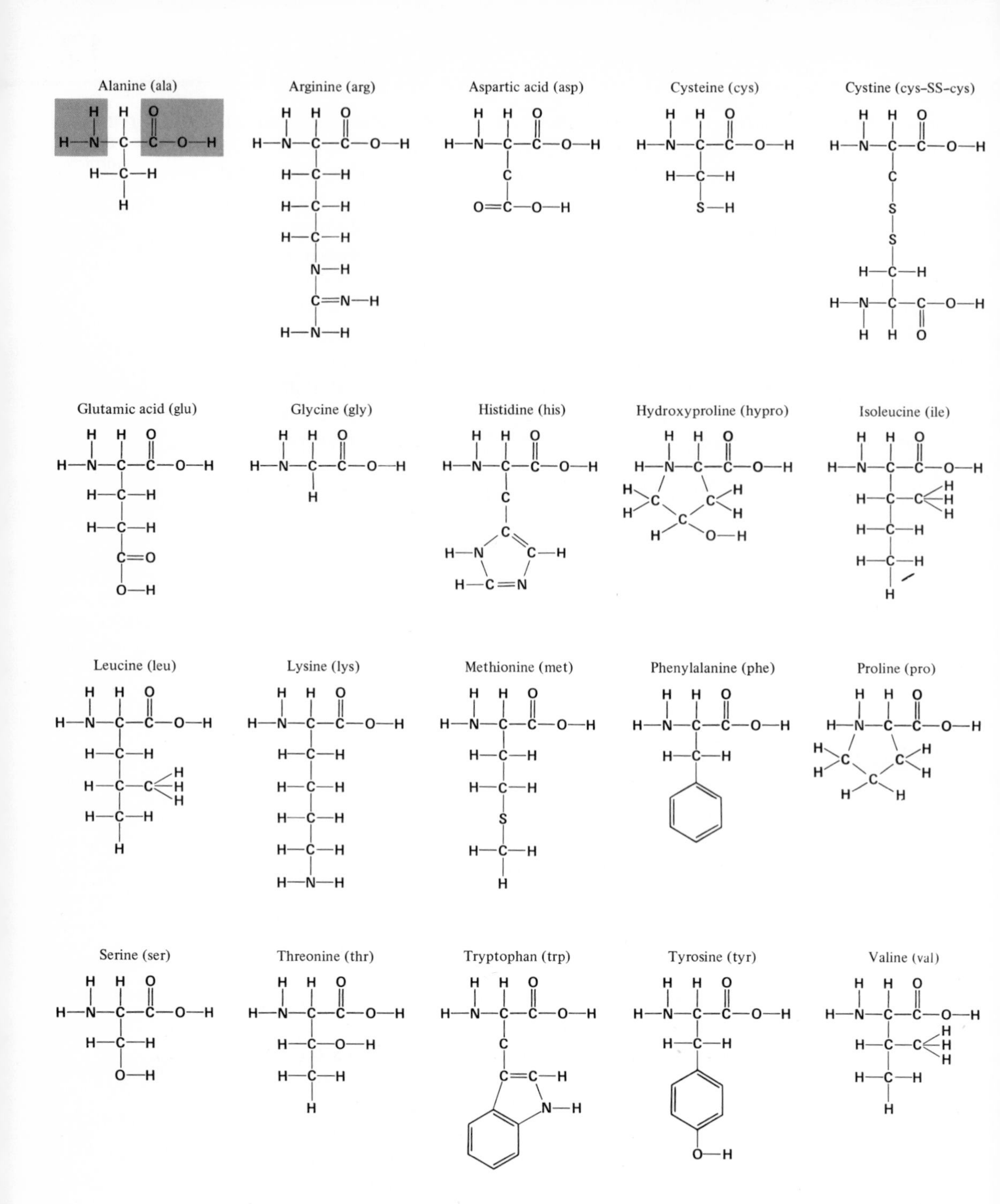

Figure 2–3. The Chemical Structure of the 20 Amino Acids Commonly Found in Plant, Animal, and Microbial Protein. Each amino acid has at one end atoms of carbon, oxygen, and hydrogen (COOH) and at the other end atoms of nitrogen and hydrogen (NH_2); see the shaded boxes on alanine. The COOH group of one amino acid joins the NH_2 group of a second amino acid to form a peptide bond (see Figure 5–7), and a molecule of water (H_2O) is given up as the reaction occurs. Peptide bonding and peptides are discussed in Chapter 5.

Amino acid position number

Whale: val(1) — leu — ser — glu(4) — gly — glu — trp — gln — leu(9) — val —
Horse: gly(1) — leu — ser — asp(4) — gly — glu — trp — gln — gln(9) — val —

leu — his(12) — val — trp — ala(15) — lys — val — glu — ala — asp — val(21) — ala — gly — his — gly —
leu — asn(12) — val — trp — gly(15) — lys — val — glu — ala — asp — ile(21) — ala — gly — his — gly —

gln — asp(27) — ile(28) — leu — ile — arg — leu — phe — lys(34) — ser(35) — his — pro — glu — thr — leu —
gln — glu(27) — val(28) — leu — ile — arg — leu — phe — thr(34) — gly(35) — his — pro — glu — thr — leu —

glu — lys — phe — asp — arg(45) — phe — lys — his — leu — lys — thr — glu — ala — glu — met —
glu — lys — phe — asp — lys(45) — phe — lys — his — leu — lys — thr — glu — ala — glu — met —

lys — ala — ser — glu — asp — leu — lys — lys — his — gly — val(66) — thr(67) — val — leu — thr —
lys — ala — ser — glu — asp — leu — lys — lys — his — gly — thr(66) — val(67) — val — leu — thr —

ala — leu — gly — ala(74) — ile — leu — lys — lys — lys — gly — his — his — glu — ala — glu —
ala — leu — gly — gly(74) — ile — leu — lys — lys — lys — gly — his — his — glu — ala — glu —

leu — lys — pro — leu — ala — gln — ser — his — ala — thr — lys — his — lys — ile — pro —
leu — lys — pro — leu — ala — gln — ser — his — ala — thr — lys — his — lys — ile — pro —

ile — lys — tyr — leu — glu — phe — ile — ser — glu(109) — ala — ile — ile — his — val — leu —
ile — lys — tyr — leu — glu — phe — ile — ser — asp(109) — ala — ile — ile — his — val — leu —

his — ser — arg(118) — his — pro — gly — asn — phe — gly — ala — asp — ala — gln — gly — ala —
his — ser — lys(118) — his — pro — gly — asn — phe — gly — ala — asp — ala — gln — gly — ala —

met — asn(132) — lys — ala — leu — glu — leu — phe — arg — lys(140) — asp — ile — ala — ala — lys —
met — thr(132) — lys — ala — leu — glu — leu — phe — arg — asn(140) — asp — ile — ala — ala — lys —

tyr — lys — glu — leu — gly — tyr — gln — gly(153)
tyr — lys — glu — leu — gly — tyr — gln — gly(153)

A

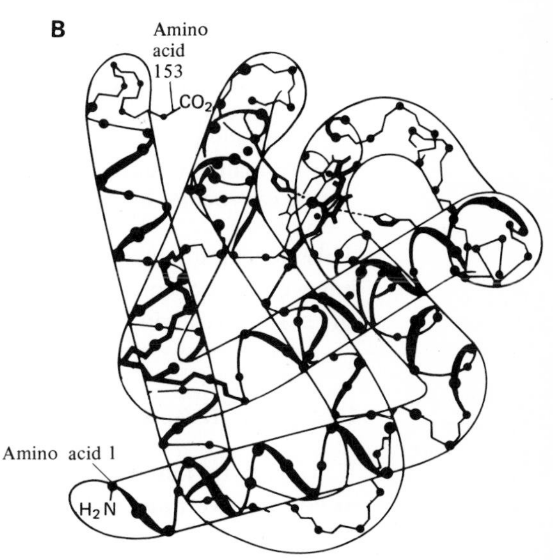

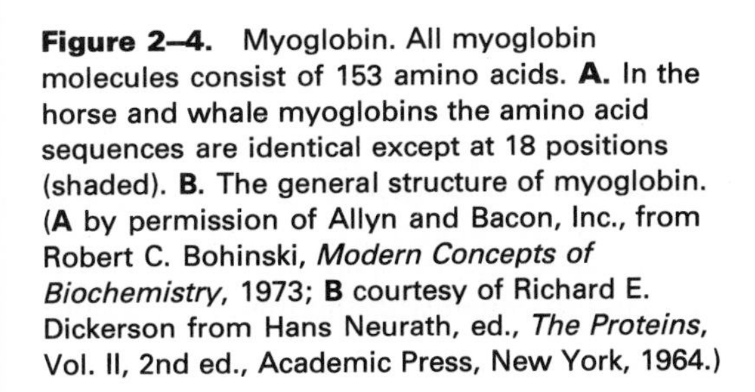

Figure 2–4. Myoglobin. All myoglobin molecules consist of 153 amino acids. **A.** In the horse and whale myoglobins the amino acid sequences are identical except at 18 positions (shaded). **B.** The general structure of myoglobin. (**A** by permission of Allyn and Bacon, Inc., from Robert C. Bohinski, *Modern Concepts of Biochemistry*, 1973; **B** courtesy of Richard E. Dickerson from Hans Neurath, ed., *The Proteins*, Vol. II, 2nd ed., Academic Press, New York, 1964.)

differences in amino acid sequences. Thus the horse and whale myoglobins (Figure 2–4) differ in 18 out of the 153 positions, although 11 of the 18 differences involve structurally similar amino acids. For example, at position 4 in whale myoglobin we find an acidic amino acid (glutamic acid) and another acidic amino acid (aspartic acid) at position 4 in horse myoglobin. Now, human myoglobin differs from chimpanzee myoglobin at only one position, whereas there are several differences between human myoglobin and the myoglobin of the bush baby (a lower primate). As the evolutionary distance increases, there is apparently a parallel increase in the number of amino acid differences even though all myoglobins have the same number of amino acids. Extensive studies of specific proteins, such as cytochrome *c*, that are found in nearly all organisms show rather convincingly that the extent of similarity correlates well with what we know about evolutionary relationships. There is less variation in amino acid sequences in proteins of organisms with a common evolutionary history than in the proteins of unrelated organisms. Thus we can use the amino acid sequence of a protein to measure, with some degree of confidence, the evolutionary distance between humans and other life forms (King and Wilson, 1975).

In terms related to the evolution of organisms, Charles Darwin (1809–1882) and Alfred Russel Wallace (1823–1913) said the same thing in a paper (1858) entitled "On the Tendency of the Species to Form Varieties, and on the Perpetuation of Varieties and Species by Natural Means of Selection" (Calvin, 1961). Both Darwin and Wallace recognized that if there are two species of, for example, birds, their ancestors were varieties of the same species and their more remote ancestors were from a single species. As suggested in the theory of organic evolution, one could continue backward to a point where there was only a single living species. Still further back is a point where it is difficult to determine if the chemical aggregates of matter are "alive." Extrapolation finally leads to an even more primitive collection of aggregates, none of which we would call alive (Figure 2–5). Darwin's thinking on the origin of life and his understanding of the gradual succession of living things from the nonliving were expressed in a letter he wrote in 1871:

> It is often said that all the conditions for the first production of a living organism are now present, which could ever have been present. But if (and oh! what a big if!) we could conceive in some warm little pond, with all sorts of ammonia and phosphoric salts, light, heat, electricity, etc., present, that a proteine compound was chemically formed ready to undergo still more complex changes, at the present day such matter would be instantly devoured or absorbed, which would not have been the case before living creatures were formed.

This statement, made over 100 years ago, contains all the basic concepts that have been the core of most "origin of life" hypotheses and experimentation since 1920 (Calvin, 1961).

the primitive earth: its atmosphere and organic compounds

If we are to find the evolutionary progress of life in a buildup of chemical molecules, we must determine that the buildup began with the materials on the

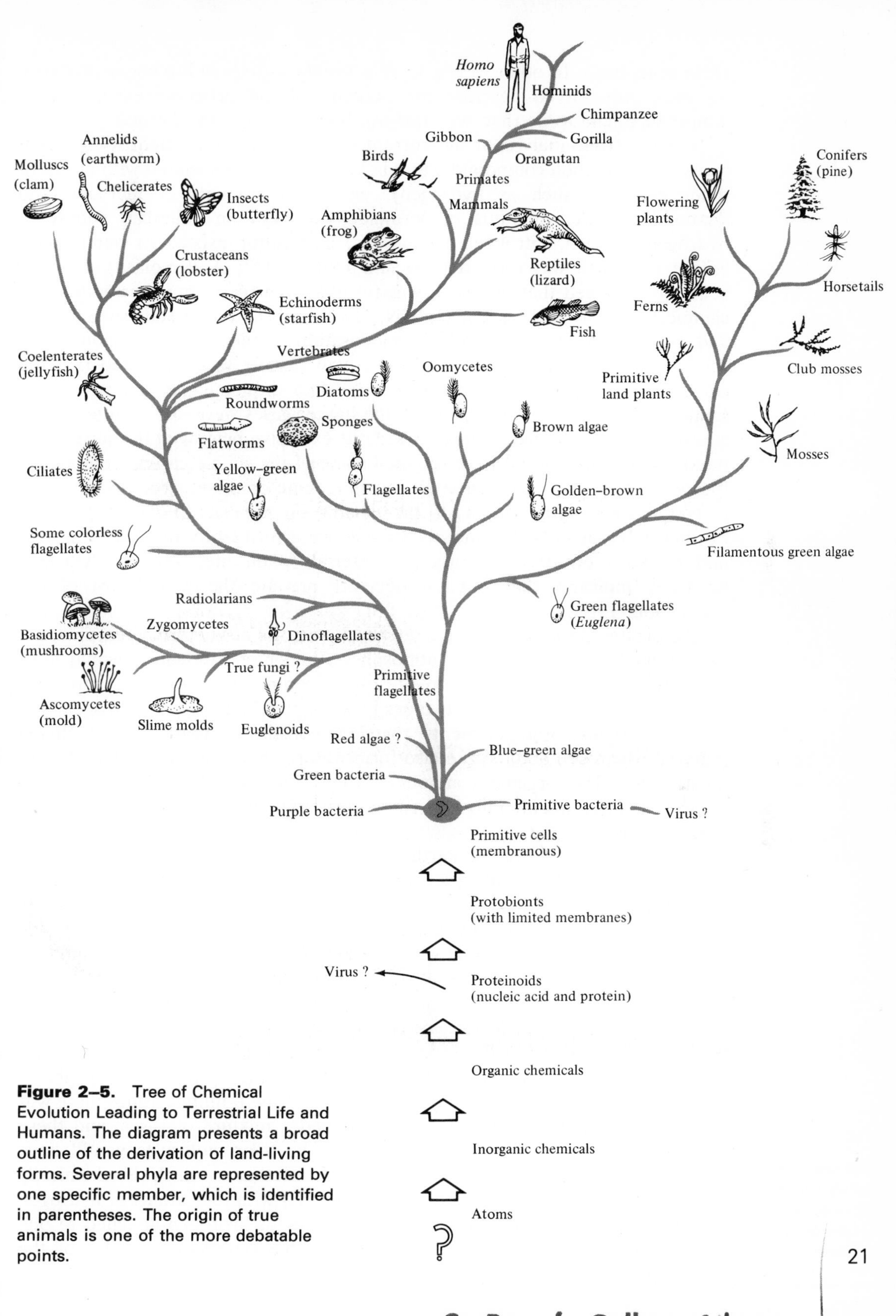

Figure 2–5. Tree of Chemical Evolution Leading to Terrestrial Life and Humans. The diagram presents a broad outline of the derivation of land-living forms. Several phyla are represented by one specific member, which is identified in parentheses. The origin of true animals is one of the more debatable points.

earth at its birth. In other words, to gain insight as to how life began on earth, we must attempt to describe the makeup of the primitive earth and its atmosphere, and to do that, we must ask how the earth was formed.

Judging from analysis of the current elements in space, scientists feel that many simple carbon compounds (that is, carbon atoms associated with atoms of other elements such as hydrogen), enormous amounts of methane and ammonia, smaller amounts of formaldehyde, and an abundant supply of hydrogen, oxygen, nitrogen, phosphorus, and sulfur existed in outer space before the earth was formed (Fox and Dose, 1972; Anderson, 1963). It is generally believed that the earth was formed from these, together with space chemical radicals such as C_2, C_3, CN, CH_3, and OH and metals such as iron and magnesium, during a gradual condensation of very large dust clouds that surrounded the sun. Urey (1952) and Levin (1959) have suggested that the earth was formed by collision of asteriod-like dust and gas clouds. As the smaller asteroid-like clouds collided, the lighter gases were lost to outer space, and the earth continued to enlarge without excessive heating at the surface. So molded, the earth would have retained some of the simple carbon compounds formed in space. It has been proposed that some of these carbon compounds played a role in the origin of the earth's oil deposits (Fesenkov, 1957). Formerly, it was held that the oil deposits were formed by the action of heat and pressure on decaying organic material (plant life). Recent evidence, however, indicates that our oil deposits predate the appearance of life (Anderson, 1963).

According to the theories of Urey and Levin, the newly formed earth would have lacked surface water, an atmosphere, and life, although it would have contained a variety of less complex organic compounds. Organic compounds consist for the most part of four types of atoms: carbon, oxygen, nitrogen, and hydrogen. These four elements together constitute about 99% of living material (hydrogen and oxygen also form water, which makes up 80 to 90% of most cells). The organic compounds found in present-day cells are very complex and fall mainly into four categories: carbohydrates, fats, proteins, and nucleic acids.

During the period of the earth's formation water is thought to have been trapped in deep geological rock formations and to have risen to the surface in the form of steam, which condensed to form rain and give rise to the early oceans. As the water flowed over the land to form the oceans, it picked up the simple organic compounds that were present on land. Over a period of millions of years, the oceans became a rich "primordial soup" of simple organic compounds, salts, various minerals and metals. Through cracks and fissures in the earth's crust and early volcanic action, outgassing of trapped volatiles from the earth's interior formed a secondary earth atmosphere. It is this secondary atmosphere that is usually referred to as the earth's primitive atmosphere. According to Oparin (1938) and Urey (1952), if the primitive atmosphere was formed as suggested, it would have contained a mixture of gases, and these gases would have been hydrogenated; that is, their molecules would have contained hydrogen. For example the gas methane, which is chemically symbolized as CH_4, has one atom of carbon (C) with four atoms of hydrogen (H) attached. Ammonia, NH_3, has three atoms of hydrogen attached to one atom of nitrogen (N). According to theory, the outgassed volatiles of the primitive atmosphere were mostly methane, ammonia, water vapor, hydrogen, nitrogen, carbon monoxide, and carbon dioxide. There was no free molecular oxygen (O_2). Such an atmosphere, containing hydrogen gas but no oxygen gas, is a reducing atmosphere. In chemical reactions involving hydrogen, atoms of hydrogen donate their electrons to other atoms, which become "reduced."

The belief that a reducing atmosphere existed is consistent with the "principle of continuity," which requires that each stage in evolution develop from the previous one in a continuous progression (Orgel, 1968). In a reducing atmosphere, a single carbon atom is reduced by hydrogen to methane (CH_4), an organic compound. Thus an organic compound can be produced not by life processes, as occurs today, but by a chemical reaction permitted because hydrogen donated its electrons to an atom of carbon.

From the initial chemical reactions that formed the first simple organic compounds, through the production of more complex organic compounds in the primitive atmosphere, chemical evolution could have proceeded to produce a continuing spectrum of more complex organic compounds, such as amino acids, carbohydrates, fatty acids, and nucleic acids. At some point an aggregate, or group of such molecules, began to "cooperate," or reproduce.

In present-day cells (Figure 2–6) amino acids are linked together to form proteins, which include some of the largest and most complex molecules known. The amino acids are strung together in chains hundreds to thousands of units long, in different proportions and sequences, and with a great variety of branching and folding. A virtually infinite number of different proteins is possible. Throughout the process of evolution, organisms seem to have

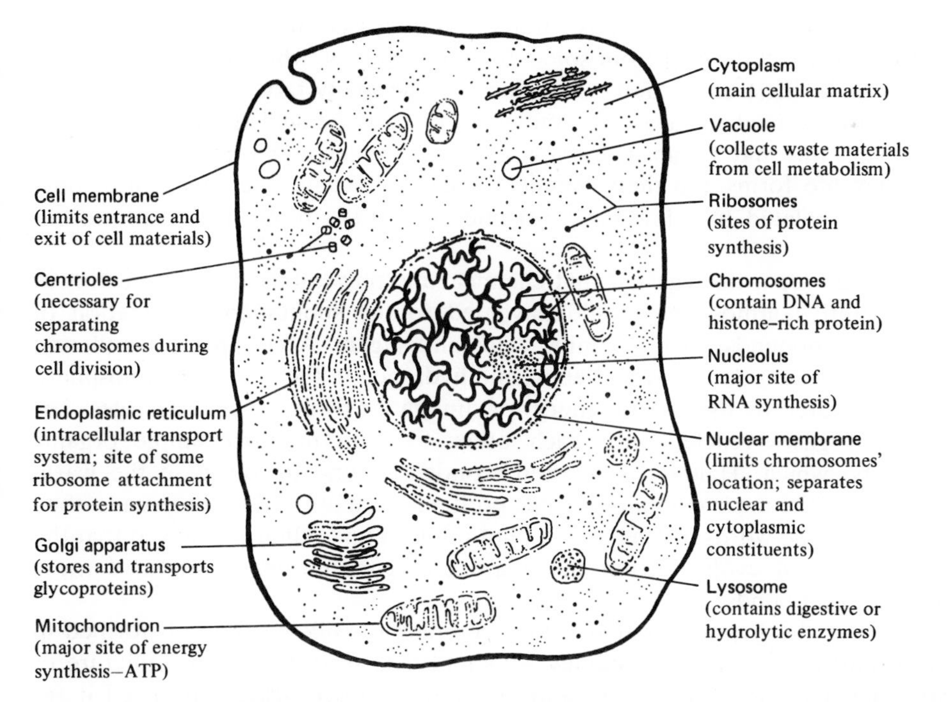

Figure 2–6. The Modern Cell. A sectional diagram of a typical animal cell as it would appear on examination with an electron microscope. Certain cellular organelles have been omitted from the diagram for the sake of clarity.

exploited this potentiality, for no two species of living organism, animal or plant, possess the same proteins. Carbohydrates and fats serve as fuel, as a source of energy, for living organisms. The nucleic acids are the main constituents of the genes, the determinants of heredity.

Are there scientific data to support the idea that the more complex organic compounds could have been produced under primitive atmospheric conditions? If scientists could reproduce the primitive atmosphere in the laboratory, perhaps they could determine whether or not the organic molecules necessary for life processes could have been formed. Organic compounds produced under these conditions would be derived abiotically (in the absence of life) in a manner similar, perhaps, to what occurred 4 to 5 billion years ago.

the synthesis of the building blocks for life

There is fossil evidence that bacteria similar to those known today existed over 3 billion years ago (Lehninger, 1975), and organic molecules have continued to evolve in complexity and variety as the organisms in which they occur evolved. Nevertheless, Lehninger states that all the various complex organic molecules are metabolic derivatives of 30 primordial organic molecules that were present in the early primordial soup (oceans). Today, for example, over 150 biologically different amino acids are known. Yet nearly all plant, animal, and microbial proteins are derived from just 20 amino acids. Similarly, all the many different nucleotides and nucleotide derivatives (the building blocks of deoxyribonucleic acid, the "chemical of heredity") contain one or more of the five major nitrogenous bases found in nucleic acids (see Chapter 5 for specific details). There are also numerous polysaccharides (sugar molecules joined together) derived from some 70 simple sugars, which in turn have their origin in a single sugar molecule, glucose. The many fatty acids found in life forms are derived from a basic molecule of palmitic acid (Lehninger, 1975). Although this may be an oversimplification, the idea is that regardless of how complex life forms and their chemical makeup become, the chemical components are related to a few basic organic molecules present in the primordial soup.

In 1924 Oparin suggested that chemical and physical processes occurring in the primitive atmosphere could have led to the spontaneous formation of organic compounds, such as amino acids and sugars. According to his theory, the methane, ammonia, water vapor, and hydrogen in the primitive atmosphere were activated by the radiant energy of sunlight or by lightning discharges. The activated gases reacted with each other to form the first simple organic products, which, Haldane suggested, condensed and dissolved in the warm oceans. As the more complex organic products accumulated in the oceans, they became, in the words of Haldane, a "hot dilute soup." Oparin believed that the first living cell evolved from the warm and concentrated solution of these organic compounds.

To test the hypothesis that rather complex organic compounds formed in the primitive atmosphere, Stanley Miller (1953) placed a gaseous mixture of methane, ammonia, hydrogen, and water in a closed vessel, sterilized it (to eliminate living cells that might been in the flask), removed any oxygen that might be present, and heated the vessel to 80°C. Then, while the gas mixture was circulating freely within the vessel, he applied an electric spark to simulate lightning (a form of energy certain to have been present under primitive atmospheric conditions) and provide energy to bring about a chemical reaction within the gas mixture (Figure 2–7).

After a week during which the spark was applied intermittently, the amino acids glycine, alanine, aspartic acid, glutamic acid, and beta-alanine had been produced, as well as several other organic acids known to occur in living systems, such as formic, acetic, propionic, lactic, and succinic acids. Thus, if Miller's apparatus truly simulated primeval conditions, such complex organic compounds as amino acids, the building blocks of protein, could have been formed under the conditions that existed 4 to 5 billion years ago. Other investigators who used a spark discharge have produced some 14 amino acids and a series of almost universal metabolic intermediates, such as acetic, propionic, lactic, and succinic acids and urea. Other energy sources tried for the abiotic synthesis of organic compounds include heat, ultraviolet light, bright sunlight, and beta, gamma, and X rays. All told, scientists have now synthesized amino acid precursors (molecules necessary to build amino acids), amino acids, polypeptides (several to many amino acids joined together), the nitrogenous bases (purines and pyrimidines) required in making up the nucleotides of DNA (deoxyribonucleic acid), polynucleotides (nucleotides linked as in a molecule of DNA), the sugars found in nucleotides, and many other nitrogen-containing organic compounds found in living systems (Keosian, 1968). Thus scientists have been able to simulate in the laboratory the proposed primitive atmosphere and use a source of energy to produce most of the 30 organic molecules believed to have been available in the primordial soup.

Now, with evidence supporting the early speculation and hypotheses concerning the makeup of the primitive atmosphere and the synthesis of the building blocks of life, we are ready to ask how the primordial organic molecules could have been organized to produce the first cell.

accumulation, combination, and the first "life"

After the return of the Apollo XI mooncraft, scientists at a conference in Houston, Texas, set the age of the earth equal to the age of the moon, 4.7 billion years. Evidence now indicates that the oldest form of life existed 3.1 billion years ago. This means that somewhere between 3.1 and 4.7 billion years ago, or within 1.6 billion years after the earth was formed, life was established. The various organic molecules were formed abiotically under conditions of a reducing atmosphere, and they accumulated in a body of water that also contained a variety of inorganic salts. This warm primordial soup contained (if we accept the results of the experiments of Miller and others) a large spectrum of amino acids, nitrogenous bases, intermediary metabolites, sugars, and even perhaps molecules of adenosine triphosphate (ATP), which today is known to be the primary energy-transfer molecule in cells. In short, the primordial soup contained the 30 primordial molecules from which, somehow, life arose.

Presumably, the first cells were much simpler than present-day bacteria and developed through a long chain of separate events, each of which involved only

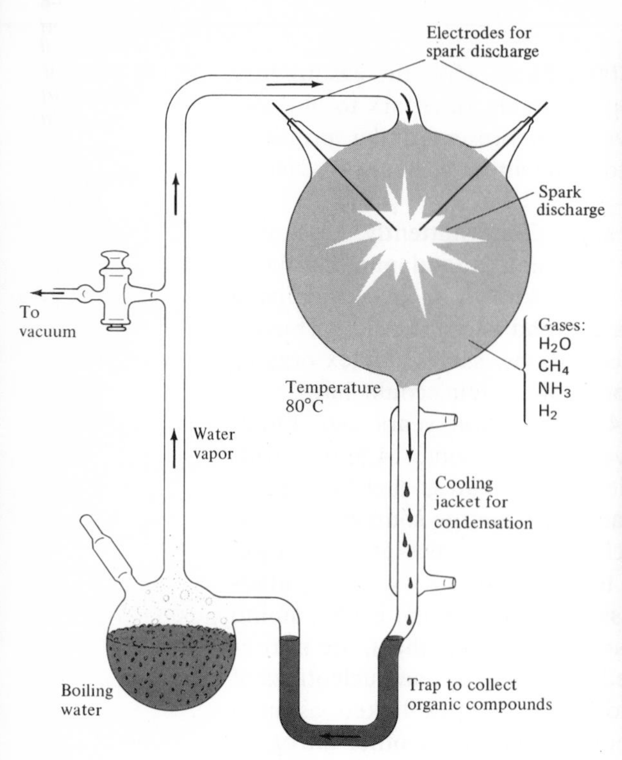

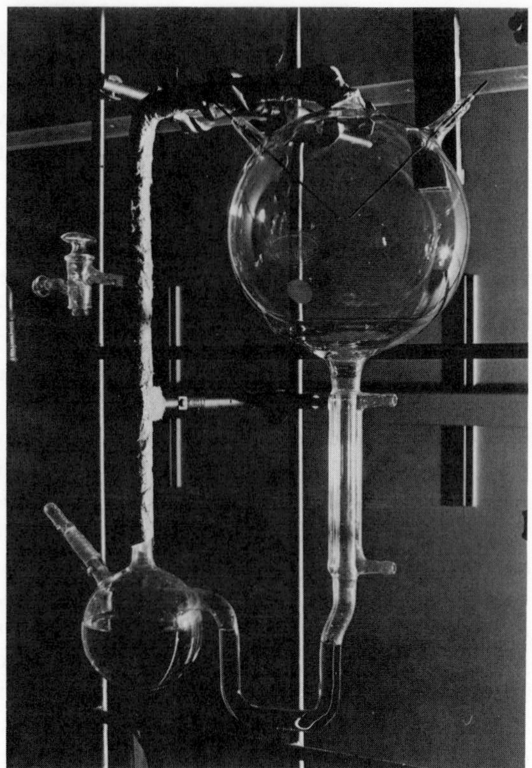

Figure 2–7. The "Miller Flask" Used in the Conversion of Simple Gases to Organic Compounds. The experiment simulated primitive earth conditions as suggested by Oparin, Urey, and others. Miller made amino acids by circulating methane (CH_4), ammonia (NH_3), water vapor (H_2O), and hydrogen (H_2) past an electrical discharge. The amino acids collected at the bottom of the apparatus and were detected by paper chromatography. Left: Stanley Miller adjusting the apparatus before starting an experiment. (University of California, San Diego, photos; courtesy of Stanley L. Miller.)

a very small change. We have seen how the building blocks of life could have accumulated in the early oceans. Calculations, however, show that the concentration of molecules was very low; that is, the solution was so dilute as to have inhibited aggregation or combination of the molecules. Yet as water evaporated, the molecules probably became concentrated in large shallow pools, ponds, lakes, and puddles. As water evaporated further from these pools, the molecules in solution became even more concentrated, until they finally came together in various combinations. Such chance meetings may have given rise to the polymerization (uniting of smaller molecules to form a larger molecule) of amino acids into protein, simple sugars into polysaccharides, and nucleotides into nucleic acids. As described earlier, much of this has been accomplished in the laboratory under "primitive atmospheric conditions."

Now to the most perplexing problem of all: In what direction did polymerization of organic compounds proceed to bring about the first living cell? Did the evolution of life start with the polymerization of amino acids to protein, and proceed through a prebiotic cell to the nucleic acids and present-day cells or through polymerization to nucleic acids, then to protein, to prebiotic cell, and finally to present-day cells? A living cell is a highly organized structure in which all elements are related to each other in a special way, so that the molecules work in concert. This very complexity causes us to wonder how the organization of a modern, or contemporary, cell developed. We know that, in contemporary cells, DNA, which is made up of nucleic acids, contains the genetic code for the synthesis of all the proteins necessary to cellular life. Some scientists, for example, Orgel, Crick, and Horowitz, think that the final assembly of a first living cell from the prebiotic organization could not have occurred without nucleic acids. (The construction of nucleic acids and their function in cellular metabolism are described in Chapter 5.) Others, such as Fox, Harada, and Kendrick, argue that the actual incorporation of informational DNA into a cell came after polymerized proteins gave rise to the prebiotic and then a biotic "cell."

Still other scientists, such as Melvin Calvin, believe that proteins and nucleic acids arose side by side in the evolutionary scheme. Calvin feels that as each macromolecule (a large molecule formed by polymerization of small molecules) of protein and nucleic acid evolved, certain structural features of the molecules gave rise to interactions between them that resulted in a cooperative order similar to what is found in modern cells. He gives an example of such cooperation in the construction of tobacco mocaic virus (viruses are considered to be very primitive). The tobacco mosaic virus (TMV) is made up of a central core of a nucleic acid covered with protein (Figure 2–8). If the protein and nucleic acid are separated, and the protein fragments placed in solution, they will "self reassemble" or reaggregate into the former TMV protein structure. On close inspection, however, the reaggregated protein rods are found to vary in length. If both the TMV nucleic acid and the disassembled protein are placed in solution, the reaggregated TMV particles are of the correct length. It is apparent that both types of macromolecules are necessary to give the correct structure of TMV. The protein structural unit can reaggregate, but interaction between protein and nucleic acid is required for the correct final construction. Nothing in the final construction of the TMV was

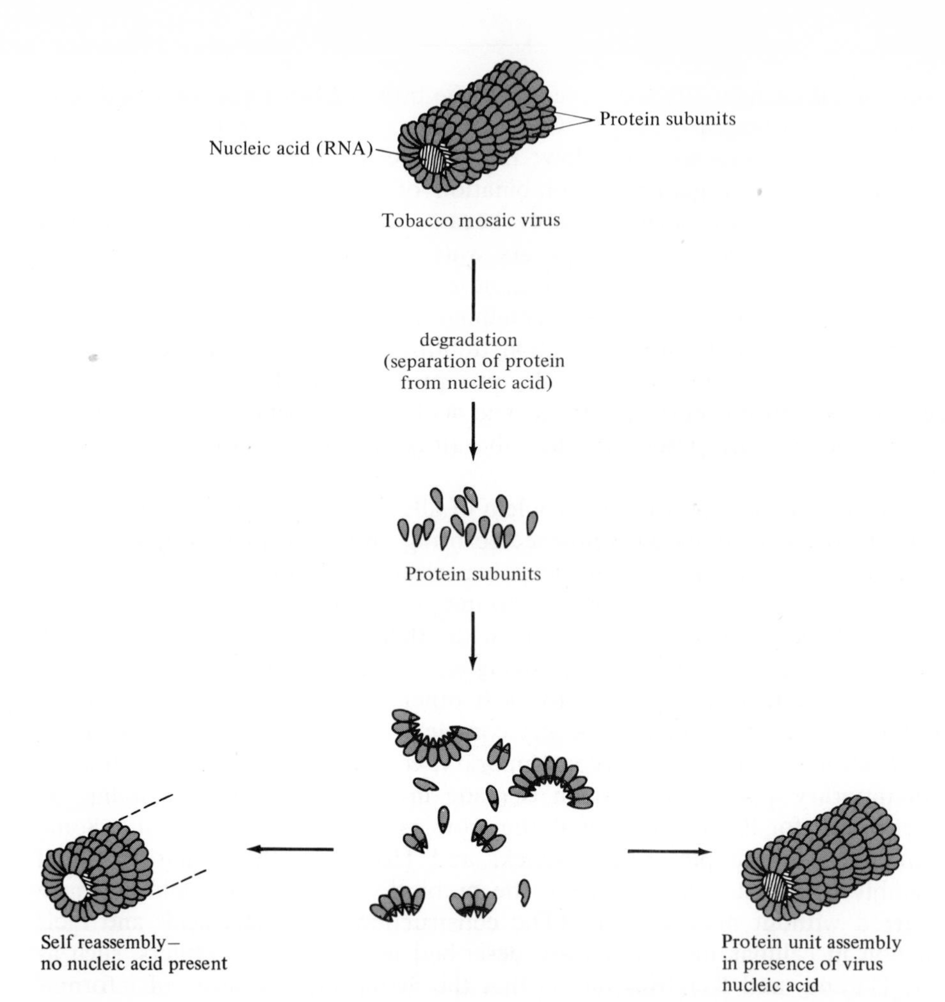

Figure 2–8. Self Reassembly, a Characteristic of Tobacco Mosaic Virus Protein. In the absence of the nucleic acid (RNA), the self-assembled protein coat of the virus will be shorter than it should be. The correct length is constructed if the nucleic acid is present with the protein during reassembly.

done, as Calvin states, by an "unknown force." There was simply an ordering of TMV parts into the correct shape (Calvin, 1961).

The question of how the first cell was formed may never be answered. But the various schools of scientific thought have presented some very interesting treatises on the possibilities. You are encouraged to read the discussions by Calvin (1961), Orgel (1968), Keosian (1968), Lehninger (1975), and Fox and Dose (1972).

the advance to more complex life

As has been discussed, life is the end product of a long, complicated, and not yet fully understood chemical evolutionary process. The earth has an age of at least 4.5 billion years. This figure is based on radioactive dating of meteorites, which were probably formed at about the same time as the earth, and radioactive materials in the earth's crust. Although the oldest Precambrian rocks on earth have been radioactively dated to 3.5 billion years, the oldest moon rock brought back by the Apollo XI crew has been dated to 4.7 billion years. Traces of what are undoubtedly primitive plants have been found in rocks 2 billion years old, and a fossilized primitive cell was found in the 3-billion-year-old Fig-tree shale. The first abundant fossils of animal life, found in rocks that are 600 million years old, are advanced invertebrates rather than primitive organisms. The Precambrian history of animals is doubtless similar to the history of plants, but few Precambrian animal fossils have been discovered. It is therefore assumed that the animals preceding the advanced forms found in the Cambrian Period had soft bodies with no hard parts. There are chemical remains of life activity in the Precambrian Period.

In Precambrian sedimentary rocks, chemical traces of complex carbon compounds have been found. Two possibilities as to their origin have been offered. First, they are of biogenic origin—that is, they were produced by living organisms—and either seeped into the sediments or were formed by later bacterial activity. Second, the chemicals were formed abiogenically, for abiogenic carbon compounds were, it is believed, common in the early history of earth. Traces of complex fats, carbohydrates, and proteins would be evidence of life, but these molecules tend to decompose after long periods of time in sedimentary rock. Many of the compounds that have been found, however, are believed to have been derived from fats, carbohydrates, and amino acids, since many of them are too complex to have been formed nonbiologically.

There are two particularly well preserved fossilized specimens of Precambrian organisms: the Gunflint chert flora, estimated at about 2 billion years, and the Ediacara fauna, which are fossilized impressions of animals over 600 million years old (McAlester, 1968). Phytane and pristane molecules, along with various isoprenoids, have been found in shale deposits that date back 3 billion years. The $-\mathrm{C}=\mathrm{C}-\overset{\mathrm{C}}{\overset{|}{\mathrm{C}}}-\mathrm{C}-$ skeleton of isoprene occurs in so many biological lipids, carotenoids, and steroids, as well as in many essential oils, that it would appear these *are* biological hydrocarbons. The phytane and pristane may have come from the hydrocarbon chain attached to the chlorophyll molecule. Lehninger (1975) states that bacteria-like organisms lived 3.1 billion years ago, and the Fig-tree shale, dated at over 3 billion years, contains organic material. If the earth is 4.7 billion years old, and these molecules are found in 2.5-to-3.1-billion-year-old shale, then life began much earlier than we formerly believed (Table 2–1).

the fossil record of human evolution

Human beings are different from all other animals in that they alone leave behind an archeological record of their behavior. They leave tools, weapons, and other artifacts, along with animal bones, at their campsites. By analyzing items found at archeological sites, investigators have been able to reconstruct

Table 2–1 Chronology of the Geological Ages

Geologic Era	*Major Period**	*Distinctive Events*	*Approx Time of Origin†*	*Geologic Samples*	? ↑
Cenozoic	Recent	Modern man	11,000		**Cultural evolution**
	Pleistocene	Early man; northern glaciation	2 million	Mud Lake, Fla.	
	Pliocene	Large carnivores	12 million		**Biological evolution**
	Miocene	First abundant grazing mammals	25 million		
	Oligocene	Large running mammals	35 million		
	Eocene	Many modern types of mammals	60 million	Green River shale	
	Paleocene	First placental mammals	65 million		
Mesozoic	Cretaceous	First flowering plants; climax of dinosaurs and ammonites, followed by extinction	130 million		
	Jurassic	First birds; first mammals; dinosaurs and ammonites abundant	180 million		
	Triassic	First dinosaurs; abundant cycads and conifers	230 million		
Paleozoic	Permian	Extinction of many kinds of marine animals, including trilobites; glaciation at low latitudes	280 million		
	Carboniferous	Great coal forests; conifers; first reptiles; insects abundant	300 million		
		Sharks and amphibians abundant; large and numerous scale trees and seed ferns	350 million	Antrim shale	
	Devonian	First amphibians and ammonites; fishes abundant	400 million		
	Silurian	First terrestrial plants and animals	425 million		
	Ordovician	First fishes; invertebrates dominant; starfish	500 million		

Table 2–1 Chronology of the Geological Ages (*continued*)

Geologic Era	*Major Period**	*Distinctive Events*	*Approx Time of Origin†*	*Geologic Samples*	
Paleozoic	Cambrian	First abundant record of marine life; trilobites dominant, followed by extinction of about two thirds of trilobite families	600 million		**Biological evolution**
				Ediacara shale	
Proterozoic	Precambrian	Fossils extremely rare, consisting of primitive aquatic plants	800 million		
		Earliest multicellular fossils	1 billion	Nonesuch shale	
		Blue-green algae (fossils)	2 billion	Gunflint chert	
				Soudan shale	
Archeozoic	Cellular (organization of various macro-molecules into cell life)	Bacteria-like organisms (microfossils?)	3 billion	Fig-tree system	**Chemical evolution**
Proarch-eozoic	Precellular (formation of organic molecules)	Formation of the earth	4.7 billion?		

*Major periods in evolution of life.

†Indicates how many years ago the events probably occurred based on a comparison of relative amounts of radioactive elements remaining in samples of rock found in different layers of the earth's crust.

‡Periods of chemical, biological, and cultural evolution probably overlapped.

the activities and cultures of people who lived long ago. Artifacts have been found recently along with fossils of early hominids.

All manlike creatures, whether ancestral to modern man or not, are called hominids. "True man," given the genus name *Homo*, is considered to be one branch of the hominid family tree. According to accepted writing on the evolution of humans, early hominids evolved from ancestral apelike creatures, and human beings evolved from hominids. For most of their past, humans have been hunter-gatherers, and the traits that distinguish hunter-gatherers from nonhuman primates are being analyzed to see whether they may have been traits of the early hominids. Evidence of behavioral patterns shared by both nonhuman primates and hunter-gatherers is now believed to have been found among early hominids (Kolata, 1975).

According to the fossil record, the direct ancestor of man may have had his beginning some 14 million years ago, if not earlier. The evidence for this consists of a few jaws with teeth that are between 10 and 14 million years old. The only clue we have that the hominid lineage may have begun to separate from the ape line is that the teeth have certain shapes found only in modern human beings and in extinct hominids.

In the Early Pleistocene, 2 to 3 million years ago, there existed in South Africa a primate known as *Australopithecus.* This animal was about four feet tall, small-brained, and with ape-sized teeth and jaws. Nevertheless, it was easily distinguishable from the fossil apes of the same period. *Australopithecus* was considered by many paleoanthropologists (persons studying human evolution) a crucial transition species incorporating traits of both ape and man. However, in the spring of 1973 Richard Leakey, a Kenyan scientist, announced the discovery of a 2.9-million-year-old form of man with a brain that was unexpectedly large for its historical age. The skull has come to be known, from its catalog number in the expedition's records, as "1470." This discovery is forcing a reevaluation of whether or not *Australopithecus* was an ancestor of man. If *Australopithecus* is not a direct ancestor of man, then the human race did not evolve simply from an isolated, linear succession of steadily advancing hominids. Rather, it would appear that modern man represents but one surviving lineage from among several lines of prehuman or nearly human creatures.

Man's development, like that of other animals, has depended on genes and the process of natural selection. Through society, however, man channels his natural urges and forces into a new style. Before Darwin's time man was considered a unique creature, "above the animals and below the angels." After Darwin, man was seen as an animal. (The impact of Charles Darwin and his theory of organic evolution is presented in Chapter 3.) In the past few decades, man has come again to be considered unique and extraordinary, for we have gained new respect for ourselves as animals. True, man is unique is some ways, but other animals are unique in other ways. Man's complexity does not set him above other animals. Certain organs of other animals are far more advanced than ours. The gorilla, for example, is anatomically more specialized than the human. Recognition of this fact should result in the social evolution of a more humble and tolerant human. But has it? Will it? What do you think?

references

Anderson, Norman. 1963. On the Origins of Life. In J. Ives Townsend (ed.), *Lectures in Biological Sciences.* University of Tennessee Press, Knoxville, pp. 1–13.

Ardrey, Robert. 1961. *African Genesis.* Atheneum, New York.

Calvin, Melvin. 1961. Origin of Life. *Annals of Internal Medicine, 54*:954–77.

Dobzhansky, Theodosius G. 1962. *Mankind Evolving.* Yale University Press, New Haven, Ct.

Fesenkov, V. G. 1957. Some Considerations About the Primeval State of the Earth. In A. I. Oparin (ed.), *The Origin of Life on Earth.* Publication House, Academy of Science of the U.S.S.R., Moscow.

Fox, Sidney W., and Klaus Dose. 1972. *Molecular Evolution and the Origin of Life.* Freeman, San Francisco.

Keosian, John., 1968. *The Origin of Life,* 2nd ed. Van Nostrand Reinhold, New York.

King, Mary-Clare, and A. C. Wilson, 1975. Evolution at Two Levels in Humans and Chimpanzees. *Science, 188*:107–16.

Kolata, Gina Bari. 1975. Human Evolution: Life-Styles and Lineages of Early Hominids. *Science, 187*:940–42.

Lehninger, Albert L. 1975. *Biochemistry,* 2nd ed. Worth, New York.

Levin, B. Y. 1959. Formation of the Earth from Cold Material and the Problem of the Formation of the Simplest Organic Substances. In A. I. Oparin (ed.), *The Origin of Life on Earth.* Pergamon, London.

McAlester, Arcie Lee. 1968. *The History of Life.* Prentice-Hall, Englewood Cliffs, N.J.

Miller, S. L. 1953. A Production of Amino Acids Under Possible Primitive Earth Conditions. *Science, 117*:528–29.

Oparin, A. I. 1938. *The Origin of Life on Earth.* Macmillan, New York.

Oparin, A. I. 1968. *Genesis and Evolutionary Development of Life.* Academic Press, New York.

Orgel, L. E. 1968. Evolution of the Genetic Apparatus. *Journal of Molecular Biology, 38*: 381–93.

Peter, W. G., III. 1970. Fundamentalist Scientists Oppose Darwinian Evolution. *BioScience, 20*:1067–69.

Urey, Harold C. 1952. *The Planets.* Yale University Press, New Haven, Ct.

Vallery-Radot, R. 1960. *The Life of Louis Pasteur.* Dover, New York.

Various theories and experimental data presented in Chapter 2 support the concept that life arose via a principle of continuity. That is, after the formation of the earth came the synthesis of small organic molecules, which evolved, through chance associations over billions of years, into more complex organic molecules. The organic molecules continued to increase in complexity, and through processes of reorganization and cooperation, they eventually became aggregates that had the ability to reproduce. During the process of chemical evolution, there must have been a process of molecular or "prebiological" selection. Those molecules most stable in a given environment would have remained, perhaps only to be eliminated later as they formed new but unstable associations with other compounds.

Such prebiological selection resulted in the formation of living cells. With the formation of the first living cells, a new but similar type of selection occurred—biological selection. Each cell type either had the potential to exist and reproduce within its environment or it ceased to exist. Thus the earth entered the time of biological evolution. Here also, as cells became more

Science is simply common sense at its best—that is, rigidly accurate in observation, and merciless to fallacy in logic.

T. H. Huxley

darwinism: the theory of natural selection / its application to organic evolution and its social impact

complex in their biochemical activities and chance associations with other cells, they may have, as a result of the increased complexity, become more susceptible to the pressures of the environment. Chemical and biological evolution did not occur progressively in the sense that living forms progressed into something "better"; they merely evolved into something that could reproduce in a given environment. As the various cell types of organized life continued to evolve, those that survived gave rise to an enormous variety of organisms (see Figure 2–5).

Naturalists have always been struck by the variety of organisms and their diverse modes of life. It was obvious that organisms were highly adapted to the various conditions in which they lived; yet no one, before Darwin, conceived that the process of organic evolution could give rise to such variety. It was precisely that variety and the adaptation of organisms to their environment that led Darwin, when writing his introduction to the *Origin of Species* in 1859, to ask "how the innumerable species inhabiting this world have been modified, so as to acquire that perfection of structure and coadaptation which justly excites our admiration." Darwin realized that the variety of species resulted from the endless variety of environments in which organisms lived, environments of both physical and biological origin. Darwin concluded that the various habitats did not directly evoke changes in an individual organism, but tended to preserve the inherited variations that were favorable in a particular habitat. Those organisms unable to cope with a particular environment did not survive. Thus natural selection perfected the organism's adaptation to its environment. Darwin, almost alone, insisted that organic change led only to increasing adaptation of organisms to their environment, not to an abstract ideal of progress defined by structural or organizational complexity. He disliked the terms *higher* and *lower* when referring to evolution. Today, many laymen still equate evolution with progress and define human evolution not simply as change, but as increasing intelligence, increasing height, or some other measure of assumed improvement.

family likeness and natural selection

Desmond Morris (1967), in his book *The Naked Ape*, quotes a sign in a zoo that says, "This animal is new to science." The African animal on display is a small squirrel with black feet. No black-footed squirrels had been found previously on the African continent. Why does this squirrel differ from the other 366 known species of squirrel? Is there anything about its way of life that makes it unique? Indeed, why are there 366 species of squirrels? They all look so similar. They literally have a family likeness. They have a family likeness because they have developed, via natural selection and organic evolution, from the same family lineage. The squirrels, according to the theory of organic evolution, all have a common ancestor, and each species of squirrel lives in its own particular environmental niche. The idea of family descent, that all species have a common ancestor, was not an original idea of Darwin's. Simple observation of *family* likeness gave rise to the idea of family descent. Yet the idea of a common ancestor, restated by Darwin in light of scientific data, created a shock that was felt around the world. Darwin made it clear to all that man had developed through organic evolution and that his closest ancestors were derived from the great apes. The idea of evolution became a family affair, and even those with limited knowledge of the subject entered the debate on whether or not man came from a "monkey."

Knowing that man resulted from the evolution of animal life was not sufficient for Darwin. He felt that evolution occurring simply along the lines of family descent did not explain how species could branch off from one another or from a common stock. He had made numerous observations during a five-year voyage on the *Beagle,* but still was unable to think of a mechanism that would allow life to progress from some common form (Darwin, 1840; Lack, 1961). Then in 1838, some 15 months after Darwin began recording and analyzing his observations, he read Thomas Malthus's *An Essay on the Principle of Population* (Irvine, 1955), an essay that had been published 40 years earlier. Darwin's observations of wild and domesticated plants and animals, along with his reading of Malthus, helped him to realize that there is a struggle for existence and that in this struggle the favorable variations of a species are most apt to survive. Those organisms with variations least favorable for a given environment are lost. Darwin followed this line of reasoning to its logical conclusion, and was struck with the idea that this struggle, or natural selection, which allowed only the more environmentally adapted to survive, could and would give rise to the formation of new species.

In his essay on population, Malthus had made two important points: (1) man seemed to be infinitely fertile and (2) the earth could sustain only a limited number of people for a given length of time. The greater the mass of people, the sooner we would deplete our world of its raw materials. But, Malthus pointed out, population increase would be kept in check by pestilence, war, and famine. Darwin realized that his observations indicated that animal populations in nature are subjected to similar forces.

To understand the processes of natural selection, Darwin asked: If pressures such as famine and pestilence restrict the size of animal populations, which members of a species survive? Why are those particular individuals spared? Darwin recognized that the variations within a species were experiments of nature, and that those of the species who were able to adapt to a given environment survived. Darwin felt that if large numbers of offspring were produced by a given species, that species would contain a sizable array of variations. He did not understand why more offspring led to greater variation, but did appreciate that greater variety gave a species a better chance for survival in any given environment. The great "struggle," envisioned by many when Darwin's theory of natural selection is discussed, is nothing more than which parents produce the greatest number of offspring. Fearful of personal bias, Darwin took four years to set down his thought-provoking ideas in writing. In 1842 he finally wrote a brief abstract of his theory concerning the forces of natural selection in the process of evolution (Bronowski, 1969).

Charles Darwin's observations and his chance reading of Malthus allowed him to synthesize a single theory of evolution by combining two previously known ideas—the idea of family descent and the idea of natural selection. Proposing the principle of natural selection as the force of organic evolution lifted Darwin from the ranks of the average scientist. Based on an analysis of his observations, Darwin not only gave a description of what probably occurred through time, but more importantly, he set up a conceptual framework not only for explaining the history of all life on earth but also for the fundamental inquiry into the history and nature of the human species.

darwin and wallace

Darwin nearly lost his right to claim the theory of natural selection. Twenty years after his discovery of Malthus's essay, he was still gathering and cataloging facts for his book when he received a manuscript from Alfred Russel Wallace (1823–1913), setting forth the same theory. Wallace also got his idea on the processes of natural selection from reading the Malthus essay. The two naturalists, working independently, but each influenced by Malthus, arrived at the theory—that animal species evolved through the process of natural selection. Wallace, ever so humble, insisted that Darwin receive full credit, since he felt Darwin had conceived of the theory before him. He also realized that Darwin's work was much more complete and that he did not have nearly as much evidence to support the theory as Darwin did. They agreed to have their papers read jointly at the next meeting of the Linnaean Society of London, which was to take place on July 1, 1858.

One year later, after pressure from Wallace and friends, Darwin published what may be *the* classic of achievement in man's search for self-knowledge, *On the Origin of Species by Means of Natural Selection* (Stebbins, 1971). The public reaction to this book was immediate and violent. The fundamentalists took immediate offense and stated that Darwin's theory contradicted the biblical story of creation. Some scientists and laymen found answers in Darwin's book to questions that troubled them, while others could not accept the concept of mutability (changeability) of species. But Darwin had prepared his manuscript meticulously, answering only those criticisms pointed directly at the theory of natural selection, and succeded in convincing many of his adversaries, religious as well as nonreligious, that organic evolution occurred. By 1872 Darwin had published six editions of this classic work. In each edition he answered criticisms, and by the sixth edition he had calmed and convinced some of his most formidable critics.

In spite of Darwin's extensive and careful work, however, his theory of natural selection had one glaring weakness. Neither Darwin nor Wallace could account for the initial appearance of a variant form or its persistence from one generation to the next. The theory did not explain how, when a change or mutation occurred in a species, that trait could be passed on to the next generation but not always, or how, still more puzzling, the "disappearing" trait could sometimes, but not always, reappear in future generations.

Still, although Darwin could not explain how, he was certain natural selection occurred. Through the study of domesticated plants and animals, Darwin knew changes occurred in all their lines. At the same time he was astute enough to realize that changes occurred by chance in domestic plants and animals and were maintained by artificial selection. That is, if the change in a plant or animal was interesting or profitable, that plant or animal was kept for reproduction.

darwin and lamarckism

Except for the new "sports" (changed or mutant varieties) he observed and could not explain, Darwin believed that gradual change occurred in the evolution of plants and animals through an inheritance of acquired characters. Darwin was influenced by Lamarck (1744–1829), a French biologist, who was the father of the first theory of biological evolution. Lamarck considered the inheritance of *acquired* characters to be the most important, if not the sole, mechanism of evolutionary change. The type of evolutionary thought in which this view is accepted is known as "Lamarckism." According to Lamarck, variations are induced in organisms in direct association with the demands of their environment. He regarded as heritable the modifications thus called into being. To Lamarck, all variations were acquired and all variations were heritable (if one demonstrates large muscles owing to weight-lifting, his children will have large muscles).

Meanwhile Gregory Mendel (1822–1884), an Augustinian monk and a contemporary of Darwin's, had determined that characters, or traits, were transmitted from parent to offspring via sets of factors we now call *genes.* Mendel postulated that one of each pair of factors, or genes, governing a specific trait came from each parent. These genes could change or mutate, but they would not be lost and could reappear. That is, the genes (and traits) could be hidden or masked in one generation and revealed in the next. Mendel recognized that his work in genetics was related to the evolution of species, for in an introduction to one of his monographs in 1854 he wrote "It [genetics] appears, however, to be the only way in which we can finally reach the solution of a problem which is of great importance in the evolution of organic forms." (Charles Darwin would not publish his *Origin of Species* for another five years.) It seems unlikely that the Austrian monk was planning a direct study of evolution. He was interested in heredity and how traits could be handed on from parent to offspring, but, as his words show, he was well aware of the evolutionary significance of his work.

If Mendel was aware of Darwin's work or at least of the idea of organic evolution, Darwin and Wallace either overlooked or ignored the principles of Mendelian inheritance, which are necessary to a complete comprehension of Darwin's theory. But, then, there is no record of anyone paying attention to Mendel's work. Mendel published his data in an obscure journal and his work (which is reviewed in Chapter 8) was promptly forgotten for the next 30 years.

darwin the man

Charles Darwin was first and last a naturalist with uncanny powers of observation (Figure 3–1). Although he was not considered a person of superior intellect, once given a set of facts he would work endlessly to organize them and synthesize their meaning. Perhaps it was his tenacity that set Darwin apart from many of his contemporaries. In view of Darwin's voluminous work, it is difficult to imagine this scientific giant as a frail, weak, and listless individual who was ill most of his life. As a result of an organic illness that sapped his

Figure 3–1. Charles Darwin. The photograph was taken during the later years of his life. (Culver Pictures, Inc.)

energy, Darwin had difficulty with his reading and writing. Yet methodically he paced and guided himself toward his moment of public controversy.

In retrospect, it is ironic that early in Darwin's career, because of his dislike for the medical arts and his sensitivity to dissection of the human anatomy, his father sent him to Cambridge to study the scriptures. Later Darwin was to alienate the clergy with his theory of natural selection. At Cambridge he observed the insect life that was a part of the Cambridge setting and developed the skill of tabulating and organizing his observations. His interest in nature led to his meeting Professor John Henslow, a Cambridge botanist. Darwin set aside ideas of the ministry and finished Cambridge in 1831. After reading Humbolt's *Travels*, he looked for an opportunity to travel to the Canary Islands (Irvine, 1955). At this point, Professor Henslow recommended Darwin for the position of naturalist of H.M.S. *Beagle*, which was about to leave on a five-year voyage to survey the coast of South America.

Darwin suffered the first attack of his lingering illness before sailing, and was told he had heart trouble. He had resolved, however, to have his dream come true regardless of personal inconvenience, and on December 27, 1831, Charles Darwin went to sea on the *Beagle*. Events following his return are well-known history.

Darwin's Marriage. As William Irvine points out in his thoroughly enjoyable book *Apes, Angels and Victorians,* "if Darwin discovered evolution by accident, he arrived at marriage by logic." Darwin himself said, "Man scans with scrupulous care the character and pedigree of his horses, cattle, and dogs before he matches them; but when he comes to his own marriage he rarely, or never, takes any such care." Darwin listed the advantages and disadvantages of marriage before he began to search for a bride. The advantages included children, companionship, and female conversation; the major disadvantage was the loss of time to work. He concluded that he did not wish to face old age alone, that he needed and would have companionship.

In the actual selection of his bride, Darwin seemed to place more importance on others' opinions than on his own. He apparently married Emma Wedgwood because his family had been marrying Wedgwoods for generations. As for his bride-elect, Emma wrote to her aunt, "It is a match that every soul has been making for us." They married on January 29, 1839. Emma found she could not compel herself to learn science, and Charles found that, true to his expectations, he spent much time caring for and loving their firstborn son, William Erasmus Darwin. As time passed, the family expanded. Darwin's health continued to decline, and a perfect wife became a private nurse to a most appreciative, loving, and continuous patient. Regardless of poor health, Darwin was a prolific, although cautious, writer (Table 3–1).

As Darwin's writing and studies continued, his health became progressively worse. After learning that his weight had increased to 336 pounds, he refused to weigh himself again. Because of his failing health he was plagued with two deep emotional anxieties: that he would not live long enough to finish his work on evolution and that his illness might be inherited by his children. (Recall Darwin believed in the inheritance of acquired characters.) As history reveals, Darwin's fears were not realized. He finished the most important biological literary work of the nineteenth century, *On the Origin of Species by Means of Natural Selection*, and his work made his later years financially secure. Darwin also lived to see his children develop into healthy individuals, displaying no symptoms of their father's illness, who were his constant and devoted companions. Charles Darwin died peacefully at 3 A.M., April 19, 1882. He was 73 years old.

Darwin rests in Westminster Abbey. He is there because he convinced people *that* evolution occurred, not *how* it occurred. At the time of his death

Table 3–1 Chronological Listing of Darwin's Major Works

The Structure and Distribution of Coral Reefs	1842
Zoology of the Voyage of the Beagle	1840
Geological Observations on the Volcanic Island	1844
Geological Observations on South America, Being the 3rd Part of the Geological Voyage of the Beagle	1846
Barnacles, Volumes I–IV	1851–1854
On the Origin of Species by Means of Natural Selection	1859
The Descent of Man and Selection in Relation to Sex	1871
Insectivorous Plants	1875
Recollections of the Development of My Mind and Character	1876
The Effects of Cross- and Self-Fertilization in the Vegetable Kingdom	1877
The Forms of Flowers	1879
The Power of Movement in Plants	1880
The Ingenious Earthworm	1881
Darwin's *Autobiography*	1881

the mechamism he had proposed, natural selection, was not accepted, and according to Gould (1975), there may not have been a single unequivocal supporter of the theory of natural selection. Wallace drew a line at man's brain—how could selection have established brains of equal size in "savages" and cultured Europeans? Sir Charles Lyell (1797–1875) felt Darwin assigned more to the power of natural selection than the theory could account for in terms of evolution. E. H. Haeckel (1834–1919) and T. H. Huxley (1825–1895), Darwin's most vociferous supporters, believed in evolution, but not in the *mechanism* of natural selection. Natural selection was not widely accepted until the science of population genetics arose in the 1920's and 1930's. A more complete acceptance came in 1937 with the publication of Dobzhansky's *Genetics and the Origin of Species.*

neo-darwinism (the new darwinism)

As previously related, *Origin of Species* presented strong evidence that organic evolution occurred, but Darwin was unable to propose an adequate theory to account for variation. Most biologists and many lay persons accepted Darwin's theory of organic evolution; yet no one could rationally determine just how species could maintain themselves over a period of time and give rise to new species. Darwin's contemporaries were unaware of hereditary mechanisms.

With the discovery, in 1900, of Mendel's work on the factoral transmission of hereditary determinants through generations, the stage was set for the marriage of Darwinism and Mendelism. Mendel's theory of heredity was the perfect complement to Darwin's idea of natural selection. First, Mendel's factors (genes) were not diluted or lost as they were transmitted between generations, and second, the mechanism of Mendelian assortment and recombination of the genes provided the means by which one could account for the production of variation essential to Darwin's theory of natural selection. Actually, many arguments ensued over the proper interpretation of Mendelian heredity before the relationship between Mendelian genetics and evolution was understood.

The conflict arose between two separate schools of thought. Naively, the naturalists in the early 1900's, believing in Lamarckism, concentrated their investigations of variation on the phenotypes (outward appearances), believing that mutant phenotypes created further phenotypic variation. The early Mendelian geneticists believed that variation came about through the changes in Mendel's particulate matter, or factors. The support that gathered for a theory of evolution through gene change, or "mutation," was unfortunate in that now there was a mutation theory of evolution that implied rapid, abrupt change in evolution, which was not in keeping with Darwin's idea of a slow, continuous evolution. Thus as Mendelism gathered acclaim between 1900 and 1915, Darwin's idea of gradual evolution through natural selection lost attention. By 1920, however, many geneticists realized that Mendelian heredity and Darwinism were complementary. We now realize that variation is due to the effect of both genes and the environment, and that natural selection, which is expressed in the phenotype (appearance), is a product of genes and their interaction with their environment. Natural selection acts on a population during all stages of development.

The resolution of the apparent conflict between genetically and environmentally caused variability in the process of organic evolution has been presented in detail in two classic books: Robert Aylmer Fisher's *The Genetical Theory of Natural Selection* (1930) and Theodosius Dobzhansky's *Genetics and the Origin of Species* (1951; first published in 1937). Fisher's work showed that the study of evolution requires an understanding of mathematics as well as Mendelism and Darwinism. The introduction of mathematics into the understanding of variability with respect to evolution had begun with G. H. Hardy and Wilhelm Weinberg.

On July 10, 1908, G. H. Hardy published a short paper in *Science* stating the proof of his equilibrium principle. The principle asserts that a gene that appears frequently in a given large population will continue to appear at that frequency in succeeding generations, if there are no new gene mutations, no large migrations into or away from the population occur, and mating occurs in a random fashion.

In 1905 a German physician, Wilhelm Weinberg, studying human genetics, used mathematics to analyze a trait recognizably inherited in man. As a result of these studies, Weinberg announced on January 13, 1908, that he had derived a general equilibrium principle for a single locus consisting of two alleles, or Mendelian factors (genes). Although Weinberg's report preceded Hardy's publication by six months, his claim was ignored, and for many years the general principle of gene equilibrium was known as Hardy's law. Today it is referred to as the Hardy–Weinberg law (Provine, 1971).

The Hardy–Weinberg law was immediately accepted by Mendelian geneticists, since Hardy's and Weinberg's work was based on the mechanisms of Mendelian principles. Accepting this work as valid indicated an understanding of what Mendelism meant to evolution. The Hardy–Weinberg law states that the frequency of genes will remain the same from one generation to the next *only if* certain conditions are rigidly fixed: if the population is large, mating is random, there is no migration, and there are no mutations, that is, if all genes are permanently "fixed" in one state. A violation of any of these conditions would allow evolution to occur. Obviously, the Hardy–Weinberg model is an ideal and cannot be a part of the natural biologically reproducing world. Humans, for example, migrate freely nearly worldwide; there is little if any control over spontaneous gene mutations; and we certainly select the partner with which we wish to produce children. Thus there is a large chance for gene exchange among people, and such exchange will lead to genetic variation and continued evolution.

The Neo-Darwin or new concept of organic evolution, like Darwin's concept, recognizes that the direction and rate of evolution are largely determined by natural selection. Nevertheless, the role of selection must be understood in terms of all forces that can and do bring about variation. Such forces include the ability of genes to change (mutate), the ability of chromosomes (on which the genes are located) to exchange genetic material and recombine with each other (recombination), and the mating of unlike parents (genetic hybridization). Neo-Darwinism incorporates the additional scientific knowledge of Mendel's work on the transmission of genes from parent to offspring into Darwin's original theory of natural selection. This understanding of the genetic basis of variability in a population strengthens Darwin's original contribution because it allows for a more complete understanding of how the forces of natural selection operate in the process of organic evolution.

the social misuse of darwin's "theory of natural selection"

A decade after Darwin published his theory on natural selection, his cousin, Francis Galton, produced *Hereditary Genius* (Galton, 1869). Galton presented considerable evidence supporting the contention that genius, talent, and morality tended to be inherited along family lines. He also suggested that Darwinian selection, if guided in a scientific manner, could improve mankind. With respect to inherited intelligence, for example, Galton argued that Negroes are, on the average, mentally inferior to European whites. Negroes are two "Galton" grades below whites, a Galton grade being equivalent to ten points on our current IQ distribution tests (Provine, 1973).

Galton concluded that an intellectually superior race, the whites, should not crossbreed with the black race. Such cross breeding, he believed, would only result in a lowering of the average intelligence and consequently reduce the number of the white geniuses available to society. After reading his cousin's book, Darwin wrote to Galton and praised his work, but Darwin thought the program of improving mankind through conscious action, or eugenics, a bit utopian. Galton, however, was not swayed from his eugenical speculation, and in 1871 he published *Gregariousness in Cattle and Men* in which he argued that we have leaders and followers or a leader and his herd. He referred to human herds as people possessing "slavish aptitudes," pointed out their "willingness for servitude," and contrasted them with leaders who were bold, original, and free of tradition. Such leaders, the Galtonian aristocrats of society, had been separated from the slavish masses through the Darwinian process of natural selection. Galton also felt that various forms of selection could eliminate people who were not really suitable to civilization.

Later, we were to hear Friedrich Nietzsche make the same case against the human mass. Both Galton and Nietzsche were *using* the *biological concept* of "survival of the fittest" (which, of course, means those capable of producing the most offspring, some of which will survive in a given environment) to gain support for their ideas on social inequality. Recently, William Shockley and Arthur Jenson have indicated that they believe there may be an IQ difference between the Caucasoid and black races, with the Caucasoid having the average higher IQ. Shockley has petitioned the National Academy of Sciences to fund basic research in order to determine if the IQ difference between the races is real or imaginary.

racial myths

We may credit Aristotle with the idea of superior and inferior people. Aristotle put forth the notion that certain people, by nature of their birth, were free, while others were to be slaves, and the concept has persisted through the ages. With the colonization of Africa and the discovery of America came an expansion of slavery, and many believed the use of slaves in colonization was justified by the writings of Aristotle. The slave, they argued, was inferior and irrational.

By the nineteenth century cotton was in such demand and so economically rewarding that the planters in the Southern part of the United States intensified their belief in the Cotton Belt Myth—that blacks were actually inferior to whites and would not be able to survive if they were left to fend for themselves. Thus Darwin's biological theory of the survival of the fittest was immediately supported by the white plantation owners, for it offered them a means of justifying their policy of expanding slavery.

spencer and social darwinism

During this same period, Herbert Spencer (1820–1903), when describing natural selection, applied Darwin's unfortunate phrase "survival of the fittest" to sociology. Consequently, use of the rifles and machine guns in war was regarded as merely the means for implementing Spencer's sociological concept of the survival of the fittest, or the replacement of the inferior by the superior. Neither the aggressor nor the victor felt any remorse because of the belief that the stronger is economically, biologically, and scientifically justified in his destruction of the weaker. The misuse of Darwin's biological fitness has also taken other forms, for example: the rich live in better homes because they are more fit to succeed; the rich should dominate the poor; the employed are more fit than the unemployed; and the Aryan should rule the world (Hitlerism of World War II).

There was and still is a tragic loss of genetic quality in warring nations, a loss that Spencer and his proponents should have recognized. For the qualities that help make armies invincible are weakened in the winning; it is the strongest, bravest, most intelligent—in short, the most fit—who die. Those who do not meet the criteria for soldiering are left to reproduce. The social and biological implications of such facts are very arguable and provide excellent topics for class discussion.

In the early 1900's two well-known social philosophers, Charles Horton Cooley (1864–1929) and Edward Alsworth Ross (1866–1951), stated theories on the value of the individual and on the idea that Social Darwinism was correct. Both authors wrote a variety of books and monographs on social inequality and "who should lead the pack." An excellent presentation of their point of view concerning man and his inherited cultural worth, from which the following excerpts are taken, is "Progressive Social Philosophy: Charles Horton Cooley and Edward Alsworth Ross" by Paul C. Violas (Karier et al., 1973).*

Both Cooley and Ross repeatedly stated that they did not consider men equal. Their belief in the basic inequality of men had important ramifications for their

* Clarence J. Karier, Paul Violas, and Joel Spring, *Roots of Crisis: American Education in the Twentieth Century,* © 1973 by Rand McNally College Publishing Company, Chicago, pp. 47–56.

social philosophy, for they used it to justify salient restrictions on the masses. Cooley confessed: "I do not know how to talk with men who believe in native equality: it seems to me that they lack common sense and observation."[1] Similarly, Ross justified elitism in decision-making, in part because "associates are unequal in capacity."[2] He described the urban migration of superior individuals from southerm Michigan and from Illinois and Wisconsin as "folk depletion," which left those areas "fished-out ponds populated chiefly by bullheads and suckers."[3]

Choosing a metaphor that likened the masses to lower species was hardly accidental. It represented a merger of nineteenth-century primitivism with Social Darwinism, which in turn yielded a view of human nature evolving along a continuum with some men still at the animal stage and an elite few as truly rational men. This concept had much currency among progressive intellectuals.

The Ross–Cooley system was keyed for change and could adapt to meet the vicissitudes of invention and fortune. It supposedly offered everyone the opportunity to develop capacities to their highest social usefulness. This system rested on the belief that men were unequal in abilities and that schools, aided by psychological tests, could fairly ascertain each individual's ability, assign him a vocational goal, and train him to become effective in it. Ideally, each individual was free to develop his capacity to its fullest as long as it was consonant with society's needs. Even this ideal, however, broke down under the weight of the authors' class, race, and ethnic biases. Their writings clearly reveal that they did not expect everyone to seize with equal dispatch all the possibilities for self-development. Cooley boldly proclaimed: "An Englishman or a German will seize upon all the opportunities in sight and demand more when an Italian or a Spaniard will perhaps make no use of those that are at hand."

Ross's writings may have influenced President Theodore Roosevelt. After receiving Ross's *Sin and Society* from Justice Oliver Wendell Holmes, the President sympathetically read all of Ross's works (Karier et al., 1973).

. . . races certainly appear to differ in the strength of those native propensities. . . . The Negro has a fiercer sex appetite than other men, . . . the Southern Italian has a bent for murder, . . . the Irishman has an uncommon taste for fighting, the Jew for money-making.[5]

Both Cooley and Ross were disturbed by what Ross had called "race suicide." The superior race, native white American, was not as prolific a breeder as the "other half." The superior race's decline in relative numbers would have frightful consequences for the nation.

"The cheap stucco manikins from Southeastern Europe," Ross lamented, "do not really take the place of the unbegotten sons of the granite men who fell at Gettysburg and Cold Harbour." Because the "Slovaks," "Syrians," and "Italians" were "undersized in spirit, no doubt as they are in body, the later comers lack the ancestral foundations of American character, and even if they catch step with us they and their children will, nevertheless, impede our progress."[6]

Both sociologists argued that the inherent inferiority of certain immigrant stocks

[1] Charles Horton Cooley, *Social Organization*, Scribner, New York, 1909.

[2] Edward Alsworth Ross, *Foundations of Sociology*, Macmillan, New York, 1905.

[3] Ross, *The Social Trend*, Macmillan, 1922.

[4] Cooley, *Sociological Theory and Social Research*, Holt, New York, 1930.

[5] Ross, *The Principles of Sociology*, Appleton Century, New York, 1938.

[6] Ross (1905), op. cit.

required new laws to exclude those who would debase the national stock. They lent the prestige of their "scientific analysis" of society to the mounting demand for legislation to restrict immigration according to racial and ethnic criteria.

The menace of foreign ethnic and racial impediments to American progress represented only one side of the problem. The other was pollution resulting from native "degenerate types." Ross agreed with Cooley's dictum: "Social improvement and eugenics are a team that should be driven abreast."[7] Cooley also stated how eugenics [the science of improving the human gene pool] should be instituted. "Scientific tests should be made of all children to ascertain those that are feeble-minded or otherwise hopelessly below a normal capacity, followed by a study of their families to find whether those defects are hereditary. If it appears that they are, the individuals having them should, as they grow older, be prevented from having children to inherit their incapacity."[7] Cooley found great eugenic value in the "upper class," which "does, after all, contain a large number of exceptionally able families."[8] Ross displayed his class bias as he argued that "from a quarter to a third of the paupers are hereditarily defective. Half or more of the chronic inebriates (drunks) are victims of a bad heredity."[9] He recommended custodial care or sterilization for what he termed "the devil's poor."[9] This is not to suggest that either Cooley or Ross agreed with all the crude eugenic and sterilization proposals formulated during their time. It does mean, however, that their writings did lend a cloak of intellectual respectability to such dehumanizing projects.

The author recommends that the student find the time to read the whole chapter.

Even though we now recognize that sociological Darwinism is a perversion of biological Darwinism, society still continues the practice. Our mistreatment of the aged and indigent stems from the misconceived notion that to be economically "fit" one cannot be idle, dependent, nonproductive, or sickly. Montaigne tells the story of Caesar, who was approached by an old soldier of his guard, who asked to be put to death. Seeing the old man's decrepitude, Caesar replied, "You think, then, that you are alive?" This remark may appear cruel, but we do no less when we think of old age as a kind of ugliness, and this attitude may be derived from unwillingness to tolerate idleness, dependence, or nonproductivity. On an even broader scale was the eugenics movement (an attempt to improve the genetic quality of man) of the early 1900's. It was based on the idea that the less fit—the poor, the mentally disturbed, the criminals, the vagrants, the physically deformed, and the economically impoverished—should not be allowed to reproduce. This movement lost support during World War II because of Hitler's eugenic policies. (The subjects of positive and negative eugenics are presented in Chapters 22 and 23.)

In keeping with Darwin's "survival of the fittest," one could say that *biologically* men are created unequal. The truth is, however, that men are not so much born biologically *unequal* as they are born biologically *different.* And therein lies the basis of Darwin's real concept of the survival of the fittest. Men are biologically and genetically different from each other. A changing environment determines who will survive, not just the individuals but the groups that will survive. Individuals do not live long enough for their survival to be of

[7] Cooley, *Human Nature and the Social Order,* Scribner, New York, 1902.

[8] Cooley, *Social Process,* Scribner, New York, 1918.

[9] Ross (1938), op. cit.

lasting biological importance. When we talk of survival, we mean the survival of a self-perpetuating group over long periods of time. Darwin's biological principle of selection was welcomed by the people of power, as they simplified it, distorted it, and adapted it to their own particular interests, but this Social Darwinism on which they based their right to social and economic privileges bears *no* relationship to Darwin's biological concept of natural selection and its role in organic evolution.

how does natural selection result in the biologically fit?

As has been noted, Darwin's theory of natural selection was, and still is in many cases, poorly understood and even distorted, especially in relation to the phrase "survival of the fittest." Darwin coined this phrase after reading Malthus's essay. Malthus stated that more children are born than live to become adults, and Darwin was quick to discern that this was true for all biological species. Next, Malthus stated that regardless of the speed of the reproductive process, a population would in time outrun its food supply, resulting in the death of some progeny in every generation as they "struggle for existence." Darwin referred to the winning of this struggle as the survival of the fittest.

The misunderstanding of this phrase stems from the fact that all species, except possibly modern man, are totally subject to their environment. Natural selection is not the annihilation of the weak. Rather, it is the survival of the greatest number of offspring because the more offspring there are, the greater the genetic variability in the population. It is partially a matter of chance which offspring survive, and in many cases the reason for their survival is not humanly comprehensible. Thus in a surviving population one can expect to find the weak as well as the strong. Humans, through the use of medicine, have increased the chance of survival for those who would have been unfit without medical aid.

Darwin showed that certain traits that appear to make a species fit are inheritable. Thus the survival of the fittest serves a noble purpose—the preservation of the adaptation that the organism has made to its environment.

Biological fitness is simply a measure of reproductive success. In any given generation, the members of a species transmit their hereditary potential to the succeeding generation. Those who survive in greatest numbers to reproduce ensure the continuation of their particular genotype or phenotype. Those genotypes and phenotypes that are less fit subsequently leave fewer reproductively competent progeny. Thus Darwin's fitness does not necessarily imply power, vigor, ruthlessness, or struggle. It implies rather that parents who are able to produce large numbers of offspring tend to ensure the survival of their kind. The Holy Bible advocated having many offspring long before Darwin, when it was written that man should "be fruitful, and multiply, and replenish the earth" (Genesis 1:28).

If matings produce no offspring (as in the case of some hybrids), the parents are said to have a Darwinian fitness of zero, regardless of their social fitness. The mating of a horse with a jackass results in a mule, which is a very strong and enduring animal. However, most mules are sterile and therefore genetically dead; they have a Darwinian fitness of zero. Conversely, reproductive fitness can be associated with certain hereditary diseases. For example, females who have diabetes or Huntington's disease or who are carriers for hemophilia, phenylketonuria, Tay–Sachs disease, or the sickle cell trait appear to be more fertile than females who are not carrying the defective genes (Firshein, 1962; Reed and Neel, 1959; Myrianthopoulos and Aronson, 1966). One who has the adult form of sugar diabetes or Huntington's disease can marry and have a large family before the defect makes its presence known. This increased fertility may actually have contributed to man's survival at an earlier period in his evolution. With a life expectancy of 20 to 30 years, prehistoric man would not have suffered much from a genetic disorder that is not expressed until the late 20's or so. A defect such as adult diabetes may also have had a further advantage. Diabetics accumulate adipose tissue, or fatty deposits, more rapidly than those having normal metabolism. Thus, a diabetic on a normal diet would accumulate a reserve of energy and a cushion from trauma and could derive sufficient energy from less food during periods of shortage. With the diabetic female more fertile and better able to survive on less food, the genes for diabetes could have been perpetuated. It would appear that what we now see as a defective gene could have been useful in the survival of early man.

We may even speculate that there would be more potential diabetics in the population today if man had not increased his life expectancy as he became agriculturally oriented and socialized. Since the expression of diabetes would have become more prevalent with increased longevity, growing awareness of the disorder may have decreased the matings of diabetics. Selection pressures thus may have been reversed relative to the gene for diabetes. Many thousands of years later we have a population where the incidence of nondiabetics is greater than the incidence of diabetics. The important point is that over great periods of time, selection favored genetic endowments that made reproductive success most probable. In general, genotypes and phenotypes that led to offspring liable to serious genetic disorders and infirmities failed to perpetuate their own kind. The process of natural selection, which bestows genetic fitness on an individual or a species, is opportunistic. It allows those exhibiting a fitness superior to others, at a given time and in a given environment, to reproduce. This fitness is an elusive factor in a population. Moreover, at any given time, a change in the environment can rapidly commit the fit to oblivion and make successes of the unfit or less fit.

When the dinosaurs died out, it made no difference that some were stronger and some were weaker or that the stronger conquered the weaker and thus obtained more food. When they perished, they perished uniformly, perhaps because size and strength made them less able to cope with a rapidly changing environment. It is not necessarily strength, ferocity, or aggressiveness but a collective "suitability" that determines which species have survival value and which disappear from history. The lowly cockroach has existed for millions of years and, some scientists speculate, may outlast man, whereas the kingly lion may not last another 100 years. In man also, the relative fitness of a given gene can change, as society or culture changes, especially if man gains control over his own evolution.

natural selection of man / why man's evolution is unique

As the angry debates over the Darwin–Wallace theory of organic evolution subsided, the theory permitted an intellectual revolution that changed man's image of how he became what he is now. To be sure, Darwin devoted little space in *Origin of Species* to the evolution of the human animal, but twelve years later, in 1871, he published his second monumental classic, *The Descent of Man and Selection in Relation to Sex,* in which he presented a detailed, in-depth study of human evolution. After Darwin's *Descent of Man,* however, nearly half a century elapsed before scientists were able to put forth a hypothesis regarding natural selection and man. The work of Fischer, Haldane, and Wright demonstrated that natural selection does occur in a human population and can be measured by devising models that can be mathematically understood.

Man is biologically unique in his capacity to think and hope. Through these processes man, alone among species, recognizes that an individual has a definite life span, that he is born to die. But while we live, we can bring about designed change. Over the generations we have relaxed many of the environmental pressures that once were primarily responsible for the condition of *Homo sapiens.* We are assured of a fairly constant environment through heating and air conditioning. We provide ourselves with the staples for required nutrition. Perhaps to our eventual demise, we have now provided ourselves with a variety of medicines and lifesaving techniques. Thus we are no longer completely at the mercy of our environment. We inhabit nearly all corners of the globe, and have multiplied to the point where biological success in reproduction threatens our very existence.

Paramount to man's survival was the evolution of a culture. It is unlikely that modern culture developed after modern man arrived as a species. Rather, modern man is probably the result of the combined simultaneous evolution of his biology and his culture. It is difficult to imagine how one could have evolved independently of the other.

We receive our biological inheritance in the form of chemical molecules of DNA, not as fixed characters or traits. Brown eyes or freckles are not transmitted directly from parents to their offspring. Rather, the parents transmit, through the DNA in their germ cells (sperm and egg), the biological potential for the child to have brown eyes or freckles, providing the environment allows the expression of this genetic potential as the fetus develops. We do not inherit a body temperature, bones that will knit when fractured, or a resistance to certain diseases. Rather, we inherit the biological potential for physiological reactions in particular environments (Dobzhansky, 1956). The physiological reactions that can or will occur are contingent on the genetic potential of the individual.

Theoretically, given a large enough population, there should be survivors regardless of the stress the natural environment places on a species. The smaller the number of members of a given species, the more likely the species will become extinct with environmental changes. Should the environmental change be so severe that no genotype in the population can resist that change the species will also perish. This is the most acceptable explanation for the disappearance of the dinosaur. Our geological strata contain the remains of thousands of species that have become extinct. Of all living species known to have inhabited the earth, only man has the intelligence to intentionally alter the course of natural selection. The real question facing man is whether our wisdom is great enough to allow us to steer a correct course as we influence our evolution. We should stop for a moment and wonder whether we would be better off evolving through the *chance* forces of natural selection without trying to alter the forces of natural selection. It is a question without answer but promotes intelligent discussion.

what forms of natural selection still operate in the human population?

Modern man, through his technological ability, is capable of altering many of the forces of natural selection that would otherwise continue to play their roles in man's evolution. Using any reasonable definition of biological success, we could say that man is the most successful product of evolution. He began his career somewhere in the Eastern Hemisphere, probably Africa, and continued to multiply and migrate until today over 3 billion members of the species are scattered over the entire earth. He has learned to domesticate plants and animals, make use of various resources for energy, and eliminate many of the predators, parasites, and diseases that plagued him. He has crossed the oceans, mountains, and deserts and even reached out into the solar system, landing on the moon and sending his machinery to the planets beyond.

Regardless of man's cultural accomplishments, including his ability to air-condition hostile environments, restore health to the sickly, and maintain his quality in life, man is still under the influence of some forms of natural selection. These include stabilization, diversification, and directionalization.

stabilizing natural selection

Stabilizing, or normalizing, selection acts to eliminate grossly deleterious genes as they arise and reveal themselves; it counteracts detrimental mutants and a wide variety of hereditary defects. It is a conservative force that favors the production of offspring fit for the current environment. If man interferes with nature's screening process (the survival of the most fit for a particular environment) by using major operations, expensive medicines, or machinery, then he assumes the moral responsibility for the increasing number of people who can transmit defective traits to their offspring.

In fact, we have already begun to neutralize the conservative force of natural selection, and, the incidence of genetically defective offspring is increasing. It has been suggested that if modern medicine continues to increase the number of defective genes in the human gene pool, we will one day reach unacceptable levels of genetic pollution. In the United States alone there are now over 6 million diabetics, and an additional 50,000 are born yearly. There are over 100,000 hemophiliacs, and 400 new hemophiliacs are born each year. Cystic fibrosis, an equally severe human defect, afflicts 1 in every 1500 newborns, resulting in some 2400 new cases per year. Some 5000 to 7000 children with Down's syndrome, a defect that causes severe mental retardation, are also born per year. And the list continues. Victor McKusick (1971), a well-known researcher in human genetics, listed some 1900 known human genetic defects in 1971; today there are over 2000. We continue to find additional hereditary defects as methods of diagnosis improve. It has been estimated that 5 to 8% of the new births in the United States involve a genetic defect. Twenty to 25% of all the beds in our institutions are occupied by persons with genetic defects. One could argue that if the severely defective either were not treated or were not allowed to reproduce, man could preserve the force of stabilizing or normalizing natural selection. Although this may be true, there do exist other trends that can help man to preserve his gene pool.

1. We have reduced the number of inbred or consanguineous (having a common ancestry) marriages, thereby decreasing the number of defective offspring in our population (Dobzhansky, 1961).
2. The number of children per family has decreased.
3. There has been an overall slowing in the national population growth.
4. Many defectives do not marry, or if they do marry, they do not have children. Some genetic defects are such that the person has difficulty finding a willing mate. Other defectives are sterile, and still others, who are mentally retarded, are institutionalized.
5. The increase in the frequency of a defective gene in the human gene pool, even with complete relaxation of selection, occurs very slowly. For example, it would take over a thousand years for the present incidence of phenylketonuria to double, that is, to go from 1 in 12,000 to 2 in 12,000 live births.
6. Although there are many possible defects, increased awareness of these defects, widespread screening detection programs, and accessible genetic counseling will help limit the number of defective humans.
7. There is hope that some day genetic manipulation will eliminate defects in the gametes or in the subsequent offspring.

Still, although the use of modern medicine will not necessarily lead to a genetic collapse, it does check the full and relentless purge of defective genes via natural selection. What this means is that the human population will never know the ideal man, that is, the man with the best constellation of genes nature could ultimately provide as a result of natural selection.

diversifying natural selection

A second force of selection presently affecting the human population leads to genetic diversification. Man lives in all geographical regions of the world and in an immense range of environments. Obviously, man has been able to adapt to the multiplicity of environments. But how does man adapt to these environments? To answer this question, we must look to genetic diversity within large populations. It is in large populations that natural selection favors diverse genetic endowments. Genetic diversity results from the various ways in which the chromosomes can cross over or exchange parts. Genes located on the chromosomes can exchange and recombine with each other as the sex cells become gametes (during the process of meiosis); sexual reproduction offers the human population an enormous opportunity for human variability (discussed in Chapter 4).

In all species in which sexual reproduction occurs, there is endless variation in the hereditary endowment of the offspring. Genetic variation comes from three principal sources:

1. Assortment of chromosomes during the formation of germ cells.
2. Recombination or rearrangement of genes within a chromosome or transfer of genes from one chromosome to another via crossing-over and translocation of chromosome parts.
3. Introduction of wholly new genetic information through mutation.

These processes produce astronomical numbers of different genetic combinations. Consider chromosome assortment alone. In man there are 23 pairs of chromosomes in the diploid set and 23 single chromosomes in the complement of each germ cell (egg or sperm). In the formation of a germ cell, 23 chromosomes can be selected, one from each of 23 pairs, in more than 8 million different ways, and since it takes germ cells from two individuals to produce a fertilized egg, the fertilized egg may have any one of more than 64 trillion (8 million times 8 million) possible combinations of chromosomes.

Now let us consider the effect of within-the-chromosome changes. Suppose the two chromosomes in a pair differ by 4 genes (most chromosome pairs probably differ by many more genes than that). The crossing-over process can form not 2 but 4^2 or 16 different possible chromosome combinations out of the original pair. If each of man's 23 pairs of chromosomes similarly differ in four gene pairs, the number expressing the possible different genetic combinations in man will have not 14 places, as in 64 trillion (64,000,000,000,000), but more than 30 (1,000,000,000,000,000,000,000,000,000,000) and will be vastly larger than the total number of people that have ever lived or will ever live. These calculations make no allowance for other chromosome changes or mutations.

Considering the enormous number of possible genetic combinations, we can more easily understand how each of 3 or 4 billion people can be genetically unique. In view of the enormous variability possible in the human species, there is no one perfect genotype for a given environment. Rather, there are many genetic variations capable of surviving in a given habitat. The interaction of a specific genotype with its environment determines its survival. Natural selection does not preclude genetic diversity, and a variety of genetic combinations can survive in a given environment.

directional natural selection

Another form of natural selection that may still be working on the human population is directional selection. Before man could control his "environment" with air conditioning, housing, and nutritional diets, he was subjected to the elements in ways modern man can hardly imagine. To survive the early hardships, he resorted to what he knew best. Early men adapted to their surroundings by responding to the challenges of the environment. Those that could adapt survived and passed on their genetic potential. With time and environmental change, genes that had once been useful but were no longer so were gradually eliminated. New mutations allowed the species to adapt to new challenges, and in time new characteristics and behavior appeared. If a change increased the reproductive fitness, or had a survival value, directional selection worked to improve and refine the trait for maximum efficiency in a particular habitat. With time whole segments of genetic potential, once necessary to survival, may have been lost because they were deleterious in a new environment. It is as though once a certain set of genes allowed survival in a given environment, other possibilities were eliminated. In this sense, after a long period of evolutionary time, there was no "turning back" (Figure 3–2). A given genotype and phenotype, always under the pressure of natural selection, gave rise to a "limited" number of new varieties. Even as these varieties were establishing themselves in a given environment, the continuing processes of gene change and genetic recombination made possible their evolution into still other species. However, each genetic combination, in a given environment at a

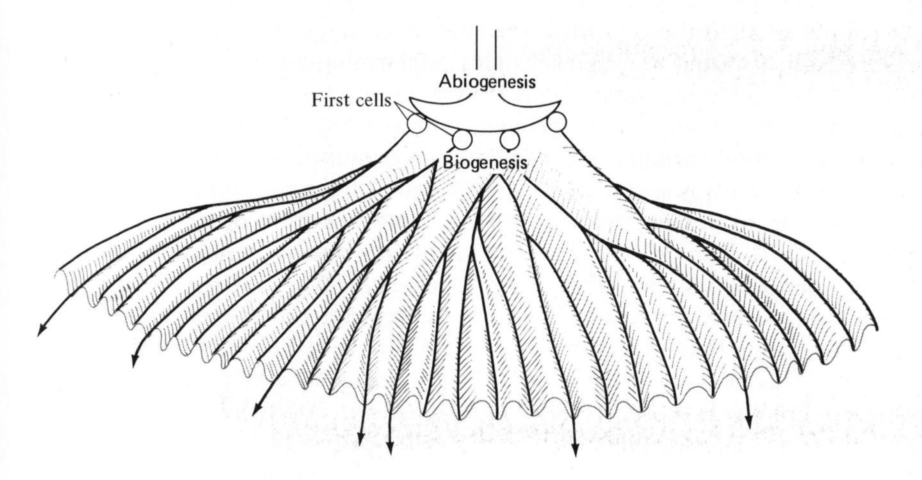

Figure 3–2. Model of the Process of Directional Selection. Many alternatives were available for the evolution of the first living cell types into simple and multicellular organisms. Cells began their evolution by following one of the various paths of development (shown as branches radiating out from the first cells). As each evolving species passed an "intersection," the number of alternative paths open to further evolution diminished. There was no turning back. Some species became extinct. Those that survived continued to evolve or be "channeled" in a "direction" as they passed each intersection of possible alternatives. With time each species takes on the appearance of stability in a given environment. (The arrows indicate that the process of directional selection is still manifest.)

given time, could only give rise to a particular species, and even by a very long-term process of genetic change a species could only give rise to a "limited" number of other species. For example, the great apes gave rise to the human family, and continued operation of directional selection led to modern man. We do not know how man will evolve in the future, but it would seem that even if natural selection were allowed to operate freely on the human species, there would always be a limited number of possibilities that humans could evolve to. Whatever evolved from our species would have been derived via directional natural selection. Our species would be "directed" toward the evolution of the next new species.

The forces of normalizing, diversifying, and directional selection promote a slow and integrated change. These processes result in the majority of a species having those traits most favorable in a given environment. The question for us is: Has man's culture eliminated the effect of directional selection? Man alters his culture much faster than changes can occur in his genetic potential. In a sense, our genes must adapt to our cultural milieu rather than our cultural environment adapting to the potential of our genes (Dobzhansky, 1961).

the social reaction to darwin's *descent of man*

According to Neo-Darwinism, units of inheritance (genes) provide the mechanism for speciation, or the creation of new species from older ones, under the influence of natural selection. In *Origin of Species* Darwin acknowledges that there must be variation for nature to act upon. He called his variants, or changed members within a species, "sports." He speculated that nature simply shuffled these "sports" around until they started off randomly in some direction, forming new species. This unalarming and rather simple idea became the real center of controversy. It was here that Darwin implied that the evolution of species came about through natural selection working at random on a set of accidental variants. This implied that evolution occurred by chance.

Further, Darwin was saying there were no goals for evolution to achieve, no direction for life to pursue, and no primary force in nature other than that of chance. In short, Darwin was saying that all life forms, regardless of size, shape, or complexity, were the work of chance. We can appreciate why religious orders were outraged. Darwin was denying the special creation of man and suggesting that he was descended from some ape or monkey-like form of primate. This was a direct attack on the unique relationship man was believed to hold with God. It was difficult, if not impossible, for religious leaders to accept the possibility that God would operate His world as though it were a game of chance, or as Albert Einstein once said, that He was "a God who plays at dice."

The clerics lashed out against the theory of evolution. A widespread and heated debate ensued over whether man was created by God, as set forth in the Bible, or had descended through the processes of evolution.

In France the opposition to Darwin's *Origin of Species* was strong, and leading statesmen referred to his work as a scientific mistake, untrue in facts and unscientific in method. In 1870 the French Académie des Sciences refused to admit Darwin as a corresponding member. In England the Royal Society of London presented the Copley Medal to Darwin, but suggested that it was for higher achievements than the writing of *Origin of Species.*

the huxley–wilberforce debate

One of the more famous debates was between Thomas Henry Huxley (1825–1895), an English biologist who supported Darwin, and Bishop Samuel Wilberforce. The debate took place before the British Association for the Advancement of Science in June 1860. This debate marked the beginning of man's seriously questioning his evolutionary relationship with God and his place in nature. The Huxley–Wilberforce debate, more than any other, left the impression that the concept of evolution was a major force in the erosion of religious concepts (Case and Stiers, 1971).

Bishop Wilberforce was well trained in mathematics and felt qualified intellectually to discuss the shortcomings of the theory of evolution. At the age of 54, he was a formidable speaker, and he knew his audience was against Darwinism. The Bishop, not really understanding Darwinism, regarded the novelties in evolution as mere absurdities and used his wit and sarcasm to denounce thoroughly the theory of evolution. As he concluded, overcome with his feeling of personal success, the Bishop leaned toward Huxley and asked with mock politeness whether it was through his grandfather or grandmother that he, Huxley, claimed his descent from a monkey. Huxley, after a moment's silence, whispered to a colleague, "The Lord hath delivered him into mine hands." As the roar of applause for Bishop Wilberforce died down, Huxley rose, and in his quiet way stated that he was there only to serve the interests of science and that the Bishop had said nothing to detract from the theory of evolution. Huxley stated that the theory of evolution was much more than a hypothesis, it was the best explanation of the formation of species ever proposed. Huxley referred to the Bishop's ignorance of science, explained Darwin's ideas very clearly, and then in the hush of the hall said he would not be ashamed to have a monkey for his ancestor, but he would be ashamed to be associated with men who used great gifts to obscure the truth (Irvine, 1955). Now the applause, previously given to the Bishop, was for Huxley. Huxley won the moment, not so much by making evolutionists or believers of the audience, but rather by relying on the value of truth telling. Huxley firmly believed there was truth in the work of Darwin.

the great monkey trial / fundamentalism versus modernism

A later renowned debate on the descent of man took place in 1925 in the small, obscure town of Dayton, Tennessee. For the first time the right to teach the theory of evolution was tested in a court of law. Laymen, religious leaders, and lawyers, secure in their fundamentalist beliefs, had sought to protect the young through enactment of legislation banning the teaching of evolution in public schools.

Although the trial took place in 1925, perhaps we can place its inception as early as 1907, when the millionaire brothers Lyman and Milton Stewart founded the Los Angeles Bible Institute. The Institute began issuing a series of pamphlets called *The Fundamentals.* These pamphlets provided the rallying point for those who were antievolution. The antievolutionists also formed a group known as the Adamists, and its membership boasted many of the influential people of the time. Perhaps the most prominent was William Jennings Bryan, a Presbyterian layman, a Sunday school teacher, an educator, and a prohibitionist. Professionally he was a lawyer-politician, a former member of the House of Representatives, who in 1908 was defeated in his third bid for the Presidency and in 1913 became Secretary of State. Successive losses in his attempt to become President and a waning political career may explain his increased concern that colleges were making unbelievers of their students in the early 1920's (De Camp, 1968). He confined his criticism of evolution to the statements that destroyed the image of man as taught in the Bible.

Bryan, who believed slogans were solutions to problems, began to tour the South, appearing before state legislatures, denouncing evolution, and demanding new laws. His highly emotional speeches before state legislatures created political and public turmoil. In Kentucky punitive legislation against teaching evolution in the public schools was defeated by only one vote. Politicians in South Carolina tried unsuccessfully to withhold public money from any school teaching the concept of evolution. Efforts were made to stop the teaching of evolution in Florida, Texas, Oklahoma, North Carolina, Alabama, Georgia, and California, although no legislation was enacted. In Tennessee, however, the farmer-legislator John Washington Butler introduced an antievolution bill in the House of Representatives, and the bill quickly passed both houses. Members of the House and Senate said they felt safe in passing the bill, since they were certain the Governor would veto it. The Governor, however, had aspirations of becoming a United States Senator and did not wish to alienate the voters. So he signed the bill into law on March 21, 1925, believing that it would never be enforced (De Camp, 1968).

The Chattanooga *Daily Times* of April 4 carried the news that the American Civil Liberties Union (ACLU) would finance a case to test the constitutionality of the Butler Act (De Camp, 1970). The following day in Dayton, Tennessee, Walter White, superintendent of schools and president of the local chamber of commerce, met with Sue K. Hicks, a young lawyer and an antievolutionist (*The South Today,* 1970). Together with George W. Rappleya, an evolutionist who managed a local, bankrupt coal company, they discussed the idea of holding the proposed test case in Dayton in order to gain national recognition and in turn attract industry to the town. After initiating the plan, they asked their friend, John Thomas Scopes, a 24-year-old, single substitute teacher, who neither understood nor taught evolution, to play the role of victim. When Scopes agreed, Rappleya, the evolutionist of the group, wired the ACLU concerning the case and received a positive reply. A warrant was served on Scopes for teaching evolution in defiance of the Butler Act. The antievolutionists retained their lawyers, Sue Hicks and John Randolph Neal, a former dean of the University of Tennessee Law School. Hicks then wrote to William Jennings Bryan offering him a place on the prosecution, and this offer turned out to be Bryan's final triumph. With Bryan's acceptance, the trial achieved national headlines. Then an offer to defend John Scopes came from America's foremost criminal lawyer, Clarence Darrow (Figure 3–3). Darrow, an agnostic, also offered the services of Dudley Malone, a divorce lawyer from New York, and they were joined by Arthur Garfield Hayes, an attorney for the ACLU.

The Great Monkey Trial, as it became known nationally, was nearly lost to Dayton as the city of Chattanooga, Tennessee, tried to rush through its version of a monkey trial for national recognition. Although Chattanooga gave up on the idea, Dayton still moved the trial up to July 10, a month ahead of the regular session.

The presiding judge, John T. Raulerson, a fundamentalist, tried to hold a fair trial despite the heat and humidity and the circus-carnival atmosphere, but he was confused by the facts of evolution and by the cleverness of the "city lawyers." In the end, the judge refused to allow scientific scholars, brought in as expert witnesses for the defense, to testify. Denied his experts, Darrow put Bryan on the stand. Bryan agreed to take the stand for an hour and 30 minutes. Darrow questioned Bryan on such subjects as the creation of man, Jonah's experience with the whale, and Joshua's power to compel the sun and moon to stand still. Bryan had to admit his lack of knowledge on these subjects and conceded that the earth might be millions of years old and not just

Figure 3–3. The Monkey Trial. The famous criminal trial lawyer Clarence Darrow, posed on the edge of the court desk, listens intently at the 1925 Dayton, Tennessee, "Scopes Monkey Trial." John Scopes is sitting at the table (fifth from the right) with his arms folded. Note the packed courtroom and the fact that the doors are open. It was a hot and humid day according to history. (The Bettmann Archive.)

thousands. This was a crucial point because the fundamentalists viewed the earth's creation and the beginning of civilization as nearly simultaneous events dating back to about 4000 B.C. (De Camp, 1968).

Finally, Darrow asked Bryan how snakes were able to walk before God made them crawl on their bellies. With this, Bryan, screaming and threatening, charged that Darrow's sole purpose was to slur the Bible. Darrow screamed back, and a near riot brought adjournment.

Bryan presented such passages from the Bible as "so God created man in his *own* image, in the image of God created he him; male and female created he them" (Genesis 1:27). Darrow, stripped of all defense, decided he was through after eight days of agitation and futile argument. Without a case, Darrow suggested a guilty verdict for John Scopes in order to appeal the case to the Supreme Court. The jury obliged, and Judge Raulerson fined Scopes $100.

With the trial over, the 100 or more journalists who had put Dayton, Tennessee, on the map left. Five days after the trial, William Jennings Bryan died in his sleep. Two years later Scopes's conviction was overturned on a technicality. He went on to a graduate degree at The University of Chicago and became a petroleum geologist. Later in life he confessed that on the day he supposedly gave his lecture on evolution he was out coaching a football team. John Scopes died in 1971.

With the victory in the Scopes trial, the Adamists continued pushing for other state legislatures to follow Tennessee. The Mississippi legislature passed the Mississippi code of 1942 prohibiting the teaching of evolution in public schools. And although the state legislature of Arkansas refused, a referendum vote passed a similar measure called the Rotenberry Act. In 1968 Justice Abe Fortas delivered the Supreme Court opinion, in the case of *Epperson v. Arkansas,* that the Rotenberry Act violated the First and Fourteenth amendments.

Nevertheless, opposition to Darwinism is not dead. In May 1963 Mrs. Jean Sumrall, leading a group of concerned California women, appeared before the California State Board of Education to protest the teaching of Darwin's theory of evolution as fact. In 1969 the California State Board of Education passed a resolution requiring teachers in California schools to teach that "scientific evidence concerning the origin of life implies at least a dualism or the necessity to use several theories." Thus the teaching of the evolutionary theory of Darwin must be balanced by the presenting of alternative theories such as creation as given in Genesis 1:27 (Peter, 1970). In contrast to the Adamists, the new advocates for including divine creation in school tests are highly educated members of the Creation Research Society. The members include those holding bachelor's, master's, Doctor of Philosophy, and M.D. degrees. They are committed to full belief in the biblical record of creation and early history.

Regardless of testimony in 1972 from members of the National Academy of Sciences, the American Chemical Society, various concerned citizens' and teachers' groups, and nineteen Nobel prize scientists, the California State School Board authorized new editions of science books for some 3 million students. A special "Creation" section will be presented alongside an altered version of Darwin's theory of evolution.

In 1973 Tennessee enacted a law calling for inclusion of the Genesis account of creation in biology textbooks, and its constitutionality was challenged in a suit filed in the U.S. District Court at Nashville. The suit was brought by the National Association of Biology Teachers (NABT) and three co-plaintiffs from Knoxville against the chairman and members of the Textbook Commission of the State of Tennessee. The 1973 act made it mandatory for biology text books to provide, whenever the theory of evolution is mentioned, an equal amount of emphasis on the Genesis account of the origins and creation of man and his world. The textbook must also emphasize that the process of organic evolution is only a theory and does not represent scientific fact. In their suit, the plaintiffs contended that:

1. The fundamental rights guaranteed by the First and Fourteenth amendments of the U.S. Constitution are being violated.
2. The Tennessee statute is an establishment of religion by the state in violation of the due process clause of the Fourteenth Amendment as that amendment incorporates the establishment clause of the First Amendment.
3. The act interferes with the free exercise of religion as that guarantee is incorporated by the due process clause of the Fourteenth Amendment.
4. The act abridges the free speech clause of the First Amendment as that clause is incorporated by the due process clause of the Fourteenth Amendment.
5. The act is a prior restraint upon the freedom of press in violation of the due process clause of the Fourteenth Amendment as that amendment incorporates the free press clause of the First Amendment.
6. The Tennessee statute violates the rights of plaintiff Wilder and Tennessee members of the National Association of Biology Teachers to academic freedom.
7. The act is void for vagueness in violation of the due process clause of the Fourteenth Amendment.

In April 1975 the U.S. Sixth Circuit Court of Appeals in Cincinnati ruled, in a two to one decision, that the Tennessee textbook law of 1973 unconstitutionally established a preference for the teaching of the Biblical account of creation over the theory of evolution. The case was called a new version "of the legislative effort to suppress the theory of evolution which produced the famous Scopes Monkey Trial of 1925," and the court pointed out that while the 1973 act of the Tennessee legislature did not directly forbid the teaching of evolution, the purpose of the new legislation was the same as that of the 1925 law.

We might conclude that the debate concerning fundamentalism and evolution is far from over. The Monkey Trial, fundamentalism, and the theory of evolution continue to command attention.

references

Bronowski, Jacob. 1969. Nature and Knowledge: Philosophy of Contemporary Science. *Condon Lectures.* Oregon State System of Higher Education.

Case, J. F., and V. E. Stiers. 1971. *Biology: Observation and Concept.* Macmillan, New York.

Darwin, Charles. 1840. *The Voyage of the Beagle.* Reprinted 1960, Bantam Books, New York.

De Camp, L. S. 1968. *The Great Monkey Trial.* Doubleday, New York.

De Camp, L. S. 1970. Evolution Still on Trial After 100 years. *Science Digest, 67*:21.

Dobzhansky, Theodosius G. 1951. *Genetics and the Origin of Species,* 3rd ed. Columbia University Press, New York.

Dobzhansky, Theodosius G. 1956. *The Biological Basis of Human Freedom.* Columbia University Press, New York.

Dobzhansky, Theodosius G. 1961. Man and Natural Selection. *American Scientist, 48*:285–99.

Firschein, I. L. 1962. Population Dynamics of the Sickle-Cell Trait in the Black Caribs of British Honduras, Central America. *American Journal of Human Genetics, 13*:233–54.

Fisher, Ronald Aylmer. 1930. *The Genetical Theory of Natural Selection.* Clarendon Press, Oxford.

Galton, Francis. 1869. *Hereditary Genius.* Reprinted 1952, Horizon Press, New York.

Gould, Stephen Jay. 1975. *Apes, Angels and Victorians: Darwin, Huxley and Evolution.* McGraw-Hill, New York.

Karier, Clarence J., Paul Violas, and Joel Spring. 1973. *Roots of Crisis: American Education in the Twentieth Century.* Rand McNally, Chicago.

Lack, David. 1961. *Darwin's Finches.* Harper & Row, New York.

McKusick, Victor. 1971. *Mendelian Inheritance in Man: Catalogs of Autosomal Dominant, Autosomal Recessive and X-Linked Phenotypes,* 3rd ed. Johns Hopkins Press, Baltimore.

Morris, Desmond. 1967. *The Naked Ape.* Dell, New York.

Myrianthopoulos, Ntinos, and S. Aronson. 1966. Population Dynamics of Tay–Sachs Disease I: Reproductive Fitness and Selection. *American Journal of Human Genetics, 18*:323–27.

Peter, W. G., III. 1970. Fundamentalist Scientists Oppose Darwinian Evolution. *BioScience, 20*:1067–69.

Provine, William B. 1971. *The Origins of Theoretical Population Genetics.* University of Chicago Press, Chicago.

Reed, T. E., and J. V. Neel. 1959. "Huntington's Chorea." In C. J. Bajema (ed.), *Natural Selection in Human Populations.* Wiley, New York.

The South Today, 1970. Scopes: This is Now, 1970. Old Law Debated in Mississippi. *1*(10):1.

Stebbins, G. Ledyard. 1971. *Processes of Organic Evolution,* 2nd ed. Prentice-Hall, Englewood Cliffs, N.J.

Individual humans live for a certain period of time and then die. If the human species is to survive on this planet, reproduction must occur somewhere between birth and death. Each generation in its turn must produce another, and the number of new individuals in each generation is very important. If fewer than the replacement number are produced, there will be a decline in population with a subsequent loss of genetic diversity. Yet with uncontrolled reproduction, although it would certainly add genetic variation, many individuals would die as a result of excessive competition for limited resources. At present we do not know how many humans are necessary to provide the genetic variability required for our survival as a species or the number necessary to sustain that variability. Man has circumvented many processes of natural selection, through technology, medicines, and nutritional awareness, without fully assessing the impact of his actions on future generations. Those individuals who once would have succumbed to various diseases and developmental malformations have survived and reproduced their kind. It is very difficult to say how this will affect the survival of our species. Yet the suggestion has already been made we face a genetic collapse, meaning that one day the majority of the people entering their reproductive years will be genetically defective.

And God said, Let us make man in our image, after our likeness: and let them have dominion over the fish of the sea, and over the fowl of the air, and over the cattle, and over all the earth, and over every creeping thing . . . . So God created man in his own image . . . male and female created he them . . . . and God said unto them, Be fruitful, and multiply, and replenish the earth, and subdue it. . . . And the Lord God planted a garden eastward in Eden; and there he put the man whom he had formed.

Genesis 1 : 26–28, 2 : 8

human reproduction, cell division, and genetic diversity

the continuum of life in a sexually reproducing species

In all sexually reproducing species, there are equivalents to the human sperm and egg. As with other species, it is through these sex cells that the human transmits the history of his species and promotes the survival of his species. Human sexual reproduction is the union of two separate cell types, the egg and sperm, and the subsequent division of the united cell, or zygote, into a new individual. The new individual, like every other, is equipped with special organs called gonads. In males the gonads are testes; in females they are ovaries. It is the gonads that give rise to the gametes (sperm and egg). The union of the sperm and egg is referred to as fertilization. The act of fertilization produces the first cell of the new generation, the zygote. The zygote contains, in structures called chromosomes, the hereditary instructions to form a new human. If the zygote is situated in a suitable environment, it undergoes "cell division," a process by which one cell divides to give rise to two genetically similar daughter cells. The first cell divides into two, those two into four, four into eight, eight into 16, etc. If the original genetic instructions are correct and followed, the result is a series of developmental changes referred to as the egg phase, the embryo phase, and the fetal phase. After nine months of normal gestation, birth occurs (Figure 4–1).

the process of cell division

All sexually reproducing multicellular organisms depend on cell division. Although each organism may start out as a single-celled zygote, enlargement and multiplication of this single cell are required to produce the final form of a given species. A complete cell division includes two separate processes: nuclear division, or karyokinesis, and division of the cytoplasm, or cytokinesis.

Two types of nuclear division are known to occur in sexually reproducing species: mitosis and meiosis. In mitosis the nuclear materials divide so that the resulting daughter cells are genetically identical. Mitosis occurs in all cells except those destined to become gametes or sex cells. Meiosis occurs only in conjunction with the formation of gametes. Meiosis can account for the variation among humans via the shuffling of the parental genetic material during the formation of gametes. How are the genetic messages of the DNA in the zygote passed on to the trillions of cells that make up the human adult? The mechanism for transmitting identical genes to all daughter cells derived from the original single-celled zygote is mitosis. Mitosis is sometimes referred to as asexual cell division, in contrast to meiosis, which is the process of cell division involved in sexual reproduction.

mitosis

In mitosis the genetic material in the nucleus is duplicated precisely ($2 \times$ DNA is duplicated to $4 \times$ DNA), and a complete set of the material is

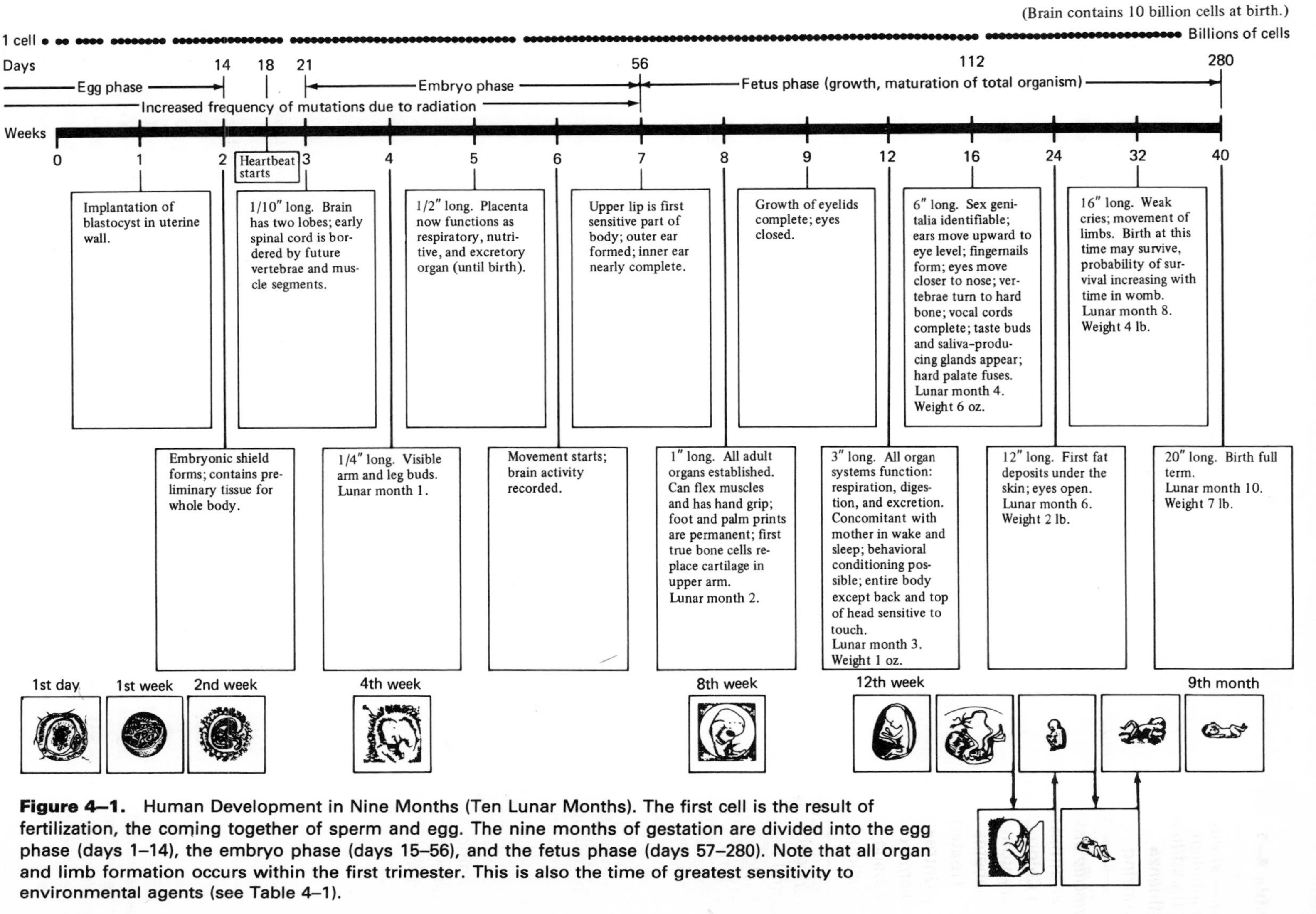

Figure 4–1. Human Development in Nine Months (Ten Lunar Months). The first cell is the result of fertilization, the coming together of sperm and egg. The nine months of gestation are divided into the egg phase (days 1–14), the embryo phase (days 15–56), and the fetus phase (days 57–280). Note that all organ and limb formation occurs within the first trimester. This is also the time of greatest sensitivity to environmental agents (see Table 4–1).

Table 4–1 Environmental Agents Affecting Development During Gestation

General environmental agents associated with birth defects	
Alcoholism	Malnutrition
Drug addiction	Vitamin A deficiency—eyes and soft parts
Influenza	Vitamin B_2 or B_6 deficiency—skeleton
Smoking	Vitamin D deficiency—bowing of long bones

Specific environmental agents associated with birth defects	
First trimester	Second trimester
p-Chlorophenylalanine—mental retardation	Syphilis
Cryptomegalovirus	Third trimester
Diabetes	Tetracycline
Treated diabetic mother—hyperglycemic, ketotic	
Untreated diabetic mother—large pancreas, severely hyperglyemic	
Diamox (acetazolamide)	
Rubella	
Thalidomide	

Drugs dangerous to pregnant women	
Central nervous system	Hallucinogen
Analgesics	Lysergic acid diethylamide (LSD)
Chlorpromazine	Metabolic drugs
Meperidine (pethidine)	Diphenylhydantoin (Dilantin)—cleft palate, skeletal and cardiac malformation, coagulation problems
Mephenytoin	Iodine
Phenantoin	Vitamin A
Phenobarbital	Narcotics
Phenothiazine	Diacetylmorphine (heroin)
Phenylbutazone	Methadone
Phenytoin	Morphine sulfate
Primidone	Stimulants
Promazine (Sparine)	Imipramine
Quinine	Phenmetrazine
Salicylates (aspirin)	Reserpine
Tolbutamide	

distributed to each daughter cell. Mitosis is usually regarded as occurring in five more or less distinct phases: interphase, prophase, metaphase, anaphase, and telophase (Figures 4–2A, 4–3, and 4–4).

Mitosis can be described for a hypothetical organism in which each cell contains four chromosomes, as diagrammed in Figure 4–2A. The chromosomes are the rodlike bodies that carry the cell's genetic information (DNA). Two of the chromosomes, a long one and a short one, were inherited from the mother and two from the father. The two long chromosomes are similar to each other, as are the two short chromosomes. Matched members of a pair of chromosomes are called homologous chromosomes. The four-chromosome cell has two pairs of homologous chromosomes. At interphase, the first phase of

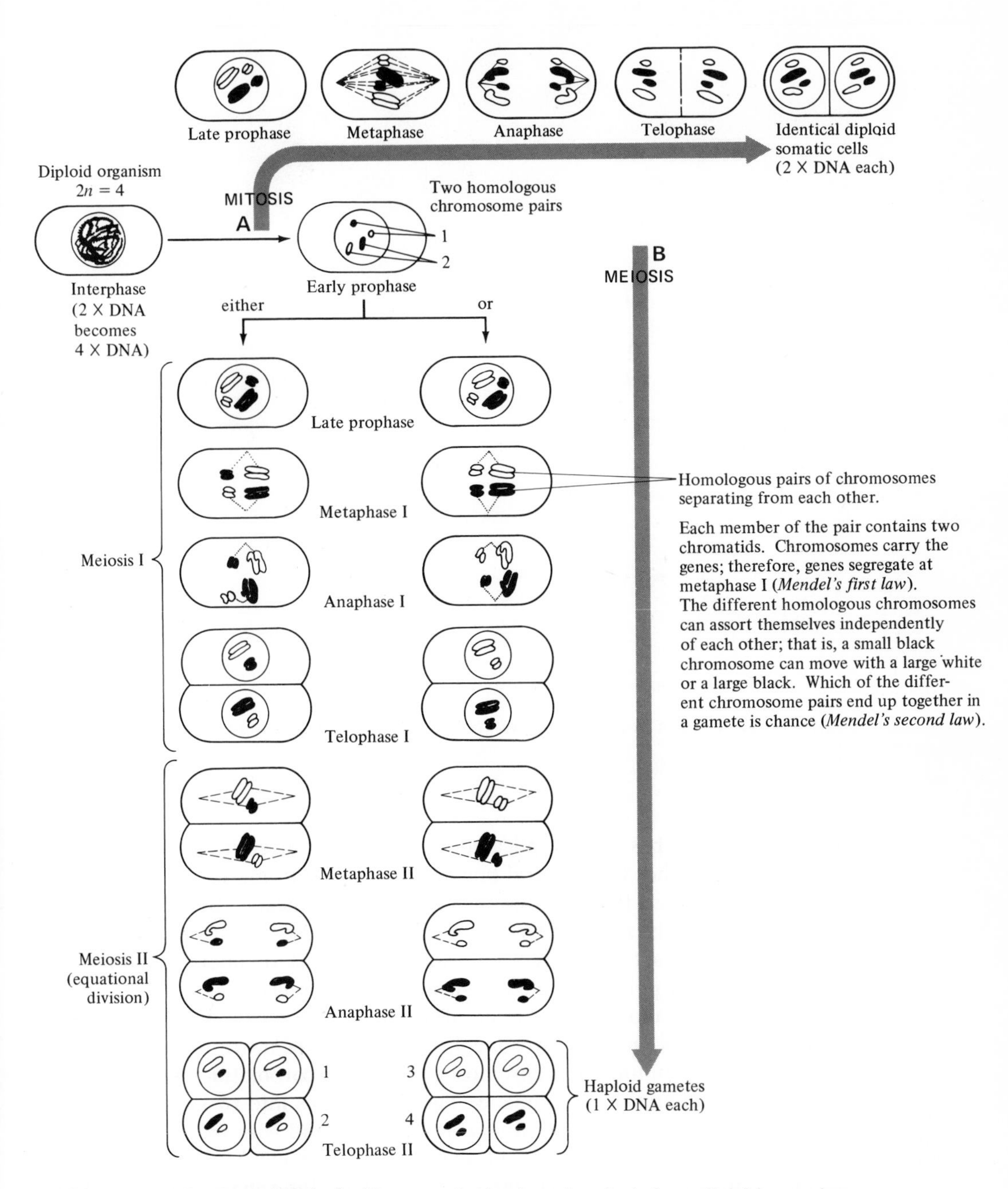

Figure 4–2. Mitosis and Meiosis. Diagram of mitosis and meiosis for a diploid organism with only two pairs of chromosomes, $2n = 4$. **A.** Through mitosis one cell gives rise to two daughter cells, each genetically identical to the first cell. **B.** Four different genetic combinations are possible from meiosis, as shown by the arrangement of the black and white, long and short chromosomes. In humans the shuffling of the 23 pairs of male and female chromosomes at meiosis results in $2^{23} \times 2^{23} = 64 \times 10^{12}$ (64 trillion) different possible genetic combinations. Meiosis offers variability; mitosis offers stability. (Adapted from G. J. Stine, *Laboratory Exercises in Genetics*, Macmillan, 1973.)

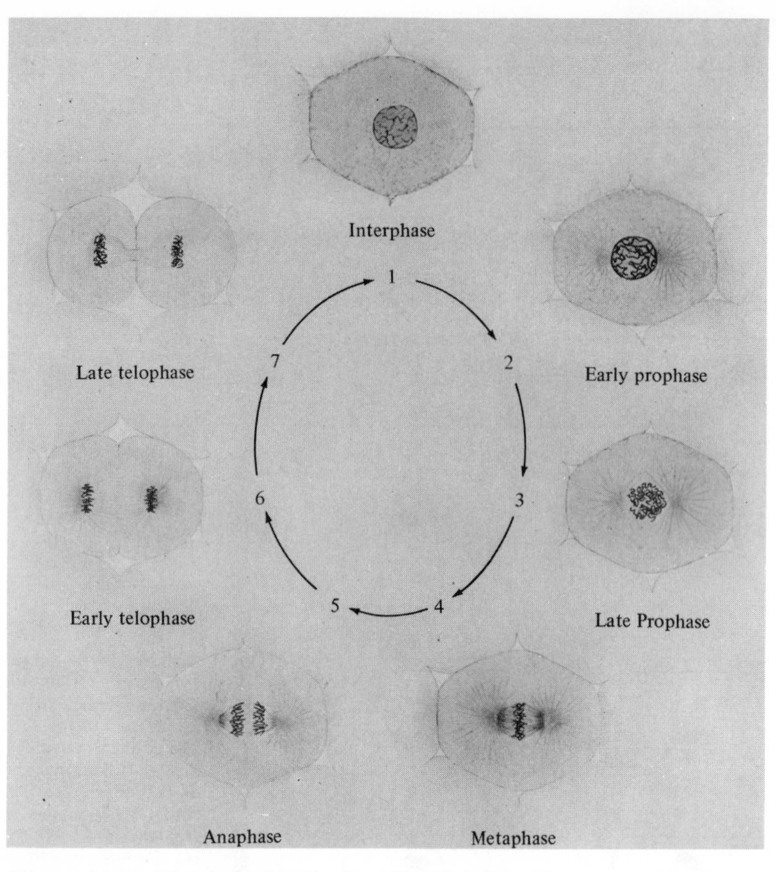

Figure 4–3. Mitosis in Developing Whitefish Egg. (Courtesy of the Sargent-Welch Scientific Company.)

mitosis, the DNA has been duplicated, and each chromosome has divided into strands held together by a central body called the centromere.

In early prophase the chromosomes shorten and thicken. By late prophase, they can be clearly seen under the microscope. Also in late prophase, the mitotic spindles start to form. These spindles consist of fibers radiating out from the opposite poles of the cell. During the metaphase stage of mitosis, all the chromosomes move to the equatorial plane of the cell. In anaphase the centromere divides, and the two daughter centromeres move toward opposite poles of the cell, pulling their chromosome strands (chromatids) with them. In telophase the separation is complete. The cell divides into two daughter cells, each with a nucleus containing four chromosomes and with the same complement of genetic information as in the original ($2 \times$ DNA). This completes mitosis until one of the daughter cells enters interphase and its DNA duplicates. The process of mitosis then begins again.

The Significance of Mitosis. Mitosis is very precise and does not give rise to the amount of variation that meiosis does. It is the exactness of DNA replication and equal chromosome distribution that ensures that one cell gives

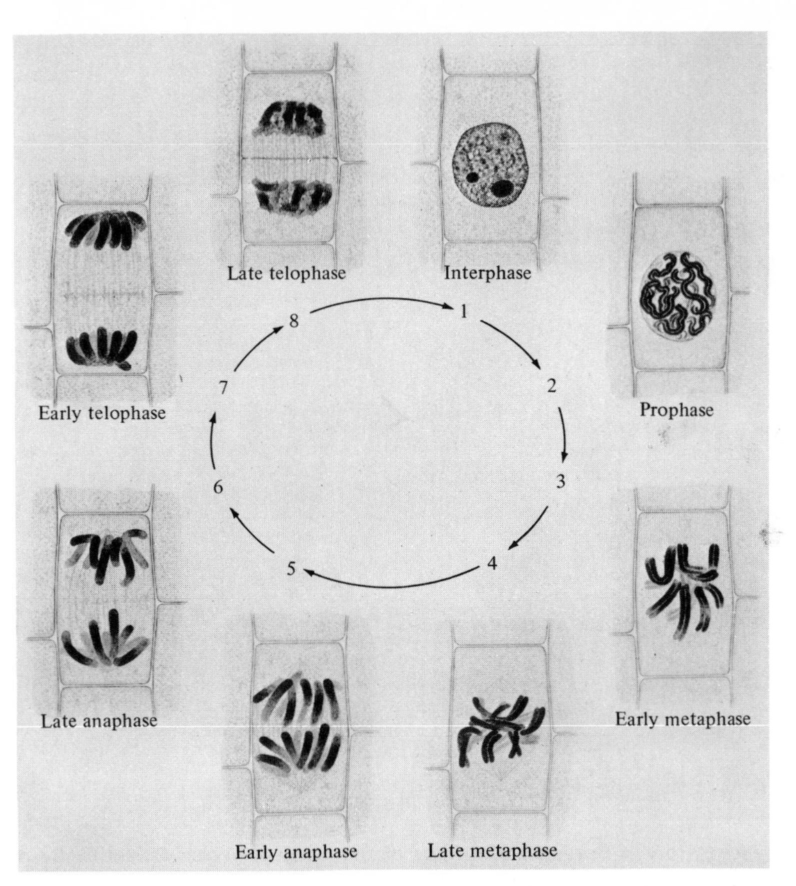

Figure 4–4. Mitosis in the Mayapple Root Tip. (Courtesy of the Sargent-Welch Scientific Company.)

rise to genetically identical daughter cells. The complement of DNA in the very first cell, the cell formed by fertilization, is reproduced and faithfully made a part of each new cell that arises. Thus in every cell in your body that contains a nucleus there is an identical set of chromosomes.

F. C. Steward (1970) at Cornell University has taken a single carrot cell and produced a new carrot plant identical to the parent. This has also been done in the tobacco plant, and a slight variation of procedure allowed the creation of identical frogs. (These experiments are discussed in Chapter 23.) This cloning process (growing a new multicelluluar organism from a single cell) begins with a body or somatic (nonsex) cell, and the new organism is identical to the parent plant or animal from which the cell was taken. This is because the cells of the cloned organism reproduced asexually through mitosis so that all chromosomes are identical. Imagine what this would mean to us if we could take any cell of a human and make a new human identical to the one from which the cell came! This fascinating area is presented in Chapter 23.

Mitosis and Human Cancer. Cancer is commonly defined as the uncontrolled process of cell division. Perhaps if scientists fully understood the

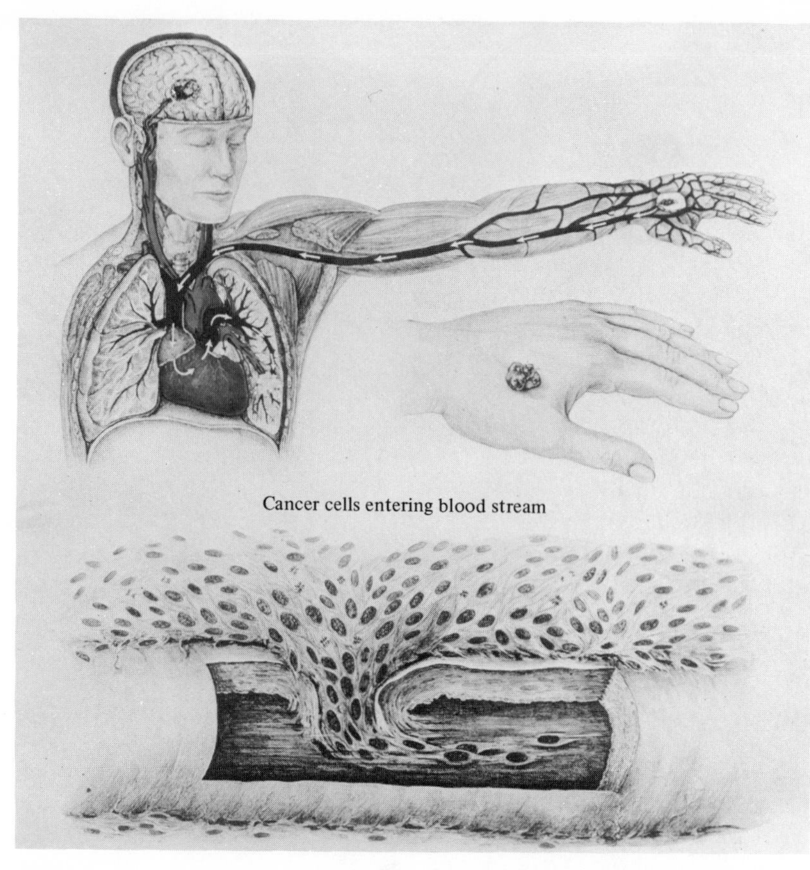

Figure 4–5. Cancer, How It Spreads. (Courtesy of the Sargent–Welch Scientific Company.)

relationship of the chromosomes to their control of mitosis, we would be able to control cancer. All body tissue—skin, blood, bone, etc.—is subject to cancer. The change in a cell that allows for uncontrolled cell division is apparently transmitted to daughter cells through the process of mitotic cell division. Different cells, with their own rate of cell division, account for the spread of cancer; one cell gives rise to two, two to four, etc. Cancerous cells accumulate via cell division. Then they migrate (metastasize) to the bloodstream and are carried to new locations, where they settle and continue to divide (Figure 4–5).

meiosis

Meiosis, although it may sound like mitosis, is a very different process. Meiosis is a special form of nuclear division that occurs only during the maturation of the gametes (sex cells).

A fundamental event in the life cycle of a sexually reproducing plant or animal is the union of gametes. Studies of chromosome behavior before and after the union of gametes clearly indicate that the chromosomes contributed by each gamete retain their separate identities in the nucleus of the zygote.

Since each gamete contributes a set of chromosomes to the nucleus of the zygote, the zygote contains twice the number of chromosomes that each gamete contains. Because each gamete is carrying only one of each kind of chromosome in its nucleus, the gamete is haploid with respect to its chromosome number. When two haploid gametes unite, a diploid nucleus is formed in the zygote. The zygote then contains matching pairs of homologous chromosomes; that is, each member of a given pair is the homolog of the other. Each pair of homologous chromosomes will in general appear physically distinct from any other pair of homologs in the nucleus. Each chromosome of a homologous pair is physically identical to the other. This statement applies to all of the chromosomes called autosomes, those chromosomes that are not associated with the sex of the individual. The sex chromosomes of a human female (XX) and male (XY) are distinct from the autosomes, and the X and Y chromosomes do not form a homologous pair. (Chromosomes, autosomes, and sex chromosomes are discussed in Chapter 6.)

Because the chromosome number of a species remains constant from one generation to the next, continuous doubling of chromosomes is impossible. Indeed, sexual organisms with continuous doubling of chromosome number would soon have large, unbalanced cells with unwieldy nuclei containing perhaps more DNA than cytoplasm. At some point in the history of life, sexually reproducing organisms evolved a mechanism that results in gametes with half the usual number of chromosomes. If a cell nucleus is diploid, it contains two sets of homologous chromosomes. The sex cell derived from that diploid cell, through the process of meiosis, contains one of each of the homologous chromosome pairs and is haploid (Figure 4–2B). Thus meiosis ensures that a sexually reproducing species maintains its number of chromosomes generation after generation. In humans the diploid number is 46 chromosomes in every generation.

There are two nuclear divisions in meiosis: meiosis I and meiosis II. The steps leading to the nuclear divisions are divided into the same five phases as in mitosis (Figure 4–2). However, two important events occur between interphase and metaphase in meiosis I that do not occur during the same interval of time in mitosis. One event is the reduction by one half of the number of chromosomes so that a diploid species can maintain its diploid number. The other is the production of variability or differences among individuals. This variability is a result of crossing-over and recombination of homologous chromosomes.

Prophase—The Period Between Interphase and Metaphase During Meiosis I. Because prophase of meiosis I is highly specialized and takes a comparatively long period of time, scientists have subdivided it into five stages: leptotene, zygotene, pachytene, diplotene, and diakinesis. Each stage relates to the activities of the chromosomes. (Figure 4–6 illustrates the process of meiosis in a cell having two chromosome pairs.) In leptotene the diploid number (two pairs) of chromosomes appears in each nucleus in the form of long, apparently single, slender threads. The four homologous threads then come together in synapsis. The threads continue to contract and become shorter and thicker in the early pachytene stage. Each chromosome then separates lengthwise into two halves, or chromatids. The four homologous chromatids thus formed remain united in a four-strand structure, or tetrad, in the diplotene stage. During the diplotene stage or late in prophase, homologous chromosomes

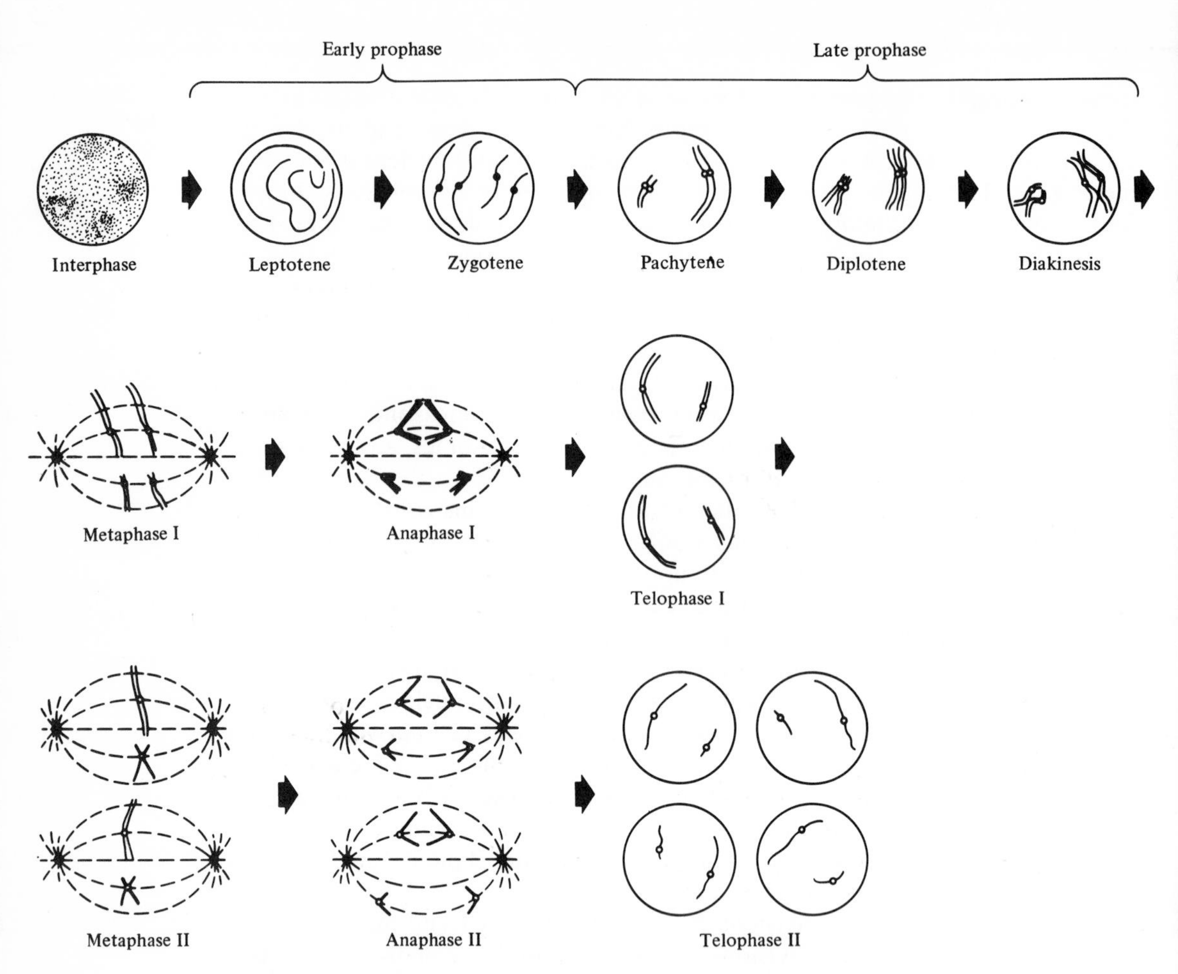

Figure 4–6. Meiosis. A general scheme of meiosis showing two pairs of chromosomes in a diploid nucleus undergoing a reduction to haploid nuclei.

tightly pair lengthwise, so that any given physical point on one chromosome is matched by the corresponding physical point on the homologous chromosome. At this time actual chromosome exchange may occur.

Now the intertwined chromosomes begin to separate as homologous centromeres pull apart in a process designated as diakinesis. At this stage the presence of one or more bridges joining together homologous points on the chromatids can be observed. A bridge, or chiasma (pl. chiasmata), may be seen between any of the chromatids, but physical exchange of chromatids is most easily detected between nonsister chromatids or those chromatids that are associated with the separate centromeres.

Crossing-over, the exchange of chromosome or chromatid segments, plays an important role in reassorting genes into new combinations (Figure 4–7). Because crossing-over permits a continual reshuffling of genes, it is believed to be crucial in providing for new and sometimes advantageous combinations of genetic material. And new genetic material is essential in natural selection. Were it not for crossing-over, the combination of genes in a particular

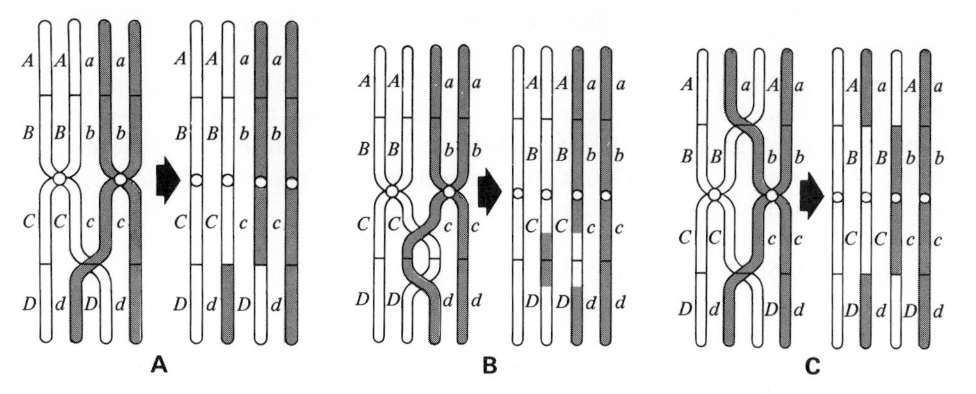

Figure 4–7. Possible Types of Crossing-over in Homologous Chromosomes. Each chromosome is divided into four regions: *A, B, C,* and *D*. **A.** A single genetic exchange of material in region *D*. **B.** A genetic exchange between regions *C* and *D*. **C.** Single region genetic exchanges in regions *A* and *D*, one region on each side of the centromere. As the number of genetic regions or "genes" on a given homologous pair of chromosomes increases and the number of homologous chromosomes per nucleus increases, the possible combinations of genetic exchange become almost limitless.

chromosome would remain constant indefinitely, except for an occasional mutation. Since most mutations are deleterious, chromosomes gradually would accumulate mutations and eventually become incompatible with normal life.

Diakinesis is followed by metaphase I, in which the still-paired chromosomes align in the equatorial plane. There is still a diploid number of centromeres and chromosomes, although each chromosome consists of two distinct chromatids. Metaphase I is followed by anaphase I, during which homologous chromosomes migrate to opposite poles of the spindle, creating two "daughter" cells, each with half the original number of chromosomes. The first meiotic division is sometimes referred to as a reduction division, since each "daughter" cell has one chromosome of each homologous pair of chromosomes.

Following anaphase I, the chromosomes do not unwind and become invisible as in mitosis. Rather, they undergo a second meiotic division—meiosis II. In each of the "daughter" cells of the first division, a new spindle forms, and the chromosomes align in the new equatorial planes, completing metaphase II. The centromeres divide, and the pairs of homologous chromosomes (formerly called chromatids) separate, migrating to opposite poles during anaphase II. Since the number of chromosomes per cell does not change, the second meiotic division is sometimes referred to as an equatorial division.

The two meiotic divisions result, then, in the formation of four nuclei, each carrying one chromatid from each tetrad, that is, the haploid number of chromosomes. In terms of DNA meiosis can be regarded as

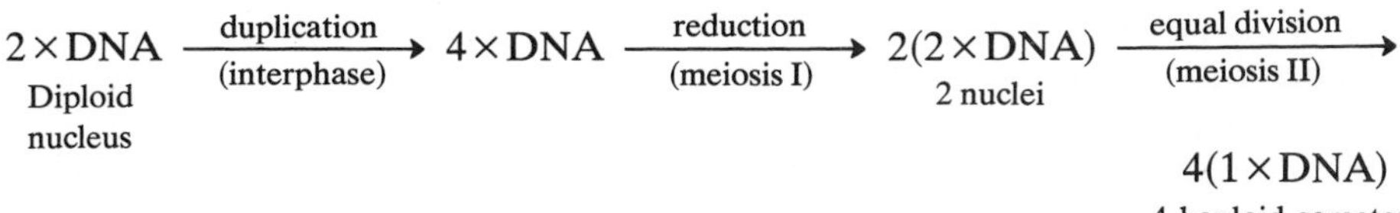

When crossing-over occurs, however, the chromosomes resulting from meiosis

are no longer like the maternal and paternal chromosomes that underwent pairing at the pachytene stage. Instead, the chromosomes coming out of meiosis may be composed of sections of the maternal and paternal homologues. Furthermore, the different (nonhomologous) maternal and paternal chromosomes undergo segregation independently of each other. As a consequence, a gamete will virtually never include an entire maternal or paternal chromosome; usually, it will contain various mixtures of maternal and paternal elements (Figure 4–2B).

The Significance of Meiosis. At least two important aspects of meiosis are related to evolution.

1. Chromosome reduction is necessary in order for a species to maintain the diploid chromosome number.
2. Variation among members of a species is essential for the operation of natural selection.

As indicated in Chapter 3, random chromosome assortment and crossing-over during meiosis lead to a fantastic number of possible combinations of genetic material. The chances are at least 1 million trillion trillion to 1 that no two independently fertilized zygotes would have the same genetic material. Given this enormous potential for variation, we can readily understand that a human female, giving rise to some 480 eggs in her reproductive lifetime, will hardly produce two eggs with the same genetic potential. And even though the male is capable of making perhaps 1000 billion sperm in his reproductive lifetime, the number of sperm is still so small, compared to the potential for variation, that one should not expect any two sperm to have the same genetic content (Stern, 1973). Thus the mechanism of meiosis ensures great variation while at the same time maintaining the diploid number of chromosomes.

the male gamete—the sperm

The sperm was first seen in 1677 by Johan Hamm (Hancock, 1970). His observations were reported to Anton van Leeuwenhoek, who then described the sperm's morphology in detail to the Royal Society of London. It took about the next 200 years, however, to establish the role of sperm in the process of sexual reproduction. In 1784 Lazzaro Spallanzani (the defender of biogenesis, mentioned in Chapter 2) filtered dog semen to separate the sperm from other material. Using the separate fractions, he found that sperm was necessary to make a female dog pregnant. Perhaps his experiment was the first recorded example of artificial insemination, a technique that is rather common today in animal husbandry.

The human male reaches puberty between the ages of 12 and 16. The onset of puberty is gradual and is brought about by the secretion of certain hormones called gonadotrophins produced by the pituitary gland. The amount of gonadotrophins secreted increases slowly until it is sufficient to promote changes in skeletal, vocal, and gonadal activity. Gonadal activity results in striking morphological changes, including the transition of a diploid germ cell into a sperm (Figure 4–8). The nucleus of the cell becomes the major constituent of the sperm head. The centriole, a minute body in the cytoplasm, produces a bundle of fibers that later become the tail; the rest of the original cytoplasm degenerates.

Inside the seminiferous tubules, which are located in the testes, many millions of diploid cells are continuously undergoing this change. It takes about 64 to 72 days from diploid cell to finished sperm, but since new cells enter the process daily, a fertile male is never in short supply for very long. In humans, after sperm are released into the region of the uterus, there is a 24-hour period in which the sperm are maximally capable of penetrating an egg. After that time this capability declines rapidly, even though the sperm may live three or four days. The fertilization time span for most mammals is less than for humans, usually six hours. An exception is the bat. The male bat deposits his sperm in the female in the fall, and the sperm winters over until the spring, when the female ovulates.

A discharge of human semen holds an average of 250 to 500 million sperm. When the number of abnormal sperm exceeds 25%, the male is usually sterile. About four out of ten infertile marriages are primarily due to a low sperm count or poor sperm morphology, that is, defective sperm. Although all the sperm in a discharge are ejaculated into the upper reaches of the vagina, only a few thousand reach the oviduct, and only a few dozen reach the egg in the fallopian tube (Austin, 1970).

In the human male there is no fluctation of hormones equivalent to the menstrual cycle in the female. A healthy male continuously produces sperm in

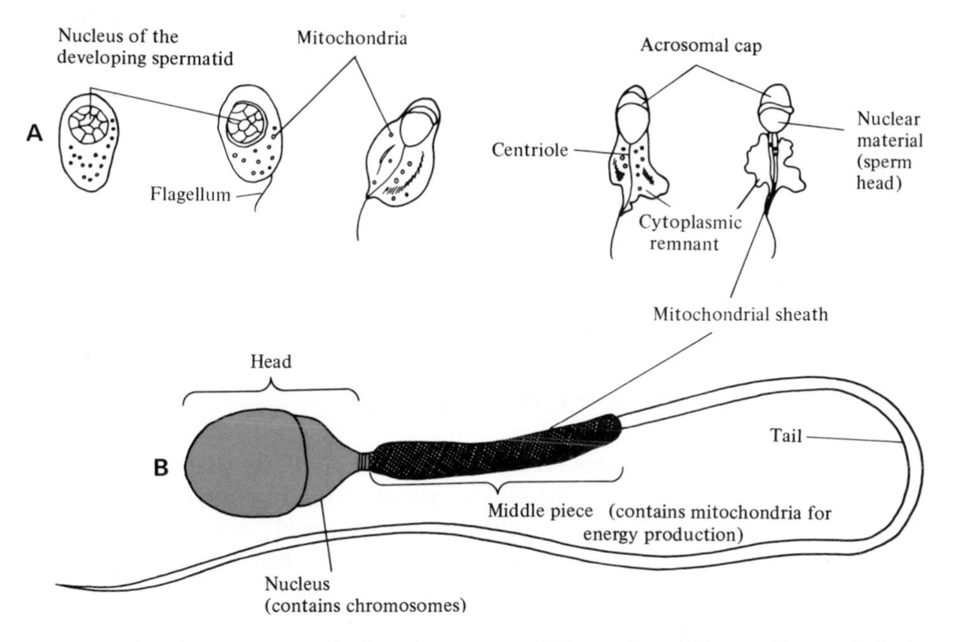

Figure 4–8. Spermatogenesis: Development and Maturation of Human Sperm. **A.** Reduction in cytoplasm and formation of the flagellum. **B.** Mature sperm: head, 3–5 microns by 1.8 microns; middle piece, 3–6 microns by 1 micron; tail, 30–50 microns by 1 micron. Total process of sperm development takes from 64 to 72 days.

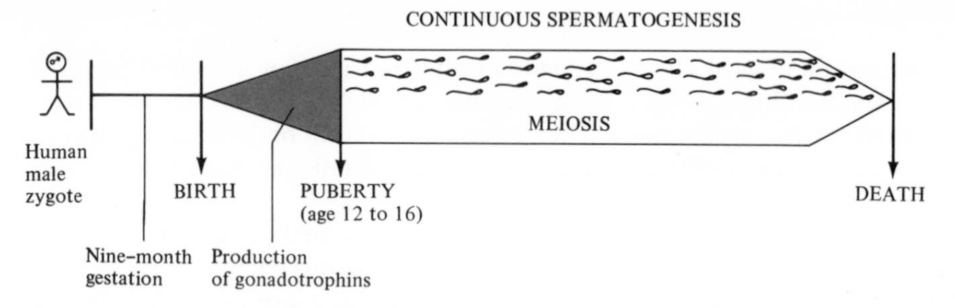

Figure 4–9. Chronology of Spermatogenesis. The diagram depicts the chronological events in the reproductive life of the human male. From the onset of puberty the male is always in the act of producing sperm.

large numbers. His maximum level of production occurs at about the age of 16, but the healthy human male continues to produce sperm until death (Figure 4–9). The continuous production of sperm comes from a reservoir of diploid cells set aside, during embryonic developent, specifically for the production of sperm.

The sperm, produced in the seminiferous tubules, are held in the epididymis, where they are bathed in a fluid secreted by specialized cells of the collecting tubules. Before entering the epididymis, the sperm are incapable of fertilizing an egg. They are "activated" as they pass through the epididymis. Sperm can remain viable for weeks or months in the epididymis, although they have a much shorter survival time once they are ejaculated (Hancock, 1970).

For spermatogenesis (the production of mature sperm from a primary diploid cell) to occur, the testes must maintain a temperature three to four degrees below the normal body temperature of 37°C (Grollman, 1969). During periods of warm weather the scrotum is relaxed and extended from the body; in winter it contracts, holding the testes closer to body heat. Our modern dress habits, some people suggest, may create higher than normal scrotal temperatures and result in lower sperm counts.

During ejaculation the sperm pass through the vas deferens, picking up various additional secretions of the Cowper's and prostate glands and seminal vesicles (Figure 4–10). Severing the vas deferens is commonly referred to as a vasectomy, one form of male sterilization.

the female gamete—the egg

The mammalian egg was discovered in 1827 by Karl Ernst von Baer. Sixteen years later Martin Barry recorded the penetration of the egg by sperm (Hancock, 1970). The human egg is a spherical cell approximately 0.14 millimeter or $\frac{1}{175}$ inch in diameter. Its weight has been estimated at 0.000004 gram or one twenty millionth of an ounce (Stern, 1973). Compared to other body cells, however, the egg is one of the largest. It contains a selection of the maternal genes, enzymes, ribosomes, mitochondria, and other elements necessary to begin a new life once the egg is fertilized.

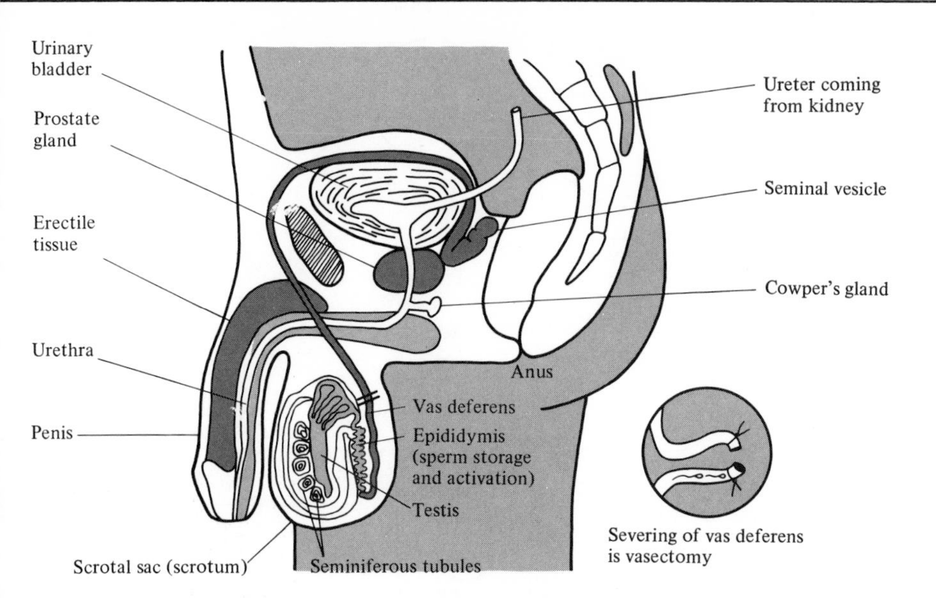

Figure 4–10. Midsagittal Section of the Male Reproductive Tract. The sperm are produced in the seminiferous tubules and pass out through the epididymis and the vas deferens. Secretions from the seminal vesicles and prostate and Cowper's glands join the sperm, and this mixture is referred to as semen. Erectile tissue in the penis fills with blood, causing it to stiffen during sexual excitement. Continued excitation causes the release of semen. Note that cutting out a segment of the vas deferens is called a vasectomy.

Like the sperm, the egg is derived from specialized primordial diploid cells, which are set aside early in the embryo's development. A few of the primordial cells migrate through the forming tissues of the early embryo and locate themselves in a region that is later referred to as the genital ridge. Due to the rapid division of these cells, the genital ridge, which is present in all human embryos, becomes distinct early in development. Then, depending on the environment and genetic message, it gives rise to the future gonads—either testes or ovary. Normally, if the diploid primordial cells carry two X chromosomes, female development takes place and the two ovaries are formed. Cells that will become the future eggs, or ova, gather in the developing ovaries. Meiosis, when completed, results in the haploid number of human chromosomes per gamete, 46 chromosomes being halved to 23. The process of meiosis is important in that when a haploid egg with its 23 chromosomes is fertilized by a haploid sperm with its 23 chromosomes, the human chromosome number of 46 is reestablished in the zygote (Figure 4–11). Mitotic cell division of the zygote gives rise to a human composed of diploid cells except, of course, for the gametes the new individual will produce.

In the female, meiosis in the developing ova, or eggs, is arrested during prophase. The egg remains at this stage, as a primary oocyte, until the female reaches puberty some 12 to 16 years later. Even then, only one oocyte at a time (usually one about every 28 days) proceeds through meiosis to become a secondary oocyte. This means that in a 45-year-old female a primary oocyte that is 45 years old is entering the second phase of meiosis I (Figure 4–12). It is believed that with the progressing age of the primary oocyte, there is an

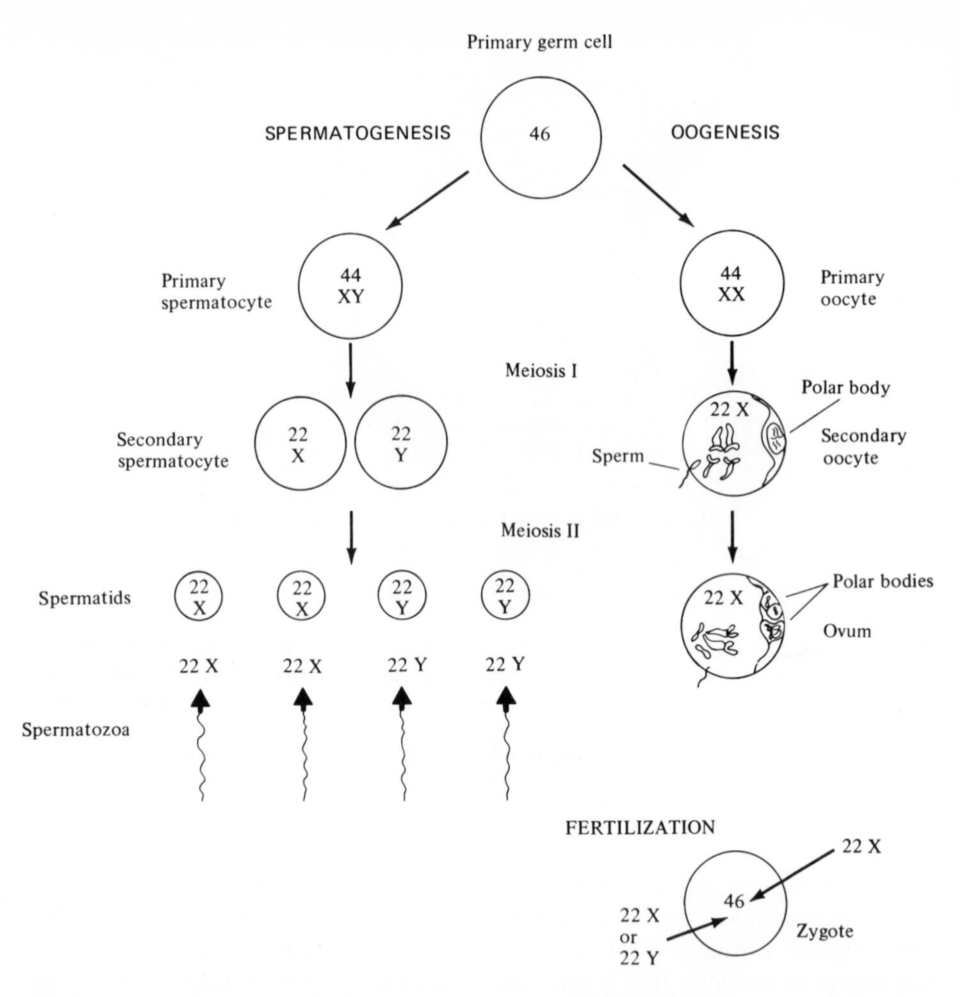

Figure 4–11. A Comparison of Spermatogenesis and Oogenesis, Starting with a Primary Germ Cell (sex not indicated). Primary human sperm cell or egg contains 46 chromosomes, 44 autosomes and 2 sex chromosomes, XX being female and XY male. The diploid cell is reduced to haploid gametes containing 23 chromosomes, 22 autosomes plus one X or one Y. In oogenesis meiosis I is completed upon entrance of the sperm; when meiosis II is completed, there are three polar bodies and one functional haploid egg. Spermatogenesis results in four functional haploid sperm.

increased chance that chromosomes will attach to each other, causing a misdivision of the chromosomes. The attachment of two homologous chromosomes to each other such that both chromosomes pass into the same nucleus during meiosis is called nondisjunction (see Figure 6–3). Nondisjunction is believed to cause the severe human abnormality of Down's syndrome, or mongolism. (Down's syndrome and other chromosome-related human abnormalities are described in Chapter 6.)

Ovulation. With the onset of puberty, the female reproductive organs begin a 28-day sexual rhythm, the menstrual cycle, that normally continues for the next 30 to 40 years (Figure 4–12). A female who reaches puberty at age 12

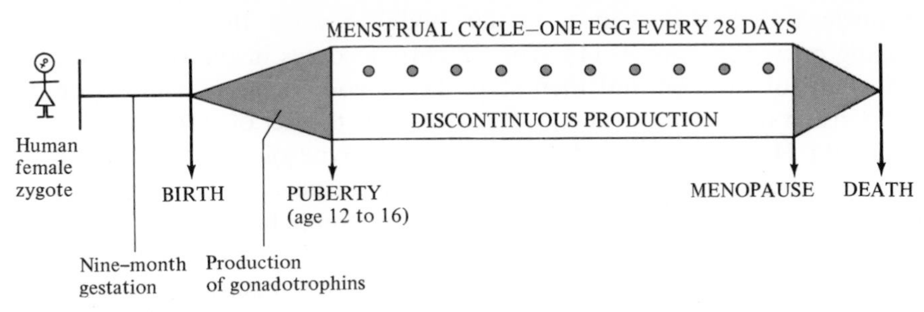

Figure 4–12. Oogenesis. The diagram depicts the chronological events occurring in the human female. Starting at puberty, the female produces a single egg every 28 days. The egg produced when she is 45 years old develops from a primary oocyte that is also 45 years old, since it was set aside, in its first phase of meiosis, before she was born.

and ovulates 12 mature eggs per year (one every month) for 40 years requires 480 oocytes (12 × 40). Estimates are that a female is born with 500,000 to 1 million primary oocytes. Those that are not used degenerate during menopause.

The 28-day female reproduction cycle terminates with menstruation. The menses is the sloughing off of the endometrial lining of the uterus, which had been prepared to receive the fertilized egg, through the vagina. Approximately midway between the beginning and the end of the menstrual cycle, the female ovulates, releasing an egg from the ovary. The ovaries normally alternate in egg production (Figure 4–13). Once in a while, however, both ovaries ovulate during the same menstrual cycle, or one ovary may produce more than one egg. In either case, multiple births may result.

Just prior to each ovulation, the process of oogenesis (maturation of the egg), which was suspended prior to the female's birth, is resumed. The primary oocyte now undergoes its first division, which results in two nuclei, each

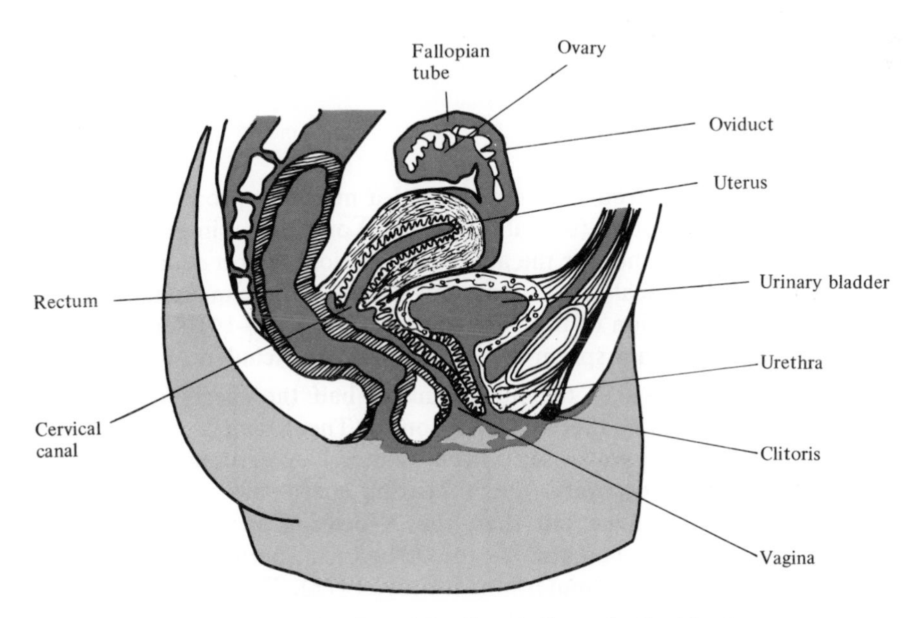

Figure 4–13. Midsagittal Section of the Female Reproductive Tract.

containing 23 chromosomes. One nucleus remains in the oocyte, while the other nucleus is enclosed in a "polar body" toward the outside of the developing oocyte (Figure 4–11). The 23 chromosomes in the oocyte begin a second phase of division and then stop. The second stage of meiosis or the second division of the chromosomes must wait for the act of fertilization.

Motoyuki Hayashi, Head of Obstetrics and Gynecology at Japan's Toho University School of Medicine, recently used a specially constructed tube equipped with a light source and lens to photograph human ovulation. His excellent color photographs appeared in the June 24, 1974, issue of *Time* magazine. Based on these photographs and their accompanying legends, the process of ovulation can described as follows:

> The follicle cells which surround the developing egg create a bulge on the outer wall of the ovary just prior to ovulation. When pressure is sufficient, the ovary ruptures with volcanic-like action spewing forth follicle fluid and finally the egg. The released egg is engulfed by finger-like fimbriae which move the egg into the fallopian tube. The egg, once in the fallopian tube, starts its five- to seven-day journey covering approximately 5 inches before it reaches the uterus. In case of pregnancy, menstruation stops until birth.

The important event, from a reproductive standpoint, in the complete menstrual cycle is the maturation and release of the egg. Menstruation ceases if the ovaries are removed. However, simply tying off the fallopian tube (tubal ligation), to keep the egg from reaching the uterus and the sperm from reaching the egg, does not interfere with the monthly period (the hormonic balance is not altered).

When the female reaches an age of 45 to 50, the level of hormones in her body changes, and she normally stops ovulating. Consequently, menstruation ceases. This termination of menstruation is referred to as menopause and marks the end of the female's reproductive capacity. Like puberty, menopause comes about gradually and perhaps irregularly.

fertilization

Once ovulation has occurred in a female and a male has produced sufficient quantity of sperm, copulation (sexual intercourse) or artificial insemination can bring sperm and egg together for the act of fertilization. Sperm deposited in the vagina pass into the cervical canal, and a few dozen of these make their way to the egg in the upper portion of the fallopian tube (Figure 4–14). Once a sperm penetrates the egg, all other sperm are physiologically excluded.

As a result of meiosis in the human male, half the sperm carry the Y chromosome and half carry the X chromosome. (The X and Y chromosomes are referred to as the human sex chromosomes.) According to Landrum Shettles of Columbia University, the Y-bearing sperm has a smaller, more rounded head and a longer tail than the X-bearing sperm. Because the X chromosome is larger than the Y, the X-bearing sperm carries approximately 4 to 5% more genetic material (McCary, 1973). The smaller Y-bearing

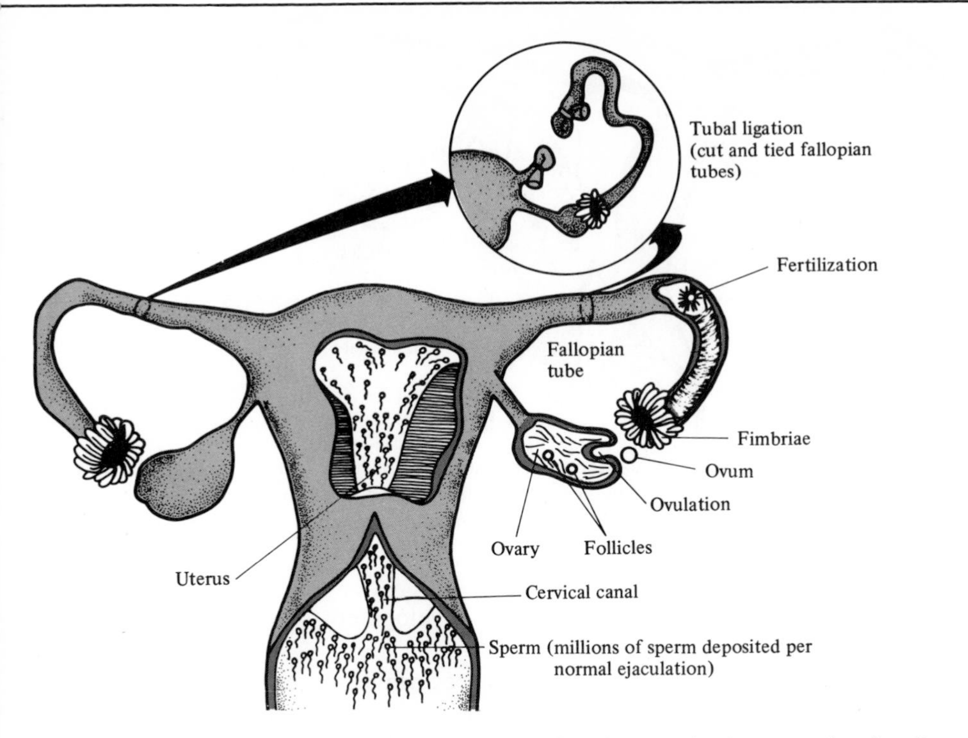

Figure 4–14. Fertilization. A diagram of human female reproductive tract showing the movement of sperm toward the fallopian tubes. Only a few dozen of the millions of sperm deposited in the uterus actually reach the egg in the tube. Arrows mark the areas where the fallopian tubes are cut and tied as one procedure for female sterilization. Notice that fertilization occurs in the fallopian tube.

sperm travels faster but is viable for a shorter period of time than the X-bearing sperm. It also contains a unique set of antigens not found in the X-bearing sperm (Lappe, 1974). The two kinds of sperm have been found to migrate in opposite directions in an electric field (Hancock, 1970), and this knowledge might be used to help persons who wish to select the sex of their children. Thus the husband's sperm (or those of another male) could be separated into X- and Y-bearing fractions and the female could be artificially inseminated with only X-bearing sperm if a female child is wanted or with only Y-bearing sperm for a male.

When the egg is penetrated by an X- or Y-bearing sperm, it is induced to complete the meiotic process, which results in an ovum with a haploid nucleus (Figure 4–11). The sperm nucleus now fuses with the egg nucleus, and the result is a single-celled zygote. These events take approximately 12 hours.

During fertilization the sperm and the egg contribute equal amounts of hereditary material; each donates its haploid complement of 23 chromosomes to the zygote. During the early hours of the zygote's existence, DNA is synthesized. Upon completion of the new DNA synthesis the zygote cell undergoes the first cell division. After several hours a second division occurs, producing four cells, and so it continues; at the hollow-ball stage, called the blastocyst, the embryo begins making its way down the fallopian tube to become implanted into the wall of the uterus. The trip takes from five to seven

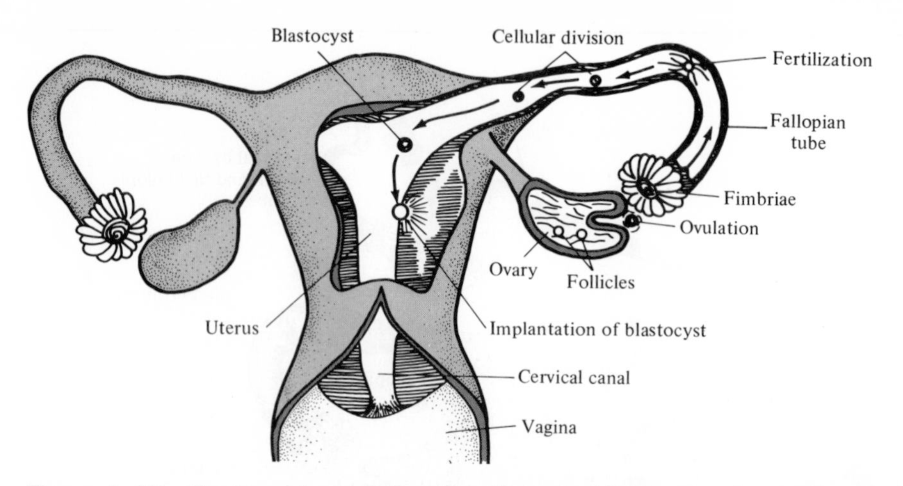

Figure 4–15. Blastocyst Implantation. The diagram of human female reproductive tract shows the movement of the developing zygote toward implantation in the wall of the uterus. It takes an egg or a blastocyst five to seven days to reach the uterus.

days (Figures 4–14 and 4–15). Further development of the embryo and the fetus is shown in Figure 4–1.

If the sperm does not penetrate the egg while the egg is in the fallopian tube (oviduct), the egg journeys on to the uterus, disintegrates, and is passed out with the menstrual sloughing of the uterine lining.

Our world population has now reached 4 billion. If we calculate the volume of an egg cell and multiply by 4 billion, we find that 4 billion people have come from a volume of about five quarts of female ova. Similar calculations indicate that the male equivalent, 4 billion sperm, is a mass smaller than an aspirin tablet (Stern, 1973).

the use of artificial insemination in humans

Artificial insemination is a procedure for depositing semen in or near the cervical canal of the woman's uterus to bring about pregnancy (Figure 4–16). When the husband's sperm are used, the procedure is referred to as homologous insemination; the use of donor sperm is heterologous insemination. Most artificial inseminations are heterologous. The husband may be infertile because of a low sperm count in his semen; less often, his sperm may contain a severe Rh incompatibility factor (see Chapter 12), or he may be carrying a dominant gene for a severe hereditary disease. As noted earlier, artificial insemination had been achieved in dogs in 1784 by Spallanzani. Later in the eighteenth century, John Hunter used artificial insemination in humans (McLaren, 1973).

Human babies conceived by artificial insemination show *no* increase in neonatal mortality or in congenital defect. This may be because the donor is screened in many cases for observable abnormalities before his sperm are taken and is a man who has previously produced only normal children. Also,

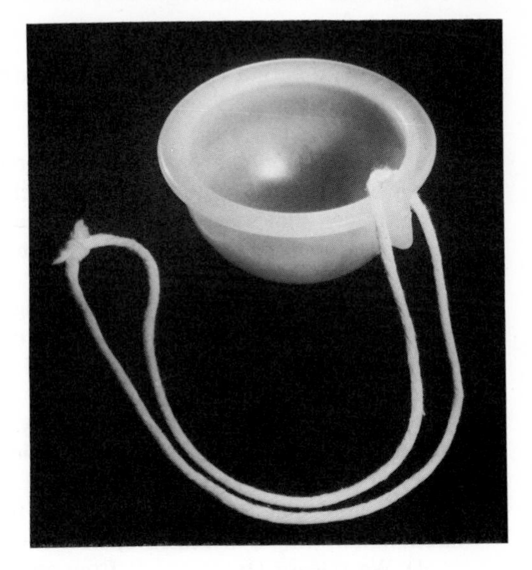

Figure 4–16. Artificial Insemination Cup. After a pelvic examination and aseptic preparation of the cervix, a semen-filled glass syringe is used to deposit a few drops of semen into the cervical canal. The remainder of the semen is placed into the cervical cup. The cup is eased through the vaginal canal to the cervix. At the cervix the cup is flipped into position, where it remains for 24 hours. The string on the cup enables one to remove the cup. (Photo by Ed Lada for *contemporary ob/gyn*.)

the likelihood of fertilization occurring at the optimum time, or just after ovulation, is greater. However, follow-up studies on the development of children from artificial insemination are limited. One study (Iizuka et al., 1968) showed that of 54 children conceived by artificial insemination, all were normal in mental and physical development.

donor selection

Donor selection is an important problem that can give rise to some anxiety. Who selects the donor and why is that person selected? A doctor may exercise a degree of positive eugenic selection, since he uses only donors who are physically and mentally sound, have a good medical record, and have semen of high quality, and he may introduce his own value judgments as to which personality types (for example) would make the best fathers (see Chapters 22 and 23). Many doctors try to match the donor to the husband as closely as possible; others inseminate with a mixture of semen from different donors so that the actual father of any given baby is unknown. However, the effect of such selection would be trivial over the whole population, unless the number of progeny from an individual donor were to become very large.

Another issue of concern is the possibility that children who were conceived from the same donor and are therefore half-brothers and half-sisters might unwittingly marry and have children. The progeny of first-cousin marriages show a higher than normal mortality and malformation rate; half-brothers and half-sisters are more closely related genetically than first cousins, so their children would be at a still greater disadvantge. (Consanguinity is discussed in Chapter 14.)

Because of the highly sensitive nature ot artificial insemination, doctors have made up forms for the prospective parents to fill out. These forms absolve the donor and doctor of legal responsibility. Below is an example of a portion of one such form:

> "Request and Permit for Artificial Donor Insemination," used by Dr. _______.
>
> To induce Dr. _______ to render the services herein requested, we, and each of us, agree that:
>
> 1) Under no circumstances shall we require that the name of the donor of the semen be divulged to us or to anyone else, and we accordingly forever waive all rights, if any, that we may have as to the name, identity or any information of any kind concerning the donor(s). We agree to rely upon the discretion of Dr. _______ in the selection of qualified donors.
>
> 2) We agree to refrain from bringing legal action of any kind, and to refrain from aiding or abetting anyone else in bringing legal action for or on account of any matter or thing which might arise out of the artificial insemination herein contemplated.
>
> 3) We do individually and mutually recognize the moral and legal responsibilities for the financial, social and emotional care of any resulting issue, and accept such responsibility as identical to that for our own naturally occurring issue. We do further agree that legal procedures for the adoption of any viable issue of such insemination will be instituted upon the live birth of such issue.
>
> 4) We release Dr. _______ from any and all liability and responsibility of any nature whatsoever which may result from complications of pregnancy, childbirth or delivery or from the birth of any infant or infants abnormal in any respect, or from the heredity or hereditary tendencies of such issue, or from other adverse consequences which may arise in connection with or as a result of the artificial insemination herein authorized.
>
> 5) We agree to indemnify the doctor for any attorney's fees, court costs, damages, judgments, or any other losses or expenses incurred by him or for which he may be responsible with respect to any claim, legal action, or defense thereto, arising out of the artificial insemination herein contemplated.

timing the act of artificial insemination

Timing is critical. The sperm have a limited life span in the female reproductive tract. Since there is no method for establishing the exact moment of ovulation in humans, insemination may be tried on several successive days. On occasion, artificial inseminations have been continued for as long as a year. Data from several centers indicate that successive inseminations result in 70 to 75% of the women becoming pregnant within three to four months after the start of treatment.

We are more successful at artificially inseminating animals, since we can more accurately determine their time of ovulation. In the United Kingdom, for example, 10 million cows had been successfully inseminated by 1950. In the United States during 1969 and 1970, 2,645,000 cows, or about 60% of all the cattle mating, were successfully inseminated (Milk Marketing Board, 1969–1970).

preservation of the sperm

Sperm perish in a matter of hours at room temperature, but they can be kept for varying periods of time if pretreated with 10% glycerol and stored in containers placed under liquid nitrogen, −196°C (Figure 4–17). Tests indicate that the pregnancy rate is no more than 50% if the semen has been stored in deep freeze for several months compared to 70 to 75% reported for the use of fresh semen (McLaren, 1973). Long-term storage of human semen has the advantage of ease of transport and administrative convenience. Administrative convenience is important, particularly because of the need to carry out inseminations during the female's short fertile period. Many couples want a second or even a third child by artificial insemination, and often they prefer to use the same donor. This may be hard to arrange unless several semen samples were deep frozen initially. In the United States commercially run banks for frozen semen are common. Many of their clients are men about to undergo a vasectomy; stored semen provides insurance for the future should the man change his mind about not having children. (Various aspects of artificial insemination are discussed in Chapters 22 and 23.)

accidents of fertilization and birth defects

David Carr (1970), a professor of anatomy at McMaster University, Ontario, states that of all eggs fertilized via copulation, 50% fail to produce a living child. If sperm does not penetrate the egg within 24 hours after ovulation, there is evidence that an aged egg will give rise to various malformations when fertilized. One common anomaly associated with the fertilization of an aged egg is the condition of dispermy (Carr, 1970). Dispermy occurs when two sperm enter one egg, and the result is a triploid condition in which the zygote has three sets of chromosomes instead of two. Triploidy is a common cause of early spontaneous abortions in humans (Carr, 1970). Few triploid children have actually been born alive; those that were, died within hours or days.

Another condition, tetraploidy, occurs if the chromosomes duplicate themselves during the first cell division of the zygote, but then the cell fails to divide. The single cell then has 92 chromosomes rather than the normal diploid number of 46. Spontaneous abortions resulting from tetraploidy occur earlier in human development than those caused by triploidy. It would appear that the addition of whole sets of chromosomes is detrimental to man.

Abnormalities also occur when certain individual chromosomes are represented more than twice. That is, through the process of nondisjunction, a person may have three of a particular chromosome but be normal for all others (as in trisomy 21, or Down's syndrome). Such conditions are referred to as aneuploidy and result in a variety of severe birth defects. Birth defects resulting from chromosome aneuploidy are discussed in Chapter 6.

Birth defects are as old as man himself. Egyptian paintings and sculptures 5000 years old show dwarfs, and a cleft palate has been found in a mummy. Babylonians catalogued more than 60 deformities of the ear, nose, mouth, sex organs, and limbs (Hurley, 1968).

When people see something unusual and do not understand its cause, they often make up vague explanations. Birth defects, in particular, have given rise to a rich lore of superstition and myths. In the past when a child was born with a harelip, its mother was thought to have been frightened by a hare during

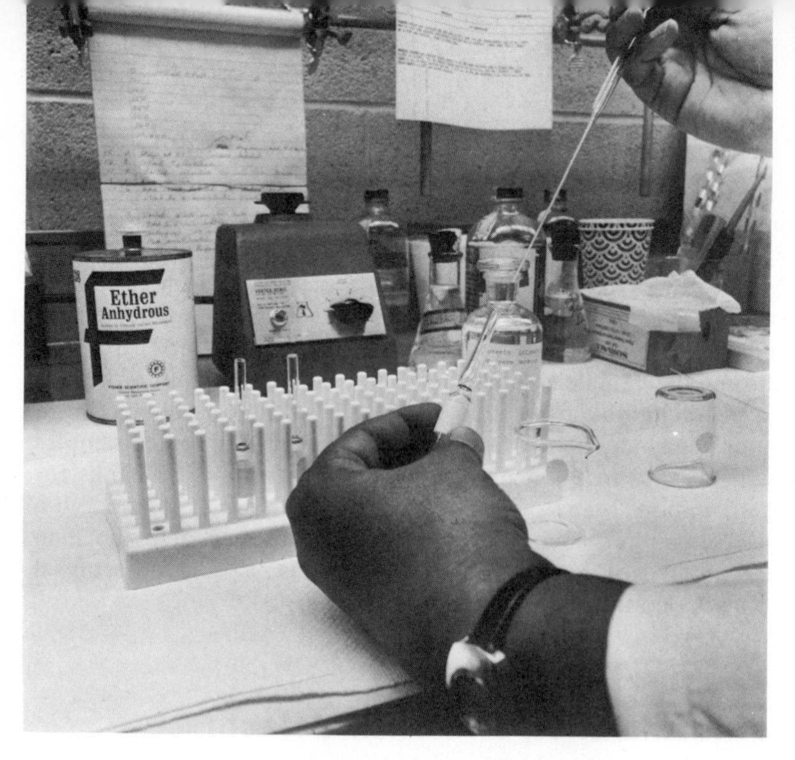

A

Figure 4–17. Deep Freezing Human Semen. **A.** Semen is separated into standard vials (1.5 to 3 ml) that will be placed into a freezing container. **B.** The vials containing fresh highly motile sperm are dipped into liquid nitrogen for storage at −196°C. (Courtesy of Marvin A. Yussman, M.D., University of Louisville School of Medicine.)

B

pregnancy. Often the birth of a deformed child was interpreted as God's punishment for a transgression. The guilt and shame often felt by parents of a defective child are a direct reflection of our past sentiments. Parents are often overwhelmed by the fear they have been responsible for what has happened to their child.

the incidence of inherited and noninherited birth defects

Today, congenital malformations, that is, malformations present at birth, are considered one of our most important public health problems. In the United States a child is born with a birth defect every two minutes. Unfortunately, congenital birth defects are not as amenable to preventive measures as are many public health diseases. Prevention of congenital defects requires a complete knowledge of human development. A recent report states that birth defects cause twice as many deaths per year as does cancer (*Laboratory Management*, 1973).

According to the U.S. Department of Health, Education and Welfare, there are at least 15 million Americans with one or more birth defects that are considered serious because they directly affect the individual's daily life (DHEW, 1974). For example, 3 million persons suffer from mental deficiency of prenatal origin, 1 million are congenital cripples, 750,000 have hearing impairments, 500,000 are affected with visual defects, and 100,000 have speech defects. Millions more have a variety of inborn birth defects of the nervous, digestive, endocrine (hormone), circulatory, and other systems. Of these 15 million birth-defect cases, about 3 million are due to environmental causes, and 9 million to a combination of hereditary and environmental factors; the other 3 million result from known inheritable defects. Each year some 250,000 babies are born with birth defects, 50,000 as a result of a known inheritable defect and 150,000 from a combination of hereditary and environmental factors interacting with the living system. These figures are conservative, since some genes do not express themselves until later in life and are not included in data on birth defects. For example, Huntington's disease, which causes the deterioration of the brain, is usually expressed after the age of 25. Diabetes, gout, and many other diseases are also expressed long after birth.

significance of birth defects

At the present time, *over* 7% of all births are affected by some type of congenital defect. The impact of congenital defects is severe, but their implications for the quality of man's genetic makeup may be even more important. "Is man still at a relatively primitive stage of evolution, intensely sorting out his genetic background through selective survival—or is his genetic makeup relatively stable but he is unable to cope with the new environmental conditions which he is encountering through his efforts to organize his society?" (Janerich, 1972).

The importance of Janerich's question can be judged by our limited knowledge of the causes of congenital diseases. Less than 20% of all congenital diseases are understood. Where the cause is known to be environmental, preventive measures have been developed. We employ population surveillance to help us recognize unusual clusters of malformations and eliminate the cause. Laboratory testing can screen new compounds before they are recommended for human use, thus avoiding human consumption of teratogenetic agents (agents that cause malformation in the embryo). In the case of rubella, which can cause the birth of malformed infants, women who intend to have children can be vaccinated to lessen the chance of their having rubella while pregnant. However, if a disease is genetic in origin, and its occurrence is inherited according to a recognized pattern, the only preventive measure, at the moment, is genetic counseling.

Neither abnormal chromosomes nor a defective gene can explain the occurrence of many common birth defects. Yet family studies strongly suggest that there is a genetic predisposition. The concept of polygenic inheritance has been proposed to explain the inheritance of such defects. This concept suggests that the cause of many congenital malformations can be found in the interaction of a number of genes or gene products. Polygenic inheritance has been suggested as the causal mechanism in anencephaly (absence of all or part of the brain), spina bifida (open spine), cleft lip and cleft palate, clubfoot, and other common congenital malformations. Another idea is that many birth defects may be caused by one gene whose action is modified by a series of other genes.

Birth defects are known to cluster, that is, to appear more often in certain seasons of the year and in patterns recurring every five to seven years. For certain birth defects the number of cases reported in a given time period has been high enough to be considered an epidemic. The most dramatic case of congenital malformation ever recorded was an epidemic of anencephaly and spina befida that occurred in the northeastern United States from 1920 to 1950 (MacMahon and Yen, 1971) and reached its peak in the early 1930's. It dwarfed the thalidomide and all rubella epidemics by several orders of magnitude. Scientists, using epidemiologic studies, have not been able to identify environmental factors that explain more than a small portion of human congenital malformations. Thus many scientists conclude that the explanation of congenital diseases or birth defects lies in genetic factors (Janerich, 1972). If the dominant factor in congenital diseases is genetic, it implies that man's hereditary material contains a large number of genetic defects and that we may have only limited success in developing measures to prevent birth defects in the near future. If, on the other hand, the causes of birth defects are environmental, we should be able to develop adequate preventive measures. Many examples of congenital genetic defects will be presented with respect to sex, time of expression, and ethnic grouping in later chapters.

references

Austin, C. R. 1970. The Egg and Fertilization. *Science Journal*, *6*:37–42.
Carr, D. H. 1970. Heredity and the Embryo. *Science Journal*, *6*:75–79.

DHEW. 1974. *What Are the Facts About Genetic Disease*? Publication No. NIH 74-370. Department of Health, Education and Welfare, Washington, D.C.

Grollman, Sigmund. 1969. *The Human Body*, 2nd ed. Macmillan, New York.

Hancock, J. L. 1970. The Sperm Cell. *Science Journal*, *6*: 31–36.

Hurley, S. 1968. The Consequence of Fetal Impoverishment. *Nutrition Today*, *3*: 3–10.

Iizuka, R., et al. 1968. The Physical and Mental Development of Children Born Following Artificial Insemination. *International Journal of Fertility*, *13*: 24–32.

Janerich, Dwight. 1972. The Impact of Major Malformations on Society: Environmental Versus Genetic Factors. In Ian H. Porter et al. (eds.), *Heredity and Society*. Academic Press, New York.

Laboratory Management. 1973. Fetal Statistics, *11*(10): 16.

Lappe, Marc. 1974. Choosing the Sex of Our Children *The Hastings Center Report*, *4*: 1–3.

McCary, James L. 1973. *Human Sexuality*. Van Nostrand, New York.

McLaren, Anne. 1973. Biological Aspects of AID. In *Law and Ethics of AID and Embryo Transfer* (Ciba Foundation Symposium 17). American Elsevier, New York.

MacMahon, B., and S. Yen. 1971. Unrecognized Epidemic of Anencephaly and Spina bifida. *The Lancet*, *1*: 31–33.

Milk Market Board. 1969–1970. *Report of the Breeding and Production Organization*, No. 20.

Stern, Curt. 1973. *Principles of Human Genetics*, 3rd ed. Freeman, San Francisco.

Steward, F. C. 1970. From Cultured Cells to Whole Plants: The Induction and Control of Their Growth and Morphogenesis. *Proceedings of the Royal Society (London)*, *B175*: 1–30.

the chemical nature of genetic material

On a Sunday in 1513 Juan Ponce de León landed in Key West, Florida, and began his search for the Fountain of Youth. He and the members of his company were ignorant of the fact they carried with them a very necessary ingredient for the elixir of that youth, which was the constant production of sperm waiting in constantly renewed abundance for the chance opportunity to penetrate an egg. This would once again establish what Ponce de León would never find—static youth. "Youth" is renewable by the act of fertilization. The cycle of youth, maturity, old age, and then death is a fact of life for all living organisms.

In discussing the creation of a new life, we are really referring to the ways that various types of life reproduce. All life forms must replace themselves through reproduction if they are to survive as a species. When reproduction is uncontrolled and the reproductive rate is higher than the required replacement rate over an extended period of time, there is an excess of the species. We currently acknowledge that there is an excess of humans, that we have a population problem. When the reproductive rate falls below that required for replacement over an extended period of time, the species becomes threatened with extinction. In the United States at this time our reproductive rate is less than the replacement requirement. For the world as a whole, however, reproduction exceeds replacement. The human population records a birth every 2.4 seconds, that is, 144 babies per minute or 8666 babies per hour. By the year 2000, 7 billion people are expected to populate the earth.

Another requirement for survival of our species is sufficient variation among individuals to withstand occasional environmental stresses. In a homogeneous (all alike) population all humans might succumb to the same

The cell has a purpose, to express its inherited potential as coded in its DNA.

the identification of our hereditary material and its function

stress at the same time and become extinct. Thus the process of reproduction must produce occasional progeny (offspring) that differ in various ways from all other members of that species. The differences may be anatomical (differences in structure) or physiological (differences in body function).

As stated in Chapter 3, if a change is even slightly advantageous to the new progeny over a period of time, future offspring continue to inherit the change. Slowly the feature becomes prevalent in the species. If the changes are slight, members of the species with the new and the old characteristics may live in close proximity. Eventually, especially if there is a physical or biological barrier, the new and the old can become separate and distinct species, as evidenced in Darwin's finches of the Galápagos Islands (Lack, 1953). Thus the redistribution of our hereditary material, DNA, through reproduction is essential for organic evolution.

the origin of hereditary material

We know one cell mitotically gives rise to identical daughter cells. Once cell life had started, there must have been some stable biochemical (a chemical made in a living system) capable of transmitting the cell's biological potential for survival. Yet, according to modern evolutionary theory, this stable biochemical molecule must also have had the ability to undergo change with time. Since we believe modern man developed through organic evolution, we must presume that this biochemical material retained certain messages of its past history and yet allowed for the acceptance of change. Not all changes in this biochemical material were satisfactory, and those organisms with unsuitable changes were selected against (died out). Those with changes that were advantageous in their particular environment survived and transmitted the successful biochemical molecule to the next generation. In this manner we received our biological inheritance.

Since the earth's first atmosphere contained more hydrogen gas than oxygen, the first species to appear would have been suited to living and reproducing in a reducing atmosphere. With the development of photosynthesizing plants, the atmosphere slowly changed from the reducing state to the present atmosphere, which contains a relatively high concentration of oxygen. One can reasonably speculate that as the oxygen concentration slowly increased over millions of years, its presence became poisonous to those plants and animals that had evolved in its absence. Except for a few species occupying exceptional habitats, life that was totally dependent on the reducing atmosphere perished. But some members of these populations could tolerate some oxygen because of small biochemical changes in their hereditary material. As evolution continued, various life forms appeared showing still greater tolerance for oxygen, until today we not only tolerate oxygen, but are totally dependent on the present levels of oxygen in our atmosphere. Thus gradual increases in free oxygen over millions of years resulted in the development of our current manner of cellular respiration. Our type of cellular respiration, as well as all other traits of our species, is transmitted to each new generation through a message encoded in a molecular descendant of that stable biochemical molecule originating in the warm oceans many millions of years ago.

the chemical nature of genetic material

Our human biological inheritance makes us similar as a species, yet unique as individuals, and also separates us from other species. Each species has its own inherited message encoded into the same kinds of biochemical molecule. In other words, all living species are made up of cells; but the cell of a human differs from that of an elephant, both of these differ from a bacterium cell, and so on. Why is it that when humans mate, they can expect to give rise to another human and not an elephant? Further, why do certain cells within us function as kidney cells or heart cells or brain cells when life begins as a single cell? These questions are intriguing when we consider that every species uses the same biochemical molecule (deoxyribonucleic acid or DNA) as its genetic contract with the next generation and every sexually reproducing species starts out as a single cell and ends up as a highly complex network of tissues, organs, and biochemical synthesis.

In 1947 Irving Chargaff determined that DNA contains equal amounts of two types of nitrogen-containing bases, the purines and pyrimidines. That is, the ratio of adenine plus guanine (the purines) to cytosine plus thymine (the pyrimidines) is close to 1.00, regardless of the organism used as the DNA source (Table 5–1). Chargaff also showed that the amount of adenine always equaled the amount of thymine, and the amount of guanine equaled the amount of cytosine. Thus Chargaff established a purine–pyrimidine relationship. By 1952 we knew that a gene was made of DNA. However, questions concerning the amount of DNA required for a single gene and the precise structure of DNA remained to be answered.

Table 5–1 Base Composition of DNA from Various Organisms

	Adenine, %	*Thymine, %*	*Guanine, %*	*Cytosine, %*	$\frac{A+G}{C+T}$
Human (sperm)	31.0	31.5	19.1	18.4	1.00
Cattle (sperm)	28.7	27.2	22.2	22.0	1.03
Chicken	28.8	29.2	20.5	21.5	0.97
Salmon	29.7	29.1	20.8	20.4	1.02
Locust	29.3	29.3	20.5	20.7	0.99
Sea urchin	32.8	32.1	17.7	17.7	1.02
Yeast	31.7	32.6	18.8	17.4	1.00
Tubercle bacteria	15.1	14.6	34.9	35.4	1.00
Escherichia coli	26.0	23.9	24.9	25.2	1.04
Vaccinia virus	29.5	29.9	20.6	20.3	1.00
Bacteriophage T2	32.6	32.6	18.2	16.6*	1.03

*5-Hydroxymethylcytosine.

can we define hereditary material more clearly?

Understanding of the mechanisms of heredity came long before we knew the chemical nature of the genetic material involved. Gregor Mendel, the father of genetics, described the methods of heredity and formulated his famous Mendelian laws of heredity about 1865. Mendel based his laws on his observations of certain traits that appeared and disappeared as he crossed garden pea plants. Mendel spoke of pairs of "factors" responsible for each trait. Some 30 years later Wilhelm Johannsen called these factors *genes.* (Mendel and his work are discussed in Chapter 8.)

An English bacteriologist, Fred Griffith, provided an early clue to the identity of the genetic material of living organisms in 1928. Griffith injected living avirulent (not pathogenic or disease-producing) pneumococci along with heat-killed virulent (disease-causing) pneumococci into mice. The mice soon died, and an examination of their tissue revealed the presence of live virulent pneumococci. Controlled experiments indicated that revival of the dead cells was unlikely, and Griffith concluded that some unexplainable transformation must have occurred (Griffith, 1928). By 1944 Avery, MacLeod, and McCarty (1944) had extracted the transforming agent of heat-killed pneumococcus and identified it as DNA, deoxyribonucleic acid. Transformation of one cell type into another through the use of naked DNA from the second cell type implied that a gene was a fragment of DNA, since only molecules of DNA brought about a permanent transmissible change in the cell.

In 1953 a scrap metal model of DNA was put together that offered an explanation of the structure of DNA, its mode of duplicating itself, and perhaps the size of a gene. Like Darwin's *Origin of Species*, Mendel's laws, and the advent of the atomic bomb, this model changed man's thinking about and understanding of his world.

This work was done by James Watson, an American biologist, Francis Crick, an English biochemist, and a woman everyone seems to forget, Rose Franklin, who was an X-ray technician in the laboratory of Maurice Wilkins, an English physicist.

Before we discuss the structure of DNA, let us consider what properties the genetic or hereditary material should possess:

1. It must be a stable molecule; yet it must have a means of giving rise to members of a species with subtle differences.
2. It must be able to reproduce itself with great fidelity, or in minute detail.
3. It must have a means of containing the information for the many thousands of traits that make up the individual.
4. It must be universal in that all life should possess it, since all living systems show a means of heredity.

the chemical nature of genetic material

DNA turned out to be the only biological chemical molecule to possess all these properties.

cellular location of DNA

DNA is located in two areas within the typical animal cell (see Figure 2–6). The majority of DNA is in the chromosomes in the nucleus. A relatively small amount is found in the mitochondria. Organelles, such as mitochondria of plants and animals and chloroplasts of algae and higher plants, contain not only DNA but also RNA (ribonucleic acid), ribosomes, and other necessary machinery for the synthesis of protein. There is a distinct difference between the DNA located in the nucleus and that found in the cell's mitochondria or chloroplasts (Goodenough and Levine, 1970). Unlike nuclear DNA, the DNA of the organelles does not associate with histone protein and exists as a naked DNA fiber. DNA fibers in mitochondria have been isolated as circular molecules of DNA. All bacteria and blue-green algae that have been studied contain DNA similar in morphology to that found in mitochondria and chloroplasts. Maybe, somewhere in time, primitive bacteria or blue-green algae became resident in cells and in this rather protected state underwent a type of symbiotic evolution. Whatever the situation, both plant and animal cells contain about 1000 times more DNA than a bacterial cell (Figure 5–1).

Let us go back for a moment to the cell nucleus. In 1869 Friedrich Miescher found in the nucleus a substance he called nuclein. Years later it came to be known as nucleic acid because of its acidic nature. By the time of Miescher's discovery, Mendel had performed his genetic experiments and Charles Darwin had written his classic work on natural selection. Yet it was not until 1953 that all of these discoveries were brought together.

In the early 1940's scientists thought that nucleoproteins (proteins found in the nucleus) were the genetic material. In fact, in 1944 Erwin Schrödinger

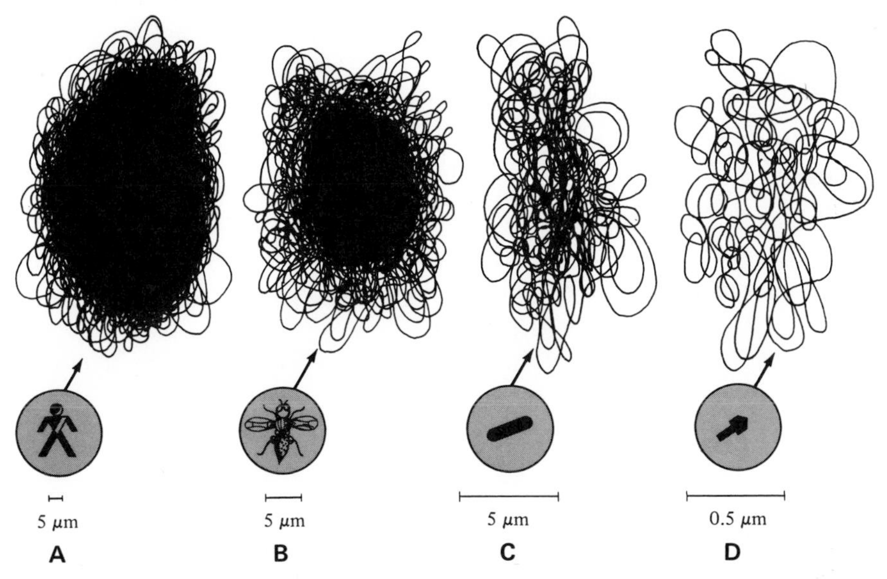

Figure 5–1. Relative Lengths of DNA. **A.** Human, about 1 million μm. **B.** The fruit fly, *Drosophila melanogaster*, about 16,000 μm. **C.** The bacterium *Escherichia coli*, about 1100 μm. **D.** Bacteriophage T4, about 68 μm. (Redrawn from F. W. Stahl, *The Mechanics of Inheritance*, ©1964, Prentice-Hall, Inc., Englewood Cliffs, N.J.).

wrote a book entitled *What Is Life?* in which he stated that genes are special kinds of proteins. We know now that the amount of nucleoproteins containing protamine or histone varies among cells in the same organism and among organisms of the same species. This lack of constancy rules out the possibility of nucleoprotein as genetic material.

In contrast, when the nucleic acids are separated out of the nucleus, the amount of DNA is constant in all cell nuclei of the same organism, regardless of cell size. However, the sex cells or gametes have only half the DNA content of somatic cells because of meiosis; through meiotic division the diploid cell with 46 chromosomes produces a sperm or egg with only 23 chromosomes, or with only 50% of the diploid cell DNA. The constant proportion of DNA in all cells led to the belief that DNA is *the* genetic material.

the structure of nucleic acids (DNA and RNA)

There are two biologically important nucleic acids: deoxyribonucleic acid (DNA) and ribonucleic acid (RNA). By learning more about the components of these two compounds, you can better understand why you are you and why you can be sure that, when you become an expectant parent, you will have another human and nothing else.

The nucleic acids are made up of nucleotides. A nucleotide can be broken down into a sugar, a phosphate, and a nitrogenous base. These substances can be further degraded into atoms of nitrogen, phosphorus, oxygen, carbon, and hydrogen.

Two different kinds of sugar molecules, a single kind of phosphate molecule, and five different nitrogenous bases are used in making up the nucleic acids DNA and RNA (Figures 5–2 and 5–3). The sugar deoxyribose, the phosphate, and one of the four bases (A, T, C, or G) attached to each

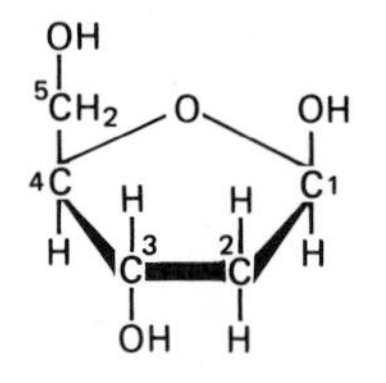

Deoxyribose
The sugar molecule of DNA

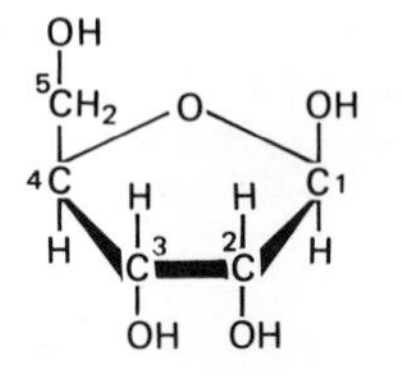

Ribose
The sugar molecule of RNA

A

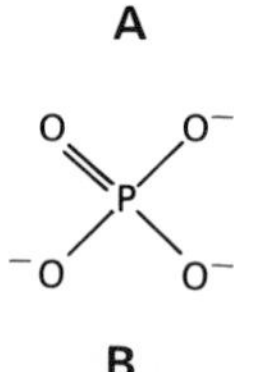

B

Figure 5–2. Components of DNA. **A.** The sugars. Note the difference between the sugars of RNA and DNA. The oxygen (O) that is present in the OH radical associated with the number 2 carbon of ribose is missing from deoxyribose. **B.** Phosphate. The same phosphate molecule occurs in RNA and DNA.

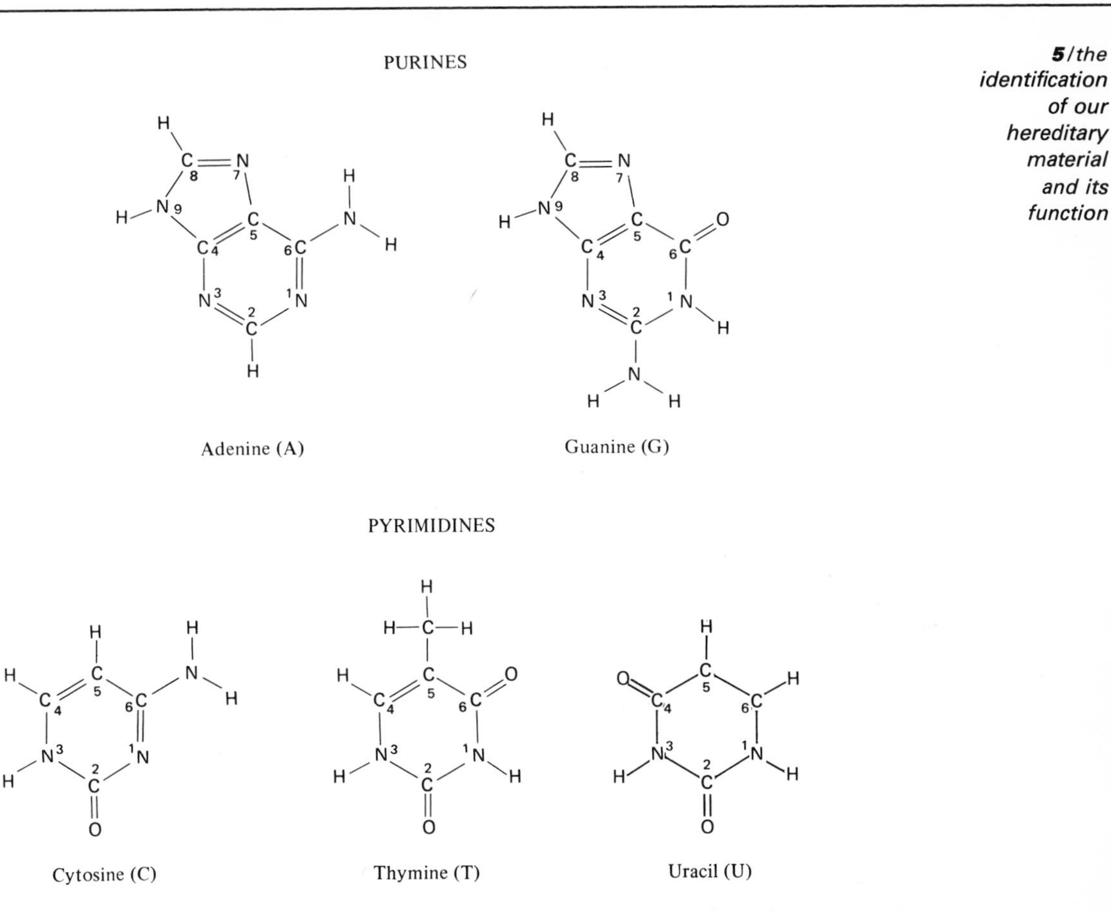

Figure 5–3. The Common Nitrogenous Bases. DNA contains the purines adenine and guanine and the pyrimidines thymine and cytosine. RNA contains the purines adenine and guanine and the pyrimidines cytosine and uracil. It is conventional to symbolize adenine by the letter A, guanine by G, thymine by T, cytosine by C, and uracil by U. This technique makes it easier to portray strands of DNA and RNA and the genetic message concisely.

other make up a nucleotide, a unit of DNA. The formation of a nucleotide containing the base adenine is represented in Figure 5–4. A polymer (chain) of nucleotides containing the four nitrogenous bases in a given sequence, as determined by evolutionary history, is a species' genetic contact with its past and contract for its future. In other words, it is the sequence and number of nucleotides in the DNA that provide the information for making the thousands of different protein molecules found in the cells.

putting it all together

In 1962 Watson, Crick, and Wilkins shared a Nobel prize in physiology and medicine for their research and their recognition of DNA as a double helix that could replicate (duplicate itself) by complementary base pairing. From 1959 to 1975 seventeen scientists were awarded Nobel Laureates for their work and achievements with DNA and its function within cells (see Table 5–2).

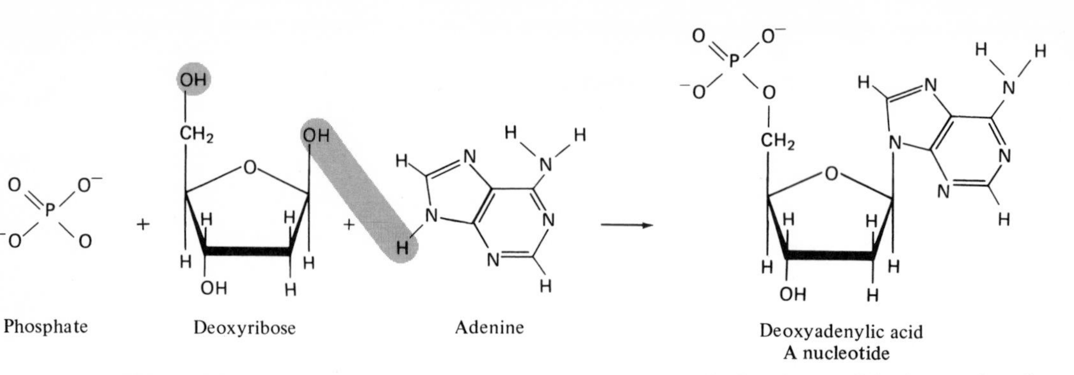

Figure 5–4. Formation of a Nucleotide. With the base adenine the result is deoxyadenylic acid.

From X-ray diffraction data provided by Maurice Wilkins and Rose Franklin, Watson and Crick (1953a) concluded that the DNA molecule has a double-stranded helical structure. They further hypothesized that the two-stranded helix of the parent DNA molecule becomes untwisted for replication; then each parent strand serves as a template for the synthesis of a new strand complementary to itself (Watson and Crick, 1953b). DNA replication occurs in a semiconservative manner, each daughter molecule consisting of one strand from the original parent molecule and one newly synthesized strand (Figure 5–5).

Table 5–2 Recipients of the Nobel Prize in Physiology and Medicine for Work Related to DNA

Year	*Name*	*Investigations*
1959	Arthur Kornberg Severo Ochoa	For studies of the chemistry of DNA and RNA
1962	James D. Watson Francis H. C. Crick Maurice H. F. Wilkins	For elucidation of the intimate structure of DNA
1965	François Jacob André M. Lwoff Jacques Monod	For discoveries concerning genetic control of enzyme and virus synthesis
1968	Marshall W. Nirenberg H. G. Khorana Robert W. Holley	For "cracking the genetic code" and elucidating the means by which a gene determines the sequence of amino acids in a protein
1969	Max Delbrück Alfred D. Hershey Salvador E. Luria	For studies of mechanisms of virus infection in living cells
1975	Renato Dulbecco Howard Temin David Baltimore	For studies of the interaction between tumor viruses and the genetic material (DNA) of the cell

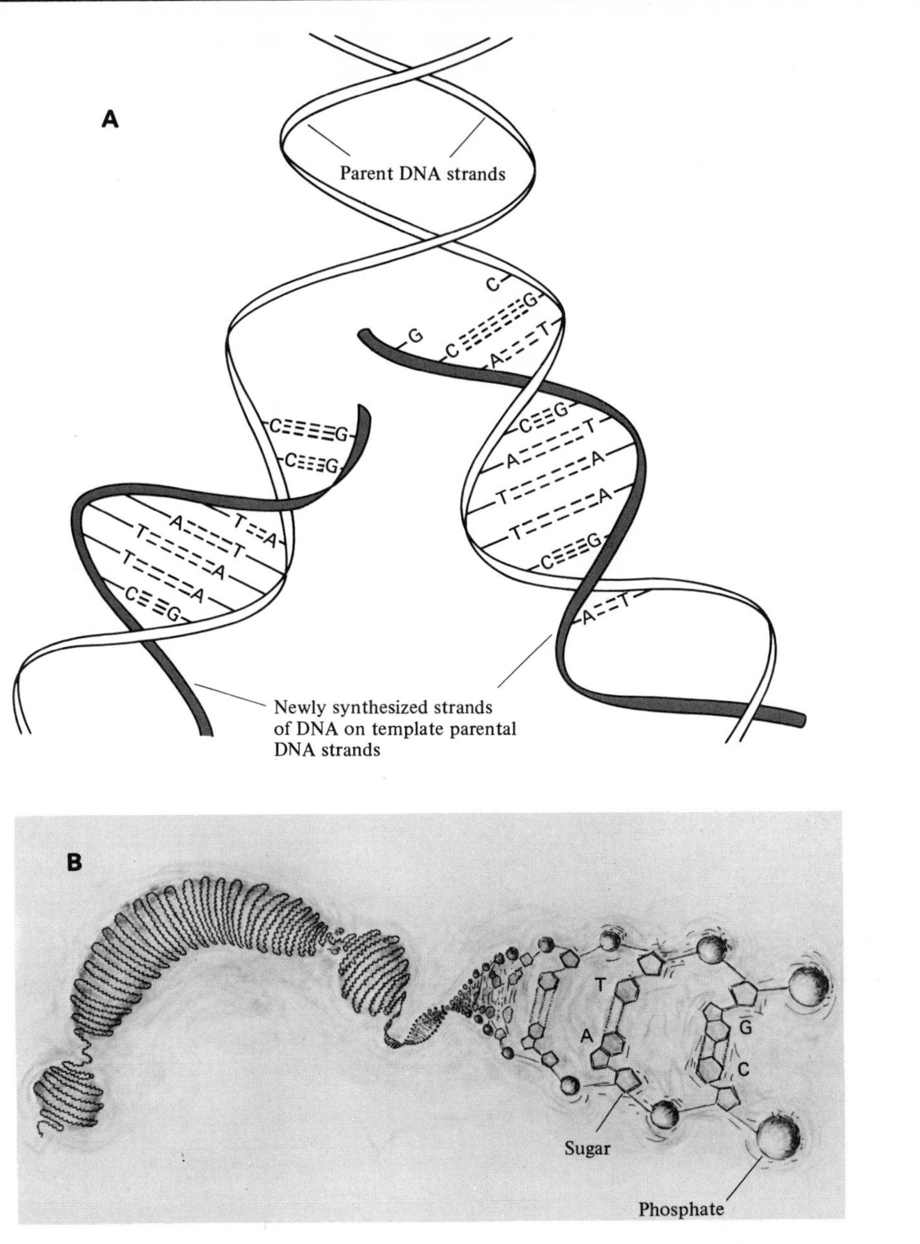

Figure 5–5. DNA Replication. **A.** Semiconservative replication of DNA through complementary pairing of bases, A with T and C with G. **B.** Hypothetical unwinding of chromosomal DNA in preparation for replication. (Photo courtesy Carolina Biological Supply Company.)

Elegant experiments performed by Taylor (1957; Taylor, Woods, and Hughes, 1957), Meselson and Stahl (1958; Meselson, Stahl, and Vinograd, 1957), Cairns (1963), and Kornberg (1968) support this theory of replication. For example Taylor et al., using radioactive thymine, followed the synthesis and replication of DNA and showed that the DNA residing in the chromosome of the plant *Vicia faba* (the Italian broad bean) is reproduced and distributed to daughter cells in a semiconservative manner.

The DNA molecule can be envisioned as a twisted ladder in which the two sides, or risers, are made up of long alternating units of deoxyribose and phosphate, and the cross-linking rungs are pairs of nitrogen-containing purine and pyrimidine bases (adenine, thymine, guanine, and cytosine) held together by hydrogen bonds (Figure 5–6). A strict system of complementary base pairing, in which adenine is always paired with thymine and guanine with cystosine, enables each strand to serve as a template for the exact reconstruction of the other strand.

This system of complementary base pairing, with A pairing only with T and G only with C, indicates that the four bases A, C, T, and G will pair with each other only in certain spatially and chemically defined ways. Thus if one strand of DNA consists of AACC, the opposite strand will be TTGG. Each strand of DNA, then, is complementary to the other. This is a crucial point because the nature of complementary base pairing is what ensures that a particular species remains that species. It is the reason humans give rise only to human infants.

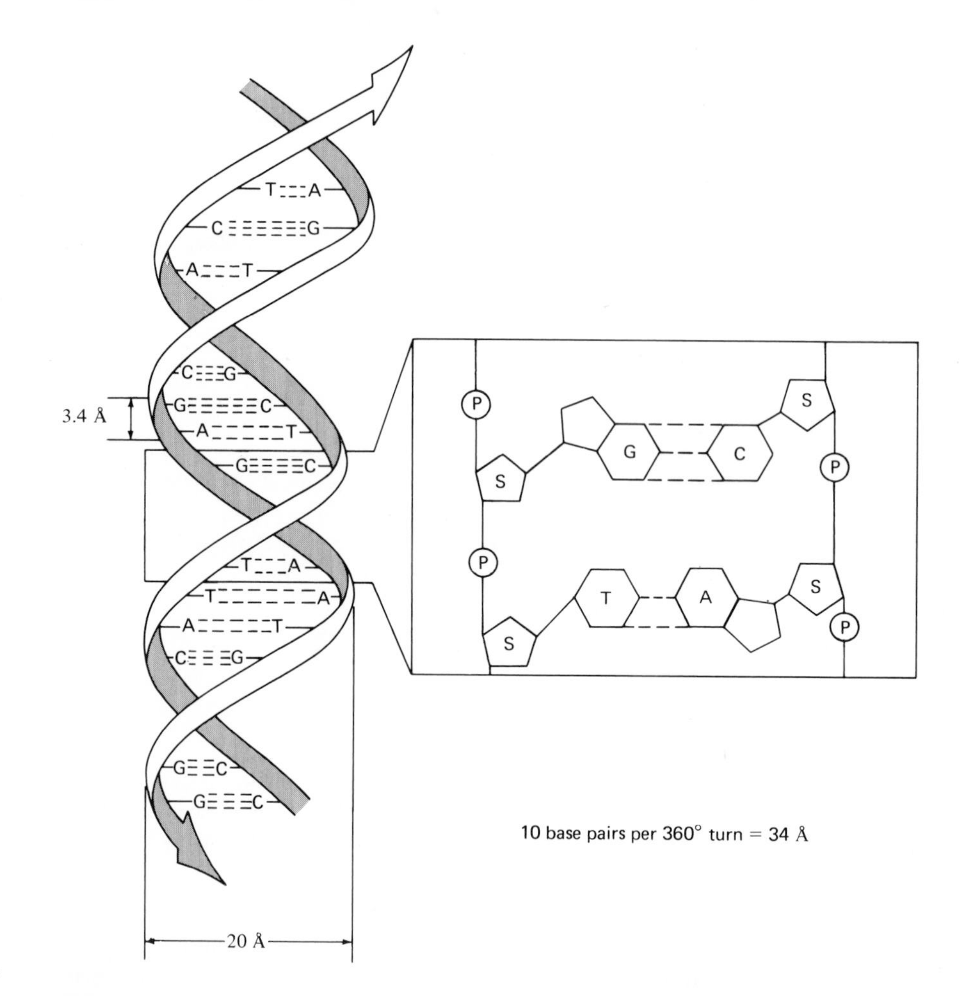

the chemical nature of genetic material

Figure 5–6. Base Pairing and Polarity. Molecules of DNA showing opposite polarity (indicated by the direction of the arrows) of the double-stranded spiral helix. Physical measurements in Ångstrom units (10^8 Å = 1 cm).

Specifically, a certain sequence and number of nucleotides give rise over time to human characteristics, which are passed from generation to generation. Slight, subtle changes in the human DNA and interaction with the environment account for the variations among humans and the uniqueness of each individual.

Base pairing is very exact. If mispairing occurs during DNA replication, the substitute nucleotide may later cause a change in some protein made by the cell. If nucleotide substitutions occur at crucial points along the DNA molecule, a new species may, in time, be derived from the old, since the information in the changed DNA may be very different from the information in the ancestral cell that existed thousands of years before. This possibility, together with directional natural selection, gives us a theory of molecular evolution that can account for the great diversity of species we know today. At the same time, stabilizing natural selection accounts for the continuation of a given species once it has evolved (see Chapter 3).

Genetic material is very stable, and humans give rise only to other humans. Yet, over millions of years, alterations in the DNA, if *not* selected *against* or if selected *for*, can accumulate in our DNA. At some point we may be making enough different proteins to subtly change the human phenotype. Continued occurrence of advantageous changes in the DNA or a shift in the environment to favor the changed DNA and eliminate the unchanged DNA could result in a new species of man.

the genetic code and protein synthesis

Our DNA contains the information for organizing and constructing the complex human body, but how does it accomplish these results? How, for example, does a single cell make the variety of protein molecules necessary for life? Proteins are essential, for various structural and functional proteins are involved in nearly all cell processes. Structural proteins are found in membranes, contractile fibers, and intracellular reinforcing filaments. Functional proteins, the enzymes, are required to catalyze the synthesis of the various chemical compounds required for life. Thus, if and when new species evolve from older species, the difference between the old and new is found in the presence or absence of certain variations of particular structural and functional proteins (see Chapter 1 for examples).

By studying cell physiology, scientists recognized that although DNA is located in the nucleus of animal and plant cells, protein synthesis actually occurs some distance from the nucleus; specifically, it occurs on the ribosomes located in the cytoplasm. Transmission of the cell's genetic message to produce a protein in the cytoplasm involves another nucleic acid, RNA. Ribonucleic acid, RNA, is different from DNA in at least three respects:

1. RNA is in general a single-stranded molecule.
2. The ribose sugar in RNA has the OH group at the number 2 carbon (see Figure 5–2).
3. RNA contains the base uracil instead of the base thymine (see Figure 5–3).

There are three types of RNA, each with its own role in the synthesis of proteins: messenger (mRNA), ribosomal (rRNA), and transfer (tRNA). The blueprint for their production is carried in the series of nucleotides in DNA. DNA is the master molecule for hereditary memory.

How is a protein synthesized and how can a cell make so many different proteins with only four bases coding the necessary information? Before answering this question, let us consider the nature of protein.

what is a protein?

A protein is made up of a series of amino acids. Two amino acids linked together are referred to as a dipeptide, and many amino acids so joined are referred to as a polypeptide. A polypeptide containing a sufficient number of amino acids to achieve a molecular weight near 10,000 is a protein and has a given function to perform. By definition, then, proteins contain various numbers of amino acids connected one to another. Dipeptide and polypeptide formation involves the joining together of amino acid molecules through the peptide bond. A dipeptide formation is represented in Figure 5–7.

A polypeptide, then, is a series of amino acid residues, joined by peptide bonds, and having an amino end (NH_2) and a carboxyl end (COOH). One such is the functional protein beef insulin (Figure 5–8). Beef insulin was the first protein in which the complete amino acid sequence was determined. It is made up of two separate (A and B) chains of amino acids cross-linked or connected to each other by the disulfide bonds formed between two cysteines as shown.

Some amino acids are large and complex, others small and simple. Some, such as lysine and arginine, carry a positive charge; others, such as glutamic

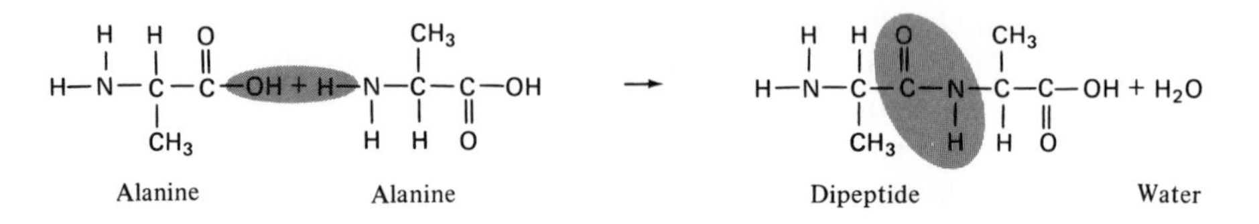

Figure 5–7. Formation of a Dipeptide from Two Molecules of Alanine. The OH and the H that are removed and the peptide bond that forms are shaded.

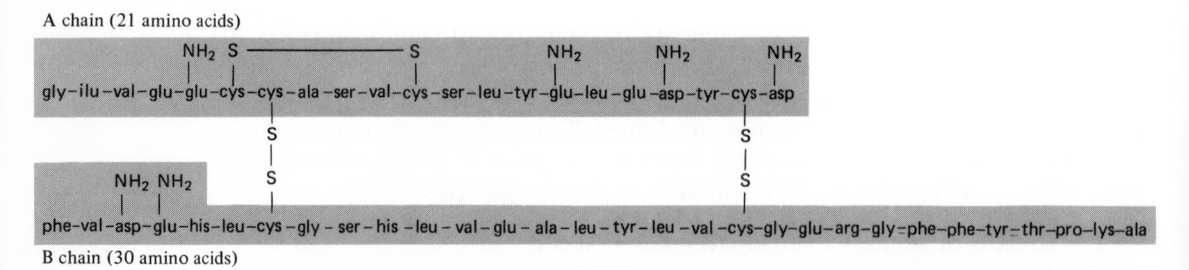

Figure 5–8. The Amino Acid Sequence in Beef Insulin. Note the disulfide bond links between the two chains of amino acids. This cross-linking is a common property of proteins.

acid and aspartic acid, carry a negative charge; but most carry none. Differences in bulk and electric charge along a polypeptide, plus an intricate coiling and folding of the protein molecule, impart much of the specificity of enzyme–substrate and antigen–antibody relations. The structural and functional characteristics of a protein, then, are based on the number and kind of amino acid residues. Although only 20 different amino acids are found in the makeup of typical plant and animal proteins (Table 5–3, Figure 2–3), there is an almost infinite variety of possible proteins. The type actually formed depends on the messages in the DNA.

Since an infinite variety of protein molecules is possible, we must ask how DNA, with only four bases, can provide the instructions so that certain amino acids will be joined together to give a cell the specific protein required. And why is a particular protein made only when the cell demands it and only in the amount the cell demands?

understanding the genetic code

In 1968 Marshall Nirenberg, H. G. Khorana, and Robert Holley shared the Nobel prize for their work in "cracking the genetic code," that is, explaining how a gene controls the arrangement of amino acids into a specific protein. In other words, cracking this code meant determining the number and sequence of nucleotides involved in the exact positioning of each amino acid in a growing peptide chain. Knowing the number and identity of nucleotides required for the placement of each amino acid, we can one day, theoretically, build any protein synthetically simply by creating an artificial message on which the protein-synthesizing machinery of the cell can act. This newfound knowledge has exciting possibilities and implications for the fields of human development, medicine, and agriculture.

Table 5–3 The Twenty Amino Acids Commonly Found in Plant and Animal Protein

Amino Acid	*Abbreviation*	*Amino Acid*	*Abbreviation*
Alanine	ala	Leucine*	leu
Arginine*	arg	Lysine*	lys
Asparagine	asn	Methionine*	met
Aspartic acid	asp	Phenylalanine*	phe
Cysteine	cys	Proline	pro
Glutamic acid	glu	Serine	ser
Glutamine	gln	Threonine*	thr
Glycine	gly	Tryptophan*	trp
Histidine*	his	Tyrosine	tyr
Isoleucine*	ileu	Valine*	val

*Ten amino acids must be provided in the human diet because the body cannot synthesize them. See Figure 2–3 for individual amino acid chemical structures.

To break the code, Nirenberg and co-workers synthesized an artificial "messenger RNA." They chemically linked together, or made, a polymer of the ribonucleotide containing the base uracil, the base that distinguishes RNA from DNA. They placed this poly-U (U—U—U—U—U— etc.) in a cell-free

First base in triplet	Second base in triplet is: Uracil		Cytosine		Adenine		Guanine		Third base in triplet
	UUU	phe	UCU		UAU	tyr	UGU	cys	Uracil
Uracil	UUC		UCC	ser	UAC		UGC		Cytosine
	UUA	leu	UCA		UAA**		UGA**		Adenine
	UUG		UCG		UAG**		UGG	trp	Guanine
	CUU		CCU		CAU	his	CGU		Uracil
Cytosine	CUC	leu	CCC	pro	CAC		CGC	arg	Cytosine
	CUA		CCA		CAA	gln	CGA		Adenine
	CUG		CCG		CAG		CGG		Guanine
	AUU		ACU		AAU	asn	AGU	ser	Uracil
Adenine	AUC	ileu	ACC	thr	AAC		AGC		Cytosine
	AUA		ACA		AAA	lys	AGA	arg	Adenine
	AUG*	met	ACG		AAG		AGG		Guanine
	GUU		GCU		GAU	asp	GGU		Uracil
Guanine	GUC	val	GCC	ala	GAC		GGC	gly	Cytosine
	GUA		GCA		GAA	glu	GGA		Adenine
	GUG*		GCG		GAG		GGG		Guanine

* A chain initiator. ** A chain terminator.

Figure 5–9. The Assignment of Codons. The amino acids for which the triplets code are shown in the shaded boxes. The first base in each triplet is given at the left, the second base is one of the four across the top, and the third base is at the right. Thus the first codon consists of three uracils, UUU; the second of uracil, cytosine, and uracil, UCU; and so on. AUG codes for *N*-formylmethionine and methionine. Discrimination between the two forms occurs through their different binding to an initiation factor necessary in translation. Normally GUG codes for valine, but is a "start" codon for the "A" protein of MS2 virus. UAA, UAG, and UGA do not correspond to any amino acid; they are responsible for terminating translation of a given protein.

extract capable of synthesizing protein. The message in this case was such that the poly-U and the system created a protein made up of only one amino acid, phenylalanine. Other single-base polymers directed the incorporation of other amino acids into polypeptides. Following this lead, but using mixed-base polymers as synthetic messenger RNA, Nirenberg and his colleagues were able to decipher the complete code. They were able to identify the messenger RNA base sequences necessary for the incorporation of the individual amino acids into protein.

Now, how many nucleotides code for the actual placement of an amino acid in a protein. That is, does a "codon" consist of one, two, three, or more nucleotides? To answer this question Nirenberg and other scientists selected nucleotide polymers at random and used them to establish that the four bases in various combinations could code for the placement of each of the 20 amino acids. More than one base is required for a codon because 4 is less than 20, and even a two-base sequence is only enough for 16 ($4^2 = 16$) combinations. That is, using only two-nucleotide sequences would code for 16 of the 20 amino acids found biological proteins. Using three bases at a time in all possible combinations ($4^3 = 64$) yields an excess of combinations, and using four at a time ($4^4 = 256$) results in an even greater excess of combinations! Experiments showed it takes three bases to account for the insertion of a given amino acid into a protein (see Figure 5–10). Three bases is the codon, the smallest genetic unit required to code the insertion of an amino acid into protein. Further research showed that more than one codon can code for the same amino acid and that some of the codons are not used for the incorporation of amino acids into protein. Some codons act as starting points, or initiators; others act as stopping points, or terminators, which regulate the length of a protein (Figure 5–9). Thus a given number of nucleotides in a specific order will specify a given protein, and that number of nucleotides will be equal to a gene! (This is only one of several definitions of a gene.) In a sense, the three-base codons are used similarly to the 26 letters in the English alphabet. The usefulness or meaning depends on how they are ordered. Further, the same letters (the same codons) can be utilized to form different words (incorporate amino acids into different proteins). For example, if we take

AUC = h = isoleucine

ACU = a = threonine

GUA = t = valine

GAU = e = aspartic acid

we have the word *hate.* If we rearrange the codons,

AUC = h = isoleucine

GAU = e = aspartic acid

ACU = a = threonine

GUA = t = valine

we have the word *heat.* We have used the same codons to transcribe two very different messages. Rearrangement of the codons also changed the placement or sequence of the amino acids. The arrangements of the codons in the RNA determine the structure and function of each protein that is synthesized by determining the identity and sequence of amino acids.

transcription: the transfer of a specific message for protein synthesis

Parallel to the studies on the genetic code, studies relating to the involvement of RNA in protein synthesis revealed that RNA is synthesized along the strands of DNA in the nucleus. The bases in this RNA are complementary to the bases in a given strand of DNA. For example, if the DNA sequence is GCTCGA, the complementary RNA is CGAGCU (recall that uracil occurs in RNA, thymine in DNA). The strand of RNA is said to be transcribed from the DNA. Each sequence of three nucleotides in the RNA specifies the placement of one amino acid in a protein when this particular RNA reaches the ribosomes in the cytoplasm. This RNA, which carries the transcribed message of the gene, is referred to as mRNA, messenger RNA. Once transcribed, the mRNA moves to the ribosomes located in the cytoplasm. The ribosomes, organelles made of protein and a different kind of RNA called ribosomal RNA (rRNA), provide for the translation of the message into a specific protein by making it possible for the mRNA to interact with a third type of RNA, transfer RNA (tRNA). This tRNA is responsible for picking up a single specific amino acid. Thus there are at least 20 different kinds of tRNA molecules in the cytoplasm. Also, since a separate amino acyl synthetase (an enzyme) is involved in the attachment of each amino acid to its particular tRNA, there must be at least 20 separate amino acyl synthetases. The tRNA–amino acid complexes also move to the ribosome surface, where the assembly of amino acids into protein occurs. A tRNA molecule with its specific amino acid is aligned with its complementary codon on the mRNA molecule. The ribosomes and mRNA then appear to move relative to each other, and an amino acid is added to the growing protein each time the ribosome passes a new group of three mRNA nucleotides.

A group of three unpaired bases of a tRNA molecule (the anticodon) pairs briefly with three complementary nucleotides of mRNA (the codon). For example, only a tRNA with the anticodon CGA can insert its amino acid at the location of the MRNA codon GCU. A second tRNA then arrives to pair briefly with the next codon of the mRNA. Immediately, the first amino acid is separated from its tRNA and linked with the next incoming amino acid by means of a peptide bond. The first tRNA is thereby released from mRNA and freed to function again in the same manner. The process continues in this way until a specific polypeptide chain, composed of a particular sequence of amino acids, is formed (Figure 5–10).

Protein synthesis is repeated over and over again during the life of your cells. Each time the system initiates a protein some 140 factors (various components required for protein synthesis) are called into action. If each performs correctly, a specific protein is produced just like the one made in your parents' cells and in their parents' cells. The genetic system, you can see, must

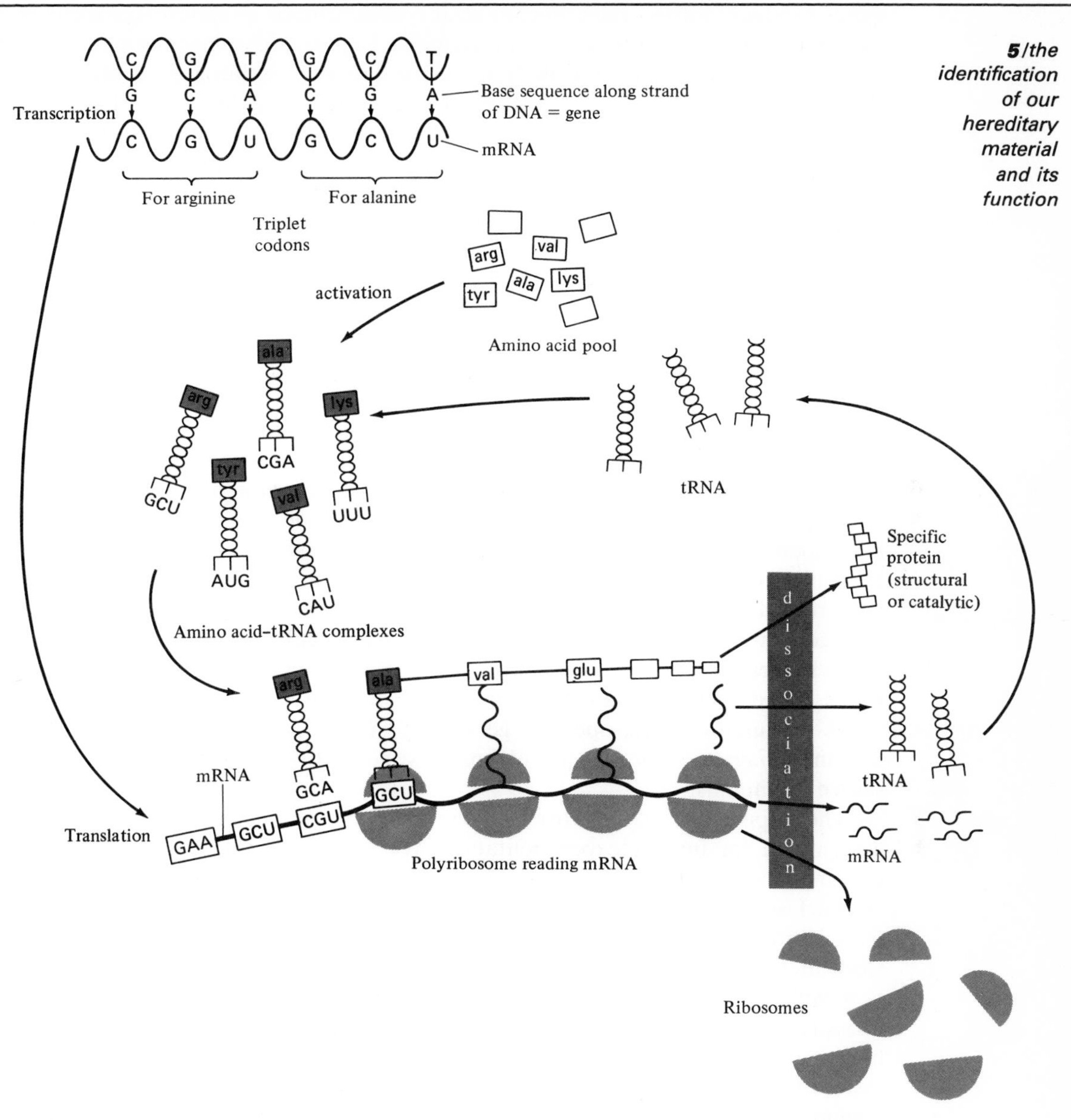

Figure 5–10. Gene Action in Protein Synthesis. The sequence of contiguous nucleotide pairs determines the genetic information available and the type of enzyme (catalytic protein) or structural protein synthesized. mRNA = messenger RNA, tRNA = transfer RNA.

have great fidelity, for a change in a single nucleotide in a single codon can change the message and result in a new or different protein, which may or may not work.

Thus far in our consideration of the gene and its role in protein synthesis we have seen that a gene is a region of DNA, located in the chromosome, that is responsible for the synthesis of a specific polypeptide. The gene consists of a linear sequence of mutable (potentially changeable) nucleotides. Identical genes produce polypeptides with identical amino acid sequences; nonidentical genes produce polypeptides with at least one difference in amino acid identity or sequence. The sequence of nucleotides is the molecular basis of one's genotype (or genetic composition), and the protein synthesized according to the particular genotype is one aspect of one's phenotype (or physical appearance). Changes in the genotype are a major concern because most of them are detrimental. An altered genotype gives rise to an altered phenotype, and in many cases the altered phenotype is recognized as a genetic disease. Of the 2000 or more genetic defects or diseases now known, only about 10 to 20% have been shown to be directly associated with an altered protein or the loss of a specific protein. In every case, however, there must be an undetected protein alteration.

gene regulation and the production of protein

We know that a gene is a sequence of nucleotides capable of creating a protein and of retaining the message of life for each succeeding generation. Genes that code for the production of the proteins used in construction of cell membranes, ribosomes, and other cell organelles and the catalytic proteins (the enzymes) are called structural genes. Primarily through the work of two Frenchmen, François Jacob and Jacques Monod (1961), who investigated the process of gene regulation of protein synthesis in the bacterium *Escherichia coli,* we have learned how at least some genes are themselves regulated in the transcription of their messages. The transcription of structural genes is regulated by other genes called regulator and operator genes. There are at least three classes of genes—regulator genes, operator genes, and structural genes—known to function in *Escherichia coli,* but this elaborate system of gene control has not been shown in animals.

The importance of the work on gene regulation in bacteria is that it serves as a model for further experimentation. We know that the genetic code is a universal language, that is, that all living cells use DNA for their genetic contract with the future and use similar systems for protein synthesis. The model proposed by Jacob and Monod is known as the *operon theory.* For this work they were awarded the Nobel prize in 1965. The concept of the operon is now widely accepted. Experimental exploitation of the model has led to the formulation of new concepts and a better understanding of gene control during human development. According to the operon model (Figure 5–11), there is a system for turning structural genes "on" and "off" as the cell requires. This system explains why some enzymes are not always present in cells. If man could learn how to bypass the mechanism of gene control, he might have the ability to restore lost limbs, eyes, and other body parts. As a price for biological complexity and specialization, we have lost most of the ability to regenerate. Since the same DNA, or set of genetic messages, is present in all our cells (except mature red blood cells, which have no nucleus), we have much to gain by learning to switch genes on and off at our pleasure.

Many lower animals can regenerate lost body parts through a process called dedifferentiation. Cells in the immediate area of an injury, although specialized in function at the time of injury, can turn on genes that have been inactive for a long period of time. The RNA in those cells gives off new messages for the synthesis of proteins that were made earlier in the organism's life, when the particular body part was first undergoing development. A planarian, for example, can regenerate at least 50% of its body if it is cut in half. The starfish can regenerate any of its five appendages. The salamander can regenerate a leg, tail, or most any part of its body, including as much as 30% of its heart.

Cloning experiments (see Chapter 23) give support to the theory that every cell contains the genetic information for making a whole organism and that, during the process of development, some genes are silenced, while others continue to turn on and off throughout the life of the organism. It is speculated that in the adult human, as well as in other mammals, at least 90% of a cell's DNA or genes are shut down or turned off. Recently, it has been shown that variously sized nonhistone protein molecules are able to turn genes on (Stein, Stein, and Kleinsmith, 1975). (Histone proteins are rich in basic amino acids and lack the amino acid tryptophan.)

Hormones have also been shown to initiate gene activity. For example, hormone-initiated gene activation occurs when the adrenal cortical glands produce the hormone cortisone. The cortisone enters the bloodstream and is carried to the liver where it turns on some 12 enzymes that were not being made previously. Perhaps the hormones that bind in the nucleus change the composition of nonhistone protein and thereby initiate gene activity. Whatever the mechanism, the result is the equivalent of pushing a reset button in the genetic program of a given cell. If we can successfully achieve programmed regeneration in humans, we will be able to take cells from "cold storage," place them in the correct medium, and grow a new heart, lung, or kidney. If we can develop the ability to grow new organs from our own cells, then perhaps the ever-present problem of graft rejection when tissue is transplanted can be circumvented.

Perhaps when we can program cells, we will also be able to deal with senescence, which is brought about by the loss of nerve cells making up the brain. According to James Bonner, a human is born with some 10 billion nerve cells in the brain, but, beginning at about age 15, loses them at the rate of 100,000 a day. Since brain tissue does not regenerate, one ultimately suffers from a gradual loss of memory and often from senility. If we could learn to program the neurons so they would divide and replace the number of cells lost per day, we might be able to restore ourselves mentally. Perhaps, since newly formed neurons would not have been programmed previously, the new cells would be ready to take in new information. If this turned out to be true, we would be losing old information but maintaining the ability to store new information, as though we were continuously young. We would, in terms of our brain, never know obsolescence.

There are, of course, a number of ethical problems arising with this type of science. For example, if we gain the ability to reset the genetic machinery of human cells so they will develop into new people (a process now referred to as cloning), who will decide how many and what kind of persons should be cloned? Since blonds have more fun, would there ever be a brunette cloned? Or a person who needs glasses? Perhaps we can say of severely defective people, "Don't clone any of those." But how do we decide which of the "normal people" should be cloned? What will happen to the family, to the meaning of the term *parent*, to the spice of life, or to variability or differences among people? More will be said about this subject in Chapter 23.

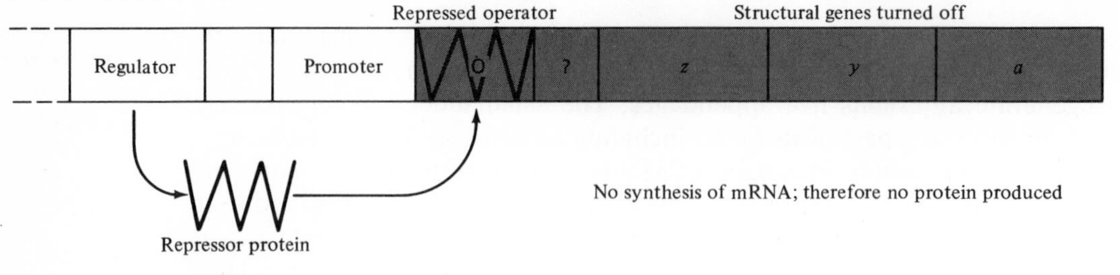

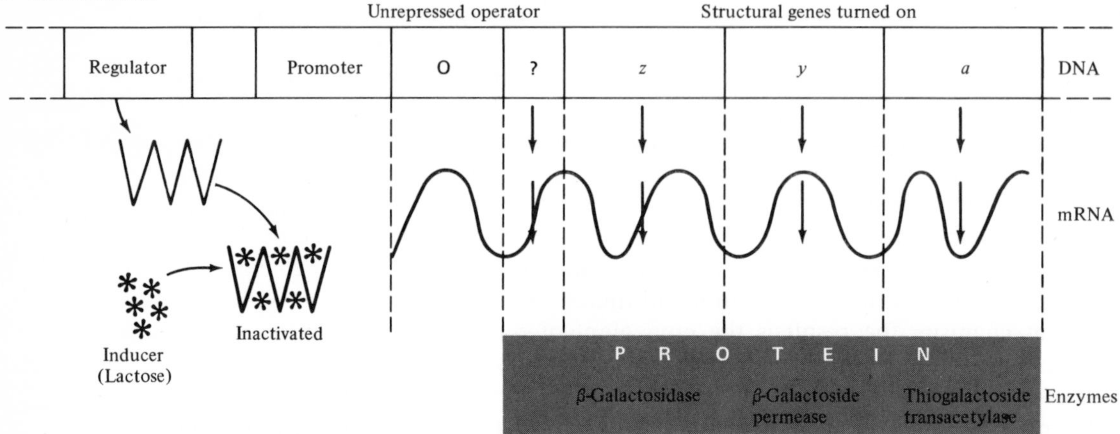

Figure 5–11. The Lactose Operon of *E. coli*. *E. coli* uses glucose as its primary source of carbon, but it has the genetic capacity to convert lactose into glucose (that is, it can utilize an alternative carbon source). If *E. coli* is placed in a culture medium containing both glucose and lactose (which is made of glucose and galactose), it will grow well until the glucose has been used up. Then growth slows or even comes to a halt. After a lag (a period of adjustment), the cells begin to utilize the lactose. They do so by synthesizing a new enzyme, beta-galactosidase (β-gal), that splits the lactose molecule:

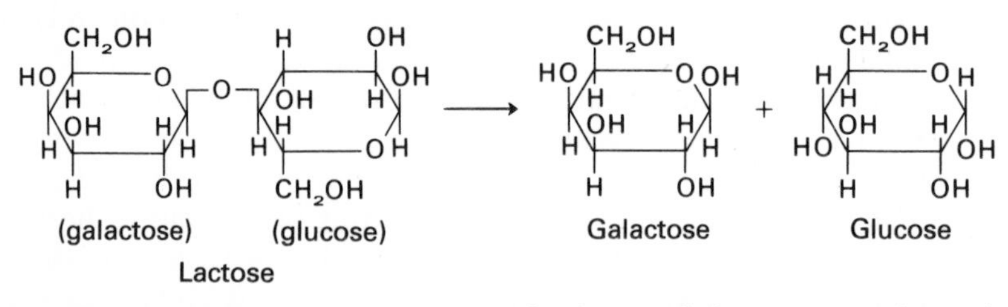

The genes that code for the enzymes necessary for the use of glucose are continuously "on"; these enzymes, which are continually synthesized, are termed constitutive. The genes that code for the production of beta-galactosidase, on the other hand, are "off" until they are activated. Such an enzyme is not constitutive. In the absence of glucose, molecules of lactose (or very similar molecules) activate the synthesis of the needed enzymes. The process is referred to as enzyme induction (an inducible system works for catabolism—the process of breaking down substances for further use).

The diagram shows the lac region of the *E. coli* chromosome. It consists of three sites—regulator, promoter, and operator—and three structural genes—*z, y,* and *a*. RNA polymerase attaches at the promoter site, but repression of the operator prevents it from transcribing mRNA. **A.** When the inducer is absent, the repressor protein made by the regulator binds to the operator site and prevents transcription of mRNA, thus preventing synthesis of the

Since the lactose operon model was devised to explain the results of several gene mutations that occur in *E. coli*, it is interesting to speculate that mutations occurring in the regulator, operator, or structural genes are involved in hereditary diseases. For example, regulator and operator genes so mutated or changed that they cannot turn on or off may be the cause of a variety of human metabolic pathologies. Genes unable to function at crucial periods in embryo, child, adolescent, or adult development will affect the individual. Perhaps it is such an inability that causes some diabetics to stop making insulin, prevents hemophiliacs from making certain blood-clotting factors, or causes under- and overproduction of hemoglobin. (In the disease hypogammaglobulinemia, not enough gammaglobulin is produced, and in hypergammaglobulinemia there is an overproduction.) Mutated genes could also cause age-dependent diseases, such as Huntington's disease, where the pathology is not usually expressed until after the individual is 25 years old.

In the inherited dominant disorder porphyria (dominant and recessive inheritance are explained in Chapters 9 and 10), the patient produces an excess of porphyrin, a molecule required in the production of hemoglobin. The overproduction of porphyrin is the result of an excessive amount of an enzyme, delta-aminolevulinic acid synthetase (ALA synthetase), that converts a precursor (a beginning substance) into porphyrin (*Laboratory Management*, 1973). Normal persons have about 25 units of enzyme, while the person with this genetic defect has about 174 units of enzyme. A similar effect occurs in hereditary oroticaciduria, where the affected person has more than ten times the normal amount of orotic acid, and in von Willebrand's disease, where varying levels of blood-clotting factor VIII are found in the patient's blood at different times. Could these situations occur because a faulty regulator gene fails to control the structural gene?

In some enzyme systems, such as those involving the synthesis of histidine and many other amino acids, as well as purines and pyrimidines, enzymes are not produced constitutively or by induction. Instead, they are repressible: their production is repressed by a specific substance that is activated during the course of metabolism. The production of the amino acid histidine, for example, is reduced when there is an accumulation of histidine, and the synthesis of the ten enzymes in the histidine pathway is reduced to a similar extent when a certain concentration of histidine is reached. This mechanism, which is called coordinate repression, serves to prevent synthesis of materials that are already present in sufficient concentration. Both repressible systems and inducible

---

enzymes for utilization of lactose. **B.** When molecules of lactose enter the cell—that is, the inducer is present—they combine with the molecules of the repressor, competitively removing them from the operator site. With the operator "unlocked," the RNA polymerase attached at the promoter site can transcribe mRNA through the series of three structural genes *z, y,* and *a*. These three genes are expressed and repressed together. The operator and the three structural genes constitute the *operon*. The (?) represents a region that is transcribed into mRNA for the purpose of initiating the translation of the message into protein. Once transcribed, the mRNA is translated into the protein necessary for the breakdown of lactose into glucose and galactose. The lag mentioned above is the time required for transcription and translation to form the enzymes needed for the utilization of lactose. Since the enzymes necessary for utilization of glucose are always present, there is little if any lag if cells using lactose are transferred to a glucose-containing medium.

systems control gene expression through repressors (negative control). The former produces an inactive repressor that is activated by a substance called a corepressor; the latter produces an active repressor that is inactivated by an inducer:

Inducible system

Active repressor + inducer → inactive repressor → permits structural gene to produce mRNA.

Repressible system

Inactive repressor + corepressor → active repressor → prevents structural gene from producing mRNA.

In addition to these examples of negative control, mechanisms of positive control over protein synthesis have also been proposed. In such systems a controlling gene produces a substance that acts on the operator gene as an activator or initiator rather than as a repressor. One example of this type is the *C* gene of the arabinose locus in *E. coli*, which has been shown by Englesberg and others to produce a substance necessary for the synthesis of certain enzymes in the arabinose-synthesizing pathway (Strickberger, 1976). These models of negative and positive controls over the process of protein synthesis form our current concepts of gene regulation.

the enzyme / its function and the effect of deficient enzymes

Of all the proteins made according to the messages of DNA, the functional proteins or enzymes may be the most numerous. An enzyme serves as a catalyst in a living system; that is, it promotes a chemical reaction without being used up in the reaction. A single molecule of a particular enzyme thus can participate many times in conversion of a given substance (called a substrate), say A, into B (Figure 5–12). Each enzyme, according to its specific genetic message, is specially constructed and uniquely suited to the conversion of one substance or of a class of structurally related substances into another. Compared to the structural proteins, however, enzymes are relatively small.

Some humans have an intolerance to lactose because they lack the enzyme lactase, which normally splits the milk sugar lactose into glucose and galactose, easily digestible sugars. In Europeans the intolerance to lactose, or milk sugar, is inherited as a recessive trait. That is, two normal parents, each having one normal and one defective gene for producing the enzyme, may expect that, on the average, one out of four of their children could react unfavorably to milk. When such children ingest milk, the result is diarrhea, vomiting, and, in severe cases, death (Segal, 1972). Numerous reports have now accumulated on milk intolerance in various non-European groups. Children in East Africa, for example, had severe cases of diarrhea after ingesting powdered milk. Scientific

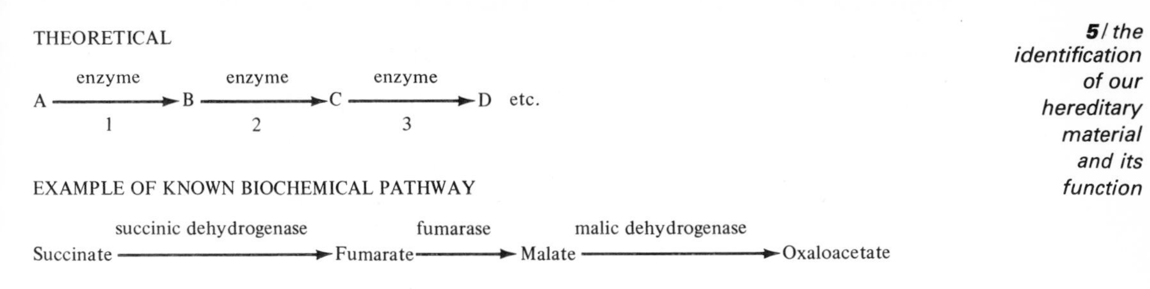

Figure 5–12. Biochemical Pathways. The theoretical pathway shows that an individual enzyme is necessary at each step of a biochemical pathway. Each enzyme is responsible for converting one specific substrate into another. Specific examples are shown in the known biochemical pathway, which is an integral part of the citric acid cycle that is essential in the production of ATP through aerobic respiration.

studies of these children showed that when they are more than four years old, their intestinal cells no longer secrete the active lactase enzyme responsible for the breaking down of lactose. Lactose intolerance has now been found in Chinese, Philippinos, Indians, and the people of Thailand and Southeast Asia in general. In all cases, there is a deficiency in lactase production beginning between the ages of one and four. Recently, research at the University of Rhode Island showed that the lactose in cow milk can be easily converted into glucose and galactose by adding lactase during the pasteurization process. If the milk is processed properly, its high nutritional content can be made available to millions of Asian and African people.

In any case, the ability of our cells to synthesize new substances from the raw materials we eat and to break down other substrates, either for their disposal as body waste or for their reuse in the building of other compounds, is carefully controlled by specific enzymes coded for by specific genes. Each type of enzyme is designed to control only one stage of a specific biological reaction (Figure 5–12). While one series of enzymes breaks down the proteins in foodstuffs to make free amino acids available for cell growth, another series of enzymes utilizes these amino acids in the synthesis of other proteins. Neither structural proteins nor the enzymes themselves could be built without other enzymes. Even DNA is attended by its own enzymes, which aid it in reproducing and repairing itself, and many different enzymes are required in the synthesis of the various types of RNA.

The enzyme is a coiled and folded structure with only two or three active sites, which are shaped to fit precisely the molecular structure of the substrate. If two molecules are to be joined, the enzyme momentarily locks into them, forcing them into such close proximity that they quickly link together. Or if molecules are to be broken apart, the enzyme strains their chemical bonds so that they separate more easily. A single biological enzyme can cleave or join molecules faster than many known industrial catalysts.

One might think of a normal person, then, as a balanced, gene-controlled, multi-enzyme system capable of development on a balanced diet. The diseased or defective system is out of balance. Sometimes additional supplements are required to reestablish balance. In diabetics, for example, the use of insulin restores a healthy balance. In many cases certain nutrients have to be

Table 5–4 Enzyme Deficiencies and Related Human Genetic Diseases

Enzyme Deficiency*	*Genetic Defect*
Acetylcholinesterase	Muscular dystrophy
Acetylcholine transferase	Muscular dystrophy
Acid lipase	Wolman's disease
Alkaline phosphatase	Hypophosphatasia
Amylo-1:4,1:6-transglucosidase (brancher)	Glycogen storage disease IV
Argininosuccinase	Argininosuccinicaciduria
Argininosuccinic acid synthetase	Citrullinemia
Aryl sulfatase A	Metachromatic leukodystrophy
Branched-chain keto acid decarboxylase	Maple syrup urine disease
Catalase	Acatalasia I and II
Ceramide lactosidase	Ceramide lactosidosis
Ceramide trihexosidase	Fabry's disease
Creatinine phosphokinase	Muscular dystrophy
Cystathionase	Cystathioninuria
Cystathionine synthetase	Homocystinuria
Deiodinase	Goitrous cretinism
2,3-Diphosphoglycerate mutase	Hemolytic anemia
DNA repair enzymes	Xeroderma pigmentosum
Fructokinase	Fructosuria
Fructose-1-phosphate aldolase	Fructose intolerance
Galactokinase	Cataract
Galactose-1-phosphate uridyl transferase	Galactosemia
beta-Galactosidase	Generalized gangliosidosis I and II
beta-Glucaronidase	"I cell" disease
Glucocerebroside oxidase	Gaucher's disease
Glucose-6-phosphatase	Glycogen storage disease I
Glucose-6-phosphate dehydrogenase	Hemolytic anemia
Glucose phosphate isomerase	Hemolytic anemia
alpha-1,4-Glucosidase	Glycogen storage disease II (Pompe's disease)
Glucuronyl transferase	Familial nonhemolytic jaundice
Glutathione peroxidase	Hemolytic anemia
Glutathione reductase	Hemolytic anemia
Glutathione synthetase	Hemolytic anemia
D-Glyceric acid dehydrogenase	L-Glycericaciduria
Hexokinase	Hemolytic anemia
Hexosaminidase A	Tay–Sachs disease
Hexosaminidases A and B	Sandhoff's disease
Histidase	Histidinemia
Homogentisic acid oxidase	Alkaptonuria
21-Hydroxylase	Adrenogenital syndrome
p-Hydroxyphenylpyruvic acid oxidase	Tyrosinemia

Table 5–4 continued

Enzyme Deficiency*	*Genetic Defect*
Hydroxyproline oxidase	Hydroxyprolinemia
Hypoxanthine–guanine phosphoribosyl transferase	X-linked hyperuricemia (Lesch–Nyhan syndrome)
Isovaleryl CoA dehydrogenase	Isovalericacidemia
Methylmalonyl CoA mutase	Methylmalonicacidemia
NAD diaphorase	Methemoglobinemia
Ornithine transcarbamylase	Hyperammonemia
Orotodine-5′-phosphate decarboxylase	Oroticaciduria
Orotodine-5′-phosphate pyrophosphorylase	Oroticaciduria
Phenylalanine hydroxylase	Phenylketonuria
6-Phosphofructokinase	Glycogen storage disease III
Phosphorylase (liver)	Glycogen storage disease VI
Phosphorylase (muscle)	Glycogen storage disease V
Phytanic acid hydroxylating enzyme	Refsum's disease
Proline oxidase	Hyperprolinemia
Pseudocholinesterase	Apnea
Pyruvate kinase	Hemolytic anemia
Sphingomyelin hydrolase	Niemann–Pick disease
Sulfite oxidase	Central nervous system disease with cataracts
Triose phosphate isomerase	Hemolytic anemia
Valine transaminase	Hypervalinemia
Xanthine oxidase	Xanthinuria

*The name of an enzyme usually ends in *-ase.*

eliminated to avoid the accumulation of metabolic products that will lead to bodily harm. For example, a person who does not produce the enzyme phenylalanine hydroxylase must be kept on a restricted protein diet (one that is low in the amino acid phenylalanine) to prevent severe mental retardation (see Figure 10–1). A diet free of galactose prevents the inherited defect galactosemia, which also leads to mental retardation, from expressing itself. Albinos lack an enzyme needed to initiate synthesis of melanin, the protein pigment giving color to eyes, skin, and hair. The hope is that such errors can be traced back through the precise biochemical pathways to their genetic source and then corrected through genetic engineering.

Many human disorders are now known to be caused by genetic mishaps that result in the absence of an enzyme (Table 5–4). However, although we now know of 62 specific diseases caused by the absence or an abnormally low level of a required enzyme, we can provide relatively little in the way of treatment and very little in the way of correction. These defects will be transmitted to the next generation if the defective person has children.

The human body is quite deficient in its ability to manufacture or convert food into vitamins, amino acids, and other essential substances. That is, our genetic code does not provide the enzymes necessary to function with the minimal nutritional requirements of such species as *Escherichia coli* or a green plant. Luckily, we have evolved a brain that has enabled us to form a society through which our nutritional needs are provided. Humans require ten amino acids in their diet, whereas the bacterial cell *E. coli* can live on a single carbon source like glucose and some minimal salts. From these simple nutrients *E. coli* can make all of the 20 amino acids, purines, pyrimidines, vitamins, fats, carbohydrates, and proteins that it requires. But humans are quite dependent on a variety of nutrients that must come preformed from the environment. We are evolutionary mutants, and in spite of our complexity as a complete organism, we are not, on the cell level, as independent as *E. coli*. The human race transmitted these genetic limitations through time and will continue to do so until we can engineer the human gene pool.

An intriguing mutation is man's loss of genetic ability to make ascorbic acid or, as it is often erroneously called, vitamin C. The lack of the enzyme necessary to create ascorbic acid leads to the serious genetic disease of scurvy. The mutation that led to our ascorbic acid deficiency must have been similar to the various mutations that led to our inability to synthesize the essential amino acids (Table 5–3), since the whole human population is deficient. The appropriate genes either are no longer in our DNA or are suppressed. Most mammals can make their own ascorbic acid, as indicated in Table 5–5. Yet the mammalian human can only be free of the disease of scurvy through diet.

Scurvy is a disease we all heard about in our early days of school. Remember the stories of the men on British sailing ships who carried crates of oranges or lime juice out to sea in order not to come down with scurvy? It has been estimated that the British navy lost 100,000 seamen to scurvy between 1753 and 1795. In man's long prehistory and during historical times, scurvy has killed more individuals, caused more human misery, and changed the course of history more than any other single disease (Stone, 1974). The disease was recognized by the early Egyptians, is mentioned in the writings of Hippocrates, and is known to have afflicted the soldiers of the Roman Legions.

Table 5–5 The Synthesis of Ascorbate in Mammals

Mammal	*Ascorbate Produced**
Rat, unstressed	4,900
Rat, stressed	15,200
Mouse	19,250
Rabbit	15,820
Goat	13,300
Dog	2,800
Cat	2,800
Human	0

* Milligrams of ascorbate synthesized per day per 70 kilograms body weight. After Stone (1974).

According to Stone, the enzyme is missing in humans because a mutation occurred in the gene that controls the synthesis of the enzyme L-gulonolactone oxidase. This mutation destroyed the cells' ability to make the enzyme necessary for synthesizing ascorbic acid (Figure 5–13). Stone states that he has been able to trace this genetic accident to a mutation that occurred in a primate ancestor during the Paleocene Epoch, about 60 million years ago. This defective gene is carried by present-day members of the primate suborder Anthropoidea, which includes man, the apes, the Old World monkeys, and the New World monkeys.

Today vitamin C, more properly called ascorbic acid, is a multimillion-ton business. It is used to improve production in bread making, to help reduce cholesterol levels in humans, to aid in bacon curing, and to combat the common cold (Lewin, 1974).

The beginning of the development of the one gene–one enzyme hypothesis is attributed to Sir Archibald Garrod. Evidence of this physician's insight into the biochemistry of human metabolic disorders is clear in his book *Inborn Errors of Metabolism* (Garrod, 1909). Garrod proposed in 1902 that alkaptonuria, a condition manifested by darkening of cartilage in the ears, melanin spots in the eyes, proneness to arthritis, and darkening of urine upon exposure

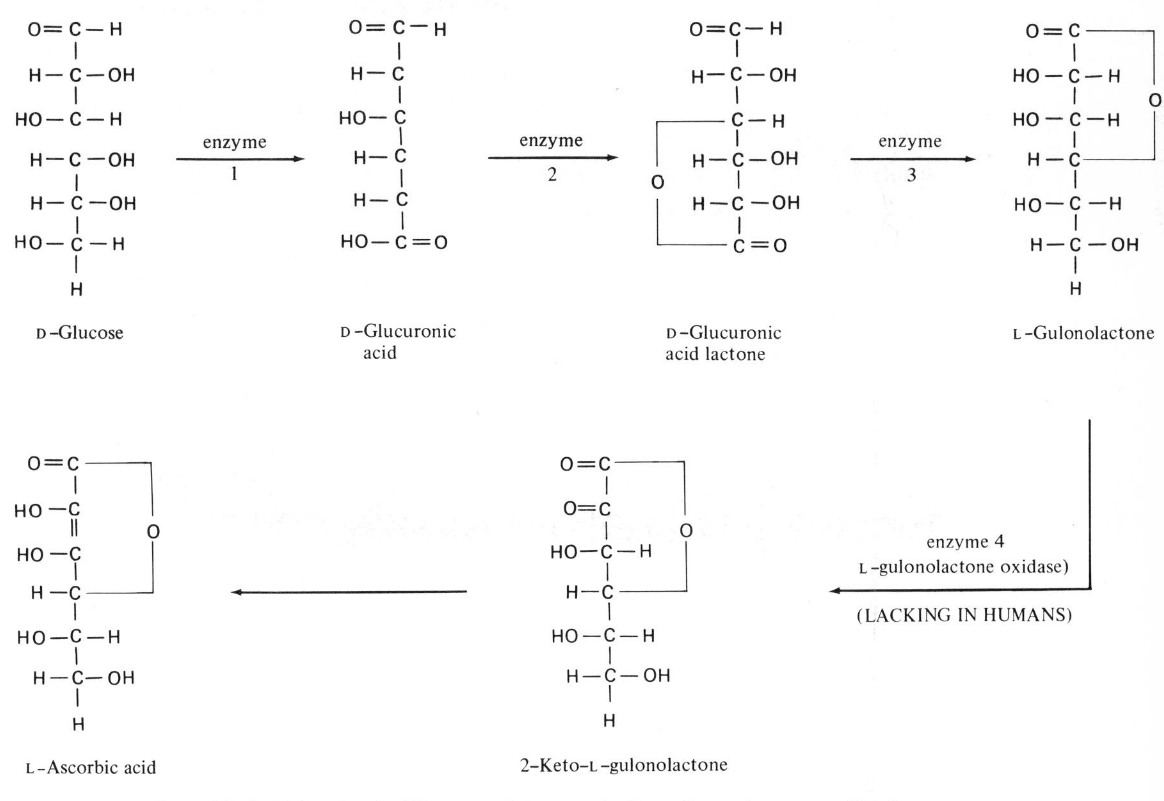

Figure 5–13. Ascorbic Acid Synthesis. Diagram of the metabolic pathway in mammalian liver synthesis of ascorbate. Nearly all mammals have the full complement of four enzymes. Humans have the first three in their livers, but lack the fourth. It is this missing fourth enzyme that prevents humans from making their own L-ascorbic acid. (After Stone, 1974.)

to air, was due to the accumulation of alkapton (2,5-dihydroxyphenylacetic acid). Garrod implied that the accumulation of alkapton was the result of an impairment of one step in a metabolic sequence. Today we understand that the steps are the points of metabolic conversion of substrates and that each is under the control of an individual and highly specific enzyme. From the time of Garrod until the early 1960's, there was a gradual evolution in the concept of stepwise metabolism through the one gene–one enzyme hypothesis to the current one gene–one polypeptide hypothesis. We know, for example, that several genes are required to code for the polypeptides that combine with an atom of iron to form hemoglobin.

Perhaps many enzyme-controlled reactions *could* occur without the enzymes, but in our present world enzyme action is required for "normal" development and life. A mutation in the DNA can throw the living system out of balance at any stage of development by producing a nonfunctional enzyme or by failing to make a required enzyme.

mutations or changes in DNA

Let us turn our attention to the concept of a mutation, the kinds of mutations, and their molecular explanation. Recall that changes in DNA and the reproduction of these changes over time have given rise to the various species that have existed.

A gene mutation is a *sudden* change in DNA. We can define a mutation as a physical or chemical change in DNA such as an addition, deletion, or substitution of a pair of nucleotides. Examples of the nucleotide sequences for a normal DNA molecule and for each type of mutation are given below.

Normal	*Addition*	*Deletion*	*Substitution*
CAT CAT CAT GTA GTA GTA	CATCCAT CAT GTAGGTA GTA	CAT AT CAT GTA TA GTA	CAT GAT CAT GTA CTA GTA

Mutation, by this definition, occurs at the level of the DNA molecule, and involves changes that are invisible even with the best electron microscope. Such submicroscopic changes in the DNA are referred to as point mutations. A point mutation affects the genetic code directly. The resulting change in the genetic code causes a variety of phenotypic changes that are due to the synthesis of abnormal protein or to the lack of synthesis of normal protein. (Agents responsible for causing mutations are discussed in Chapter 13.)

One excellent example of a point mutation and its effect comes from the pioneer work of Ingram and others investigating the changes in hemoglobin, the oxygen-carrying porphyrin pigment of red blood cells.

gene control of hemoglobin synthesis

The oxygen in the air we breathe is essential for human survival. For transport to our tissues, oxygen is reversibly bound to a special transport protein called hemoglobin. This protein is not dissolved in the blood plasma, but is contained in the specialized red blood cells, or erythrocytes, which are suspended in and transported by the blood plasma. In mammals, which include humans, cell specialization has gone so far that the mature erythrocyte has lost its nucleus. Only the small membrane containing a concentrated hemoglobin solution is left. This hemoglobin solution also contains a number of enzyme systems and nutrients, which protect the hemoglobin against destruction. During early stages in the developing embryo, long before birth, hemoglobin is synthesized in special cells; oxygen transport by hemoglobin starts with the beginning of circulation.

In humans hemoglobin is a protein containing a heme group (ferroheme) and a basic protein, globin, with a molecular weight of approximately 66,000. A complete molecule of adult human hemoglobin (Hb A) consists of four separate polypeptide chains: two identical alpha (α) chains, each composed of 141 amino acids coded by one gene, and two identical beta (β) chains, each composed of 146 amino acids coded by a second gene. Each alpha chain begins with the amino acid valine and terminates with arginine. The beta chains begin with valine and terminate with histidine. The alpha and beta chains are identical through a segment of 61 consecutive amino acids. After the polypeptide chain has been synthesized, it coils together into its normal tertiary structure. Before or during this process the heme group is attached. Four polypeptide chains together with four heme groups form one hemoglobin molecule (Figure 5–14).

During early embryonic life, two hemoglobin components, which have been named hemoglobin Gower 1 and 2, are present. This embryonic hemoglobin (Hb E) has alpha chains identical to the adult hemoglobin connected with two distinct epsilon chains. At about 12 weeks of development epsilon chain synthesis ceases (gene shutdown), and the alpha chains now combine with newly synthesized gamma chains to make hemoglobin F or fetal hemoglobin (Hb F). Fetal hemoglobin is synthesized until birth, or in some cases in trace amounts through the first years of life. At birth, another gene, which codes for beta chain hemoglobin, is activated, and from that point on both the alpha and the beta chain hemoglobins are made throughout childhood and adult life. There is supportive evidence that separate genes control the syntheses of the alpha, beta, gamma, and (probably) epsilon hemoglobin chains.

Formal genetic studies have confirmed the existence of genetic loci controlling the structures of the alpha and beta chains and a delta chain. It is presumed that the alpha and beta chains' loci are some distance apart on the same chromosome or on different chromosomes. Unfortunately, there is no formal genetic evidence as to the chromosomal sites of the epsilon and gamma loci, and it is not known which of the chromosomes carry the hemoglobin genes (Weatherall and Clegg, 1969).

The sequence of gene activation and shutdown helps to explain many of the findings in patients with various hemoglobinopathies (hemoglobin disorders). It explains the finding of both normal and abnormal hemoglobin in heterozygous carriers (persons carrying normal and abnormal states of the same gene) and the absence of normal hemoglobin in those homozygous for an abnormal hemoglobin gene. It also accounts for the observation that infants with beta chain hemoglobin variants are normal in the early neonatal period and are only

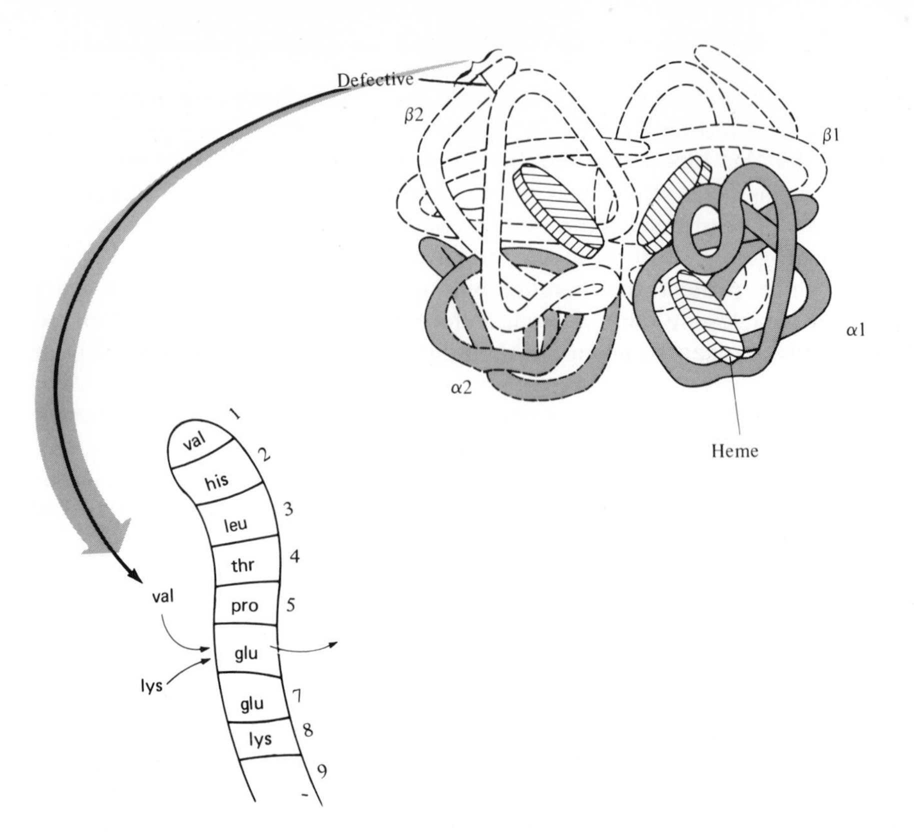

Figure 5–14. Hemoglobin Molecule. A structural diagram of hemoglobin made up of two identical beta chains (white) and two identical alpha chains (shaded). Each chain is associated with an iron-containing heme group (circular objects). The four chains fit very closely together in a sort of tetrahedral arrangement. There are about 280 million molecules of hemoglobin in one red blood cell. The replacement of glutamic acid by valine results in sickle cell anemia; the replacement of glutamic acid by lysine results in hemoglobin C disease.

affected clinically after one to four months because of extended gamma chain synthesis and lack of expected beta chain synthesis. Hemoglobins with altered alpha chains, on the other hand, are found in the fetus and in the adult. There are at least 132 structural variants of hemoglobin, some of which are discussed in Chapters 10, 14, and 15.

effect of a point mutation: base changes leading to sickle cell hemoglobin

Genetic mutations can result from the alteration of a single base in an organism's DNA. Such a point mutation can cause the incorporation of an amino acid other than the one normally incorporated and result in the production of a partially active or totally inactive polypeptide. Much information on base-substitution mutants has been obtained from studies on human hemoglobin, the protein coat of tobacco mosaic virus, rII bacteriophage mutants, and the enzyme tryptophan synthetase from the bacterium *Escherichia coli.*

One malfunctioning gene can create a medical problem that is very difficult to diagnose. Since genetic defects, occurring at the level of the DNA, are not yet correctable, we must settle for treating the symptoms (see Chapter 22).

A well-known example of hemopathology is sickle cell anemia. Persons who are homozygous defective (have two defective genes) produce only defective hemoglobin, which is severely altered with respect to its ability to carry oxygen. Further, red blood cells carrying the defective hemoglobin tend to sickle or change shape (Figure 5–15). Such sickled red blood cells block the small blood vessels (capillaries), leading to tissue death and severe pain. This process is clinically referred to as a "crisis." The genetic, social, and economic aspects of sickle cell anemia will be covered in Chapters 10, 15, and 22.

From what has been said of protein synthesis, it follows that the altered hemoglobin molecule produced by persons with sickle cell anemia has its beginning at the level of the DNA. A base change in DNA results in an altered message, which in turn leads to the synthesis of defective hemoglobin. Since the mutation occurs at the level of the bases in DNA, it is a point mutation.

Analysis of the amino acid sequence of normal and sickle cell hemoglobin has not revealed any changes in the amino acid composition of the alpha chain. In the beta chain, however, the sixth amino acid (glutamic acid) of the 146 that make up the chain is replaced by valine. In a second, less serious hemoglobinopathy, referred to as hemoglobin C disease (Hb C), glutamic acid in the sixth position is replaced by lysine (Figure 5–14). It is clear that at the level of DNA, where a three-base sequence is responsible for the exact incorporation of a specific amino acid into a polypeptide chain or protein, a single base substitution must have occurred for each abnormal beta chain (Table 5–6).

Table 5–6 Base Substitutions Responsible for Abnormal Hemoglobins S and C

Hemoglobin	*Amino Acid at Sixth Position*	*Possible DNA Base Sequences*
Normal	Glutamic	CTC or CTT
Sickle cell	Valine	CA*C or CA*T
Hemoglobin C disease	Lysine	T*TC or T*TT

* Base substitutions in DNA that lead to an amino acid substitution in the protein hemoglobin. Codons resulting from base triplets can be found in Figure 5–9; for example, C T C in DNA gives rise to the mRNA codon G A G, and C T T to G A A.

Thus a single base substitution can cause the incorporation of the wrong amino acid and subsequent hemoglobinopathy. Geneticists consider the amino acid substitution in hemoglobin C disease a transitional mutation. That is, one pyrimidine, cytosine, is replaced by a second pyrimidine, thymine. In sickle cell

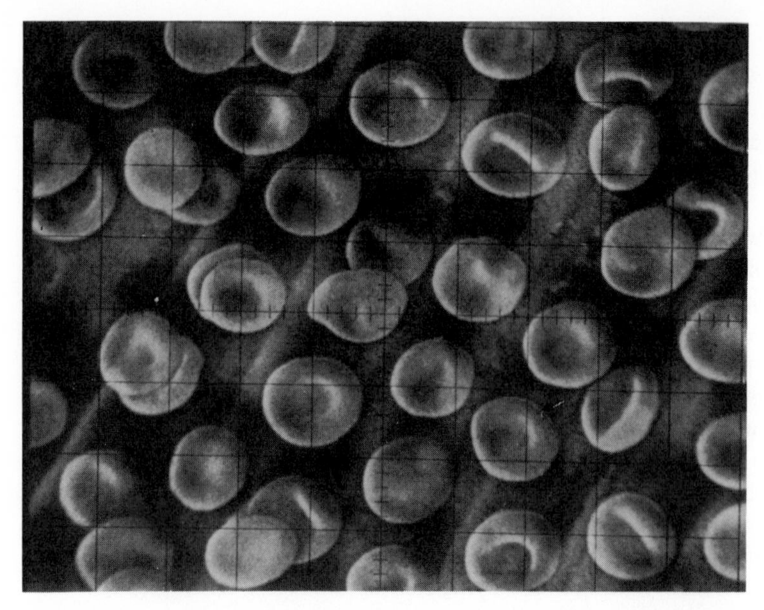

A

B

Figure 5–15. Normal and Sickle-Shaped Red Blood Cells. **A.** Normal red blood cells are round cells. **B.** Red blood cells in various states of sickling. (Photos courtesy of Philips Electronic Instruments, Inc.)

hemoglobin there is a base transversion. A pyrimidine, thymine, is replaced by a purine, adenine. These transitions and transversions lead to point mutations. Point mutations can be caused by a variety of environmental agents, such as radiation (X rays and ultraviolet), certain chemicals, and natural accidents in the normal DNA replication or repair process (discussed in Chapter 13). The wide range of effects a single base substitution can have on a phenotype is emphasized by the fact that the change from glutamic acid to valine in the sickle cell hemoglobin causes no less than 22 physical and/or medical problems for the individuals affected (Figure 5–16).

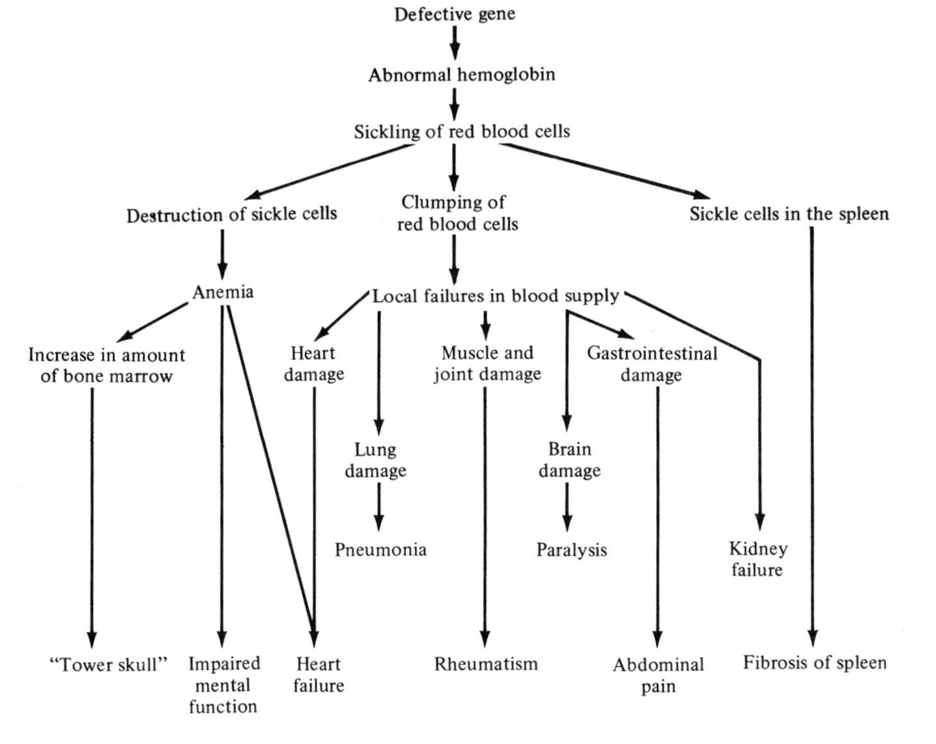

Figure 5–16. Genetic Pleiotropy. There are at least 22 physical and mental defects associated with a single change of a base pair that results in the replacement of one amino acid in the hemoglobin molecule. (Adapted from M. W. Strickberger, *Genetics*, 2nd ed., Macmillan, 1976.)

references

Avery, O. T., C. M. MacLeod, and M. McCarty. 1944. Studies on the Chemical Nature of the Substance Inducing Transformation of Pneumococcal Types. *Journal of Experimental Medicine*, *79*: 137–58.

Burns, George W. 1976. *The Science of Genetics*, 3rd ed. Macmillan, New York.

Cairns, J. 1963. The Bacterial Chromosome and Its Manner of Replication as Seen by Autoradiography. *Journal of Molecular Biology*, *6*: 108–213.

Garrod, A. E. 1909. *Inborn Errors of Metabolism*. Oxford University Press, London.

Goodenough, U. W., and R. P. Levine. 1970. The Genetic Activity of Mitochondria and Chloroplasts. *Scientific American*, *223*(5): 22–24.

Griffith, F. 1928. Significance of Pneumonial Types. *Journal of Hygiene*, *27*: 113–59.

Jacob, F., and J. Monod. 1961. Genetic Regulatory Mechanisms in the Synthesis of Proteins. *Journal of Molecular Biology*, *3*: 318–38.

Kornberg, A. 1968. The Synthesis of DNA. *Scientific American, 219*(4): 64–78.
Laboratory Management. 1973. Will Enzyme Assays Help End Misdiagnosis? *11*:55–57.
Lack, David. 1953. Darwin's Finches. *Scientific American, 188*(4):66–72.
Lewin, Roger. 1974. And Now It Takes over the Brain. *New Scientist, 62*:110.
Meselson, M., and F. W. Stahl. 1958. The Replication of DNA in *Escherichia coli. Proceedings of the National Academy of Sciences USA, 44*:671–82.
Meselson, M., F. W. Stahl, and J. Vinograd. 1957. Equilibrium Sedimentation of Macromolecules in Density Gradients. *Proceedings of the National Academy of Sciences USA, 43*:581–88.
Segal, Harold L. 1972. The Role of Societies and Institutions in the Maximization of the Human Value of Science. *Annals of the New York Academy of Sciences, 196*(4):256–62.
Stein, G. S., J. S. Stein, and L. J. Kleinsmith. 1975. Chromosomal Proteins and Gene Regulation. *Scientific American, 232*(2):46–57.
Stone, Irwin. 1974. Humans, the Mammalian Mutants. *American Laboratory, 33*:32–39.
Strickberger, Monroe W. 1976. *Genetics*, 2nd ed. Macmillan, New York.
Taylor, J. H. 1957. The Time and Mode of Duplication of the Chromosomes. *American Naturalist, 91*:109–221.
Taylor, J. H., P. S. Woods, and W. L. Hughes. 1957. Organization and Duplication of Chromosomes as Revealed by Autoradiographic Studies Using Tritium-Labeled Thymidine. *Proceedings of the National Academy of Sciences USA, 43*:122–28.
Watson, J. D., and F. H. C. Crick. 1953a. Molecular Structure of Nucleic Acids: A Structure for Deoxyribose Nucleic Acid. *Nature, 171*:737–38.
Watson, J. D., and F. H. C. Crick, 1953b. Genetical Implications of the Structure of Deoxyribonucleic Acid. *Nature, 171*:964–67.
Weatherall, D. J., and J. B. Clegg. 1969. Disorders of Protein Synthesis. In Cyril Clark (ed.), *Selected Topics in Medical Genetics.* Oxford University Press, New York.

the physical basis: the transmission of hereditary material through the chromosomes / some social implications

In 1879 Walther Flemming named the easily stainable material within a cell nucleus *chromatin.* In early stages of nuclear division, chromatin appears as long threadlike structures; in later stages it appears as the short rod-like bodies that Waldeyer named *chromosomes.* The number of chromosomes in each cell is a characteristic of a species (Table 6–1), but the number of chromosomes in itself is not necessarily sufficient for species identification, since several species may have the same number of chromosomes. As can be seen in Table 6–1, rice, tomato, pine, white oak, and grasshopper have 24 chromosomes, yet they are clearly separate species. All normal members of the same species have the same number of chromosomes; humans, for example, regardless of race, color, or geographical location, have a haploid number of 23 chromosomes. The haploid number of chromosomes for different species may vary from 2 in a few arthropods and the flowering plant *Haplopappus gracilis* to 510 in the fern *Ophioglossum petiolatum* (Whitehouse, 1969).

Studies of the structure, shape and movement of chromosomes (cytogenetics) have helped to unravel the path through which various species may have evolved. And though the hereditary material of evolution is DNA, the chromosome number represents the number of packages into which the DNA of a species is divided. Therefore, although different species may have the same number of chromosomes by chance, they do not necessarily have the same type and amount of DNA, and at least some of the base sequences are different. Otherwise, they would not be separate species. Change in the number of chromosomes may be reflected in changes in the organism's phenotype.

In 1902, two years after the rediscovery of Gregor Mendel's research on the inheritance of "factors" (Mendelism is discussed in Chapter 8), W. S.

The causes of events are even more interesting than the events themselves.

Cicero

human chromosomes and chromosome abnormalities

Table 6–1 Diploid and Haploid Chromosome Numbers Found in Some Species of Plants and Animals*

Common Name	Species	Chromosome Number	
		Diploid	*Haploid*
Plants			
Garden onion	*Allium cepa*	16	
Rice	*Oryza sativa*	24	
Corn	*Zea mays*	20	
Tomato	*Lycopersicum esculentum*	24	
White oak	*Quercus alba*	24	
Garden pea	*Pisum sativum*	14	
Pine	*Pinus* species	24	
Potato	*Solanum tuberosum*	48	
Animals			
Man	*Homo sapiens*	46	23
Cat	*Felis domesticus*	38	19
Dog	*Canis familiaris*	78	39
Horse	*Equus calibus*	64	32
Rat	*Rattus norvegicus*	42	21
Golden hamster	*Mesocrictus auratus*	44	22
Rabbit	*Oryctolagus cuniculus*	44	22
Frog	*Rana pipiens*	26	13
Grasshopper	*Melanoplus differentialis*	24	12

* Excerpted from Strickberger (1976).

Sutton suggested that the transmission of Mendel's factors (genes) might be associated with the behavior of the chromosomes during meiosis. Sutton's hypothesis was correct, and it became known as the *chromosome theory of heredity:* that genes are located in and remain with the chromosome during the formation of daughter cells (mitosis) or gametes (meiosis).

the morphology of chromosomes

The description of chromosome morphology is limited to chromosomes of cells in the process of nuclear division (see Figures 4–2, 4–3, and 4–4). During the interphase and mitotic prophase, the chromosome appears to show near-maximum extension and consists of two chromatids loosely wound around each other. As prophase progresses, the number of coils holding the chromatids together increases, and the chromosome becomes shortened and thickened. As a result of this coiling process, the individual chromosomes become more distinct.

The sister chromatids (identical halves of one chromosome, Figure 6–1A), after completely uncoiling from each other, are held together by a common centromere. The centromere is an area of constriction located at a given point

Twisting and turning of sister chromatids

Sister chromatids

Satellite

Centromere

Secondary constriction

A

Centromeres

Metacentric

Submetacentric

Acrocentric

B

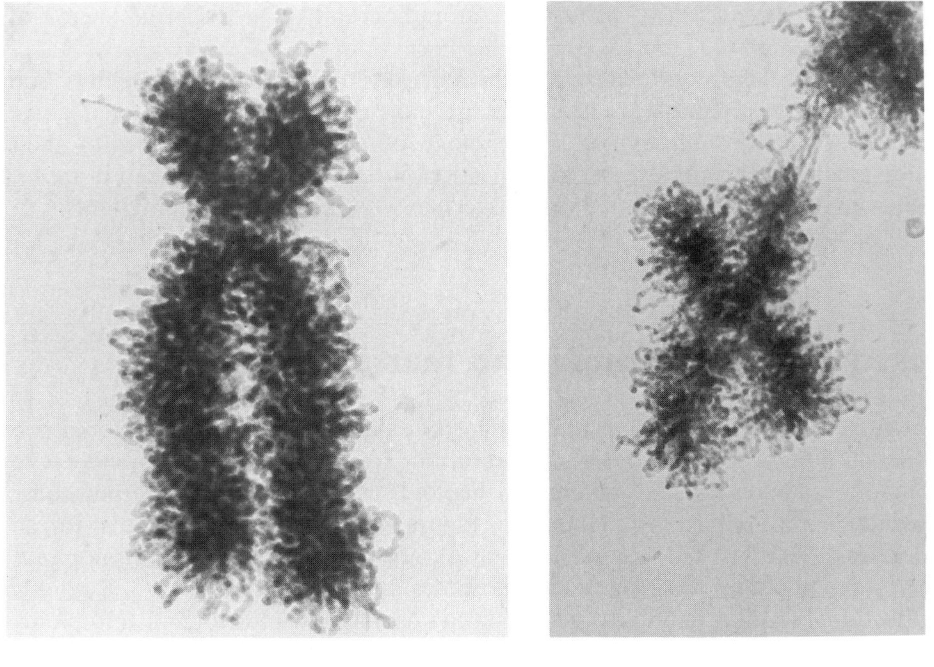

C **D**

Figure 6–1. Chromosome Morphology. **A.** Chromosome morphology is determined by locations of centromeres, satellites, and secondary constrictions. **B.** A chromosome with its centromere located at its center is metacentric; if the centromere is to one side of the center, the chromosome is submetacentric; and if the centromere is near one end of the chromosome, the chromosome is acrocentric. **C.** Electron micrograph of a human #12 chromosome ($\times$21,000) with submetacentric morphology. The total dry mass for the two chromatids is 13.2×10^{-13} gram, 74% of which is in the long arms. **D.** A human chromosome of the 19–20 group ($\times$21,800) with metacentric morphology. The total dry mass for two chromatids is 7.4×10^{-13} gram. (**C** and **D** courtesy of E. J. DuPraw from *DNA and Chromosomes,* Holt, Rinehart and Winston, New York, 1970.)

along the length of the chromosome, where the chromatids meet. The constricted area is characterized by a clear zone containing a small granule. The centromere (also sometimes called the kinetochore) is the site to which the cell's spindle fibers attach for the movement of the individual chromosomes during mitosis and meiosis.

Chromosomes are classified into at least three types by their shape. The position of the centromere in the maximally contracted chromosomes of metaphase and anaphase determines shape. The longest human chromosome at metaphase is 0.5 by 10 microns (Ris and Kubai, 1970). *Acrocentric* chromosomes are rodlike with the centromere located near the end. *Submetacentric* chromosomes have unequal "arms"; the centromere is located off center. One arm is longer than the other, and the chromosome is J- or L-shaped in anaphase. In *metacentric* chromosomes the centromere is located in the middle, giving rise to arms of equal or nearly equal length. This chromosome is V-shaped in anaphase. The location of the centromere is usually referred to as the primary constriction. Secondary constrictions, located at other places along a chromosome, other than at the centromere, are very useful in chromosome identification (Figure 6–1B). Other aids to chromosome identification during mitosis are size, number of secondary constrictions, satellite formation, and location of the nucleolus, as well as autoradiography (De Robertis, Saez, and De Robertis, 1975).

The size, length and width, and the locations of centromeres, satellites, and secondary constrictions are constant for individual chromosomes of a given species. The constant overall morphology of chromosomes allows for their individual identification. Such identification in humans has been used in recent years to determine fetal sex and to ascertain whether the developing fetus is or will be defective.

chromosome number and rearrangements

In general, complex multicellular animals are diploid; that is, they contain two sets of homologous chromosomes resulting from the random union of two haploid gametes. Each gamete is haploid as the result of chromosome reduction through meiosis (shown in Figure 4–2B). The general term for an organism with its full complement of chromosomes is euploid. Monoploids carry one genome (n) or set of chromosomes; diploids carry two ($2n$). Euploids with three or more complete sets of chromosomes—triploids ($3n$), tetraploids ($4n$), pentaploids ($5n$), hexaploids ($6n$), and so on—are called polyploids. Polyploidy appears to have evolved more successfully in plants than in animals. Many of our commercial ornamental plants, as well as hybrid fruits and grains, are polyploid. Wheat, for example, is hexaploid, and strawberries are octaploid. Complete polyploidy in humans is rare, and the few cases that have been reported were stillborn or died a few hours after birth (Schindler and Mikamo, 1970). The condition of polyploidy offers the geneticist an opportunity to study gene–dosage relationships. A diploid has two genes per trait, a triploid three, tetraploid four, etc. The question is: How do the additional genes affect the final phenotype?

Sometimes a variation in chromosome number does not involve a complete set, but only one or a few chromosomes of a set. This condition is called aneuploidy. A numerical prefix and the suffix "somic" indicate the kind of aneuploidy involved. For example, "monosomic" means a diploid organism missing one chromosome of a single pair, expressed $2n-1$. In animals the loss of one homolog of a homologous pair of chromosomes usually shifts the genetic balance in such a way as to increase mortality, reduce fertility, and alter the phenotype of an individual in a number of other undesirable ways. In humans there may be mental retardation and social misbehavior.

A trisomic contains the diploid number and one extra chromosome, expressed $2n + 1$. That is, one of the diploid chromosome pairs has an extra homolog. Studies of the effects of trisomy in man show, for example, that an extra chromosome 21 results in a child with Down's syndrome (mongolism). In general, trisomics (and monosomics) result from the failure of a pair of chromosomes to separate during meiosis. This failure is called meiotic nondisjunction and was first observed in *Drosophila* (Bridges, 1916). If a gamete that is a product of meiotic nondisjunction of chromosome 21 (homologous pair) unites with a normal gamete, the result is a trisomy 21, or Down's syndrome (Figure 6–2). There are a number of other somic abnormalities: tetrasomic, $2n + 2$; double trisomics, $2n+1+1$; and nullosomics $2n - 2$. Of these only rare tetrasomics are fully viable in man.

The chromosomes of the cell nucleus are very stable structures. However, when they are subjected to a variety of environmental agents, such as virus,

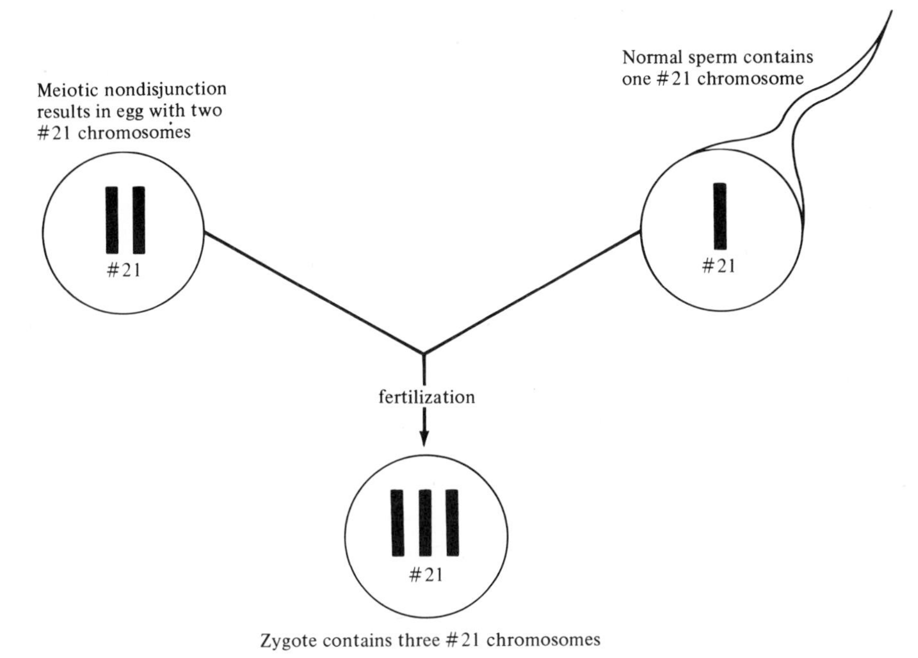

Figure 6–2. Down's Syndrome due to Nondisjunction or Failure of Chromosome 21 to Separate. The ovum (egg) that develops contains two #21 chromosomes. Fertilization with a normal sperm results in a zygote containing three #21 chromosomes (trisomy 21). The process of nondisjunction also occurs in males but at a much lower frequency than in females. This defect increases in frequency with the age of the mother.

radiation, and chemical mutagens, point mutations and chromosome breakage may occur.

Many changes in chromosome number or chromosome morphology have been detected and characterized through examination of their subsequent effects. We shall deal with the effects of chromosomal rearrangements, additions, and deletions, as they are known to influence the phenotype (physical appearance) of man, after briefly considering the actual changes in chromosome morphology: translocations, inversions, deletions, and duplications.

translocations and inversions

Translocations and inversions alter the arrangement of genes, but not necessarily their quality or quantity. Organisms carrying either type of rearrangement may appear normal unless the new gene position affects genes in neighboring regions or vice versa. Both translocations and inversions in diploid organisms become evident during meiosis. A nonreciprocal translocation is an unequal exchange of chromosome segments, or when only one segment of a chromosome unites with a nonhomologous chromosome. For example, a segment of chromosome 22 may attach to chromosome 9. Reciprocal translocations, or the exchange of equal chromosome segments, can also occur and produce the cross-shaped chromosomal figures observed cytologically in meiotic prophase (Figure 6–3A). The cross-shaped configuration results from the pairing of homologous chromosomal segments. Some translocations are known to be associated with semisterility, the alteration of gene linkage relationships, and the modification of gene expression.

In humans a nonreciprocal chromosome translocation *appears* to result in a cell with only 45 chromosomes. Actually, there has been a loss of part of one or both chromosomes. In Figure 6–3B, perhaps owing to a centric fusion, there appears to have been a translocation between one of the number 13 and one of the number 14 human chromosomes. Although this fusion produced a clinically normal person, the person has a high risk of producing abnormal offspring (Redding and Hirschhorn, 1968). A similar and common translocation, which causes a form of Down's syndrome, occurs between chromosome 21 and one of the number 13, 14, or 15 chromosomes. The translocation most often referred to with respect to Down's syndrome occurs between chromosome 21 and chromosomes 14 and 15 giving rise to 14/21 and 15/21 translocations (Figure 6–4).

Both reciprocal and nonreciprocal translocation carriers can be clinically normal, but their offspring may be defective. If the mother is the carrier of the chromosome 21 translocation, there is a 20% chance a child born to her will have Down's syndrome, whereas, for unknown reasons, there is only a 2% chance if the father carries the same translocation (Dow, 1973).

Recently, two other chromosome translocations that are associated with serious disorders have been recognized. First, the missing portion of chromosome 22, associated with chronic myelogenous leukemia, is attached to

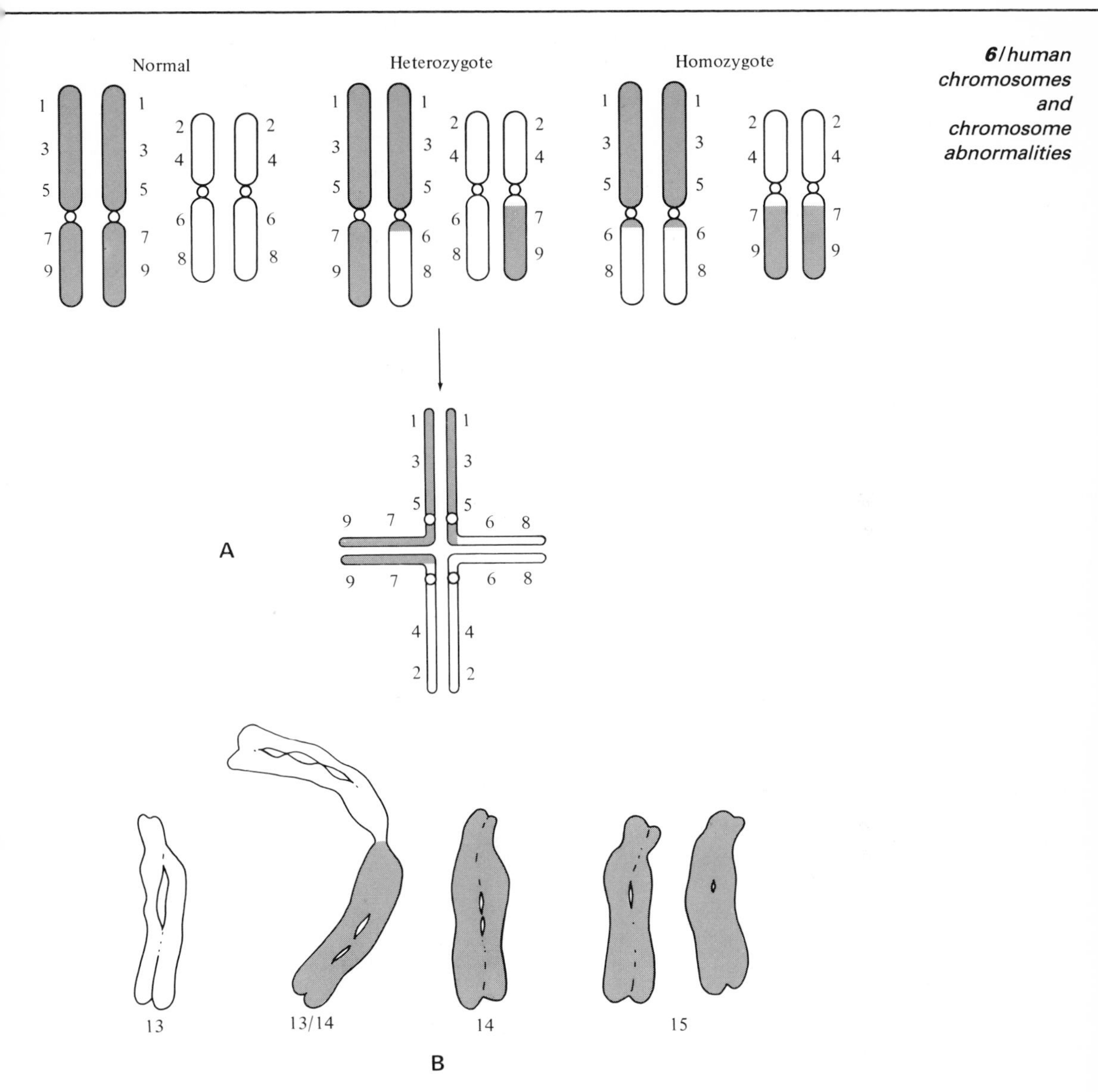

Figure 6–3. Translocation Heterozygote for Genetic Material 6,8/7,9 in Prophase of Meiosis I. **A.** The cross-shaped figure results from forces involved in homologous pairing during meiosis. This type of figure does not occur with a reciprocal translocation homozygote because in that case regular pairing occurs during meiosis. **B.** A translocation in human chromosomes. One chromosome 13 and one 14 have been reassociated in a centric-fusion type of translocation. There would be only 45 chromosomes in this particular karyotype. The person would be clinically normal, but would be a translocation carrier heterozygote and would have a high risk of having abnormal children. (From G. J. Stine, *Laboratory Exercises in Genetics*, Macmillan, 1973.)

chromosome 9 (*Bio Medical News,* 1973). Second, in a Y to X translocation, the Y chromosome is attached to the X of a female, causing habitual abortions (Ris and Kubai, 1970). In general, translocation of the sex chromosome to an autosome in humans is rare. Only 14 X/autosome translocations have been

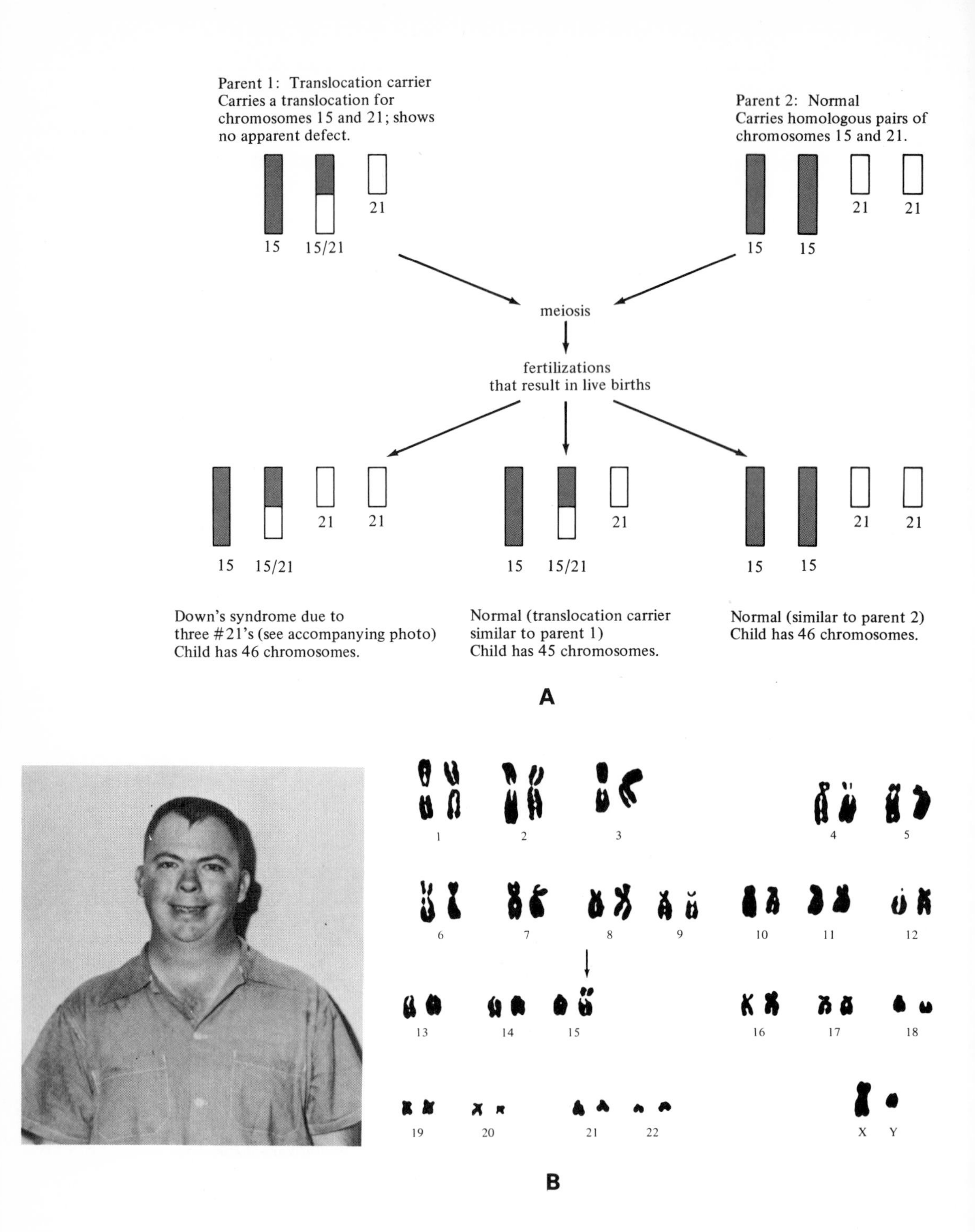

Figure 6–4. Translocation of Chromosomes 21 and 15. **A.** A diagram showing the carrier with only 45 chromosomes but the equivalent of three #21 chromosomes. The translocation type of Down's child is usually born to younger parents than the trisomy 21 Down's child. **B.** A 15/21 translocation Down's syndrome patient. Notice that his karyotype reflects the 15/21 translocation. (**B** by permission of Charles H. Carter, M.D.)

documented, and only a few Y/autosome translocations have been described (Khudr et al., 1973).

In chromosomal inversions a given chromosomal segment is reinserted into the same location on the chromosome, but the segment is turned 180 degrees. For example, the gene segment ABCDE of a chromosome with an inversion for CD gives ABDCE. The phenotypic results or genetic consequences of an inversion in diploid organisms are similar to those of a translocation. That is, only in certain cases does an inversion result in an abnormality. In one case an inversion in chromosome 13 is known to have been carried in a family with no recorded phenotypic effects in the first several children. When an abnormal child was born, cytogenetic studies revealed that a piece of the inverted section of chromosome 13 had been lost (Kutay et al., 1973).

deletions and duplications

Deletions and duplications, in contrast to translocations and inversions, are changes (by loss or addition) of the normal quantity of genetic materials. In general, these types of chromosomal aberrations are apparent during cytological investigation because of deviation from the expected chromosome morphology. A duplication of a single gene is not observable, but the presence of an extra homolog is easily identified. An extra homolog, as in trisomy 21, means that every gene located on chromosome 21 is present three times instead of just twice. Such gene duplication can also occur as a result of translocated segments of chromosomes, as discussed in 14/21 and 15/21 chromosome translocations. In any case, a gamete ends up with duplicate genes for a given trait rather than just a single gene.

A deletion may be as small as a nucleotide pair (although such a short segment could not be seen cytologically) or as large as whole chromosomes. Genetically, deletions are distinguished from gene mutations because they cannot back mutate; if a gene change is due to a loss from the chromosome, the DNA is no longer available for further change. Phenotypically, the loss of a large segment of genetic material is usually lethal, owing to genetic imbalance. Cytologically, deleted chromosome sections can be detected in prophase of meiosis I. At that stage a loop appears in one of the paired chromosomes because there is no corresponding homologous segment in the deleted section of that chromosome. (See the effect of deletions as displayed in Figures 6–5 and 6–6.) Currently, over 300 numerical and morphological aberrations have been described (World Health Organization, 1972).

the number of human chromosomes per cell nucleus

In 1882 Walther Flemming published a drawing of human chromosomes, and in 1891 David Paul van Hansemann attempted to count the number of human chromosomes per cell. In 1912 Hans von Winiwarter reported that the human

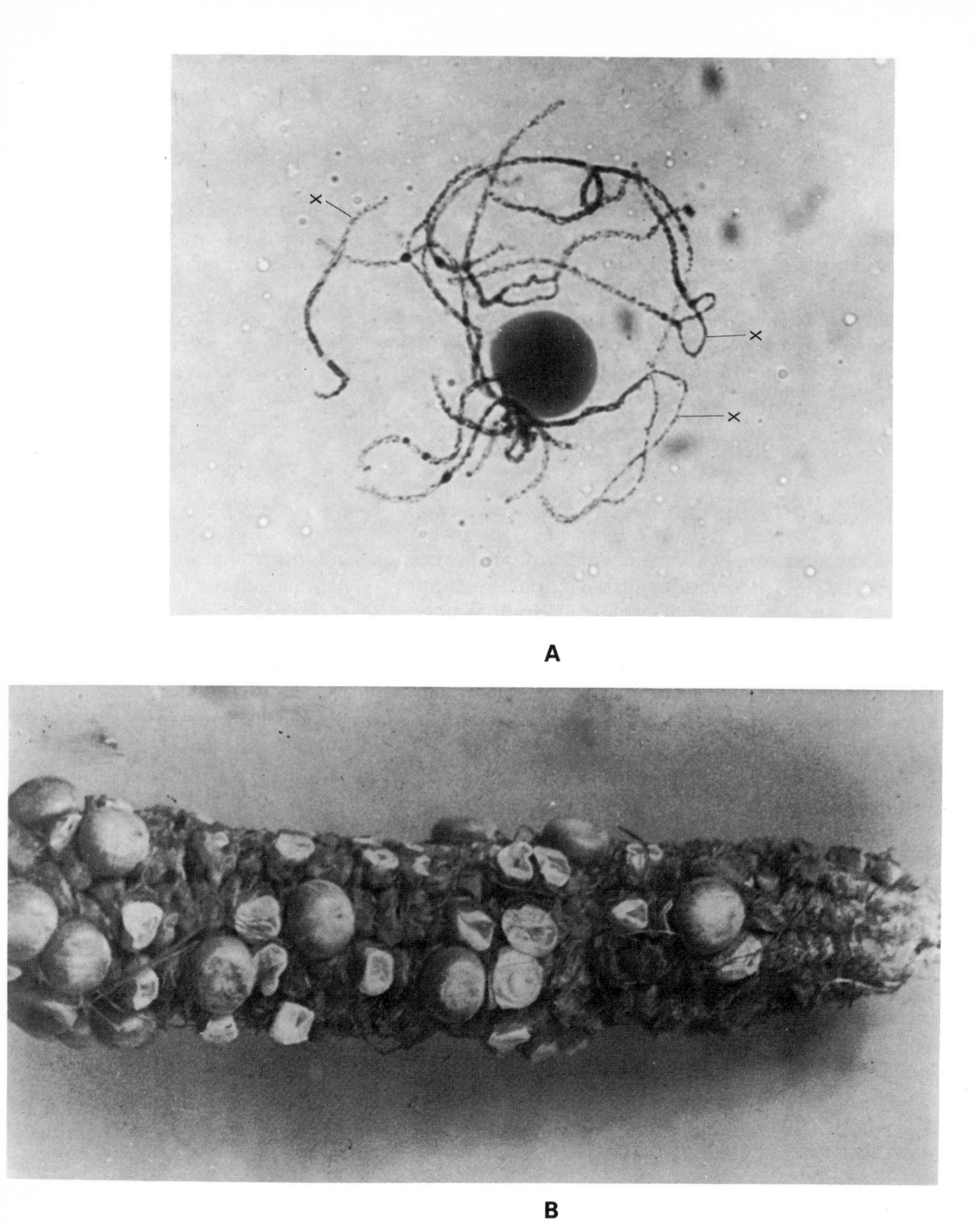

Figure 6–5. Asynaptic Homologous Chromosome Pairing in *Zea mays* (Corn). **A.** Structural asynaptic pairing of pachytene chromosomes. In the regions of asynaptic pairing, indicated by the crosses, homologous chromosomes do not pair with each other. Malformed ear and kernels result. **B.** The defective phenotype that results from asynaptic pairing during meiosis I. (**A** courtesy of O. L. Miller; **B** reprinted by permission of the Crop Science Society of America from M. G. Neuffer et al., *The Mutants of Maize*, 1968, p. 8.)

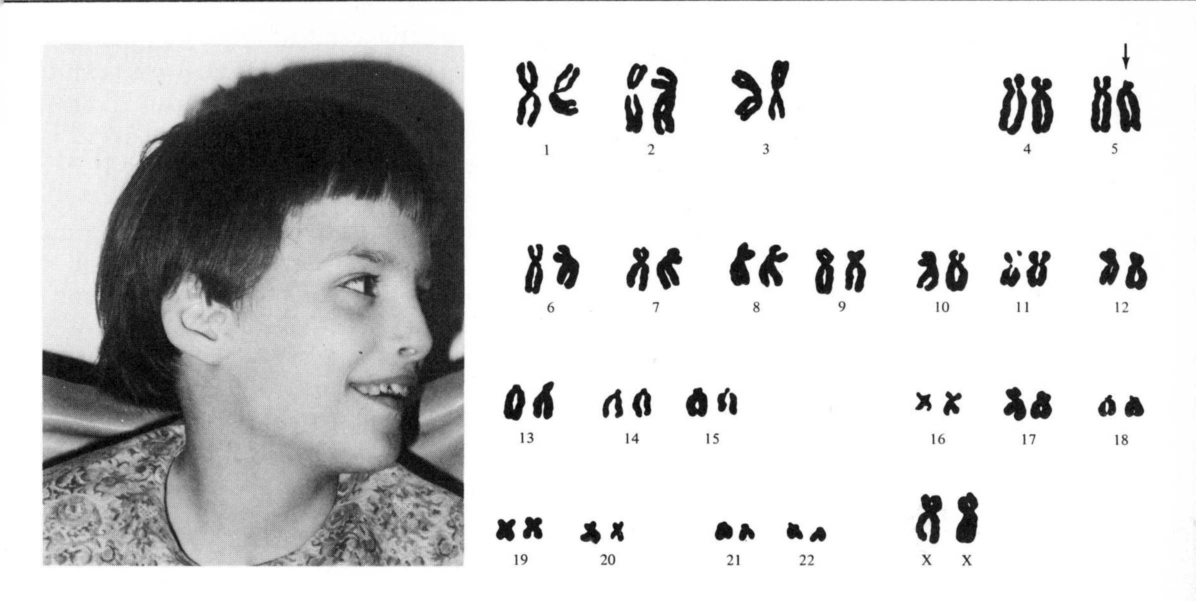

Figure 6–6. Cri-du-Chat Syndrome. This defect is associated with a partial deletion of the short arm of chromosome 5. An affected infant demonstrates a microcephalic head, round or moon face, small chin, and prominent beaklike nose. A major diagnostic feature of the young patient is a mewing or catlike cry instead of a normal child's cry. Autoradiographic studies have shown that some children with similar symptoms are missing a piece of the short arm of chromosome 4. However, these children do not exhibit a catlike cry. In rare instances cancer of the eye has been associated with the #4 chromosome deletion. (By permission of Charles H. Carter, M.D.)

male nucleus contained 47 chromosomes and the female nucleus contained 48. He stated that the female contained two X chromosomes and the male one X. The Y chromosome had not yet been found. In 1917 von Winiwarter reported that the male had X and Y chromosomes and that there were thus 48 chromosomes in both sexes. In 1921 F. S. Painter reported a variation of 45 to 48 chromosomes in different males. He later suggested that the real number of human chromosomes per cell was 48. Ernest Chu (1960) noted that the males studied by Painter in 1921 were from a mentally defective population (inmates of a state mental institution). In light of modern knowledge we can guess that they may have had chromosome aneuploidy. From a combination of studies through 1955, it was generally accepted that the male did carry the Y chromosome and the human diploid cell carried 48 chromosomes. This belief was short-lived, as S. H. Tjio and A. Levan (Human Chromosome Study Group, 1960) reported in 1956 that they had counted only 46 chromosomes in cells taken from human lung tissue in culture. These investigators were the first to use tissue culture techniques for human chromosome studies. Further use and refinement of tissue culture techniques confirmed that human diploid cells did indeed contain 46 chromosomes.

It took 65 years (from 1891 to 1956) to establish 46 as the correct number of chromosomes in a human cell, but after that progress was rapid. In 1959 Jerome Lejeune and colleagues reported the important discovery of a definite association between chromosome aneuploidy and Down's syndrome. The aneuploidy was a trisomy for chromosome 21 (the cell had three #21 chromosomes instead of the normal two). This finding was important in that it was the first direct association made between human chromosomes and human disease. Further studies revealed that because of this anomaly alone, approximately 1 in every 600 live births is severely mentally retarded (German, 1970). Since 1959 a number of chromosome-associated diseases have been recognized.

Of the 46 chromosomes found in human somatic cells (cells other than the sex cells or gametes), 22 pairs or 44 are referred to as autosomes. The X and Y chromosomes, the twenty-third pair, are the sex chromosomes. A normal female has two X chromosomes, and a normal male has both an X and a Y chromosome. The process of meiosis provides males with equal numbers of X- and Y-bearing sperm, while all ova carry one X chromosome. The union of two X-bearing gametes gives rise to a female child, and the union of an X-bearing ovum and a Y-bearing sperm gives rise to a male child. Each child has 46 chromosomes. Most numerical or structural chromosomal anomalies occur or are cytogically evident during the period of meiosis.

the human karyotype

All the human chromosomes have been classified according to size, shape, location of centromere, and various patterns of banding produced by certain dyes. The karyotype (karyo means cell nucleus), or classification, was first developed in 1966. The current classification of chromosomes is the result of three major conferences held in Denver in 1960, in Chicago in 1966, and in Paris in 1971. A diagrammatic representation of the human karyotype (complement of 46 human chromosomes) was the result of these conferences (Figure 6–7). Each chromosome is separated into segments, which begin at the centromere and are labeled *p* and *q*, as shown for chromosome 1. The letter *p* designates the short arm (the shorter piece of chromosome extending out from the centromere), and *q* the long arm. Different lengths of the long and short arm can be seen quite clearly in the Y chromosome and chromosomes 4, 13, and 21. Regions, or recognizable areas along the arms of the chromosome, are numbered consecutively outward from the centromere in the larger arabic numbers; on chromosome 1 there are regions 1, 2, 3 on *p* and regions 1, 2, 3, and 4 on *q*. Adjacent bands, in the light- and dark-stained areas, which are easily distinguished, are numbered in smaller arabic numbers close to the arms of the chromosome. Each region has its own set of numbered bands. The centromere is the most consistent morphological feature of a chromosome, and its discovery has been called the landmark in chromosome research (Paris Conference, 1971). The autosome numbers were assigned on the basis of physical size starting with chromosome 1, the longest, but we now know that chromosome 22 is actually larger than 21. However, changing the chromosome designations would only lead to confusion, especially with respect to Down's syndrome. Therefore, the original numbering of these chromosomes is retained in setting up human karyotypes.

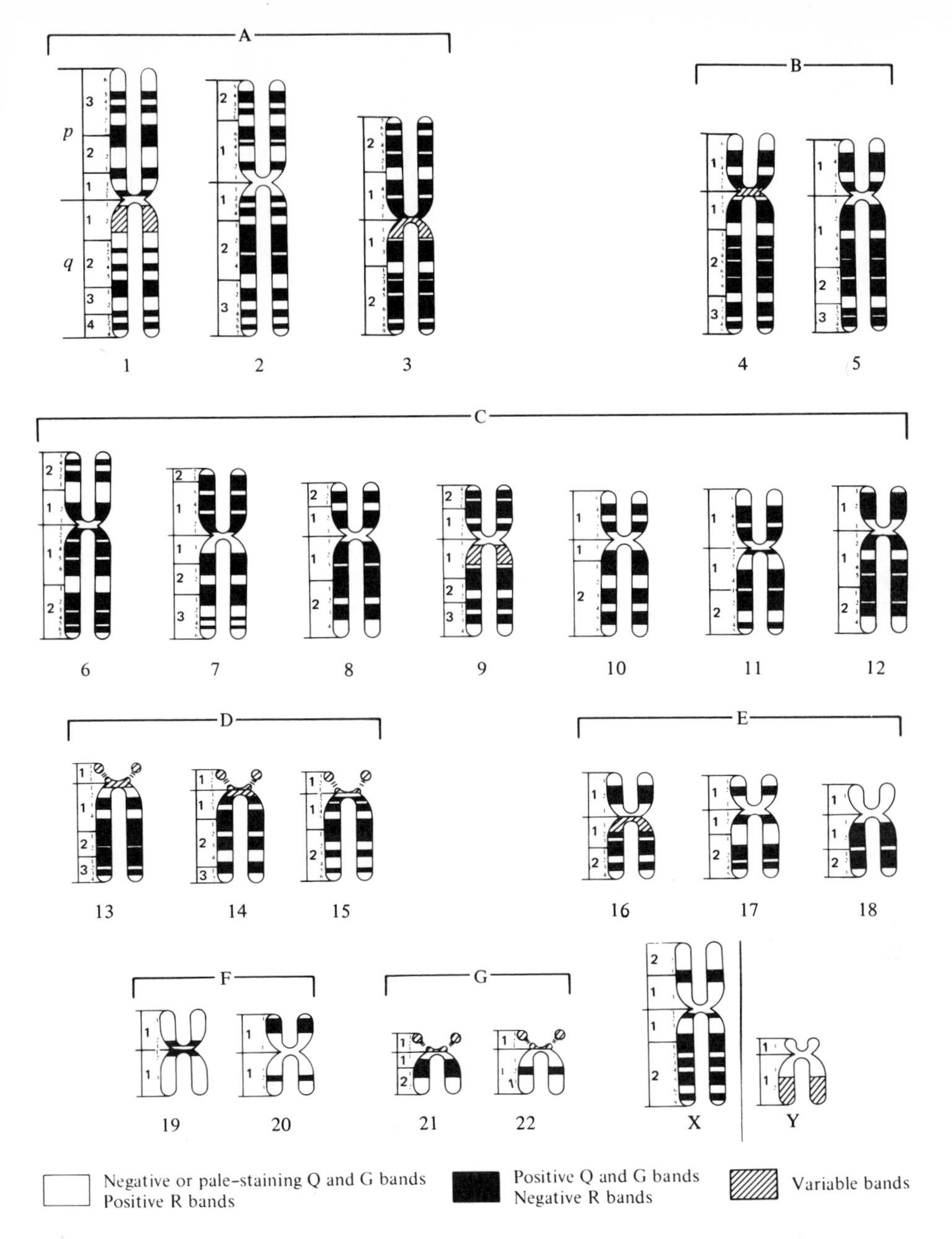

Figure 6–7. Diagrammatic Representation of a Human Male Karyotype According to the Denver, Chicago, and Paris Conventions for the Standardization of Chromosome Nomenclature. The chromosomes are aligned according to size and shape, location of the centromere, autoradiographic characteristics, satellites, and the Q-, G-, and R-stained bands, as outlined at the Chicago and Paris conferences. Q bands, which are fluorescent, result from use of quinacrine mustard stain; G bands are formed when Giemsa dye is used to stain the chromosome; and R bands result from the use of Giemsa dye with a reverse (R) staining technique, which gives banding patterns that are the reverse of the G bands. The numbers 1 through 22 are assigned to the autosomes, and X and Y represent the sex chromosomes. The seven letters A through G represent the Denver-system grouping, which is based mainly on the location of the centromere. Note the satellite chromosome segments extending from chromosomes 13, 14, 15, 21, and 22 and the size difference between the X and Y chromosomes. (From Paris Conference (1971): Standardization in Human Cytogenetics. In *Birth Defects: Orig. Art. Ser.*, ed. D. Bergsma, vol. VIII(7), 1972. Published by The National Foundation/March of Dimes, White Plains, N.Y.)

describing a karyotype

In describing a karyotype (Figure 6–7) by the standardized international terminology one states first the total number of chromosomes, then gives the sex chromosomes and a description of any anomaly. For example:

1. Normal female would be written as 46 XX.
2. Normal male would be written as 46 XY.
3. Female missing an X would be written as 45 X.
4. Female with an extra X would be written as 47 XXX.
5. Male with an extra G group chromosome would be written as 47 XY G+.
6. Male with a missing G group chromosome would be written as 45 XY G−.

The + or − sign following the group letter means an extra or missing chromosome and can be the equivalent of a chromosome or a piece of a chromosome, depending on whether some other symbolism is used along with the + or −.

Deletion Symbolism: In deletion symbolism the letters *p* (= petite) and *q* are used as follows:

p = short arm deletion (or piece of chromosome missing from the short arm).
q = long arm deletion (or piece of chromosome missing from the long arm).

For example, a defective cri-du-chat male would be described as 46 XY B*p*−. If the long arm were missing, the designation would be 46 XY B*q*−.

Translocation Symbolism. People with balanced translocations can have children with unbalanced translocations, as in Down's syndrome. A man or woman may have a balanced D/G translocation chromosome, made up of the long arms of one D and one G chromosome. If the person is female, her karyotype is 45 XX D− G− t (D*q*G*q*+). An unbalanced female Down's 46 translocation would be karyotyped 46 XX t (D*q*G*q*).

chromosome banding

Although the chemistry of the staining reaction is not fully understood, standard staining techniques can be relied on to produce bands in given regions of the chromosome. Band stains are valuable in the search for chromosomal abnormalities not readily apparent with other procedures, especially the Q, G, and R bands, which indicate the location of euchromatin or the "active portions of chromosomal DNA." Q staining receives its name from the use of quinacrine mustard, or quinacrine dihydrochloride, to produce fluorescent bands. G staining uses the Giemsa dye mixture, and the R staining uses a reverse Giemsa method, which results in differently stained bands, or R bands.

autoradiography

Autoradiography is the use of reagents emitting low-level radiation to produce photographic images in silver halide film emulsions. In chromosome research, radioactive nucleotides (recall that nucleotides make up DNA) are presented to dividing cells. During preparation for cell division, the cells incorporate the radioactive nucleotides into the new DNA, which in turn is incorporated into the chromosome. These chromosomes are now referred to as "hot." That is, they emit low-level radiation from portions of the chromosome where radioactive nucleotides have been incorporated. If the chromosomes are placed adjacent to photographic film, the radiation from the chromosomes exposes the film, and the hot chromosomes take their own pictures. The high-resolution (fine-detail) chromosome analysis made possible with autoradiography enables us to distinguish chromosomes 4 from 5; 13, 14, and 15 from each other and 17 from 18. Also autoradiography is the best method for detecting which one of the two X chromosomes is late in replicating its DNA, and has also served in the identification of certain features of the Y chromosome.

the process and costs

The procedure for producing human karyotypes is as follows. A blood sample of the subject is cultured so that particular cells, leukocytes or lymphocytes, in the sample begin to divide. A cell division inhibitor, colchicine, is added to hold the chromosomes in metaphase (Figures 4–2, 4–3, and 4–4). These cells are then subjected to mild physical pressure or to a hypotonic solution to disrupt the cell and disperse the chromosomes. A photograph is then taken of the chromosome spread (Figure 6–8), and the individual chromosomes are cut out of the photograph and pasted on a blank karyotype analysis sheet (Figure 6–9). After all of the chromosomes have been cut and pasted (Figure 6–10), the karyotype is checked for chromosome abnormalities.

Figure 6–8. Representation of a Human Male Metaphase Chromosome Preparation.

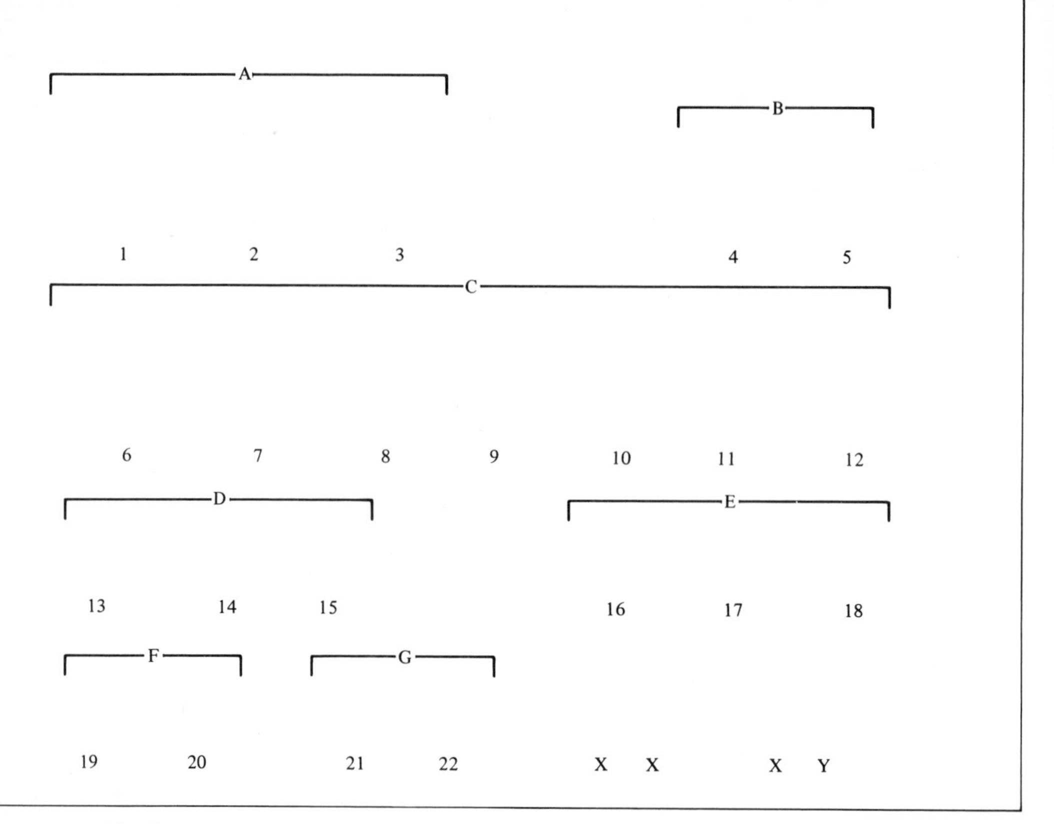

Figure 6–9. Karyotype Analysis Sheet. Chromosomes are individually cut out of a photograph and pasted on the page according to their size and shape and the locations of centromere, satellites, and bands where proper staining has been applied.

The manual labor involved in karyotyping is tedious and relatively expensive. Approximately two thirds of the cost of karyotyping is for technician labor, but the equipment is also quite costly.

A person referred to a laboratory providing a karyotyping service can expect to pay approximately $250 to $350 if the cells to be examined are to be obtained by amniocentesis (the withdrawing of amniotic fluid from the amniotic sac containing a developing fetus). The cost for karyotyping blood leukocytes is approximately $100 to $150.

The charge per case includes not only the cost of the karyotype, but also the complete diagnosis. In a prenatal diagnosis the charge of $300 would include all services from the amniocentesis through diagnosis and genetic counseling. Patients are usually billed after the final diagnosis. If a problem occurs with cell cultures and additional amniocenteses are required, there is no additional charge. Some laboratories maintain technicians exclusively for their cytogenic laboratory. When funds are available, one technician deals exclusively with blood leukocyte cultures, and another is responsible for all tissue cultures.

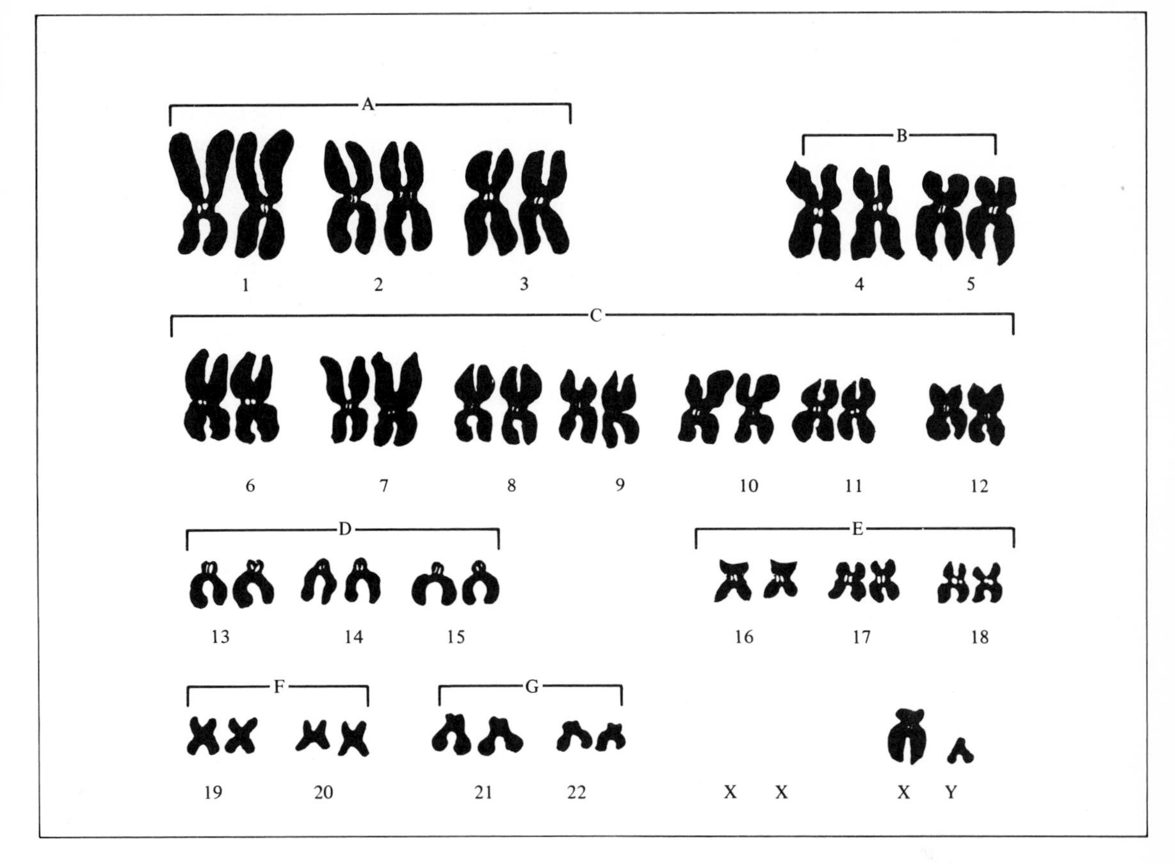

Figure 6–10. Normal Male Karyotype Analysis. Chromosomes are arranged in descending order of size and separated into the seven groups A, B, C, D, E, F, and G. Females have 22 paired autosomes and two X (sex) chromosomes; males have 22 paired autosomes and the sex chromosomes X and Y. The X chromosomes are difficult to distinguish from the C group, and some laboratories place them beside group C. The Y is acrocentric, similar to group G, and placed to one side of that group. According to convention standards, the sex chromosomes are placed in the lower right-hand corner of the karyotype.

general application of karyotype analysis

Karyotype analysis is generally limited to cases where an abnormality is expected. This conservative use is due to the expense, the limited number of trained personnel, a slight potential hazard in amniocentesis, and the fact that three to six weeks are required to culture the cells. In many cases clinical evidence is such that the karyotype is not really necessary and would serve only as supportive evidence. Karyotype analysis is quite effective, however, when the counselor believes there is a high probability of a defective birth due to a chromosome abnormality. For example, when it is known that a parent is a translocation carrier for Down's syndrome, karyotype analysis is most beneficial (Figure 6–4). If an inexpensive, routine method for screening prenatal or newborn karyotypes were available, certain disorders could possibly be detected before they occurred. Early detection and more frequent analysis of large populations would provide new information, and perhaps spin-offs from the various cases would allow for new methods of early and effective treatment. If, for example, we could prenatally detect all the chromosomal conditions that lead to mental retardation, we would be in a position to counsel prospective parents more effectively. The screening studies might also serve as an effective sensor in assessing the quality of our environment and subsequent effects with respect to chromosomes damage and genetically caused birth defects. In brief, karyotype analysis is the single most important test we have in screening for possible cytologically, "genetically" defective offspring and for the recognition of potential genetic congenital malformations resulting from chromosome abnormalities. Karyotype analysis can make possible better counseling of prospective mothers, before or during pregnancy. Feelings of guilt and the ensuing emotional stress involved when a malformed or mentally retarded child is born could be eliminated in many cases.

automation of human karyotyping

As mentioned, the cutting, matching and pasting, and inspecting of random metaphase chromosome cells is extremely time-consuming. Castleman and Melnyk, two California scientists, are working to alleviate this situation by developing a microscope-computer combination that turns out, in minutes, pictorial karyotypes from randomly spread chromosomes. The automated karyotyping system, under development at Caltech Jet Propulsion Laboratory, accepts a conventionally prepared microscope slide and produces a high-resolution pictorial karyotype in standard format (Figure 6–11). The system includes an automated light microscope, a digital computer, an image-processing system, and a high-resolution picture output device. Starting with a regularly prepared metaphase spread of chromosomes, the automated system scans, identifies, and classifies the chromosomes. The computer orients and arranges the chromosomes into the regular karyotype pattern. The investigator, by viewing the video screen, can select the metaphase spreads most suitable for further computer processing of the chromosomes (Castleman and Melnyk, 1974; *News and Features,* 1974).

prenatal diagnosis of chromosome abnormalities / amniocentesis

Amniocentesis is a prenatal procedure permitting the physician to perform tests on the fluid (amniotic fluid) that bathes the developing fetus within the amniotic sac. The information available from study of the amniotic fluid and the fetal cells within the fluid gives the parents the option of aborting the fetus rather than giving birth to a malformed or mentally retarded child. In 1975 there were 251 centers offering amniocentesis compared to 121 centers in 1971 (Lynch, 1974). This increase illustrates the growing interest in this field. The National Genetics Foundation (250 West 57th Street, New York, N.Y. 10019)

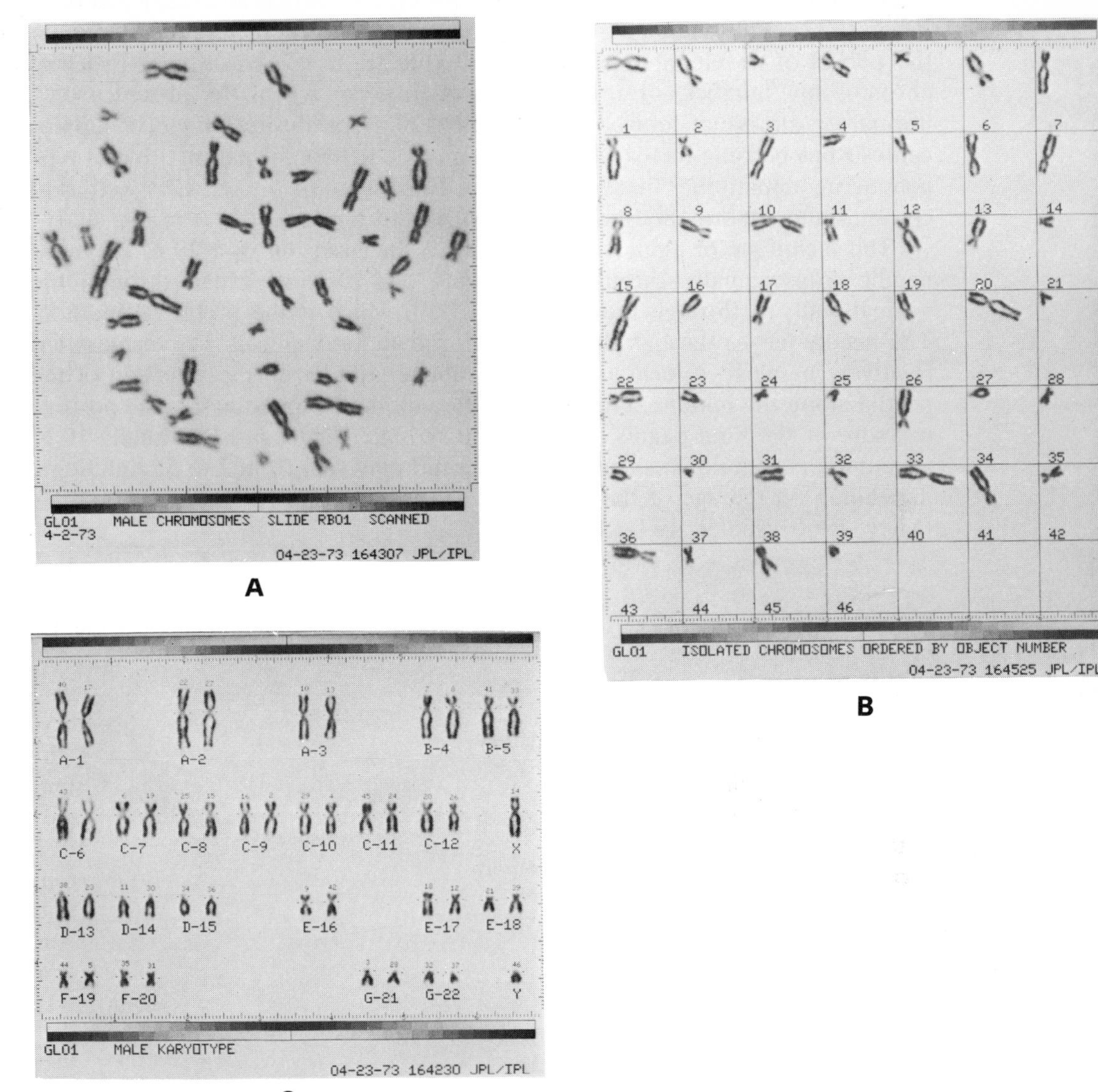

Figure 6–11. Automated Karyotype Analysis. **A.** Random spread of 46 chromosomes. **B.** The new automated light microscope system scans and sorts each chromosome. **C.** The computer orients and arranges the chromosomes into the traditional karyotype pattern. The number above each chromosome is its position in the scanning order shown in **B.** The group and number classification appears below each chromosome. Two California scientists, who have a contract with the National Institute of Child Health and Human Development, are developing the system to speed the preparation of karyotypes. (Photos courtesy of Jet Propulsion Laboratory, California Institute of Technology; reproduced with permission.)

and the National Foundation/March of Dimes (1275 Mamaroneck Avenue, White Plains, N.Y. 10605) maintain listings of diagnostic and consulting centers.

Interest in the composition of amniotic fluid began over 100 years ago (Nitowsky, 1971). Studies have shown that half the total amniotic fluid volume is ingested or passed through the fetus every 24 hours (*Laboratory Management*, 1973). The numbers of fetal cells and biochemical factors fluctuate over the period of development. Detection of changes in cell constituents, such as chromosome number, or in fluid components can reveal the presence of a normal or abnormal fetus. Over the past 20 years the technique of amniocentesis has become increasingly important in detection of parental blood type genetic incompatibility that may result in Rh disease of the fetus, as well as of chromosome abnormalities such as Down's syndrome.

The technique of amniocentesis requires the insertion of a 20 to 23 gauge needle transvaginally through the vagina and cervical canal, through the vaginal wall, or through the abdominal wall, which is the preferred method. The needle passes through the midline of the abdominal wall. The technique is relatively painless; a local anesthetic is not required with the insertion of the needle along the midline. Once the needle enters the amniotic sac, the positive pressure of the fluid begins to fill up the syringe (Figure 6–12). Usually 10 to 20 milliliters of fluid is taken, but the amount may vary from 2 to 55 milliliters, depending on the age of the fetus. The fluid is used for cell and fluid analysis (Liley, 1960; Nadler, 1971).

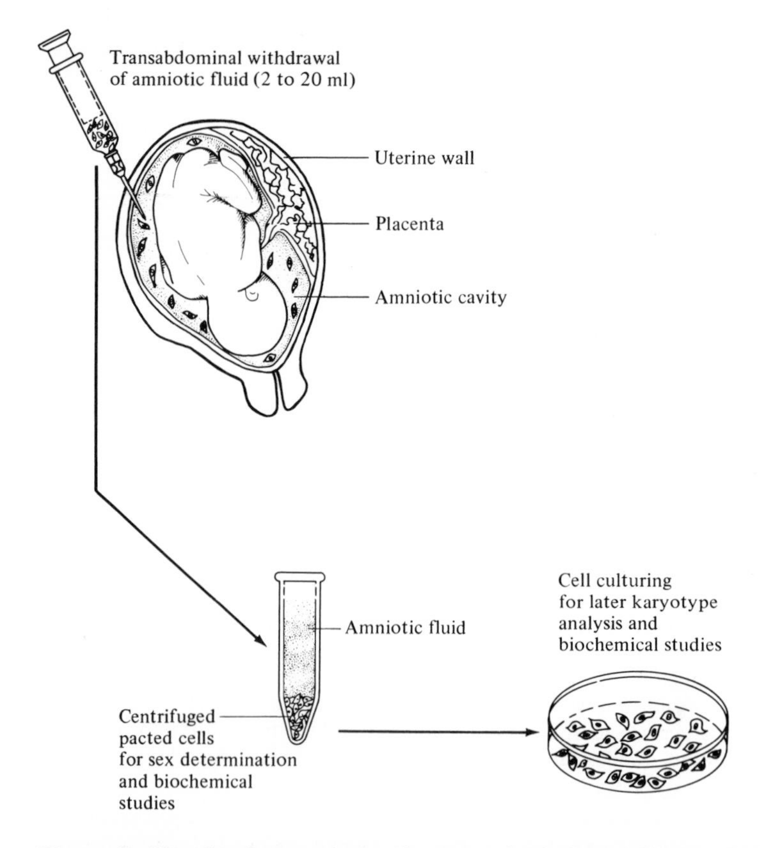

Figure 6–12. Amniocentesis for Karyotype Analysis and Biochemical Studies. The drawing represents the procedure of amniocentesis in order to obtain fetal cells and amniotic fluid for analysis. Sex determination can be performed on the pacted cells or later if necessary after the cells are ready to be processed for karyotype analysis.

The most frequently employed method for diagnosing fetal genetic abnormalities, combining amniocentesis with karyotype analysis, involves the in vitro cultivation of amniotic cells and a subsequent biochemical or cytologic study of the cell populations. After the cells and supernatant fluid are separated, the cells can be analyzed for sex and chromosome count, or they can be dispersed in a nutrient medium and 'allowed to multiply. Initially, the cells vary in morphology, although the majority are epithelial or polyhedral in appearance. Sufficient cell growth should occur in 10 to 14 days, at which time chromosome preparations may be made. At this time a fibroblast type of cell (cells found in fibrous tissue) predominates in the culture. Frequently, up to six weeks may elapse before sufficient numbers of cells are available for biochemical or enzymatic studies.

Speed is a critical consideration with amniotic cell cultures because diagnosis must be rendered in time to permit selective abortion. With a three- to six-week incubation period for amniotic cells, amniocentesis should be performed no later than 14 to 16 weeks into pregnancy to allow for therapeutic termination no later than the 24th week of gestation (*Laboratory Management*, 1972). This should provide sufficient time even if the procedure has to be repeated due to a culture failure or unclear results.

In addition to chromosome analysis of cultured and uncultured cells, the cultured cells and fluid can be examined for their enzymatic and metabolic properties. A number of enzyme activities can be readily be measured in cell cultures, and deficient activity of a specific enzyme is evidence of a fetus with an inborn metabolic disorder. (Metabolic defects caused by a single defective gene are covered in Chapter 5.) The National Foundation/March of Dimes advocates amniocentesis in order to improve understanding of normal and abnormal metabolic patterns and to develop prenatal treatment for defects so that a healthy child will be born.

how early in pregnancy can amniocentesis be of diagnostic value in detecting chromosome abnormalities?

Technically, the physician should remember that the fetus is the patient, even though the counseling must be given to the parent. The parent decides whether to have amniocentesis, and after learning the results, decides on the fate of the developing fetus.

Pregnancy begins with the implantation of the blastocyst (a hollow ball of dividing cells produced following fertilization) into the wall of the uterus. The obstetrician calculates the beginning of pregnancy from the first day of the last menstrual period, which usually precedes conception by two weeks and blastocyst implantation by three to four weeks (Fuchs, 1971).

In general, except for the determination of Rh and fetal lung maturity, which are preferably ascertained in late pregnancy, the optimal time for amniocentesis is at 14 to 16 weeks into pregnancy. This time period is considered optimal because there is a sufficient volume of amniotic fluid (about 125 milliliters) so that withdrawal of 2 to 20 milliliters will not endanger development and there should be an ample number of viable cells available for culturing and chromosome studies. In fluid samples taken before the 12th week, the volume that can safely be withdrawn is too small, and in many cases too few cells are available to be utilized. The number of cells in the fluid increases from 2000 per milliliter in the 12th week to 10,000 per milliliter in the 20th week, and these cells have a 50% viability rate (that is, if cultured, 50% of the cells will divide). Beyond the 20th week, the fluid continues to increase, but cell viability decreases. A 97% success rate has been achieved in culturing cells taken from amniotic fluid during the 14th to 20th week. Only a 14 to 33% success has been achieved in culturing cells from the first trimester, or first three months, of pregnancy, and a 10 to 15% success rate in the third trimester (Hahnemann, 1972). Theoretically, only one viable cell is necessary to start a culture; but culturing is a demanding technique, and success depends upon the number of viable cells available initially. The fewer viable cells, the less chance for a viable culture. In addition, as already noted, up to six weeks may be required for culturing before the karyotype can be done, and time must be allowed for repeated sampling if necessary, without risking undue delay in the decision for elective abortion. Law and social sanctions restrict a person's decisions concerning abortion in the third trimester.

maternal and fetal risks of amniocentesis

In the United States in 1973, there were 3,141,000 live births (*Vital Statistics,* 1974). According to the World Health Organization (1972), 1% of these children, or 31,410, were born with a chromosome abnormality, and 13% of those or 4383, were born to women over the age of 35. Over half the children with Down's syndrome, some 2,500 to 3,000 children, are born to women 35 and older (Milunsky, et al., 1970). Although diagnostic amniocentesis would be desirable for all pregnant women over 35, the large number of such pregnancies would overwhelm the presently limited laboratory facilities. Most investigators advocate amniocentesis for women 40 or older and when a chromosomal aberration has been previously detected in the family. The second category include "high-risk" mothers, those who are known to be carriers of balanced translocations, though they appear normal, and those who are known mosaics, that is, who carry two or more cell types. There were 3,141,000 children born born in 1973, but less than 1000 of their mothers submitted to amniocentesis and karyotype analysis. There is a 96% chance that if the cells had been cultured for the identification of trisomies and translocations in high-risk pregnancies, the abnormality would have been correctly diagnosed (Nadler, 1971). Although amniocenteses have proved reliable in most cases, it should be noted that results that are negative for a chromosome disorder do not guarantee that the fetus will be completely healthy. The fetus still has the same risk as other members of the general population of inheriting other gene-related disorders. It is estimated that everybody carries between five and eight defective genes. This lack of guarantee should be stressed to the prospective parents before amniocentesis is undertaken.

The risks of hemorrhaging, infection, and trauma are minimal to the mother and fetus. Data compiled for 1971 through 1975 support this assertion. With proper gynecological examinations and the use of sterile techniques, the chance of complications when amniocentesis is conducted within the optimum time period is less than 1 in 1000. However, there is concern about long-term risks involved in amniocentesis (Miller, 1971). For example, Miller postulates that the removal of a large amount of fluid during the period from 14 to 16 weeks may produce a decrease in IQ of 25 points. He states that studies to date have no way of determining this kind of potential damage. Recognizing the far-reaching social importance of amniocentesis, he encourages greater attention to assessing the effects of its use before it becomes widespread with no one realizing the more subtle inherent dangers.

other prenatal diagnostic techniques

Amniography. A water-soluble dye is injected into the amniotic sac. The dye mixes with the fluid and provides contrast in order for the X-ray to outline the fetus.

Ultrasonography. Ultrasound is used to locate the placenta and fetal head prior to amniocentesis. Normal fetal specimens are expected to elicit sound patterns that differ from those of a dead or grossly malformed fetus (Merut et al., 1973).

Fetoscopy. This is a procedure whereby the physician can actually see the developing fetus with the use of a fiberoptic endoscope. Most potentially diagnosable inherited defects escape in utero due to the absence of morphological or biochemical findings in amniotic cells. With fetoscopy it may become possible to obtain tissue, particularly fetal blood, and to expand the field prenatal diagnosis (Valenti, 1972).

the personal benefits of prenatal diagnosis

The topic of prenatal diagnosis would be incomplete without a discussion of attitude. Most people find little harm in submitting to the analysis to learn the sex of a child when sex-linked disorders have been found in the family or for information about suspected chromosome abnormalities. However, these same people often object violently to the alternatives proposed after learning their child will be mentally retarded or malformed.

The Catholic Church is absolutely opposed to therapeutic abortions for genetic reasons, and there are approximately 50 million Catholics in the United States—that is, one in every five persons—and some 600 million in the world. The opposition from this group alone can be a considerable force. Other groups, such as the nationwide Right to Life group, also oppose the elective abortion. Despite the efforts of these groups to influence legislation and individual decisions, it is, in the end, the parents' choice to act in a manner in keeping with their personal beliefs and code of ethics.

Table 6–2 lists most chromosomal abnormalities, both genetic and nongenetic, presently detectable by prenatal methods. Amniocentesis lets the parents know medically where they and the unborn child stand. Whether to abort the fetus or allow it to be born is still the choice of the parents in consultation with a competent physician. The question is: How long will the decision remain a right of the parents?

In brief, amniocentesis and subsequent analysis can provide the following benefits.

1. It serves as a diagnostic tool, enabling the physician to make a more definite diagnosis of the condition of the fetus.
2. If the results prove negative, it can reduce or perhaps dispel the anxiety and emotional strain on families who have a high risk of producing a child with a genetic disorder. Without this procedure such people might choose to have a therapeutic abortion rather than risk having an abnormal child. Amniocentesis can rule out certain cytologically observable abnormalities.

Table 6—2 Prenatal Human Karyotypes and Related Clinical Conditions in Live Births

Recognized Karyotypes	*Mental Retardation in at Least 50% of the Cases*	*Clinical Condition*	*Incidence at Birth*
46 XY		Normal male	
46 XX		Normal female	
Trisomies			
13 D	×	Patau's syndrome, multiple congenital anomalies	1 in 6000–7000
18 E	×	Edwards' syndrome, multiple congenital anomalies	1 in 4000
21 G	×	Down's syndrome	1 in 600–700
21 G (?)	×	Cat's eye syndrome (non-Down's), severe mental retardation, muscle hypertonia	Very rare
22 G	×	Schizoid, mental retardation	Very rare
Monosomy			
21 G	×	Only known living monosomy except for X0; death usually occurs in first year due to multiple severe malformations	
Sex chromosome complement			
X0	×	Turner's syndrome	1 in 2500 females
XXX	Less than 50%	Triple X; few, if any, problems	1 in 1000 females
XXXX	×	Tetra X, mental retardation	Very rare
XXXXX	×	Penta X, mental retardation	Very rare
XXY	×	Klinefelter's syndrome	1 in 400 males
XXXY	×	Modified Klinefelter's syndrome	Very rare
XXXXY	×	Modified Klinefelter's syndrome	Very rare
XYY		YY syndrome (super male)	1 in 250 males
XXYY		YY syndrome (super male)	Less often than XYY

Table 6–2 continued

Recognized Karyotypes	*Mental retardation in at Least 50% of the Cases*	*Clinical Condition*	*Incidence at Birth*
X0/XY	Less than 50%	Mosaicism	Common (?)
XX/XY	Less than 50%	Mosaicism	Common (?)
Y0		Never observed	
Translocations			
B/C, B/G, and B/D unbalanced	×	Cri du chat	Rare
C/E balanced		Translocation carrier parent (genetic)	Rare
D/D unbalanced	×	D1 syndrome	Rare
D/D balanced		Translocation carrier parent (genetic)	Rare
D/G unbalanced	×	Down's syndrome (genetic)	Rare
D/G balanced		Translocation carrier parent (genetic)	Rare
C/G (9 and 22)	?	Philadelphia chromosome (leukemia)	Rare
G/G unbalanced	×	Down's syndrome (genetic)	Rare
G/G balanced		Translocation carrier (genetic)	Rare
Y/X		Repeated spontaneous abortions	Very rare
Ring chromosomes†			
B (5)	×	Cri du chat	Very rare
D (13)	×	Severely retarded, lower spine abnormalities	Very rare
E (18)	×	Severely retarded, lower spine abnormalities	Very rare
G (either 21 or 22)	×	Mental retardation and developmental problems	Very rare
Partial deletions (sections missing)			
B (in short arm 4)	×	Most die in a few months due to heart failure and multiple malformations; retinoblastoma (cancer of the eye)	Very rare
B (in short arm 5)	×	Cri du chat	Rare
D (in long arm 13)	×	Mental retardation, other congenital malformations	Rare
D (in short arm 13)	×	Mental retardation, other congenital malformations	Rare
E (in long arm 18)	×	Severe mental retardation, others	Rare
E (in short arm 18)	×	Severe mental retardation, others	Rare
G (in short arm 22)		Variable refractory dysplastic anemia	Very rare
Y (in long arm)		Some cases of infertility	Rare
Y (short Y deletion)		Associated with several blood cell malignancies	Very rare

*Rare indicates that the abnormality occurs at least once in every 100,000 live births; very rare means it occurs at least once in 1 million live births.

†A ring chromosome implies the loss of both ends of a chromosome so that the raw ends attach to each other forming a ring.

3. If the results are positive and the parents choose therapeutic abortion, it can reduce the occurrence of genetic disorders that impose a serious emotional and financial burden on parents and society.
4. If handled properly, it could lead to a decrease in the pathological gene pool of autosomal dominant disorders (in which one defective gene causes an abnormality), such as Huntington's disease. We must be careful, however, since the elimination of fetuses with recessive disorders (where two defective genes are required for the abnormality to be expressed) could lead to an increase in the frequency of the defective gene. Through amniocentesis heterozygous parents (who carry one normal and one defective gene) may elect to have more offspring since they can be assured of phenotypically normal children. But two thirds, or 67%, of these surviving offspring would in turn be heterozygous, thus increasing the defective gene in the population (Prescott, Magenis, and Buist, 1972). The increase in the defective gene through the use of elective abortion based on information provided by amniocentesis can be more clearly understood as follows:
   a. Two heterozygous normal parents mate: $Aa \times Aa$.
   b. Each parent gives rise to gametes A and a

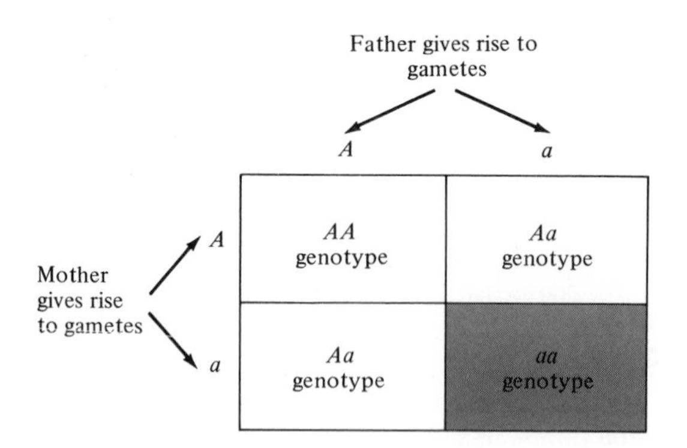

   c. On fertilization the mother may carry a fetus with any of the genotypes shown above.
   d. If the recessive defective *aa* genotype is detected and aborted each time, the parents will have only children with genotypes *AA* and *Aa*. Thus the probability is that two out of three normal births will be heterozygous like the parents.

chromosome aberrations and their effect on the phenotype

the physical basis: transmission of hereditary material through the chromosomes

Chromosome abnormalities have been known to occur in plants, lower animals, and mammals for a long time. In 1927, for example, Gates found that a chromosome abnormality was associated with a behavioral problem in mice (Gates, 1927). Subsequently, a variety of chromosome abnormalities were found to be associated with behavioral changes in humans (see Chapter 7).

As can be ascertained from Table 6–2, chromosome abnormalities in man vary from monosomies to quintosomies. Conditions where aborted fetal material was found to be triploid or tetraploid were not included in Table 6–2 because no polypoid live birth has ever been known to survive beyond a few days. In fact, no autosomal trisomies for chromosomes other than 13, 18, 21, and 22 have been found in live births. The loss of the X chromosome in males (Y0) and of autosomes other than chromosome 21 is also lethal, for examples of these have not been found in live births. Most translocations, ring chromosomes, and partially deleted chromosomes must also be lethal, since very few of these have been found among live births. Such data establish that through human evolution the chromosomes have reached the point where those remaining in the species have been carefully balanced, so that the normal diploid complement results in socially and physiologically meaningful humans. A deviation in this balance means a change in the amount of DNA left to code for cell function. Such changes are most often detrimental to the cell, or to the species if they are transmitted to offspring.

The fact that changes in chromosome morphology or number are detrimental does not imply that they do not occur. On the contrary, collectively they occur quite often (Table 6–3). For example, 10% of all zygotes have a chromosome defect. Approximately 17 to 20% of all recognizable pregnancies end in spontaneous abortions. About 15% occur before the completion of the first trimester, and 60% of these spontaneously aborted fetuses possess a chromosome abnormality (Carter, 1969). Of the spontaneously aborted fetuses with chromosome abnormalities, Graham states, 5% are due to tetraploidy, 23% to triploidy, 47% to trisomies, 22% to Turner's X0 condition, and 3% to other abnormal chromosomal situations.

Table 6–3 Frequency of the More Common Chromosome-Associated Human Abnormalities in the United States*

Syndrome	Chromosome Number	Estimated Incidence	Calculated Number Born in 1973†	Calculated Number in the United *States*‡
Down's (trisomy 21)	47	1 in 600–700	5,000	350,000
Patau's (trisomy D)	47	1 in 6,000–7,000	450	30,000
Edwards' (trisomy E)	47	1 in 4,000	800	50,000
Triple X (XXX)	47	1 in 1,000 females	3,000	200,000
Turner's (X0)	45	1 in 2,500 females	1,300	85,000
Klinefelter's (XXY)	47	1 in 400 males	8,000	530,000
Double Y (XYY)	47	1 in 250 males	12,000	850,000
		Total*	30,000	2,100,000

* This list is confined to those chromosome abnormalities with high incidence. Thus, the number 30,000 is not the total for all live births with chromosome problems.

† Calculated number based on live births for 1973 (*Vital Statistics*, 1974).

‡ Calculated number based on total United States population of 211,438,000 as of March 1, 1974 (U.S. Bureau of Census figures for 1974).

Of the pregnancies that end in live births, 1% produce offspring with a chromosomal abnormality compatible to some degree with life. There were 3,141,000 live child births in 1973. Thus, 31,410 of these children have some type of chromosome abnormality. Of these, approximately 40%, or 12,564, have autosomal defects, and 60%, or 18,846, have sex chromosome abnormalities. Among the 12,564 live births displaying an autosomal anomaly associated with human defects, the most frequent is the trisomy for chromosome 21, which is associated with Down's syndrome, or mongolism (Figure 6–13). The second most frequent autosomal disorder is associated with chromosome 18, Edwards' syndrome (Figure 6–14), and the third most frequent is Patau's syndrome (Figure 6–15).

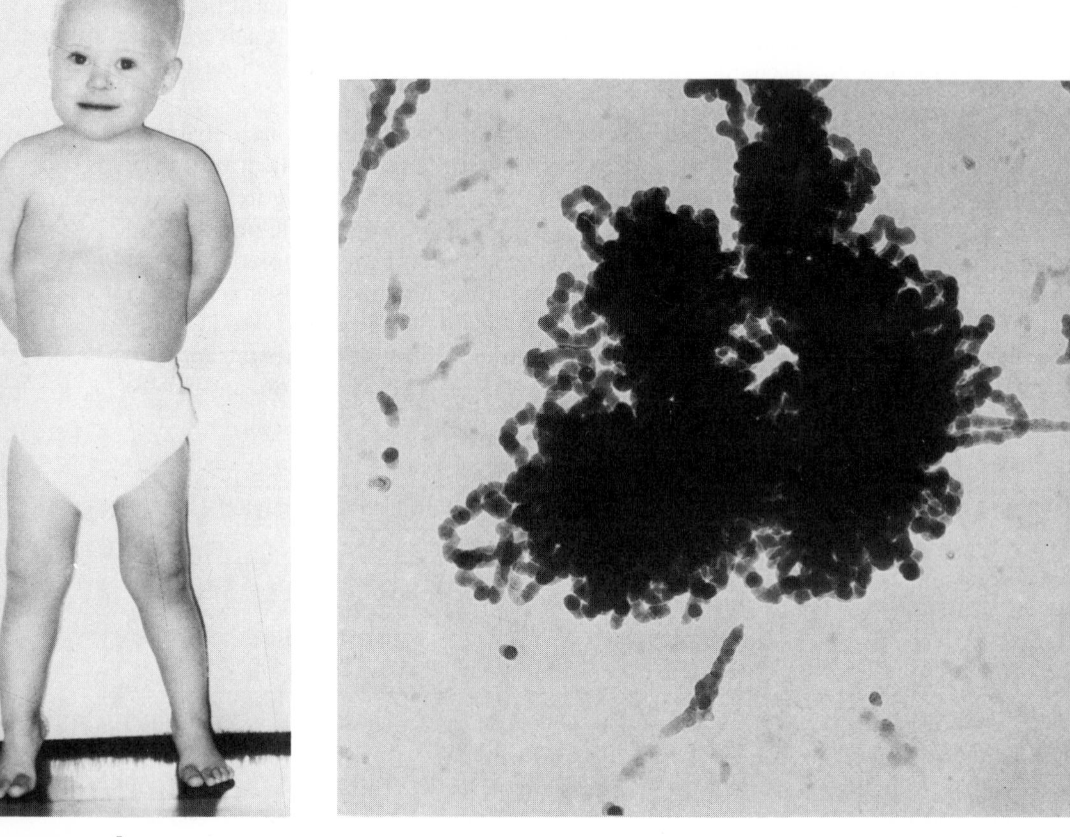

A B

Figure 6–13. Down's Syndrome (Trisomy 21). **A.** This disorder is of high incidence, occurring in approximately 1 in every 600 to 700 live births. Generally, the surviving adult is of short stature, mentally retarded, and more susceptible to infection and leukemia than normal. Many characteristics are associated with the syndrome, although only a few may be present in any one case. Ninety-five percent of all cases of Down's syndrome are associated with trisomy of chromosome 21. **B.** Electron micrograph of a human chromatid pair from chromosomes 21–22 (×35,300) with a total dry mass of 6.3×10^{-13} gram. (**A** by permission of Charles H. Carter, M.D.; **B** courtesy of E. J. DuPraw from *DNA and Chromosomes*, Holt, Rinehart and Winston, New York, 1970.)

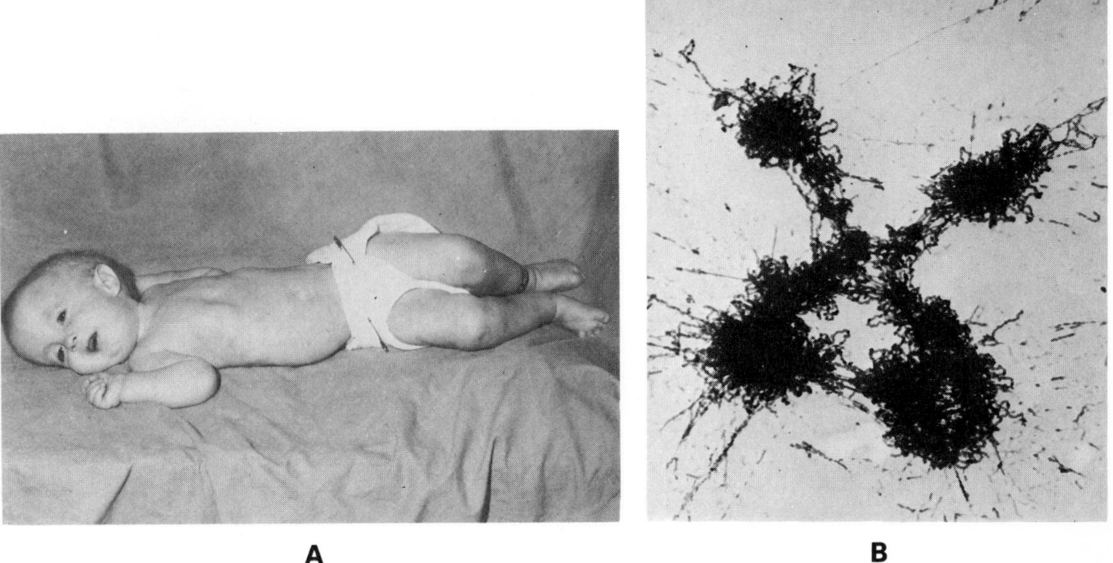

Figure 6–14. Edwards' Syndrome (Trisomy 18). **A.** In this congenital defect the patient has low-set ears, stunted growth, and often cleft lip and/or palate. Many patients show fusion of the fingers and a peculiar finger-folding pattern. They are difficult to feed and usually die at a young age. **B.** Electron micrograph of a human chromosome 17–18 (×8200) with a total dry mass for two chromatids of 11.1×10^{-13} gram. (**A** by permission of Charles H. Carter, M.D.; **B** courtesy of E. J. DuPraw from *DNA and Chromosomes*, Holt, Rinehart and Winston, New York, 1970.)

The most common abnormality of the sex chromosomes is the double Y (XYY), but the associated human disorder, if any, is mild (Figure 6–16), as is also true of the triple X (XXX). The second most frequently found is Klinefelter's syndrome, XXY (Figure 6–17). The fourth most frequent sex chromosome anomaly involves a missing sex chromosome, monosomy X0; this is known as Turner's syndrome (Figure 6–18). Other common sex chromosome aneuploidies found in humans are the various mosaics that result from errors in cell division (Figure 6–19). A significant number of mosaics now being found are listed in Table 6–4.

Live birth mosaics start out as a single sex. But early in development, an error in cell division gives rise to two daughter cells that differ in their sex chromosomes. For example, during division of an XY cell, if the Y should lag behind in anaphase, for whatever reason, and not become enclosed in the nucleus of one of the daughter cells, there would be only one cell XY; the other would be X0. From that point on, every time those cells divided, each would give rise to identical daughter cells. The XY cells would make more XY cells, and the X0 more X0 (Figure 6–20). Thus one person can have various tissues containing two, three, or more cell types, depending on when in development the error occurred. Mosaics can also occur as a result of nondisjunction, as shown in Figure 6–20. Table 6–4 shows the results of mitotic nondisjunction and anaphase lag in XX, XY, and XXY cells. These phenomena in X0 cells would result in XX cells, more X0 cells, and cells with no sex chromosome at all, which would be expected to die. In XXX cells, these

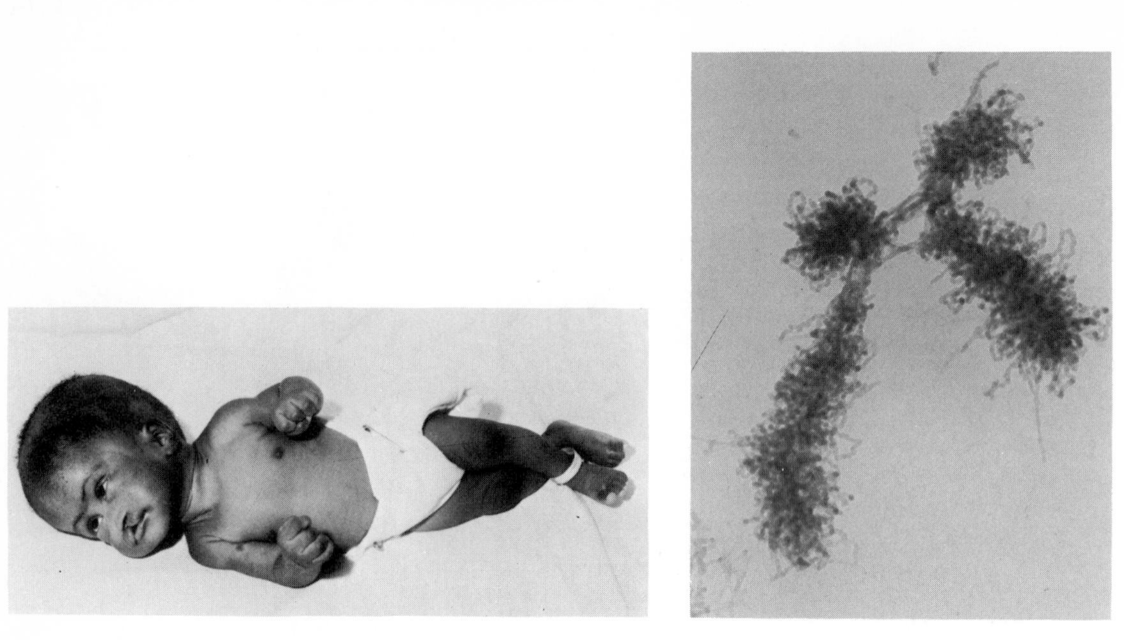

A **B**

Figure 6–15. Patau's Syndrome (Trisomy 13). **A.** In general, the patient has retarded growth, microcephaly, cleft lip and/or palate, a small chin, and most often polydactyly. Some of these patients also have cyclopia or the complete absence of eyes. They have low-set ears and a short neck, and hernias are common. Nearly all males have undescended testes. They are difficult to feed and have a short life span. **B.** Electron micrograph of a human chromosome 13–15 (×24,000) showing acrocentric morphology. Total dry mass for two chromatids is 9.9×10^{-13} gram, of which 84% is in the long arms. The length of fiber in each chromatid is 218 ± 11 μm, and the length of DNA double helix is about 3 cm. (**A** by permission of Charles H. Carter, M.D.; **B** courtesy of E. J. DuPraw from *DNA and Chromosomes*, Holt, Rinehart and Winston, New York, 1970.)

phenomena would produce more XXX cells, XXXX cells, and some normal XX cells.

Kemp (1951) states that chromosome mosaics give rise to the formation of a rare human gynandromorph. The gynandromorph in insects may be bilaterally symetrical, with one side and/or one half (top or bottom) of the body

Table 6–4 Chromosome Mosaics Expected by Proposed Mechanisms*

Anaphase Lag		*Mitotic Nondisjunction*	
Original Cell Type	*Mosaic Expected*	*Original Cell Type*	*Mosaic Expected*
XX	XX/X0	XX	XXX/X0/(XX)
XY	XY/X0	XY	XXY/X0(XY)
XXY	XXY/XX		XXY/(Y0)/(XY)
	XXY/XY	XXY	XXXY/XY
			XXYY/XX

* From Redding and Hirschhorn (1968).

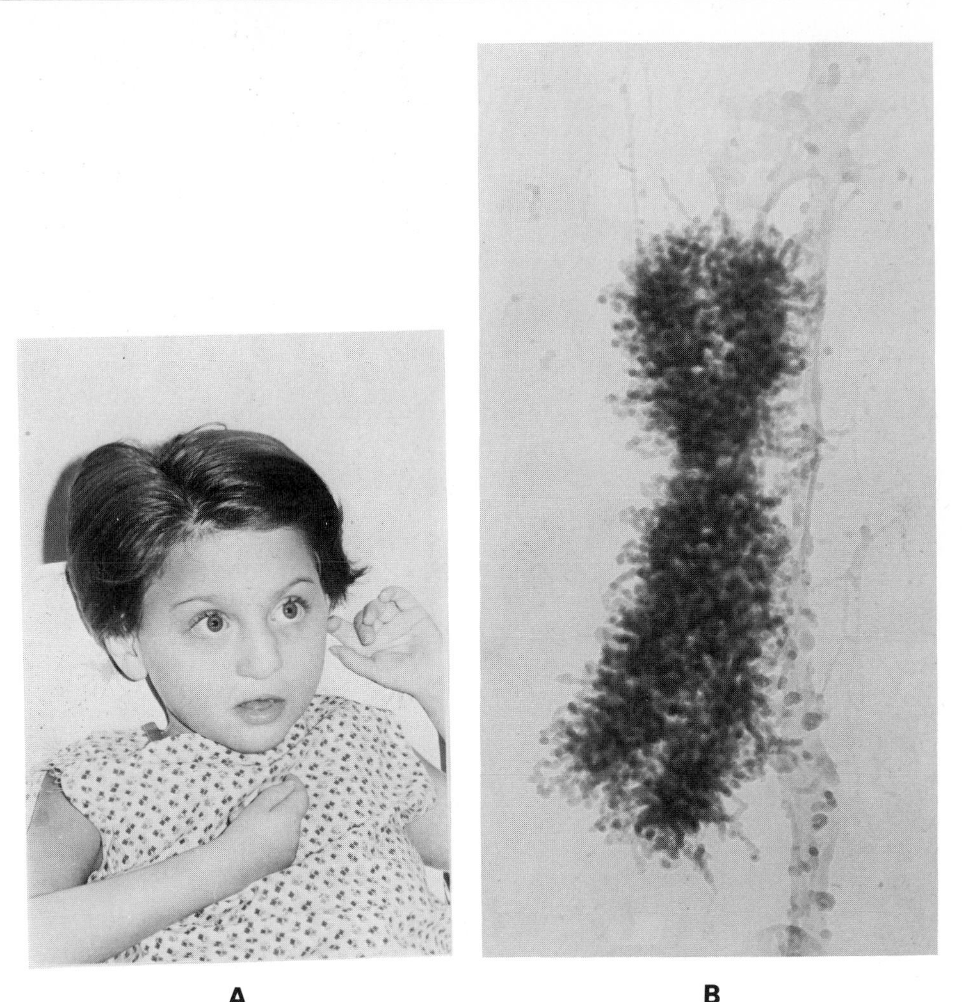

Figure 6–16. Triple X Female (XXX). **A.** A triple X female has three X chromosomes instead of the normal two and demonstrates two Barr bodies, as expected. Disadvantages of this abnormality appear to be limited, in keeping with the Mary Lyon hypothesis of sex chromosome inactivation, although some investigators have reported instances of low IQ. In general, a triple X female is located by chance in screening rather than because of some demonstrable disorder. **B.** Electron micrograph of a human chromosome tentatively identified as the X chromosome (×24,200) from its weight and attachment to the nuclear envelope. Total dry mass for two chromatids is 15.4×10^{-13} gram, 63% of which is in the long arms. (**A** by permission of Charles H. Carter, M.D.; **B** courtesy of E. J. DuPraw from *DNA and Chromosomes*, Holt, Rinehart and Winston, New York, 1970.)

developing as a male and the other side or half as a female. In humans, as Stern (1973) points out, a genetic setup for gynandromorphism would not lead to clear-cut sexual mosaics since the development of certain tissues and organs is not solely determined by their sexual genotype. Sex type in humans also depends on hormonal influences emanating from other tissues and organs. However, human gonadal mosaics have been discovered with testicular and ovarian tissues or organs that have developed side by side and with secondary

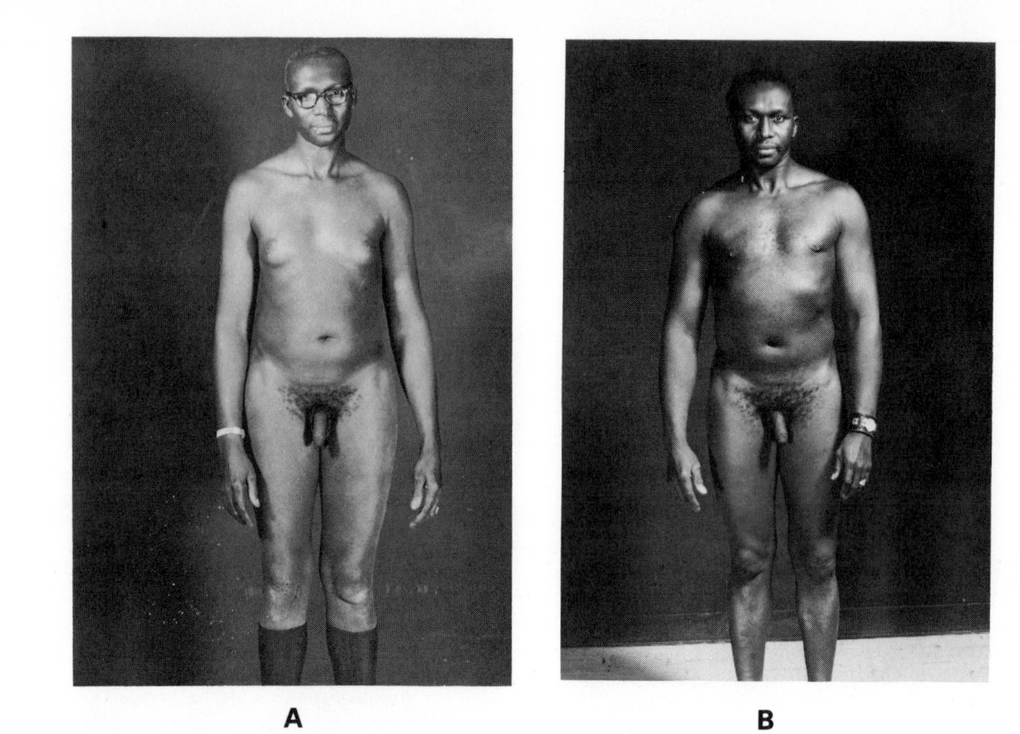

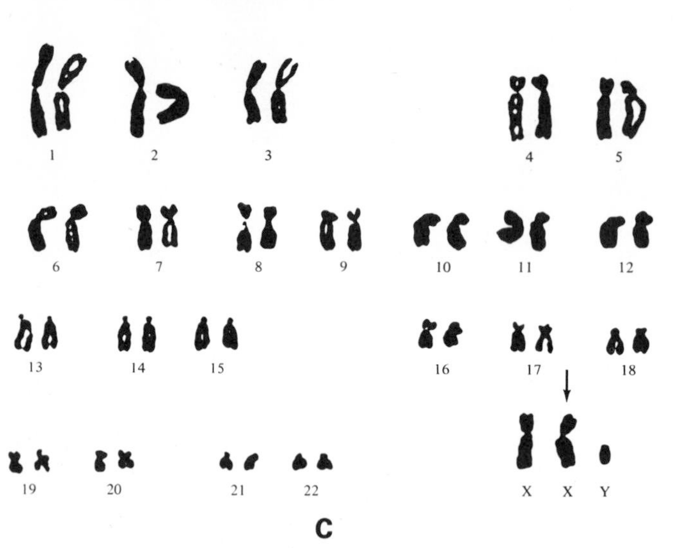

Figure 6–17. Klinefelter's Syndrome (XXY). The cause of the XXY male is believed to be nondisjunction of the sex chromosomes during the formation of the sex cells. The extra X can be either paternal or maternal in origin. The patient has a normal life expectancy, but he may have difficult social adjustments to make. About 75% of Klinefelter males have breast development. Most have a normal IQ. **A.** Patient aged 43 with Klinefelter's syndrome (XXY) prior to breast surgery and testosterone treatment. Notice the light skin and feminine-type musculature. **B.** Patient after surgery and three years of testosterone therapy. Notice the skin pigmentation, development of muscle, and change in body shape. **C.** Karyotype of a patient demonstrating the XXY syndrome. (**A** and **B** from Kenneth L. Becker, M.D., *Fertility and Sterility, 23*:568–78, ©1972 The Williams and Wilkins Co., Baltimore, reproduced by permission; **C** by permission of Charles H. Carter, M.D.)

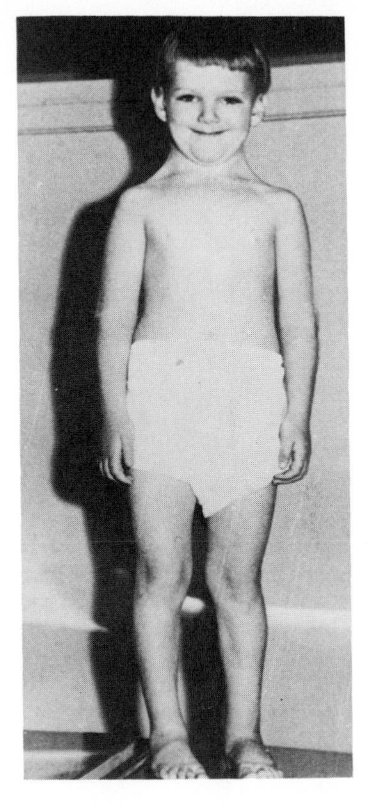

Figure 6–18. Turner's Syndrome (XO). The phenotypically female patient is missing one of the X chromosomes. She demonstrates sexual infantilism, short stature, a webbed neck, a small chin, and a broad or shield chest. In general, these patients have less than normal IQ and an increased incidence of nerve deafness. They are sterile and do not demonstrate feminine development or menarche at puberty. (By permission of Charles H. Carter, M.D.)

sex characters that give a mosaic appearance. For example, they may have one male and one female breast or both bearded and nonbearded facial areas.

In general, the causes of the various chromosome abnormalities are not known. Certain findings, such as those reported by Robinson at the American Pediatric Society meeting in 1973, indicate that sex chromosome abnormalities are found more frequently in children born between the months of May and October. Also, the incidence of Down's children born to mothers under 35 was two to three times higher in May to October births. These data are in agreement with earlier reports that out of 7000 live births examined in Denver, Colorado, the highest frequency of abnormal sex chromosomes was found to occur during the five months of spring and summer. Certain months of the two reports overlap. And in a study reported in Australia, Down's syndrome increased periodically, peaking every five to seven years. Further, authors of the report were able to predict an increase in the incidence of Down's syndrome based on a virus current in the population. It would appear that infectious agents must be considered a potential cause of many of our known chromosomal abnormalities. It has also been established that the occurrence of Down's (trisomy 21) syndrome increases with the age of the mother (Table 6–5). The chance of a woman 16 to 29 years old having a trisomy 21 child is 1 in every 1500 live births. Should this same woman reach the age of 45 before becoming pregnant, her chance of having a trisomy 21 child is 1 in 60. There is a drastic increase in the probability, as the female ages, of this particular abnormality occurring.

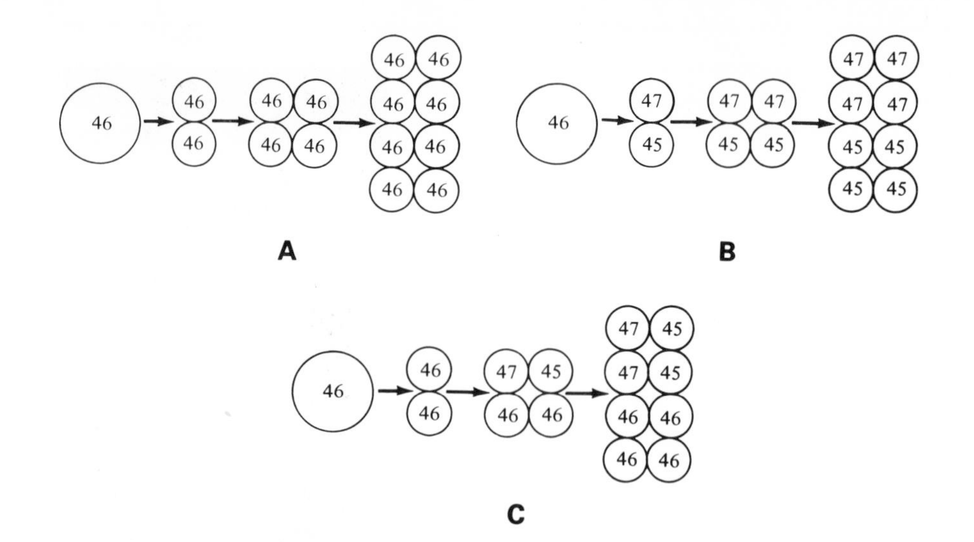

Figure 6–19. Relationship of Timing of Division Error to Type of Mosaicism. **A.** Normal cell division. All resulting cells contain 46 chromosomes. **B.** Error occurring during first cell division after fertilization. Two types of cells are present in the developing individual, half containing 47 chromosomes and half containing 45. **C.** Error occurring later in cell division. Three different cell lines are present. Half of the cells have the normal 46 chromosomes, one fourth have 47, and one fourth have 45. (After A. Redding and K. Hirschhorn, Guide to Human Chromosome Defects (1968). In *Birth Defects: Orig. Art. Ser.*, ed. D. Bergsma, vol. IV (4), 1968. Published by The National Foundation/March of Dimes, White Plains, N.Y.)

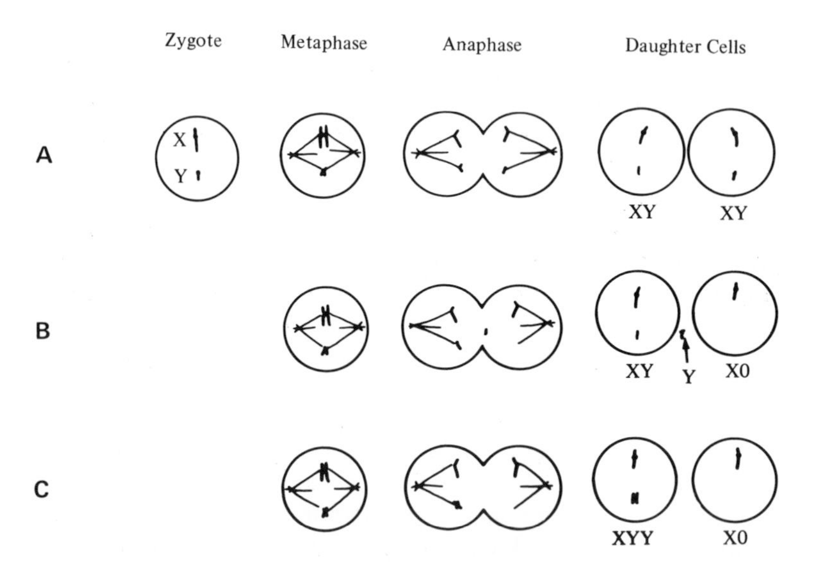

the physical basis: transmission of hereditary material through the chromosomes

Figure 6–20. Formation of Mosaicism (Illustrated for X and Y Chromosomes). **A.** Normal mitosis. **B.** Anaphase lag. **C.** Mitotic nondisjunction in representative male cells. Anaphase lag and mitotic nondisjunction can occur in the autosome as well as in the sex chromosomes. (After A. Redding and K. Hirschhorn, Guide to Human Chromosome Defects (1968). In *Birth Defects: Orig. Art. Ser.*, ed. D. Bergsma, Vol. IV (4), 1968. Published by The National Foundation/March of Dimes, White Plains, N.Y.)

Table 6–5 Maternal Age and Its Relationship to Down's Syndrome Associated with Trisomy 21*

	Down's Syndrome in Newborn	
Maternal Age	*First Risk*	*Recurrence Risk*
16–29	2 in 1500	1 in 500
30–34	1 in 750	1 in 250
35–39	1 in 600	1 in 200
40–44	1 in 300	1 in 100
45 and over	1 in 60	1 in 20

* Adapted from Redding and Hirschhorn (1968).

autosomal-chromosome-associated defects

down's syndrome (47, trisomy 21 G)

Down's syndrome, or mongolism, is an extremely old human disease, having occurred among the Saxons (Brothwell, 1960). The disease was first described by John Langdon-Down in 1866. Some 93 years later, Lejeune, Gautier, and Turpin (1959) published a report that persons with Down's syndrome had 47 chromosomes instead of 46 and that the extra chromosome was a third chromosome 21. We now know that a number of chromosomal abnormalities are associated with Down's syndrome. Ninety-five per cent of all Down's patients have trisomy 21; 4 to 5% have translocations; and about 1% are mosaics. Newer techniques of chromosome staining using fluorescent stains have shown that a partial trisomy for chromosome 21 also causes Down's syndrome (Uchida and Lin, 1973). Down's syndrome is one of the most common malformations known. When the various chromosome abnormalities associated with this disease are added up, Down's syndrome is found in 1 out of every 600 Caucasian births. Thirty per cent of Down's children die within the first year, 50% within five years; very few reached age 20 *before* antibiotics were used.

A child with Down's syndrome ages rapidly, and has characteristic epicanthic folds (slanted eyes), a flat facial profile, and a broad "saddle" nose. The ears are small and round; the tongue is large and flat. The child has a protruding lower lip, a broad short skull, thick short hands with stubby fingers, and thick short feet and body. Down's children usually have light-colored specks in the iris of their eyes, referred to as Brushfield spots, and nearly always have a simian crease, an unusual crease extending across the palm of the hand (Figure 6–13). Heart trouble, hernias, and a marked susceptibility to infections are associated with the syndrome. With the use of antibiotics, some Down's children live past their twentieth year. Possibly as a result of this life extension, we now have documented about two dozen cases where Down's females have become pregnant and delivered live children. Half of these children were Down's and half were normal (Thompson, 1961).

In general, one can count up some 18 to 20 distinguishable characteristics associated with Down's syndrome, but any one child may have only four or five of them. Thus the age-old myth that "all mongoloids look alike" is just that, a myth. The majority have characteristic features, but each looks different from the other. In general, Down's children have a very low IQ. However, their IQ's may range from severe retardation to IQ's of 50 to 70. Also, they are usually retarded in physical development. For example, a six-month-old baby with mongolism often acts like a baby of about two months. He will be slow to sit up, crawl, and walk. Sometimes the child does not talk until he is six years old or older, and some never develop speech. Down's patients are also more apt to be diabetic and to develop leukemia than persons in the general population.

Most persons with Down's have a cheerful, affectionate disposition, so they may do well at home, particularly during their early years. An increasing number are now attending special education classes, and working in sheltered workshops and other sheltered work situations in the community. Those wishing more information concerning mental retardation or the various chromosome abnormalities discussed herein can write to the The National Association for Retarded Citizens, 2709 Avenue E East, Arlington, Texas 76011 (a voluntary organization) or to The National Foundation/March of Dimes, 1275 Mamaroneck Avenue, White Plains, N.Y. 10605.

edwards' syndrome (47, trisomy 18 E)

In 1960, one year after Lejeune and co-workers reported that an extra chromosome is associated with Down's syndrome, Edwards reported the association of an extra E group chromosome with multiple congenital defects. Follow-up studies using radioactive labeling of the chromosome showed that the extra E group chromosome is an additional number 18.

Trisomy 18 occurs once in every 4000 live births and is three times more frequent in females than in males. Thirty percent die during the neonatal period (first 28 days after birth), 50% by the end of the second month, and 90% within the first twelve months.

The child born with trisomy 18 is severely retarded in mental and physical development. The child has a diminished response to sounds, low-set ears, a small mouth, micrognathia (an abnormally small jaw), inguinal or umbilical hernia, a small pelvis, abnormal genitalia, and diseases of the heart (Figure 6–14).

patau's syndrome (47, trisomy 13 D)

Approximately 1 person in every 6000 to 7000 live births has an extra number 13 chromosome. About 50% of these infants die within 28 days; 70% fail to survive six months; and 80 to 90% die within the first twelve months after birth. Those surviving are severely retarded. In 1973 the oldest known patient was about 36 (Howell, 1973).

The characteristic clinical features of trisomy 13 include seizures, jitteriness, apnea (cessation of respiration), moderate microcephaly (small head), wide-spaced eyes with a sloping forehead, eye defects, beaked nose, clefts of the lip with or without cleft palate, absence of the nasal septum, poorly differentiated low-set ears, and scalp defects. Trisomy 13 victims display a simian crease, as in Down's syndrome, have polydactyly (extra fingers), congenital heart disease, malformed genitalia, and unusual persistence of embryonic and/or fetal type hemoglobin (Figure 6–15). They also show kidney disorders (Heller, 1969; Howell, 1973).

The incidence of all three trisomies—21 G, 13 D and 18 E—appears to increase with the age at which pregnancy occurs. The faulty distribution of chromosomes in the egg is hypothesized to be due to the failure of homologous pairs of chromosomes to separate at the first division of meiosis (Figure 6–2).

other trisomies

Other trisomies have been reported. One, the cat's eye syndrome, is a trisomy 21 or 22 that does not lead to Down's syndrome but to persons with congenital urinary and heart problems. Another trisomy 22 has been associated with a behavioral change called schizophrenia. And trisomy 17 patients demonstrate problems that are similar to those of patients with trisomy 18 (Heller, 1969).

monosomy (45 X and 45 −21 or −22)

The loss of a complete chromosome is generally considered lethal, since living examples are extremely rare and until recently occurred only in the sex chromosome complement, the familiar Turner's syndrome X0. We now know of at least *one* autosomal monosomy, a missing number 21 or 22 chromosome (Warren and Rimoin, 1970). Ten patients have been described, and they appear to have many of the symptoms associated with Down's patients as well as considerable genitalic malformation.

translocations (45 and 46)

Translocations are the rearrangements of chromosomes such that a part of one chromosome becomes attached to another. If no DNA is lost during the rearrangement process, the cell is still genetically balanced. Individuals carrying such balanced translocations generally appear normal (Figure 6–21). Problems appear, however, when the translocation carriers produce gametes. Through the random chance process of meiosis, some sex cells will be normal, some will be missing the translocated chromosome and thus will be deficient, and still others will have too many chromosomes (see Figures 6–3 and 6–4). Until recently, the only known unbalanced translocation carriers contained chromosome translocation combinations of the D/G groups (Down's syndrome). With the new techniques for labeling chromosomes with dyes and establishing various banding patterns, we now recognize D/D and G/G unbalanced

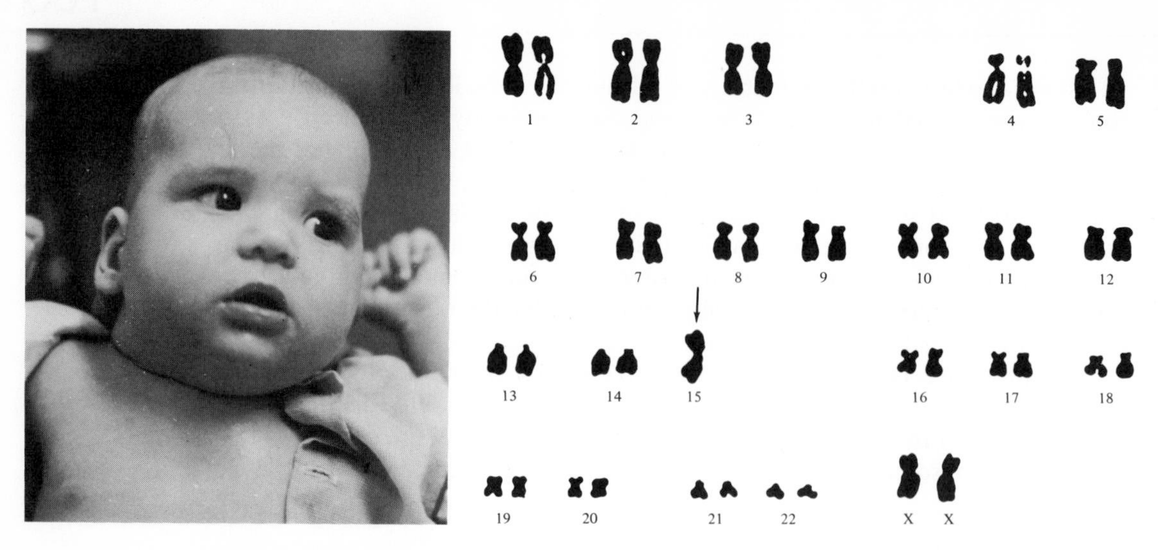

Figure 6–21. Balanced D/D Translocation. In balanced translocations the carrier, as shown by the photograph, does not appear to exhibit any abnormalities. This is in contrast to unbalanced translocation, such as shown in Figure 6–4 where the patient carries an imbalance of chromosome 21 and demonstrates Down's syndrome. (By permission of Charles H. Carter, M.D.)

translocations associated with Down's syndrome. Also, it has been established that unbalanced B/C, B/D, and B/G are associated with cri du chat, a defect with severe mental retardation. A deletion of 40% of the long arm of chromosome 22 was once thought to be associated with leukemia. Recently, however, this "deletion," or missing piece of chromosome 22, was found translocated to chromosome 9. Thus, the former long arm chromosome 22 deletion is a G to C, 22 to 9 translocation. We have also recently learned that a translocation of the Y chromosome to the X, when transmitted to the female, can lead to spontaneous abortions (Khudr et al., 1973).

ring chromosomes

The occurrence of a ring chromosome in live births is very rare. A ring chromosome is formed by the accidental loss of both ends of a chromosome. The remaining segments of the chromosome attached to the centromere fuse together to form a ring. Ring chromosomes have been found in group B, chromosome 5; group D, chromosome 13; group E, chromosome 18; and group G, either 21 or 22. In all cases the surviving child was severely retarded, mentally and physically.

partial deletions

In 1963 Lejeune and his colleagues reported a clinically recognizable syndrome associated with a deletion of a specific region, a segment of the short arm of chromosome 5 (German, 1964). The child having this specific deletion at birth had a cry resembling a suffering cat, and the disorder was accordingly called the cri-du-chat (cry of the cat) syndrome. Although the number of cases in the United States is not known, 1% of those patients having an IQ of less than 20 are cri-du-chat (Howell, 1973). As can be seen in Table 6–2, nearly every chromosome abnormality, expecially the partial deletions, affects mental development. Other clinical features of cri-du-chat are low birth weight, frequent seizures, cleft palate, small head with scalp defects, severe physical retardation, low-set ears, and micrognathia (small jaw). Approximately 10 to 15% of the cri-du-chat cases have a partial deletion of chromosome 5. Deletion of a portion of the short arm of chromosome 4 presents manifestations similar to the deletion of the short arm of chromosome 5. These manifestations are found less often, however, and are associated with the formation of eye cancer (Figure 6–6).

Short- and long-arm deletions of chromosomes 13 and 18 (Figure 6–22) are very similar, and the variable expressions of the different deletions make the cases difficult to distinguish. Deletions of chromosome 13 usually have more severe manifestations than deletions of chromosome 18, but in either case most

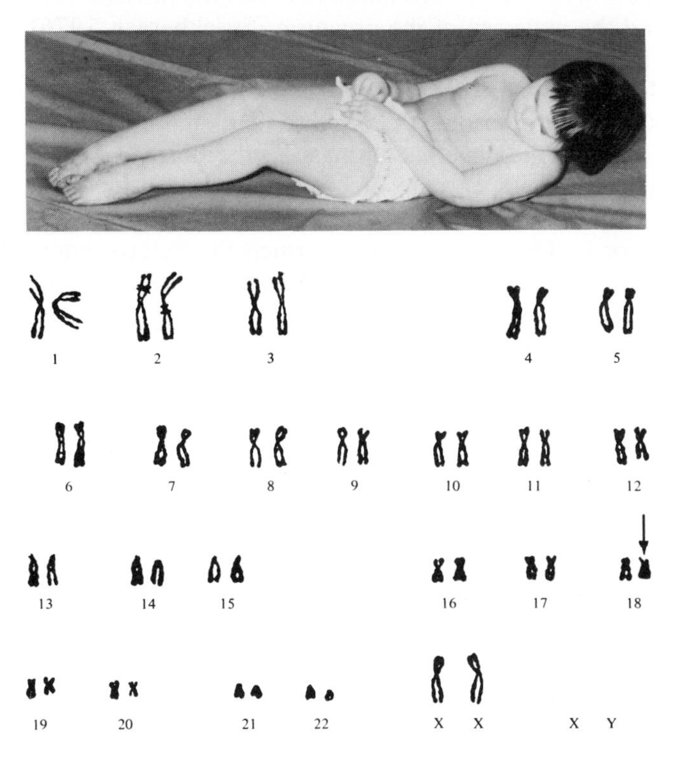

Figure 6–22. Partial Deletion of the Long Arm of Chromosome 18. In contrast to a trisomy of chromosome 18, a deletion of the long arm results in microcephaly. The face in many cases appears to be underdeveloped with a prominent chin. The sexual organs are underdeveloped, and general growth is stunted. Some patients exhibit a congenital heart disease and deafness, and most patients are severely mentally retarded. Most cases are sporadic with either the mother or the father being a balanced translocation carrier. The symptoms for a partial deletion of the long arm of chromosomes 13 are very similar. (By permission of Charles H. Carter, M.D.)

patients have congenital heart disease, microcephaly (small head), conductive hearing loss, and usually underdeveloped genitalia. These deletions are easily identified by karyotyping analysis, however.

sex determination, barr bodies, and sex-chromosome-associated defects

In early history men attributed sex determinations to the phases of the moon, time of fertilization, the direction of the wind, the testis involved, and the physical condition of each partner at the time of conception. Avicenna, an Arabian physician of the Dark Ages, believed that semen desposited in the left side of the uterus gave rise to a girl, in the right side to a boy, and in the middle to a hermaphrodite (Ballantyne, 1902). However, Gregor Mendel suggested, in a letter to Karl Nägeli, that sex might be an inherited trait with a mode of transmission similar to that of the other traits he had studied. Shortly after the rediscovery of Mendel's work in 1900 by Tschermak, De Vries, and Correns, McClung observed that there was a close correlation between the presence of a particular chromosome and the male sex of the organism (Strickberger, 1976). In 1905 Wilson and Stevens independently showed that the "accessory" chromosome found by McClung was present in both male and female *Protenor*. Wilson called the "accessory" chromosome X, because of its unknown function, and showed that in *Protenor* the male had a single X chromosome and the female had two, XX. Only a single X chromosome was observed in the male, but researchers found two kinds of gametes resulting from meiosis—those with the X and those without, or 0. Thus the male was termed the heterogametic sex. Since all of the gametes produced by the female contain an X, she was named the homogametic sex.

Following these findings, scientists discovered a number of different sex chromosome patterns that governed sex determination in various organisms. In some cases the female was found to be X0 and the male XX, as in the members of the order Lepidoptera. In general, however, the most common mechanism of sex determination for diploid organisms is that of the X and Y type, where XX results in a female and XY results in a male of the species. But here also there are unusual circumstances. In *Drosophila melanogaster*, for example, the Y chromosome does not influence maleness or femaleness, since a single X results in a male and XX is female even when a Y is present. In man, however, the Y chromosome is necessary for maleness, and the X is essential for his viability (Painter, 1921). Although the sex chromosomes influence maleness and femaleness, the subsequent phenotypes resulting from a given sex chromosome genotype can vary greatly, depending on the environment of the developing embryo.

The X chromosome was first described by H. Henking in 1891 as a peculiar chromatin element that lagged behind during anaphase chromosome separation and passed undivided to one spermatocyte during the second meiotic division in certain insects. When cytologists discovered that in certain species XX determined one sex and a single X the other, confusion occurred, since the two sexes appeared to be genetically unbalanced. The homogametic sex (XX) would have each gene on the X chromosome represented twice, whereas the heterogametic sex contained only a single dose of these genes. One should expect to find twice the gene product, a protein, from an XX individual as from an X0 or an XY individual. Muller, however, in his studies of the sex-linked apricot eye color of *Drosophila* in 1932, showed that the addition of an X chromosome did not increase the intensity of eye color. Also, the eye color of the single dose (XY) of the male was exactly the same as the eye color of the double dose (XX) of the female. Muller postulated that there must be a mechanism to account for the equalization of gene product in the different sexes. He named this mechanism dosage compensation.

The physical state of the sex chromosome has attracted considerable interest since it was observed that the large heterochromatic regions of the chromosomes are genetically inert. The genetic inertness of heterochromatin has recently been shown to be due to its inability to synthesize messenger RNA (Hsu, 1962; Frenster, Allrey, and Mirsky, 1963). Thus heterochromatin and euchromatin apparently represent alternate states of the chromosome related to various functions during cell differentiation. Heterochromatic regions of a chromosome are dark-staining regions in which the chromatin is tightly coiled and condensed. Such regions are usually visible during interphase when the euchromatin is dispersed to the extent that it is not visible under the light microscope. The heterochromatin is characteristically easily observed in the X chromosome of animals. Following this lead, S. Ohno demonstrated that a heterochromatic X chromosome is visible in interphase nuclei of somatic cells of female mice, but not in the male somatic cells. Most important, he reported that in the human female as well as in the female mouse one of the X chromosomes is heterochromatic along its entire length and that this same sex chromatin mass is never seen in corresponding male cells (Ohno and Hauscha, 1960; Ohno and Makino, 1961).

Mary Lyon (1961) and L. Russell (1961), working independently with mosaic mice whose coat color was mottled, reported that only one active X chromosome is necessary for normal development. Lyon based her conclusions on two particular observations: (1) the X0 mice were perfectly normal, and (2) females heterozygous for the sex-linked coat color genes were mottled. (A sex-linked gene is a gene located on the X chromosome.) The coat of hair on these females had patches of normal hair color and patches of mutant color. Through extensive breeding experiments with heterozygous female mice containing two nonallelic X-linked color mutants, Lyon was able to show that when one of the mutant genes was located on each X, only one or the other acted in each cell. Conversely, when the mutant genes were located on the same X chromosome, some cells showed the genetic expression of both genes, and other cells remained free of mutant expression. Observations such as these resulted in the formulation of the Lyon hypothesis. Briefly it proposes that

1. In each female somatic cell in mammals, one of the two X chromosomes is inactive, the inactivation occurring early in development.
2. It is a matter of chance whether the maternal or paternal X is inactivated.
3. Once an X chromosome is inactivated, that same X is the inactive chromosome in all progeny of that particular cell.

By implication, the X chromosome of the male would remain functional. However, in cases of XXY, XXXY, XXXXY males, the extra X chromosomes should become inactive. If the Lyon hypothesis holds true, the question of how the cell undergoes dosage compensation will have been partially answered.

Actually, interphase sex chromatin was first described by M. Barr and E. Bertram (1949) (although at the time it was not identified as the X chromosome). They later showed that the sex chromatin, or as it was subsequently called, the Barr body, can generally be found associated with the inner side of the nuclear membrane of female somatic cells. Note the lack of a corresponding Barr body in a male cell (Figure 6–23).

A minority of Barr bodies are found to be free in the nucleoplasm (the solution inside the nuclear membrane). This fact aids in their detection for the purposes of determining the sex of the cell. The average size of the Barr body, or X-chromatin mass, is 0.7 by 1.2 micrometers in the nuclei of human cells from the lining of the mouth and from other tissue. Recent audiographic studies leave no doubt that the Barr body is a single X chromosome that is highly condensed during interphase. It has also been shown that the Barr body is late in DNA replication, and it has become apparent from studies on human patients with 2X, 3X, 4X, and 5X conditions that the number of late-replicating Barr bodies or X chromosomes in a cell is always one less than the total number of X chromosomes in the cell.

These data fit in well with what one would expect from the Lyon hypothesis. The Lyon hypothesis also helps clarify why numerical chromosomal aberrations or deletions in man can lead to drastic consequences. For example, autosomal trisomy and complete or partial deletion of a chromosome usually produce a lethal genetic imbalance in man. Those with a condition of trisomy or deletion that do survive (trisomy 13, 18, 21 and monosomic X0 individuals) are severely afflicted. In many individuals aneuploidy of the sex chromosomes does not result in a lethal genetic imbalance. Supposedly there is complete inactivation of all but one of the X chromosomes. However, mental and physical disorders have been associated with extra X chromosomes similar to those found with the viable autosomal trisomies. It is believed that in these cases the extra X chromosome or chromosomes are only partially inactivated, leading to a genetic imbalance (Saxén and Rapola, 1969).

It is often desirable to sample the chromosomal complement of the developing embryo to determine whether it is male or female, since we know of a number of sex-linked genes that, when defective, cause severe physiological conditions, such as hemophilia (inability to clot blood), Duchenne muscular dystrophy (destruction and wasting of muscle tissue leading to death), and the Lesch–Nyhan syndrome (self-mutilation). Sex-linked genes are transmitted in the X chromosome. Few, if any, defective genes are known to be located on the Y chromosome. If a female has one defective gene for the above traits, her chance of passing the defects to her son is one in two, a very high risk. That is, one of her X's carries the normal gene and one carries the defective gene. A son, XY by virtue of fertilization, receives either the normal or the defective X by chance. Also, we know that nearly two thirds of the known chromosomal anomalies in newborn children are sex chromosome anomalies, and that these anomalies are mostly numerical aberrations. That is, there are too many or too few sex chromosomes. Thus our ability to screen families or the population at

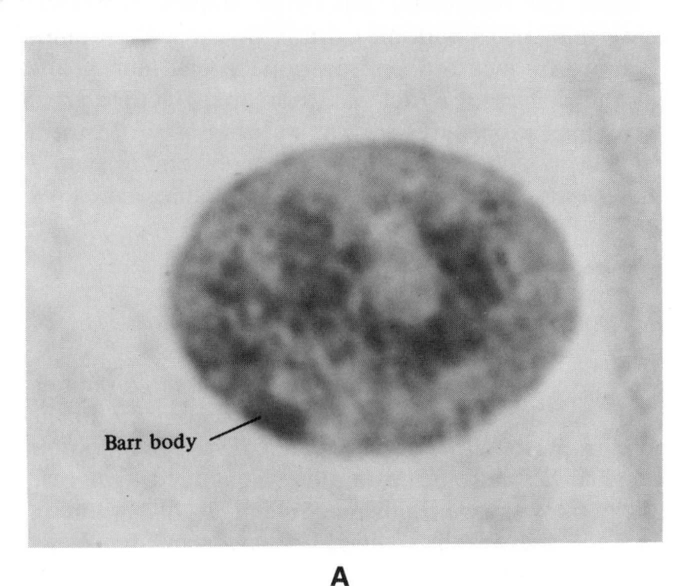

A

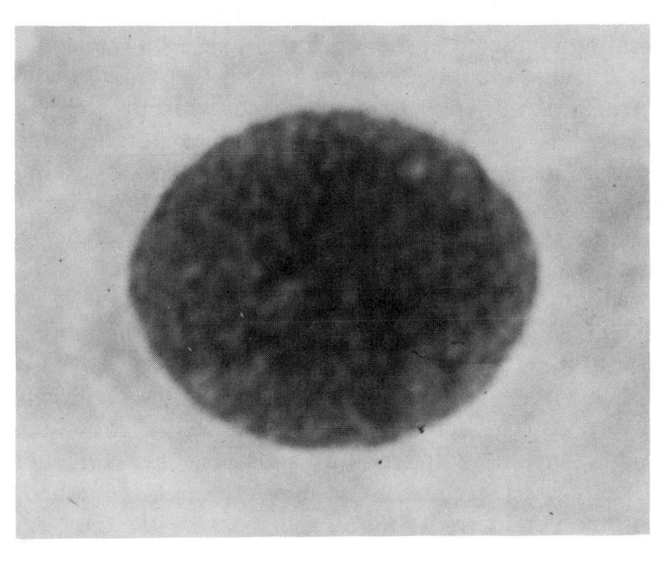

B

Figure 6–23. Squamous Epithelium Cells. **A.** Female cell containing a Barr body. **B.** Male cell with no Barr body. (Courtesy Carolina Biological Supply Company.)

large for early indications of numerical sex chromosome aberrations and screen the fetus for sex identification will allow us to offer genetic counseling when it will be the most effective. In the United States, as can be seen in Table 6–3, there are many thousands of aneuploid males and females whose phenotypes resulted from their sex chromosome constitution. There are about 750,000 combined cases of triple X, Turner's X0, Klinefelter's XXY, and double Y males (XYY).

Very little is known about the behavioral aspects of victims of genetic aneuploidy. We are learning that males with an extra X or extra Y chromosome may be represented more frequently in prison populations than in the general population. But is the degree of social risk from these people great enough to set up broad screening procedures so that we can cope with them from the time of conception? Although we are not now in a position to answer this question, we *can* determine which embryos are carrrying extra sex chromosomes.

the olympics and the barr body test

The Barr body analysis for sex identification is used by Olympic committee officials to judge whether or not those persons registering to compete in the women's events are in fact chromosomally normal women. Those females, for example, with an extra X (triple X syndrome) would be disqualified. One case in point concerns Ewa Klobukowska, a Polish coholder of the women's 100-meter dash record. In 1968 she was disqualified for having one too many chromosomes. The chromosome was not identified by the news media, but we can assume it was a sex chromosome, since in 1968 the Barr body test identified only the late-replicating X. The examination of her chromosomes and subsequent disqualification implied that she was more male than female, or at least not a normal female. What is strange to the author is that females with extra X chromosomes are generally less fit than those with the normal complement. If, in fact, Ewa was an XXY male (Klinefelter's syndrome), she would still be less than fit to compete in track competition, due to poor physical and emotional development. (See discussion of the various sex chromosome abnormalities below.) In 1968 the committee took away Ewa's title and gold medal, since she failed the sex identification test. According to Volpe (1971), Ewa's boyfriend was amazed anyone would think she was not a woman.

Since the unfortunate incident involving the Polish track star the test has become more sophisticated. Using Atebrin-stained human hair roots under a fluorescent microscope, scientists can easily detect both the Barr body (X chromosome) and the Y body (Y chromosome) (François, Matton-Van Leuren, and Costa, 1971). According to the investigators, the hair root technique offers many advantages: It is more accurate than the older method; it identifies the Y chromosome in addition to the X and thus can identify both sexes; it can be easily done on babies, mental patients, and research animals, since the hair is much easier to obtain than a scrape of cells from the lining of the cheek.

The Barr body test has also been used to determine the number of females to males (sex ratio) from conception through birth and death. Calculations based on Barr body tests show that more male than female conceptions occur, but that more male fetuses are involved in spontaneous abortions. Still 106 males are born to every 100 females. By age 20 the ratio is reduced to 100 males to 100 females, and at age 85 the ratio is 62 males to 100 females (Jacobs et al., 1959). The higher mortality rate for males following birth has not been explained satisfactorily, but it is suggested that the two X chromosomes of the female offer a considerable survival advantage over the single X found in the male. It may well be that many undetected genes associated with our survival under conditions of stress are on the X chromosome and that, in some manner yet to be understood, the presence of the double X in the female provides her with greater endurance.

sex-chromosome-associated defects

Turner's Syndrome (45, X0). Turner's syndrome, also known as female gonadal dysgenesis, is manifested by gonad impairment resulting in the loss of reproductive power. The syndrome was described by H. Turner (1938). In 1954 scientists recognized that patients with gonadal dysgenesis did not show a Barr body in their cells. In 1959, C. E. Ford and his colleagues' analysis of the chromosomes for Turner's patients revealed that the patients were missing one X chromosome, or were X0 (Ford, et al., 1959). The syndrome has also been associated with mosaicism, where the patient has cells (presumably XX/X0) of 45 and 46 chromosomes (Figures 6–19 and 6–20). As shown in Table 6–6, the incidence of the X0 anomaly is approximately 1 in 2500 live female births. In terms of the 1973 United States population with the birth of 3,141,000 children, we would expect that 1256 children were born with a missing X chromosome. Evidence gathered from spontaneously aborted fetuses indicates that over 95% of the X0 fertilizations are nonviable. If they were viable, the incidence in the population would be much higher. During the past few years, a series of X0 aborted embryos and fetuses ranging in age from five weeks to seven months has been collected. Analyses of these specimens indicate a particular pattern of development. From conception to ten weeks development appears to be normal. After ten weeks malformations are clearly recognizable. A departure from normal changes in the genitals occurs along with generalized lymphodema (swelling).

The clinical symptoms of a Turner's patient (Figure 6–18) are as follows: The adult female is relatively small in stature, usually sterile, and has wide-spaced and underdeveloped breasts. Her ovaries fail to develop, and consequently she fails to menstruate. She has sparse pubic hair and abnormal hormone secretions. These patients have low-set ears, and the more severe cases have webbed necks and low IQ's. The Turner's female should not be considered a case of sex reversal; she is an undeveloped female. Most cases of Turner's patients are believed to have resulted from nondisjunction in the mother. An egg was formed without an X chromosome and was fertilized by an X-bearing sperm, forming an X0 zygote (Figures 6–18 and 6–20).

The Triple X (47 XXX), Tetra X (48 XXXX), and Penta X (49 XXXXX) Females. Reports on females with extra X chromosomes in their makeup have appeared in scientific literature since Jacobs (1959) described her findings on the triple X (Haunden et al., 1966). Carr and colleagues (Carr, Barr, and Plunkett, 1961) reported finding two mentally retarded females, both with four X chromosomes. Since Carr's work only a dozen additional cases have been recognized. This is an indication that live births with the tetra X condition are very rare. Then Kesaree and Woolley (1963) described findings on penta X patients. To date, less than a dozen penta X conditions have been reported. In view of the very rare occurrence of tetra and penta X live births, we can assume that these conditions are usually lost in early spontaneous abortions. The triple X condition, however, is fairly common. It is estimated that 1 in every 1000 live births is triple X.

The description of triple (Figure 6–16), tetra, and penta X females is rather vague. Most of the tetra X and penta X individuals were found in the screening of mentally retarded individuals, but there may be tetra and penta X people in the population without a definite appearance to call attention to their abnormal chromosomal condition. Even in those patients known to be triple, tetra, and penta X, the behavioral and physical expressions are quite variable. In fact, without the aid of karyotype studies, identification of the individual additional X cases is almost impossible. In short, no one specific phenotype has been consistently recorded except, perhaps, for a less than normal IQ. Even this exception is a biased measurement, since most cases have been diagnosed in institutional populations. In keeping with the inconsistent clinical features, children born to triple X females are normal males and females. This fact is very surprising, since one would expect the mother to produce three kinds of eggs, X, XXX, and 0, so that fertilization by X- and Y-bearing sperm could give rise to (a) normal females, (b) normal males, (c) Y0 (lethal), (d) X0 Turner's female, (e) tetra X females, or (f) XXXY males. Yet, of all these possibilities, only normal offspring have been born to triple X females.

The triple X patient is best described as having menstrual irregularities and premature menopause. Some are sterile, and a small number of triple X patients have demonstrated less than normal IQ. This description is certainly not a picture that differs greatly from that of a group of normal people.

J. Heller (1969) stated that an extra X predisposes a person to mental deficiency, that a person having extra X chromosomes has twice as much risk of being admitted to a hospital for some form of mental illness as those with a normal complement of chromosomes, and that each additional X beyond the normal two involves a greater degree of mental retardation. Such risk factors should lead us to test mentally abnormal females for additional X chromosomes. Testing may be a way of obtaining better counseling data. Care must be taken, however, to establish reasonable control groups of people, so that those carrying extra X chromosomes are not stigmatized for this accident of nature. Incidentally, the triple X female and at times the tetra and penta X females have been referred to as "super females." In reality, there is nothing biologically "super" about being abnormal, and it is questionable that there is an overall social advantage to being abnormal.

Klinefelter's Syndrome (47, XXY). Klinefelter and his colleagues (Klinefelter et al., 1942) described nine men with gynecomastia (enlarged breasts), small testes, an inability to produce sperm, and an increased secretion of the female follicle-stimulating hormone (secreted by a female for the maturation of the ova). Jacobs and Strong (1959) reported that Klinefelter's male patients contained an extra X chromosome (XXY). We know that Klinefelter's patients are always male by virtue of having the Y chromosome, but they have variable male genital development. The number of such persons in society is relatively high. The incidence is 1 in every 400 live male births. The frequency of Klinefelter's births has, like other genetic defects, been associated with the age of the mother. The older the mother, the greater the risk (Heller, 1969). Over 12,000 such persons were born in the United States in 1973 (Table 6–3), and at that time there were 0.5 million or more in the total population. Subsequent to the finding of the XXY male, rarer cases of XXXY and XXXXY (Figure 6–24) were found, but their frequency is not known. The XXXY, XXXXY, XXYY, and XXXYY and numerous sex chromosome mosaics (XX/XY, XXX/XXY, XY/XY, XXXX/XXXXY, etc.) resemble the Klinefelter males to some extent, and these patients are considered modified Klinefelter's. The true Klinefelter's patient is sex chromatin positive with one Barr body, like the normal female, but is morphologically (physically) male.

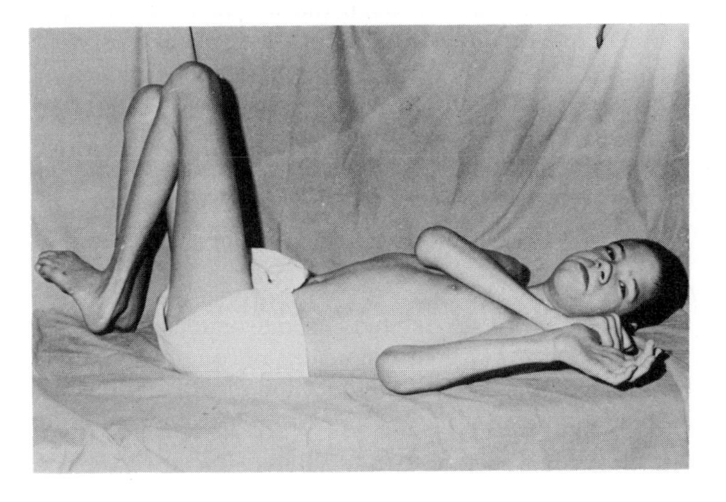

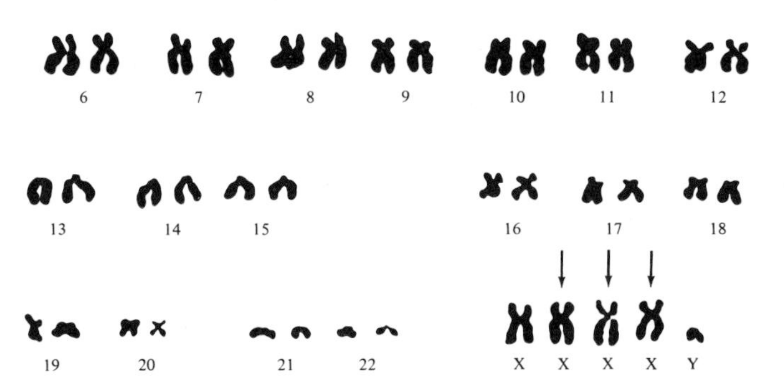

Figure 6–24. A Modified Klinefelter's Male. The presence of more than two X chromosomes accompanying a Y in a cell is regarded as a modified Klinefelter's case. In general, with each additional X, the frequency of the syndrome decreases and the chance of mental retardation increases. Modified Klinefelter patients, like the Klinefelter XXY, are tall, but generally more clumsy and definitely more retarded than the XXY. (By permission of Charles H. Carter, M.D.)

Approximately 75% of the true Klinefelter males (XXY) have an enlargement of one or both breasts. Slightly more than half (51%) have less than normal IQ, and most are sterile. Klinefelter's males usually have normal-sized penises, but small testes and prostate glands, and scarce body, pubic, and facial hair. They are slightly heavy and average about five feet, ten inches in height. A statistically significant higher frequency of mentally defective Klinefelter's males has been found in prison populations and mental hospitals than in the general population. The source of error leading to the XXY condition is believed to be nondisjunction (Figure 6–20). An XX egg is fertilized by a Y sperm to form an XXY zygote. There is a definite correlation of increasing mental deficiency with increasing numbers of X chromosomes over the XXY Klinefelter's patient.

Of the high-incidence chromosomal disorders, the Klinefelter's patient is one of the most amenable to treatment. Through the use of hormones and surgery, the typical feminine-appearing male can be converted successfully into a masculine male in most cases. In one particular case, the patient underwent surgery for removal of his breast (mastectomy). This operation was followed by three years of male hormone (testosterone) treatment. The change in his physical stature was very pronounced toward maleness (see Figure 6–17). Concomitant mental or psychological changes occur and must be dealt with professionally (Becker, 1972).

The Double Y Syndrome (47, XYY). The XYY syndrome (Figure 6–25) was kept until last in our discussion of sex-related chromosome abnormalities for two reasons. First, it occurs with very high frequency, once in every 200 to 1000 live male births (Hook, 1973a). Second, the question of whether one has a tendency toward becoming a criminal if he has an extra Y chromosome is of great social significance. The birth of an XYY individual is most likely caused

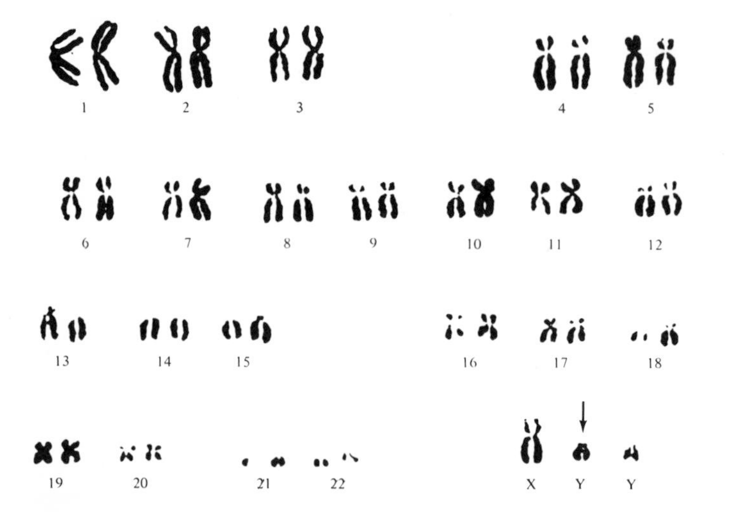

Figure 6–25. An XYY Karyotype. The XYY male appears normal; a photograph would, at most, show a tall male, about 6 feet 4 inches, who may or may not have facial acne. Research data and social problems of the XYY male are discussed in the text. (By permission of Charles H. Carter, M.D.)

by nondisjunction at the second division of meiosis, when two Y chromosomes are available for further separation into the single Y-bearing sperm. A double Y-bearing sperm fertilizing a normal egg results in an XYY zygote.

The association of criminal tendencies with males carrying an extra Y chromosome is a very controversial issue. Depending on the research papers one reads, he or she can become biased toward one of two opposite extremes. Some believe there is a statistically significant association between an extra Y chromosome and criminal behavior, and others believe there is no significance to a male having an extra Y chromosome.

Regardless of the controversy, some who have committed crimes in England and Australia, including rape and murder, have successfully used the XYY chromosome abnormality as a defense for their actions and received reduced sentences.

In reviewing the large number of conflicting reports on whether XYY males have a tendency toward criminal behavior, we can say, but with some hesitation, that the XYY male must be clinically described in behavioral as well as physical terms. There is an apparent consensus that about half of the 100 or more XYY cases reported had less than average IQ. All were described as tall (over six feet) or very tall (six feet, five inches to seven feet). Several authors say that the XYY male has unusual sexual tastes, including homosexuality (Heller, 1969). Remember, however, that most of the XYY individuals were found in prisons, where the incidence of homosexuality among all residents is high. Several reports refer to XYY males as having facial acne and increased male hormone production.

In spite of later debate concerning the double Y karyotype, the first XYY individual discovered was, ironically, a normal, happy, family man, whose wife gave birth to a Down's syndrome child. When a karyotype analysis of the parents was prepared, the husband was found to be XYY (Hook and Kim, 1971). The normal appearance of this male is not unusual. According to the research reports, many of the persons identified as XYY displayed no aggressive behavior. Further, persons with no Y chromosomes (women) have shown that they can be quite aggressive. The author does not believe there is conclusive evidence that the Y chromosome predisposes one to aggression or hostility, but rather that aggression is the result of an interplay of one's total genetic potential and the environment to which he has been exposed.

One method of assessing the genetic contribution of an extra Y to aggressive behavior would be to study XYY identical twins. However, only one such pair has been studied (Ranier, Abdullah, and Jarvik, 1972). A study by Brown (1968) covers the role of the environment in the expression of criminal tendencies by XYY males. His study showed that the XYY male tended to come from parents with high socioeconomic backgrounds.

Because of a tendency for XYY males to be taller than average, an attempt was made to correlate their unusual height with their frequency in institutions. It was hypothesized that because of their large size XYY males might develop personality patterns, such as impulsive behavior, that could be interpreted as antisocial behavior. If this hypothesis is true, it would seem logical that the mean height of males in all penal institutions should be higher than in the total population. Studies in this area showed that during late childhood large height per se is not responsible for the frequency of delinquency among XYY individuals and is unlikely to be a strong contributory cause (Hook and Kim, 1971). Further, it has been pointed out that children afflicted with giantism do not show aggressive behavior (Stephenson, Mellinger, and Manson, 1968).

There are no other phenotypic characteristics to date that might account for the frequency of XYY's in mental and penal settings, except possibly slight neurological defects, mostly motor dysfunction. And the incidence and severity of these defects in the XYY's studied are so small that they should not be considered significant. Electroencephalograms (EEG's) of XYY, used to search for brain dysfunctions, have only served to enforce the normalcy of the XYY individual (Hook, 1973b). Other studies reveal no marked increase in the frequency of clinical epilepsy in XYY individuals, and there is no documentation concerning cryptic seizures.

Endocrine abnormalities known to have an effect on behavior would be consistent with some types of brain dysfunction. Yet studies done to test the testosterone excretion rate in XYY's as compared to normal males found that the extra Y chromosome had no effect upon the excretion of the male sex hormone (Rudd, Galal, and Casey, 1968). The same results were reached concerning pituitary gonadotrophic, growth hormones and the female luteinizing and follicle-stimulating hormones. Since these tests involved either adults or adolescents, it is possible that hormonal abnormalities occurred earlier in development.

If the neural hypothesis that there is a low-level brain dysfunction is correct, then the presumed physiological factors might only be operative in some fraction of XYY's, as determined by the individual and his environment. The only support to date for the neural hypothesis is that there is no evidence against it. At present the other hypotheses and the research do not explain the apparent association of the XYY with deviance. The association itself may even prove faulty.

In spite of the lack of evidence, conflicting reports, and our inability to measure the association of the XYY and social deviance, Gardner and Neu (1972) state unequivocally that the XYY male has a definite tendency to sociopathic behavior. However, Borgankar and Shah (1974), after a comprehensive review of the XYY literature conclude, that "... the frequency of antisocial behavior of the XYY male is probably not very different from non-XYY persons of similar background and social class."

A recent study of the XYY karyotype at Harvard Medical School has led to a debate at the medical school as to whether or not the research should be done. Implicit in the XYY karyotype research are the assumptions that science can and should attempt to distinguish between the behavior of people on the basis of genetics. The XYY study program tends to focus the blame for certain antisocial behavior on the individual's genes rather than on the social-environmental conditions.

In the study the researchers are trying to determine if *any* psychopathology is associated with the extra Y chromosome. Parents whose children were found to be XYY were informed of the condition (Beckwith et al., 1975), although Valentine et al. (1971) had reported previously that giving parents the information that their child is XYY induces anxieties about the child's development and leads to behavior that might not have developed if the parents had not been told.

Screening for the XYY karyotype continues even though the investigator cannot know whether behavioral problems that arise in XYY children are due to the additional Y chromosome.

In summary, XYY individuals apparently are found more often in mental–penal settings than in the general population. Tallness is a common characteristic, and facial acne occurs less frequently. Their behavior tends to be impulsive rather than aggressive, and according to Hook, this impulsive behavior increases the risk factor of their ending up in jail (Hook, 1973).

If it turns out that the XYY chromosome abnormality does in fact modify socially acceptable patterns of behavior, the defect carries with it serious social and legal implications. Should we convict a man of murder if his actions were directed biologically? Should the mother carrying a child known to have the XYY chromosome condition be forced to abort that child? Should society allow individuals with the XYY chromosome complement to reproduce? The answers to these questions affect both the individual involved and the rest of society. To date, it appears we have no acceptable answers.

the size of the Y chromosome and behavior

We are only beginning to gather information on the size of the Y chromosome and what, if any, correlation can be made between its size and male behavior. The Y chromosome is a relatively small chromosome. However, through the use of new fluorescent staining techniques, it has been shown that not all Y chromosomes are of the same size. They range in size from a large Y with four discernible sections in the long arm to a short Y chromosome without a long arm. Persons with a short Y are male, but persons with a Y short-arm deletion develop as females (Krmpotic et al., 1972). On the basis of this evidence, it has been concluded that the genes for maleness are located in the short Y material.

There have been reports that there is a correlation between a large Y and abnormal behavior, but Urdall and Brøgger (1974) report that there is *no* such correlation.

references

Ballantyne, J. W. 1902. *Antenatal Pathology and Hygiene*, vol. 1. W. Green & Sons, Edinburgh.

Barr, M. L., and E. G. Bertram, 1949. A Morphological Distinction Between Neurons of the Male and Female, and the Behavior of the Nucleolar Satellite During Accelerated Nucleoprotein Synthesis. *Nature*, *163*: 676–77.

Becker, K. L. 1972. Clinical and Therapeutic Experiences with Klinefelter's Syndrome. *Fertility and Sterility*, *23*: 568–78.

Beckwith, Jon, et al. 1975. Harvard XYY Study. *Science*, *187*: 298.

Bio Medical News. 1973. Staining Shows Missing Portion of Chromosome. *4* (Aug.): 14.

Borgankar, D. S., and S. A. Shah. 1974. The XYY Chromosome Male-or Syndrome? *Progress in Medical Genetics*, *10*: 135–222.

Bridges, C. B. 1916. Nondisjunction as Proof of the Chromosome Theory of Heredity. *Genetics, 1*: 1–52, 107–63.

Brothwell, D. R. 1960. A Possible Case of Mongolism in a Saxon Population. *Annals of Human Genetics, 24*: 141–50.

Brown, C. M. 1968. Males with XYY Sex Chromosome Complement. *Journal of Medical Genetics, 5*: 341–59.

Carr, D. H., M. L. Barr, and E. R. Plunkett. 1961. An XXXX Sex Chromosome Complex in Two Mentally Defective Females. *Canadian Medical Association Journal, 84*: 131–37.

Castleman, K. R., and J. H. Melnyk. 1974. Automatic Karyotyping with Pictorial Output. *Medical Research Engineering* (ms. submitted).

Chu, E. H. Y. 1960. The Chromosome Complements of Human Somatic Cells. *American Journal of Human Genetics, 12*: 97–103.

DeRobertis, E. D. P., F. A. Saez, and E. M. F. DeRobertis, Jr. 1975. *Cell Biology*. Saunders, Philadephia.

Dow, E. 1973. Fetography. *American Journal of Obstetrics and Gynecology, 115*(5): 718–21.

Ford, C. E., et al. 1959. A Sex-Chromosome Anomaly in a Case of Gonadal Dysgenesis (Turner's Syndrome). *The Lancet, 1*: 711–13.

François, J., M. Matton-Van Leuren, and J. Costa. 1971. Male and Female Sex Determination in Hair Roots. *Clinical Genetics, 2*: 73–77.

Frenster, J. H., V. G. Allrey, and A. E. Mirsky. 1963. Repressed and Active Chromatin Isolated from Interphase Lymphocytes. *Proceedings of the National Academy of Sciences USA, 50*: 1026–32.

Fuchs, Fritz. 1971. Amniocentesis and Abortion: Methods and Risks; Intrauterine Diagnosis. *Birth Defects: Original Article Series, 7*(5): 18–19.

Gardner, L. I., and K. L. Neu 1972. Evidence Linking an Extra Y Chromosome to Sociopathic Behavior. *Archives of General Psychiatry, 26*: 220–22.

Gates, W. H. 1927. A Case of Nondisjunction in the Mouse. *Genetics, 12*: 295–306.

German, J. L. 1964. The Pattern of DNA Synthesis in the Chromosomes of Human Blood Cells. *Journal of Cell Biology, 20*: 37–55.

German, James. 1970. Studying Human Chromosomes Today. *American Scientist, 58*: 182–201.

Hahnemann, Niels. 1972. Possibility of Culturing Foetal Cells at Early Stages of Pregnancy. *Clinical Genetics, 3*: 286–93.

Haunden, G. D., et al. 1966. Numerical Abnormalities of the X Chromosome. *The Lancet, 1*: 398–400.

Heller, J. H. 1969. Human Chromosome Abnormalities as Related to Physical and Mental Dysfunction. *Journal of Human Heredity, 60*: 239–48.

Hook, E. B. 1972. Some Comments on the Significance of Sex Chromosome Abnormalities in Human Males. In Ian H. Porter et al. (eds.), *Heredity and Society*. Academic Press, New York.

Hook, E. B. 1973. Behavioral Implications of the Human XYY Genotype. *Science, 179*: 139–50.

Hook, E. B., and D. S. Kim. 1971. Height and Antisocial Behavior in XY and XYY Boys. *Science, 172*: 139–50.

Howell, R. R. 1973. Prenatal Diagnosis in the Prevention of Handicapping Disorders. *The Pediatric Clinics of North America, 20*: 141–49.

Hsu, T. C. 1962. Differential Rate in RNA Synthesis Between Euchromatin and Heterochromatin. *Experimental Cell Research, 27*: 332–34.

Human Chromosomes Study Group. 1960. *The Lancet, 1*: 1063.

Jacobs, P. A., and J. A. Strong. 1959. A Case of Human Intersexuality Having a Possible XXY Sex-Determining Mechanism. *Nature, 182*: 302–303.

Jacobs, P. A., et al. 1959. Evidence for the Existence of the Human "Super Female." *The Lancet, 2*:423–25.

Kemp, Tage. 1951. *Genetics and Disease.* Munksgaard, Copenhagen.

Kesaree, N., and P. V. Woolley, Jr. 1963. A Phenotypic Female with 49 Chromosomes, Presumably XXXXX. A Case Report. *Journal of Pediatrics, 63*:1009–1103.

Khudr, Gabriel, et al. 1973. Y to X Translocation in a Woman with Reproductive Failure. *Journal of the American Medical Association, 226*:544–49.

Klinefelter, H. F., Jr., C. E. Reifenstein, Jr., and F. Albright. 1942. Syndrome Characterized by Gynecomastia, Spermatogenesis Without A-Leydigism, and Increased Excretion of Follicle-Stimulating Hormone. *Journal of Clinical Endocrinology and Metabolism, 2*:615–27.

Krmpotic, E., et al. 1972. Localization of Male Determining Factor on Short Arm of Y Chromosome. *Clinical Genetics, 3*:381–87.

Kutay, Taysi, et al. 1973. Duplication/Deficiency Product of a Pericentric Inversion in Man: A Case of D_1 Trisomy Syndrome. *Pediatrics, 82*:263–68.

Laboratory Management. 1972. How to Set Up a Genetics Laboratory. *10*(Oct.):22–24.

Laboratory Management. 1973. Amniotic Fluid: A Viable Route for Fetal Nutrients. *12*(June):33.

Lejeune, J., J. M. Gautier, and R. Turpin. 1959. Etude des chromosomes somatiques de neuf enfants mongoliens. *Comptes Rendus de l'Academie des Sciences, 248*:1721–22.

Liley, A. W. 1960. Technique and Complications of Amniocentesis. *New Zealand Medical Journal, 59*:581–86.

Lynch, H. T. 1974. International Directory. Birth Defects Genetics Service. The National Foundation/March of Dimes. New York.

Lyon, M. F. 1961. Gene Action in the X-Chromosomes of the Mouse. *Nature, 190*:372–73.

Mermut, S., et al. 1973. The Effect of Ultrasound on Human Chromosomes *in vitro. Obstetrics and Gynecology, 41*(1):4–6.

Miller, O. J. 1971. Discussion of Symposium Papers. Intrauterine Diagnosis. *Birth Defects: Original Article Series, 7*(5):33–34.

Milunsky, Aubrey, et al. 1970. Prenatal Genetic Diagnosis. *New England Journal of Medicine, 283*:1370–81.

Nadler, Henry L. 1971. Indications for Amniocentesis in the Early Prenatal Detection of Genetic Disorders: Intrauterine Diagnosis. *Birth Defects: Original Article Series, 7*(5):5–9.

News and Features, 1974. Computer System Speeds Karyotype Preparation. April 10, pp. 9–10.

Ohno, S., and T. S. Hauscha. 1960. Allocycly of the X-Chromosomes in Tumors and Normal Tissues. *Cancer Research, 20*:541–45.

Ohno, S., and S. Makino. 1961. The Single X Nature of the Sex Chromatin in Man. *The Lancet, 1*:78–79.

Painter, T. S. 1921. The Y-Chromosome in Mammals. *Science, 53*:303–304.

Paris Conference (1971): Standardization in Human Cytogenetics. *Birth Defects: Original Article Series, 8,* 1972. The National Foundation, New York.

Prescott, G. H., R. E. Magenis, and N. R. Buist. 1972. Amniocentesis for Antenatal Diagnosis of Genetic Disorders. *Postgraduate Medicine, 51*:212–17.

Ranier, J., S. Abdullah, and L. F. Jarvik. 1972. XYY Karyotype in a Pair of Monozygotic Twins: A 17-Year Life History Study. *British Journal of Psychiatry, 120*:543–48.

Redding, Audrey, and Kurt Hirschhorn. 1968. Guide to Human Chromosome Defects. *Birth Defects: Original Article Series, 4*:1–16.

Rudd, B. T., O. M. Galal, and M. D. Casey. 1968. Testosterone Excretion Rates in Normal Males and Males with an XYY Complement. *Journal of Medical Genetics, 5*:286–88.

Russell, L. B. 1961. Genetics of Mammalian Sex Chromosomes. *Science, 133*:1795–1803.

Saxén, Lauri, and Juhani Rapola. 1969. *Congenital Defects.* Holt, Rinehart and Winston, New York.

Schindler, A. M., and K. Mikamo. 1970. Triploidy in Man: Report of a Case and a Discussion on Etiology. *Cytogenetics, 9*:116–30.

Stephenson, J. N., R. C. Mellinger, and G. Manson. 1968. Cerebral Gigantism. *Pediatrics, 41:130–38.*

Stern, Curt. 1973. *Principles of Human Genetics,* 3rd ed. Freeman, San Francisco.

Strickberger, Monroe W. 1976. *Genetics,* 2nd ed. Macmillan, New York.

Turner, H. H. 1938. A Syndrome of Infantilism, Congenital Webbed Neck and Cubitus Nalgus. *Endocrinology, 23*:566–76.

Uchida, Irene A., and C. C. Lin. 1973. Identification of Partial 12 Trisomy by Quinacrine Fluorescence. *Journal of Pediatrics, 82*:269–72.

Urdall, Trygre, and Anton Brøgger. 1974. Criminality and the Length of the Y Chromosome. *The Lancet, 1*:626–27.

Valenti, C. 1972. Endoamnioscopy and Fetal Biopsy: A New Technique. *American Journal of Obstetrics and Gynecology, 114*:561–64.

Valentine, G. H., et al. 1971. Genetic Disorders: Prevention, Treatment, and Rehabilitation. *World Health Organization Technical Report Series,* No. 497, pp. 1–46.

Vital Statistics. 1974. Publication No. 22, 13:1. Department of Health, Education and Welfare, Washington, D.C.

Volpe, Peter E. 1971. *Human Heredity and Birth Defects.* Pegasus/Bobbs-Merrill, Indianapolis, ch. 7.

Whitehouse, H. L. K. 1969. *Towards an Understanding of the Mechanism of Heredity,* 2nd ed. St. Martin's Press, New York.

7

As stated in Chapter 4, sexual reproduction is essential to the successful continuation of the human species as we know it. Over millions of years we have evolved with two distinct forms—male and female. Each contributes about the same amount of hereditary material to the formation of the next generation. The primary and secondary sexual characteristics, along with the reproductive organs, are reserved for a specific sex, and these are transmitted in such a manner that the sexes have been maintained.

Socially and biologically, there are few words having the overpowering connotations of the word *sex*. The dictionary defines sex as being synonymous with gender, male or female. This vague definition leaves many questions unanswered in relation to Turner's and Klinefelter's patients (Chapter 6). Psychologists state that sex can be considered in at least five different ways: appearance of the external genitals, possession of the correct number of sex chromosomes (XX or XY), production of certain hormones in certain amounts, assignment of sex at birth (what the doctor or your parents thought you were), and gender (what you think of yourself). With these interpretations of sex, the matter of defining sex becomes a very complex issue involving genetics, psychology, and physiology.

In humans, providing all developmental processes work correctly, the Y chromosome confers maleness on the developing embryo, and the lack of Y results in a female. Thus, from the very beginning, each cell receives one type

Sex has become one of the most discussed subjects of modern times. The Victorians pretended it did not exist; the moderns pretend that nothing else exists.

Fulton J. Sheen

the sex chromosomes / intersexuality and homosexuality

of nonreversible sex labeling through the sex chromosomes, and as far as the sex chromosomes are concerned there is no possibility of sexual reversal.

This chromosomal permanence was demonstrated with salamanders. In salamanders, unlike mammals, the males have the identical XX chromosomes and the females the XY chromosomes. Humphrey and Gallien, working separately, mated normal male (XX) salamanders to hormonally sex-reversed (XX) salamanders. The hormonally sex-reversed males still had their male chromosomes. As expected, all the progeny from the normal male mated with a *feminized male* were male. Thus in these amphibians there is the possibility of disassociating the genetic sex of the cells from the anatomical and functional sex of the individual.

The child with Turner's syndrome, who has only a single X chromosome, is anatomically female. Children with a single Y and more than one X chromosome are all males although abnormal; Klinefelter's syndrome XXY and those with conditions of XXXY, XXXXY, and XXXXXY, are all male. We know that two X chromosomes are necessary for the development of a normal female ovary, since Turner's syndrome (X0) patients are sterile owing to undeveloped ovaries. Studies by Hammerton indicate that the second X chromosome in the female is required for the production of a normal ovary, and genes on X chromosomes are essential for production of normal testes in the male. Thus Hammerton postulates that in the normal XY male the X chromosome is in charge of bringing about sexual differentiation and that the Y chromosome regulates the precise time the X chromosome becomes active for the development of the male gonads (Hammerton, 1966; Mittwoch, 1970).

During embryonic development the gonads of both sexes are, at first, morphologically identical, and only subsequently do they develop into testes or ovaries according to the sex chromosome constitution (Figure 7–1). Gonadal sex differentiation begins about 35 days after fertilization if the cells carry the male XY chromosome constitution. The ovarian changes in a female occur later on in the developmental sequence.

If the male induction process does not begin according to schedule, early in embryonic development, abnormalities can be expected in gonadal genitalia at birth. If no induction occurs, the embryo will continue toward female development. The correct sex chromosome complement and precise timing are both necessary for the induction processes of male development. A disturbance in this process can create varying degrees of anatomical sexual development. Such errors or accidents in development can lead to the incomplete development of one sex or the partial display of both sexes (intersex) in persons with normal sex chromosomes. The spectrum of human sexual development can be seen in Table 7–1, which presents the normal male and female and the transition through the various forms of intersexual development.

male pseudohermaphroditism

The most common cause of male pseudohermaphroditism is testicular feminization, which is an inherited sexual disorder. Those affected display a normal feminine appearance and behavior, but are genetically (XY) male. They have

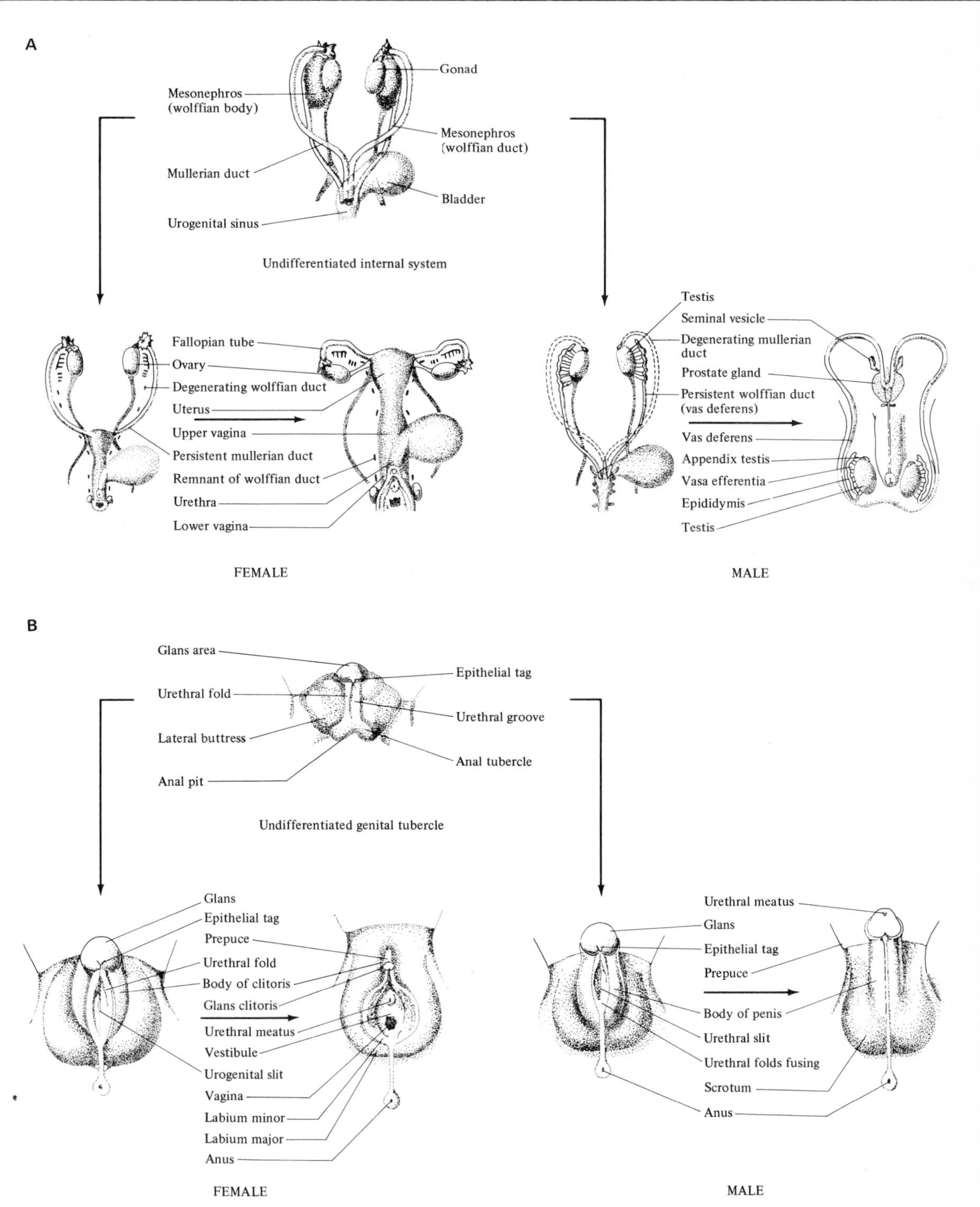

Figure 7–1. Differentiation of Sexual Characteristics. The early embryo grows without showing any sexual characteristics. The male and female sexual organs eventually develop from common basic structures. **A.** Development of the internal genital ducts from the primitive undifferentiated system. **B.** Development of external genitalia from the primitive genital tubercle. (Courtesy of Dr. Alfred D. Jost.)

Table 7–1 Human Sexual Characteristics

	Normal Female	*Adrenogenital Syndrome (Female Pseudohermaphrodite)*	*Intersex: Without Adrenal Disorder*	*True Hermaphrodite*	*Turner's Syndrome*	*Gonadal Dysgenesis*	*Intersex: Testicular Feminization Syndrome (Male Pseudohermaphrodite)*	*Klinefelter's Syndrome*	*Adrenogenital Feminism*	*Male with Hypospadias*	*Normal Male*
Sex chromosomes	XX	XX	XX	XX	X0	XY	XY	XXY	XY	XY	XY
Gonads (○female, ●male)	○○	○○	○○	●○	○○ (small)		●●	●● (small)	●●	●●	●●
Sex hormones*											
Estrogens (female)	Normal				<		Normal	Normal	<	Normal	Normal
Androgens (male)	Normal	>	Normal	Normal or low	<		Reduced	<	>	Normal	Normal

External anatomy						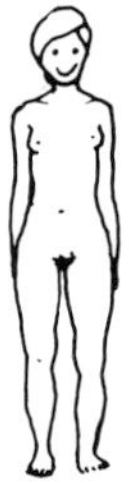			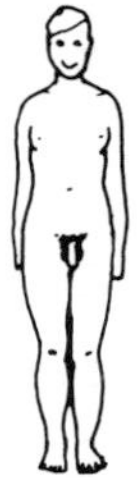	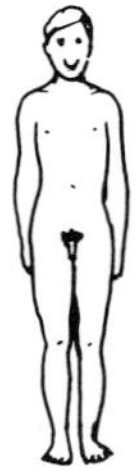	
Phallus Penis				Normal or				Small to moderate	Normal	Small (may resemble enlarged clitoris)	Normal
Clitoris	Normal	Enlarged	Enlarged	Enlarged clitoris	Normal	Normal or enlarged	Normal				
Menstruation	Normal	Sparse	Normal	Sparse	Sparse	Sparse	Sparse	None	None	None	None
Production of gametes	+	–	–	–	–	–	–	–	–	–	+

*Normal means that the amount of sex hormone expected for the sex chromosome complement is found.
>greater than normal for a person with a given sex chromosome complement.
<less than normal for a person with a given sex chromosome complement.

well-developed breasts, but do not menstruate and are sterile. The vagina ends in a blind pouch and undescended testes occupy the normal site of the ovaries. The testes produce female estrogens responsible for the secondary sexual characteristics. Presumably, a defective gene alters the ability of the Y chromosome to confer maleness on the embryo, but somehow allows for the production of the female hormone. The patient has a reduced amount of male testosterone.

Since the individual is sterile, it has not been possible to ascertain the manner of inheritance. Perhaps the defective gene somehow upsets the developmental time sequence and is transmitted via the X chromosome from the mother (Volpe, 1971).

In some cases of male pseudohermaphroditism an opening is present beneath the penis simulating a vagina. The scrotum or scrotal sac is usually small and does not contain the (undescended) testes. Many of these individuals have normal-appearing female external genitals, and may even undergo feminization at puberty, although they remain amenorrheic, that is, they do not menstruate. The majority of the male pseudohermaphrodites tested, including those with true testicular feminization, are chromatin negative. They do not show a Barr body, and have the *normal* sex chromosome complement.

A small number of male pseudohermaphrodites have sex chromosome anomalies, such as X0/XY and X0/XX mosaicism. The discovery that an individual can have both X0 and XY cells indicates that attempts to equate the sex chromosome complement (the genetic sex) with the anatomical sex of an individual are destined to failure. An individual's anatomical sex is a part of his or her phenotype, not of the genotype. The significance of this fact is that many chromosomal, genic, hormonal, and environmental factors influence the appearance of the internal and external genitals, or the anatomical sex, and may even produce so-called sex reversal.

female pseudohermaphroditism (congenital androgenital syndrome)

The female pseudohermaphrodite has the normal (XX) sex chromosome complement, but her genitalia display various degrees of phallic development. Abnormally functioning adrenal glands produce an excess of fetal androgens, which inhibit the complete differentiation of the female duct system and stimulate the development of male sex organs. There are variable expressions of genital duct development. That is, the time when the male or female ducts depart from normal depends on the kind and levels of hormones present. In these females (XX) the presence of androgens leads to development of portions of the male ducts.

true hermaphroditism

Although true hermaphroditism, the condition of having both male and female reproductive organs, is rare, at least 25 patients have been studied (Miller,

1964). They showed varying degrees of intersexual development of the genitals. A few had a chromosome abnormality in the form of mosaicism. Some did not exhibit the Barr body and were considered sex chromatin negative. However, most true hermaphrodites are chromatin positive.

In one case, recorded as a sex chromosome mosaic, the individual was reported to be chromatin positive with indistinct genitals. He displayed XX, XXY, and XXYYY chromosome complements in his leukocytes, XX and XXY cells in the testes, and XX in the skin cells (Miller, 1964). The mode of formation of such a mosaic individual is difficult to explain.

True hermaphroditism was described as early as 1601 (Sadler, 1970):

> A weird happening has occurred in the case of a Lansquenet (soldier) named Daniel Burghammer .... When the same was on the point of going to bed one night he complained to his wife, to whom he had been married by the Church seven years ago, that he had great pains in his belly and felt something stirring within. An hour thereafter he gave birth to a child, a girl .... He has then confessed on the spot that he was one-half man and one-half woman.... He also stated that ... he only slept once with a Spaniard and he became pregnant therefrom. This, however, he kept a secret unto himself and also his wife ... he has never been able to get her with child .... The aforesaid soldier is able to suckle the child with his right breast only and not at all on the left side, where he is a man. He has also the natural organs for a man for passing water .... All this has been set down and described by notaries. It is considered in Italy to be a great miracle by the clergy and is to be recorded in the chronicles. The couple, however, are to be divorced by the clergy. Paidena in Italy, the 26th day of May, 1601.

No true hermaphrodites have been documented in recent times as reproducing, and they are considered to be infertile. Self-impregnation in a true hermaphrodite would be highly unlikely on both morphological and endocrinological grounds.

the mini male

Males with the normal XY chromosome complement who have extremely small penis, scrotum, testes, and prostate have been called "mini males." According to Melicow, a deficiency of testicular hormone and possibly a pituitary failure lead to inadequate production of the male hormone (Melicow, 1971).

turner's and klinefelter's syndromes

The two most outstanding cases of intersexual development are the Turner's (X0) and Klinefelter's (XXY) patients. In many of the cases described in the literature, mental retardation was detected during infancy. As stated earlier, Turner's females are short in stature, menstruate sparsely if at all, and are sterile. The nipples of the breast are wide apart, and the typical body hair of normal females is either absent or sparse. The vagina is smaller than normal.

There may be ovaries, but the lack of the second X chromosome is associated with incomplete or no ovarian development and accordingly a variable reduction in the level of ovarian hormones.

Many Klinefelter's patients are tall, presumably because of the genetic imbalance that occurs with the extra X chromosome. These individuals have underdeveloped prostate, testes, and scrotal sac. Frequently they are sterile, although some have fathered children. According to Becker (1972), there is a possibility of breast cancer developing in the 75% of Klinefelter males who develop breasts. There is a definite feminine distribution of fatty tissue, giving the individual a female body build.

summary of the uncertainties of sex

According to the report of Jost (1970), there is a period during development when the embryo and fetus can become either male or female regardless of the normal XX or XY condition. It has been shown that in many cases the sexual role is acquired in childhood. Thus it is possible that a number of XX males and XY females exist in our society without being aware of their sex chromosome constitution.

This concept is largely based on detailed psychological examinations of persons with sexual abnormalities who frequently accepted the sex of their rearing. Even the most conservative reader must admit that it is most difficult to specify what constitutes normal human sexual behavior.

historical attempt to relate homosexuality and inheritance

"I heard from a friend that the New Jersey Medical Society was having an all-day conference on homosexuality and was looking for a gay physician to speak. So I called up and volunteered," a young physician told 700 other physicians. In the middle of his speech he stopped and said, "I really wasn't invited here as a medical scientist, but as a street homosexual. From here on in I'm going to tell you what the real problems of homosexuals are."

For 30 years Dr. Howard Brown kept his secret—from his classmates at medical school, from other professors at New York University, and from high-ranking members of New York's city government. His homosexual friends knew, but others saw him in his public roles—in class as a professor of health policy and administration, or perhaps on television as the city's director of health in 1966 and 1967.

Dr. Brown was careful lest anyone suspect. He avoided the subject of homosexuality in class. After television interviews, he made absolutely certain that every trace of makeup had been wiped off.

His straight friends and passing acquaintances must have been surprised when they picked up their *New York Times* on October 3, 1973, and found on the front page a picture of Dr. Howard Brown and the headline: EX-CITY AIDE SAYS HE'S HOMOSEXUAL (Van Dyne, 1973).

Due to the prominence society gives to the question of homosexuality, it is fitting to discuss the relationship of genetics to the possible inheritance of homosexual behavior.

Since at least 1886 attempts have been made to show a genetic basis for homosexuality. Krafft-Ebing (1886) concluded from his pedigree studies that homosexuals had multiple hereditary flaws such as alcoholism and tuberculosis. In a later study Hirschfeld reported, as evidence for a hereditary factor, that 35% of male homosexuals have homosexual brothers or other close relatives (Stoller, 1968). In the middle 1930's Goldschmidt (1938), a geneticist studying intersexuality in gypsy moths, found apparent males with female (XX) chromosomes. He thought this might be the key to human homosexuality, as did Lang (1940).

Lang wanted to show that homosexuality was inherited so that Nazi Germany could eliminate it by execution. Lang reasoned that since males and females are born in nearly equal proportion, the existence of a homosexual in a family would be equal to a "missing female." Having access to police files in Hamburg and Munich, Lang compiled a list of more than 1500 homosexuals, investigated their backgrounds, and in 1940 announced that his theory was correct: the homosexual did have a larger proportion of male siblings than the normal male in the population at large. The following year Lang's studies were repeated and confirmed (Kallman, 1952). The significance of his data remains unclear.

A different approach to establishing a relationship between genetics and homosexuality was taken by the geneticist E. Slater (1962). Because chromosomal anomalies become more frequent as the mother's age increases, Slater hypothesized that if homosexuality is a chromosomal anomaly, it should be more common among males born later in their mother's lives. Also, males born late or last among their siblings should show a higher proportion of homosexuality. Slater studied some 400 homosexuals and found that this was the case. He concluded that homosexuality is a chromosomal anomaly.

The studies that appeared for a time to solidify the genetic basis of homosexuality were those of F. J. Kallman (1952). Kallman studied identical and fraternal (nonidentical) twins. Monozygotic, or identical, twins are of the same sex, look alike, and have many similar characteristics, since both come from a single fertilized egg. Dizygotic, or fraternal, twins are formed from separately fertilized eggs, and the two children are no more alike than brothers and sisters in general. Now, since monozygotic twins share the same hereditary material, it would follow that if genes are the determining factor, and if one twin turned out to be homosexual, the other would also. When a trait is studied and it appears in both identical twins, scientists say there is concordance for that trait. Of course, several sets of identical twins are used to make the data statistically reliable. The researcher also compares the relative amount of concordance for the trait, say, homosexuality, between identical twins with that between fraternal twins. A higher rate of concordance in identical twins offers good evidence that the trait in question is genetic. Kallman reported 100% concordance for overt homosexual activities in his study of 40 identical twin pairs. Behavioral scientists still do not know what to do with the Kallman studies. Few, if any, subsequent studies have been able to support an association between genes and homosexuality. The Kallman data stand alone;

they have not been repeated. From the basic principles of genetics, we would expect that many of the fathers or mothers of the homosexual twins would also have been homosexual (providing the gene is not recessive). However, only one father of the 85 male parents involved in the Kallman studies on monozygotic and dizygotic twins was homosexual (Gadpaille, 1972).

Another attempt to associate genetics with homosexuality involved the work of L. Barr. Barr and his colleagues discovered the now famous Barr body, an inactive X chromosome, which is found near the membrane of normal female cells (see Figure 6–23). The normal XY male does not have a Barr body. When 50 homosexuals and 5 male transvestites (males who wear female clothing) were given the Barr body test, no abnormality with respect to the sex chromosome was found (Pare, 1965).

Later, when techniques for analyzing all the chromosomes of a cell were developed, homosexuals were given further tests and still did not show a chromosome abnormality. With results such as these, it would appear that recognizable chromosomal abnormalities are not involved in homosexuality.

In a recent CBS TV film, *Homosexuality,* it was stated that there are no happy homosexuals; yet in the United States the Gay Liberation movement claims over 20 million members. The Kinsey report claims that between 4% and 17% of white Americans are homosexuals or have tried homosexual relationships. At an incidence of one in six, calculating for both male and female, it would appear that there are some 35 million persons who have tried homosexual relationships or are practicing homosexuals (calculations based on 212 million Americans). Regardless of the figures one accepts, the proportion of persons engaged in homosexual activities is very large, although the real number may never be known because of actual or feared social reprisals.

In 1970 the United Presbyterian Church declared that "The practice of homosexuality is [a] sin." In 1973 the July–August issue of *Trends*, a bimonthly adult-education journal of the United Presbyterian Church, asked: "Is homosexuality a manifestation of sin? Is it a sickness?" The reply from the Presbyterian membership was no to both questions.

In mid-December 1973 the American Psychiatric Association voted to remove homosexuality from its official list of mental disorders and place it in a new category of sexual orientation disturbances. They also adopted a resolution deploring discrimination against homosexuals in employment, housing, public accommodations, and licensing, and requested a repeal of the sodomy (usually defined as copulation between males) laws still existing in 42 states and the District of Columbia (*The Sciences,* 1974). Today there are open gay bars and social groups in every major city of the United States.

a matter of definition

the physical basis: transmission of the hereditary material through the chromosomes

There are a number of definitions available for the homosexual. One, provided by scientists working in the area of homosexuality, is: "The physical contact between two individuals of the same gender which both recognize as being sexual in nature and which ordinarily results in sexual arousal" (Gebhard, 1972). This definition defines the homosexual *act*, but the term *homosexual* implies a definite preference for erotic stimulation by, or attraction to, members of the same sex (Marmor, 1965). Either definition allows a marriage between an androgen-insensitive woman (a male pseudohermaphrodite) and a normal male to be considered heterosexual. Here we are playing at what appears to be a word game, but let us pursue it.

For unknown reasons, normal XY embryonic tissue becomes insensitive to the action of the male androgen hormone during the crucial stage of sexual development and continues development to become a female. The resistance of embryonic cells to androgen-induced sexual differentiation is known to be an inherited phenomenon; most likely the gene occurs on the X chromosome and is dominant (Money, 1970). These females are androgen insensitive or male pseudohermaphrodites. Children born with this syndrome can externally appear to be perfect females. They are invariably assigned female roles. Their testes, which are usually undescended, do not make sperm. The uterus is vestigial and menstruation does not occur. These persons are usually judged defective because of menstrual failure or lumps of the testes in the groins. In some cases the vagina is too short for comfortable intercourse and requires surgical lengthening. The onset of puberty is at the usual time and is invariably feminizing, since the cells of the body are unable to respond to the normal masculine output of androgen from the testes but do respond to the estrogen produced in a male, which is enough to produce a degree of feminization so complete as to be compatible with a career as a fashion model—a characteristic of the syndrome is that the girls are tall. When such a female marries, the semantics of homosexuality comes in because genetically and gonadally both partners in a sexual relationship have testes. The married partners are male XY and XY. Morphologically, hormonally, and psychosexually, they are heterosexual in their relationship. Legally, and by common sense, they are heterosexual.

But what of those children who for reasons of accident after birth or misassignment at birth must change their genetic sex roles? Do they tend to become homosexual because of the interaction of genes and social misassignment? Some interesting cases were reported by a medical psychologist, John Money, to the American Association for the Advancement of Science in 1973. In 1963 parents of identical twin boys took them to be circumcised. During the operation, performed with an electric cauterizing needle, a surge of current burned off a baby's penis. To cope with this tragedy, the parents took the advice of sex experts: "Bring the baby up as a girl." Aided by plastic surgery and reared as a daughter, the once normal baby boy has grown into a nine-year-old child who is, at least psychologically, a girl. In another case a child was born with an underdeveloped penis. He was assigned at birth as a male, because he was XY and had testicles. Later, at 17 months, the child was reassigned as a female. Her two brothers are helping this child develop a feminine attitude.

Money also cited two hermaphrodites, each belonging to a different family. Both children were born XX and had female ovaries and a male penis. The first family raised their child as a boy who later married even though both he and his partner were XX. The second family resorted to corrective surgery and therapy. The girl reached puberty and had a steady date. Socialization was necessary, however, for her to become a female even though she was XX. We will have to wait for the outcome of these cases. It is still too soon to determine whether or not they will develop homosexual tendencies.

This question of the sex chromosomes versus sex identity should make it clear that sex assignment can be difficult. A person's sexual identity is more than just the inheritance of sex chromosomes. The sex chromosomes are not, of themselves, responsible for homosexuality. The specific causes of homosexuality remain open to question.

the human brain, male or female?

Recent and continuing investigations involving the human brain may yield new knowledge as to whether one can inherit a predisposition to homosexuality. Apparently there is an early, embryonic, hormonal influence on the hypothalamus, a portion of the brain that, among other things, regulates the menstrual periods. When radioactively labeled male hormone was administered to rat fetuses, the labeled hormone was later found concentrated in the cells of the hypothalamus. When the labeled hormone was given during the critical period of sexual differentiation of a genetic female XX, the hypothalmic cells became "masculinized." As a consequence of being treated with the male hormone, these cells were completely inhibited from beginning the cyclic female periods. The females are not normal with respect to ovulation, nor do they experience the female sexual cycle. The reverse of this situation holds when the male rat is chemically castrated by antibodies that block the male androgens from initiating certain biochemical events in the hypothalamus. However, as admitted by Money (1970 and 1972), it is a giant step from the reactions observed in a rat to those that occur in humans. Androgenized females do behave in a tomboyish manner, according to self-, parental, and peer observations—but how much of this is social conditioning?

In relation to homosexuality, these findings suggest that an otherwise normally developing embryo may have a predisposition to homosexuality based on the inherited lack of ability of certain cells, perhaps of the hypothalamus, to respond to normal levels of hormones. That is, a genetic defect may occur in certain hormone-sensitive cells necessary for normal development and, given certain as yet undefined environmental conditions, the "defect" becomes predominant in the course of biological or social development. This is in keeping with the fact that there is a higher incidence of homosexuality in families that have other homosexual persons in their background than in families that have no known history of homosexuality (*The Sciences*, 1973).

the physical basis: transmission of the hereditary material through the chromosomes

Although many claims for a definite genetic relationship have been made throughout the history of investigations into homosexuality, there is no currently acceptable evidence of an innate form of homosexuality. Nor is there any evidence that homosexuality, as traditionally defined, is exclusively *acquired.* Whatever the degree of an individual's homosexual commitment, the cause of the behavior may be in some degree hereditary and biological, and in some degree environmental, learned, and sociological.

sex identity changes

We know comparatively little of the behavioral aspects of genetic aneuploidy; for that matter, we are still having problems trying to understand the behavior of what are apparently genetically normal individuals. Even the apparently genetically normal have acute social and legal problems. For example, a former transvestite (man who dresses in women's clothes) wanted to void his 1963 marriage because he had found that his wife was still a "man," in spite of earlier surgery. He claimed they could not consummate their marriage. Nine physicians, including anatomical and psychological experts on sexual abnormalities, testified during the 16-day trial. Evidence presented showed that the sex chromosomes of Miss Ashley, a former sailor in the Merchant Marine who had worked as a model after the sex operation, were in fact male (XY). In ruling, Justice Ormrod of the British High Court had to define what a woman was before ruling on the validity of a marriage between the supposed female, Miss Ashley, and the former transvestite. Justice Ormrod stated: "The only cases where the term *change of sex* is appropriate are those in which a mistake as to sex is made at birth and subsequently revealed by further investigation." He then annulled the marriage. Ormrod suggested that in fact the woman in this case was a transsexual, "a female imprisoned in a male body." He pointed out that surgery for these people relieves the psychological distress but does not change their genetic sex. He said that the biological sex of an individual is fixed not later than birth and cannot be changed by medical or surgical means. When informed of the court's decision that she was still a male, the 34-year-old "female" said she was "absolutely shattered." She added, "It means I can't get married at the moment."

In another such case, a reserved 42-year-old geologist expressed horror and disbelief when told that his 25-year-old bride was really a male from Peterhead, Aberdeenshire, who had previously married a 30-year-old female. Further, the bride's first wife was pregnant with "his" baby. The confused geologist said he would seek legal advice on how to dissolve the marriage.

A third case involves a navy veteran, husband, and genetic father of a 15-year-old daughter, who underwent a sex change operation. A New York judge granted the new Deborah a divorce on the grounds that he had become a woman and wanted to marry a man. Questions: Who will be responsible and liable for the actions of the daughter? Will Deborah have to pay child support? Also much more important, what of the daughter's emotional problems?

The most celebrated sex change to date is Christine Jorgensen, a Brooklyn GI and a normal genetic male who underwent a surgical sex change in 1952. The newspaper headline writers had a field day; for example (Jorgensen, 1967):

Dear Mom and Dad, Son Wrote, I Have Become Your Daughter
Ex-GI Becomes Blond Beauty (front page New York *Daily News,* December 1, 1952)
Judge Wonders: Is Chris Really George?
Christine! By George?
Christine—Is She He or He She

Recently, Christine Jorgensen demanded an apology from the then Vice President Spiro T. Agnew for linking her name with Senator Charles E. Goodell, a New York Republican who left the Nixon political fold. Agnew termed Goodell "the Christine Jorgensen of the Republican Party." In March of 1973, after 21 years of womanhood, Christine was asked about marriage:

> I've been engaged two times and I've been in love two times, but I wasn't in love with the men I was engaged to and I wasn't engaged to the men I was in love with.

Miss Jorgensen also suggested that persons with sex identity problems visit a gender identity clinic like the Erickson Educational Foundation in Baton Rouge, Louisiana, or any of the 50 hospitals with gender identity boards in the United States.

In a similar case judges, lawyers, policemen, and prison guards were baffled over the sex of 29-year-old Bernadette Cassell, another example of the apparently genetically normal person with acute social-legal problems. She was born a genetic male but underwent sex surgery and was in the process of receiving hormone injections to become a woman when arrested for solicitation for the purpose of prostitution. When Judge Gerstung asked her what her sex was, she replied that since she was in the process of receiving the hormone injections, she was halfway to becoming a woman. The police booked her as a transvestite, but were afraid to perform a physical examination, since they did not know who should examine her, a man or woman. The judge tried her as a male and sent her to the male correctional institution at Jessup, Maryland. However, the authorities at the institution said she had the physical attributes of a woman and sent her to the women's correctional institution.

As a final example, a 52-year-old high-school music teacher of Bernard Township, New Jersey, was fired after he underwent a sex change to become a woman. She was fired because she posed a psychological threat to her students. He had taught for 14 years prior to the sex change operation in New York.

It should be obvious that "sex" and sex identification can mean a variety of things to different people. In a world where one's sex is of primary importance in becoming established in society, nature's roulette of natural selection through trial and error can be *socially* very cruel.

references

Becker, K. L. 1972. Clinical and Therapeutic Experiences with Klinefelter's Syndrome. *Fertility and Sterility, 23*: 568–78.

Gadpaille, W. J. 1972. Research into the Physiology of Maleness and Femaleness: Its Contribution to the Etiology and Psychodynamics of Homosexuality. *Archives of General Psychiatry, 26* : 193–206.

Gebhard, Paul H. 1972. Incidence of Overt Homosexuality in the United States and Western Europe. In John M. Livingood M.D. (ed.), *Homosexuality.* Publication No. HSM 72–9116, pp. 22–29. Department of Health, Education and Welfare, Washington, D.C.

Goldschmidt, R. 1938. *Physiological Genetics.* McGraw-Hill, New York.

Hammerton, J. L. 1966. Significance of Sex Chromosome Derived Heterochromatin in Mammals. *Nature, 219*:910–14.

Jorgensen, Christine. 1967. *A Personal Autobiography.* P. S. Ericksson, New York.

Jost, A. D. 1970. Development of Sexual Characteristics. *Science Journal, 6:67–72.*

Kallman, Franz. 1952. Comparaitve Twin studies on the Genetic Aspects of Male Homosexuality. *Journal of Nervous and Mental Disease, 115*:238–98.

Krafft-Ebing, Richard von. 1886. *Psychopathia Sexualis* (first unexpergated edition in English; original in latin). Reprinted 1965, Putnam, New York.

Lang, Theo. 1940. Studies on the Genetic Determination of Homosexuality. *Journal of Nervous and Mental Disease, 92:*55–64.

Marmor, J. (ed.). 1965. *Sexual Inversion: The Multiple Roots of Homosexuality.* Basic Books, New York.

Melicow, M. M. 1971. Proper Approach to Intersexuality in Infants and Children, *Medical Aspects of Human Sexuality, 5:*25–49.

Miller, O. J. 1964. Sex Chromosome Anomalies. *American Journal of Obstetrics and Gynecology, 90*:1078–1139.

Mittwoch, U. 1970. How does the Y Chromosome Affect Gonadal Differentiation? *Philosophical Transactions of the Royal Society of London, B259:*113–17.

Money, John. 1970. Sexual Dimorphism and Homosexual Gender Identity. *Psychological Bulletin, 74*:425–40.

Money, John. 1972. Pubertal Hormones and Homosexuality, Bisexuality and Heterosexuality. In John M. Livingood M.D. (ed.), *Homosexuality.* Publication No. HSM 72-9116, pp. 73–77. Department of Health, Education and Welfare, Washington, D.C.

Money, John. 1973. Biological Imperatives. *Time,* January 8, p. 34.

Pare, C. M. B. 1965. Etiology of Homosexuality: Genetics and Chromosomal Aspects. In J. Marmor (ed.), *Sexual Inversion: The Multiple Roots of Homosexuality.* Basic Books, New York.

Sadler, B. L. 1970. The Law and the Unborn Child. A Brief Review of Emerging Problems. *Fogerty International Proceeding, 6*:215–18.

Slater, E. 1962. Birth Order and Maternal Age of Homosexuals. *The Lancet, 1*:69–71.

Stoller, Robert J. 1968. *Sex and Gender.* Science House Publishers, New York.

The Sciences. 1973. Are Hormonal Factors Associated with Sexual Preference? *13*(Oct.):5.

The Sciences. 1974. Instant Cure for 20 Million Gay Americans. *14*(March):4.

Van Dyne, Larry. 1973. The Feeling You Have to Hide. *Chronicle of Higher Education,* December, p. 7.

Volpe, E. Peter. 1971. *Human Heredity and Birth Defects.* Bobbs–Merrill, Indianapolis.

part 4

the physical basis: transmission of units of hereditary material through the genes / some social implications

8

gregor mendel's legacy and classical genetics

It is the common wonder of all men, how among so many millions of faces there should be none alike.

Sir Thomas Browne

Gregor Johann Mendel was born July 22, 1822, in the small village of Heinzendorf in northern Moravia, at that time a part of Austria but now in Czechoslovakia near the Polish border. Johann (the additional name of Gregor was not given to him until he became a member of the Augustinian Order) was the only son of Anton and Rosine Mendel, a peasant couple. The life of a peasant in the early nineteenth century was still dominated by the hardships of a remnant feudal system whereby peasants were burdened with taxes, deprived of all democratic rights, and forced to contribute horses, carts, and their own labor three days a week for the benefit of the nobility, who enjoyed the privileges of exemption from military service, freedom from taxation, and trial in common law courts. Mendel was forced to struggle throughout much of his life under these conditions.

His interest in the cultivation and improvement of plants developed from his early years, which he spent raising crops with his father. As a student in the Heinzendorf school for elementary education, he was also taught the essentials of gardening, fruit growing, and beekeeping. Because of these early experiences, Mendel's approach to botany was practical rather than philosophical, experimental rather than conjectural.

mendel's education, life at the monastery, and teaching experience

At the age of 18, Mendel received his diploma from the Troppau high school, with special commendation from his teachers for his ability and achievement.

Although he was concerned about how he would have enough to eat, his interest in education led him to enroll at the Philosophical Institute in Olmütz. During his last year at Olmütz, Mendel became ill and realized it would be impossible for him to continue to push himself further. He began to look for a profession in which he would be spared constant anxiety about earning a living.

Mendel asked his physics professor at Olmütz, Friedrich Franz, for assistance just when Franz had been asked to recommend students to the Augustinian monastery at Altbrünn. Johann Mendel was admitted as a novice to the monastic community of St. Thomas in 1843. From then on, as was the custom, another name, Gregor, was used before his baptismal name (Iltis, 1966).

The Augustinian monastery, located a short distance from the Moravian capital of Brünn, was one of the principal centers of culture and learning in Austria. Here Mendel, for the first time free of worry about a means of survival, devoted himself completely to his studies. Among those who were most influential in his life were Pater Franz Bratranek, a philosopher and botanist, and philosopher Pater Klacel. With his excellent scholastic record and outstanding character, Mendel was ordained a priest, Pater Gregor Johann Mendel, on August 4, 1847 (Figure 8–1).

Figure 8–1. Gregor Johann Mendel (1822–1884). (From Radio Times Hulton Picture Library, British Broadcasting Corporation.)

Mendel's first teaching appointment was as supply teacher at the Znaim high school. The term *supply* meant that the position was temporary until the teacher passed the state examination for a teaching certificate (Iltis, 1966). Mendel took the examination in the summer of 1850. His essay in physics was accepted, but he did not do well in the history essay or in the oral portion of the examination. However, the examiners decided that, in view of his intelligence and character, he would be allowed a second examination after a year's time.

Mendel returned to Znaim depressed by his first failure. He was unaware that his physics examiner had written to his superiors urging that a way be found to offer him more extensive training at the University of Vienna. Impressed with the examiner's opinion of Mendel, the bishop gave his consent.

In October of 1851 Gregor Mendel began two of the most intellectually stimulating years of his life at the University of Vienna. He studied physics, under the famous Doppler, who had discovered the Doppler effect. He also studied chemistry, mathematics, zoology, entomology, botany, and paleontology. He came under the influence of the outstanding biologist Franz Unger, who took a concrete, practical view of inheritance and insisted on sticking to the experimental facts (Bronowski, 1974). From Unger Mendel was to gain a background in the use of quantitative and experimental methods in biological research. Mendel completed his stay at the university in August 1853, and shortly thereafter accepted a supply teaching position at the Brünn Modern School.

After teaching at Brünn for almost a year, he decided, at the age of 32, to take the state teaching examination again and returned to Vienna in May 1854 for a second attempt to pass the examination. As a member of the Vienna Zoological and Botanical Society, Mendel received research literature and was in a position to cite names, dates, and experimental conclusions about evolution, plant breeding, and heredity. However, these conclusions were regarded as unorthodox by many members of the examination committee, and Mendel refused to modify his opinion merely to please the examining professors. Aware of the hostility of certain members of the committee, Mendel suddenly asked that his application for a teaching certificate be withdrawn. He had learned that it was too much to expect his examiners to listen objectively to views they opposed, and so he returned to Brünn in defeat. His second struggle with the examining board did one important thing. It made him stop thinking about a teaching certificate and thus gave him time to renew his interest in plant research.

mendel's research

For a long time scientists had known that hybrids produced by crossing two varieties of a plant would not breed true. The offspring of the hybrid plants would be a mixture of plants like the parents and plants like the grandparents. Mendel knew from his own observations that hybrids always possessed a certain "likeness" to their parents and that with predictable regularity every hybrid plant of a species would turn out to be like every other hybrid plant of that species. He also observed that characteristics reappeared in the offspring of the hybrid plants with regularity, and that regularity suggested to him that certain "factors" must be present in the transfer of characteristics from one generation to the next.

Mendel seemed to be alone in his conclusions, for nowhere in the scientific literature could he find an account of the extensive numerical analysis of successive generations, bred from hybrid plants, that would demonstrate the existence of such factors. Mendel said, in fact, that (Sturtevant, 1965)

> among all the numerous experiments made [by his predecessors], not one has been carried out to such an extent and in such a way as to make it possible to determine the number of different forms under which the offspring of hybrids appear, or to arrange these forms with certainty according to their relations.

Mendel felt that if the "factors" of inheritance followed some logical distribution, then a law or pattern could be discovered by performing large numbers of crosses and counting the different forms as they appeared in the offspring. This simple, practical approach turned out to be the key to his success in deriving the basic patterns of inheritance.

Mendel carried out experiments with the garden pea, *Pisum sativum*, during the eight years 1856 to 1863, when his pea crop was destroyed by the pea weevil (Sturtevant, 1965).

Cross-pollinating the pea plant is not difficult, but it requires time and patience. Mendel took pollen from the ripened anthers of one pea plant and dusted it on the exposed stigma of a second pea plant. He then placed a small bag around the flower he had artificially pollinated to prevent unknown pollen from being introduced by wind or insects (Figure 8–2).

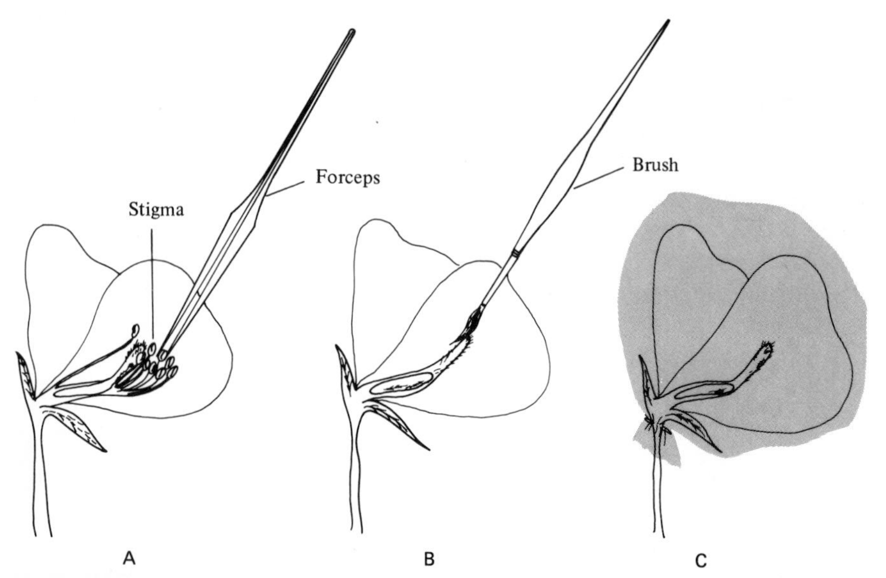

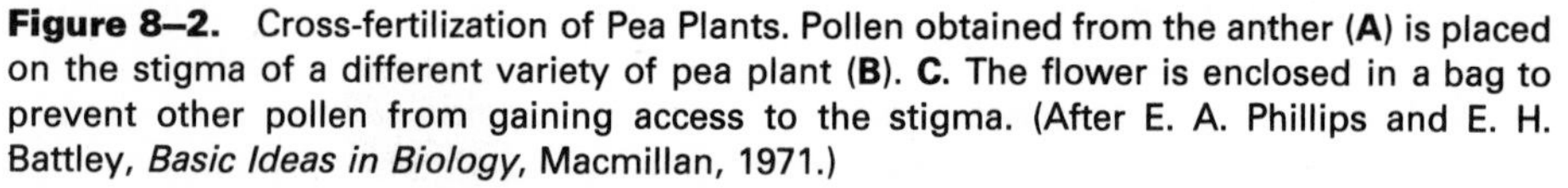
Figure 8–2. Cross-fertilization of Pea Plants. Pollen obtained from the anther (**A**) is placed on the stigma of a different variety of pea plant (**B**). **C.** The flower is enclosed in a bag to prevent other pollen from gaining access to the stigma. (After E. A. Phillips and E. H. Battley, *Basic Ideas in Biology*, Macmillan, 1971.)

Mendel knew that if the experiments were to give an accurate picture of heredity, a large number of generations would have to be bred, and many plants would have to be included in each generation. "It indeed required some courage to undertake such far-reaching labors," Mendel would later write in the introduction to his paper. "It appears, however, to be the only way in which we can finally reach the solution of a problem which is of great importance in the evolution of organic forms" (Moore, 1975).

Mendel first devoted two years (1854–1856) to testing his pea plants to be certain that they would breed true—that, for example, a self-pollinated plant with wrinkled seeds would give rise only to wrinkled seeds no matter how many generations were grown. He then selected for study seven sharply contrasting pairs of characters in order to eliminate the confusion that had overcome the work of his predecessors. The seven characters that Mendel chose to observe in his crosses can be seen in Figure 8–3.

In his first experiment Mendel made a total of 287 cross-fertilizations on 70 plants. In each area of his small garden (about 100 feet by 20 feet) he planted, next to each other, the seeds that contained one of the seven pairs of contrasting characters. In a few weeks he artificially cross-pollinated seven matched pairs of pea plants. For example, he applied pollen from a tall plant to a dwarf plant; in this case the dwarf plant was the bearer of the resulting hybrid pea seeds. To make certain that it made no difference which plant served as the seed bearer, Mendel also reversed his fertilization procedure, applying pollen from the dwarf plant to the tall plant.

It took about 80 days for Mendel's plants to "set" the hybrid seeds in which a pair of contrasting characters had been joined. These seeds were studied to gain information about the seed shape and cotyledon characteristics of the F_1 generation. Mendel recorded and summarized his data and confirmed his ideas about the paired characters in the F_1 hybrid plants. In every case one member of each pair has "overpowered" or "masked" the effect of the other. For example, the seeds from the tallness-dwarfness cross had grown into plants that were all uniformly tall, a fact that contradicted the botanists' traditional view of hybrid inheritance that the plants should have been intermediate in height. Mendel decided to call the character that was expressed in the hybrid offspring *dominant* and the hidden character *recessive.*

The next spring Mendel planted the F_1 seed, which gave rise to hybrid F_2 plants from which Mendel learned about the expression of the F_1 genes for flower position, stem length, pod color, etc. (Figure 8–4). When Mendel had summed up the results for the experiments involving the F_2 generation, he arrived at an average dominant-to-recessive ratio of 2.98 to 1, or approximately 3:1. In the preceding F_1 generation the recessive characters had not been present; they seemed to have disappeared completely, and all the F_1 plants had been of the dominant type. Now in the F_2 generation, the recessive character had suddenly reappeared in 25% of the offspring, after being hidden for a generation. The other 75% displayed the dominant character. Thus the distribution of the factors of inheritance showed a mathematical relationship, as Mendel had hoped to find. From these results, Mendel was able to say that the appearance of an organism does not always reveal its actual hereditary factors. "It can be seen," said Mendel, "how rash it may be to draw from the external resemblances conclusions as to their internal nature" (Moore, 1975).

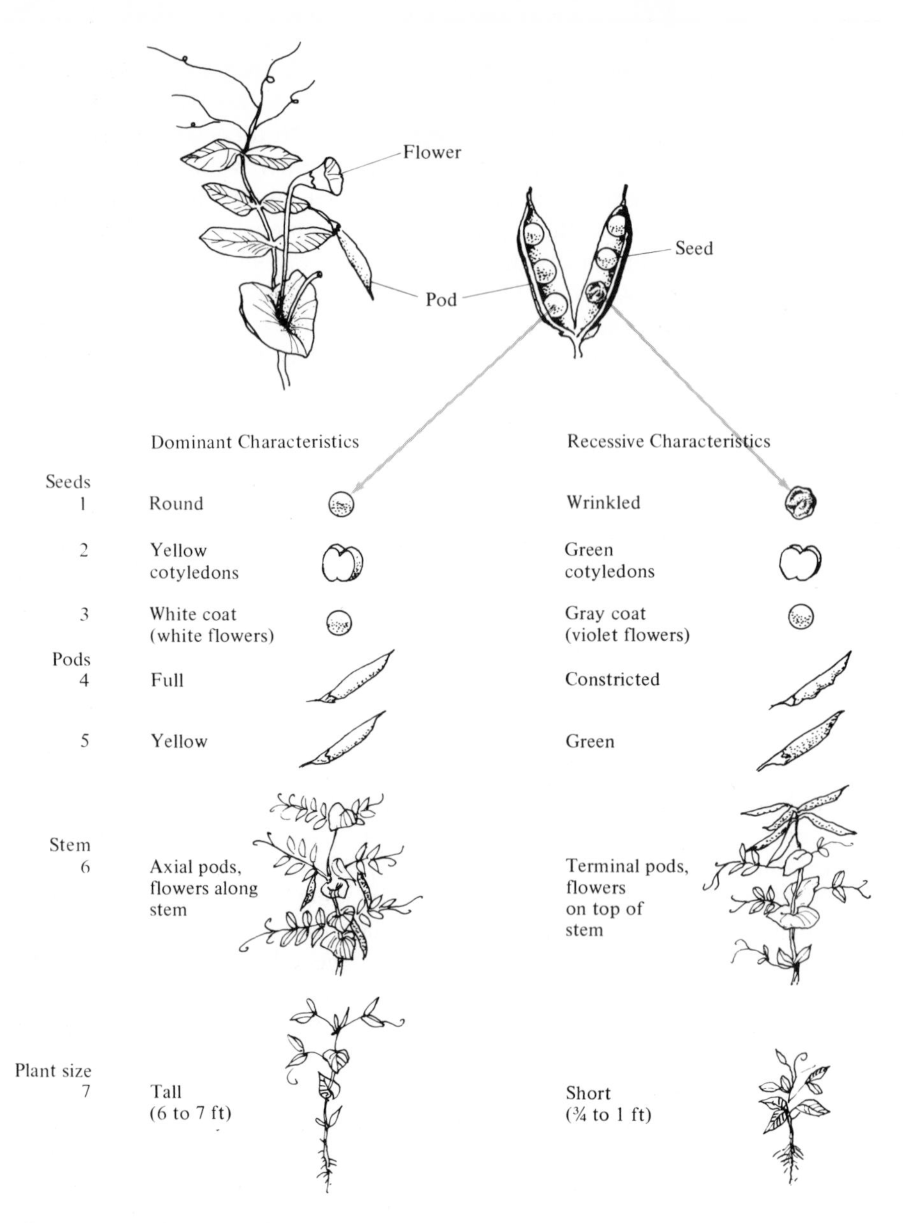

Figure 8–3. Pea Plant Characteristics. The seven traits observed, counted, and published by Mendel. (Modified from M. W. Strickberger, *Genetics*, 2nd ed., Macmillan, 1976.)

the physical basis: transmission of units of hereditary material through the genes

Mendel's explanation for the mathematical consistency he had found is best appreciated if considered against the background of the scientific knowledge at that time. It was known that parental traits were transmitted to the offspring by single specialized reproductive cells, and that these cells, or gametes, united to form the fertilized egg from which the new plant or animal would develop. What was not understood was the mechanism by which these reproductive cells were formed and how they were different from body cells. Mendel knew nothing of chromosomes or meiosis; it was 15 years later that staining

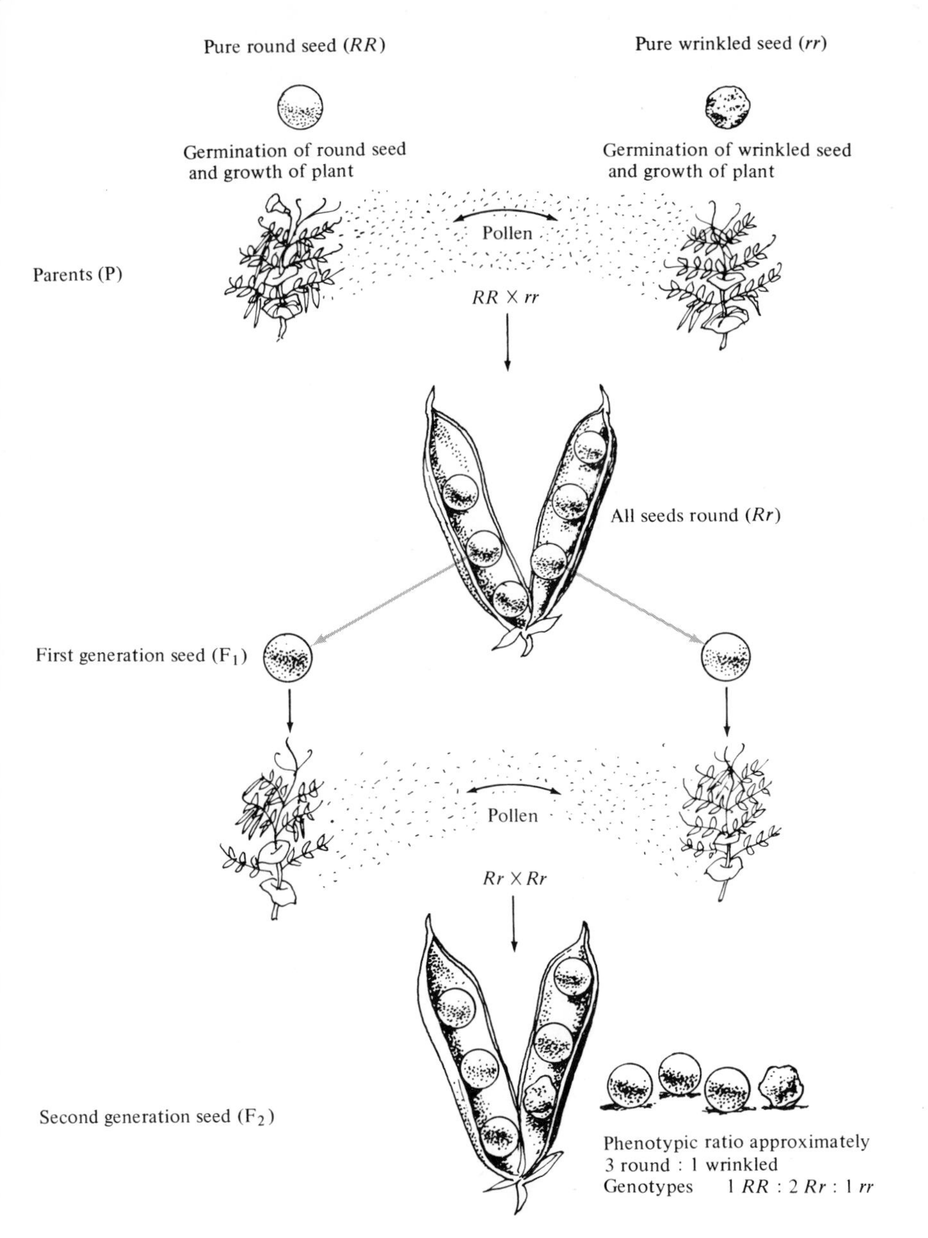

Figure 8–4. Transmission of a Single Trait in Pea Plants. It took Mendel's plants about 80 days to mature. The seeds resulting from the P_1 cross (the F_1 seeds) are hybrid and contain the hereditary information necessary to produce round and wrinkled seeds, but round is dominant to wrinkled so all F_1 seeds are round. In the F_2, however, the recessive wrinkled seed reappears.

techniques made it possible for Strasburger to see and describe the rodlike forms (chromosomes) appearing in the nucleus at cell division.

In order to explain the results of his crosses, Mendel visualized units of heredity, which he called factors, within the cells of plants; later these units came to be known as genes, were located on the chromosomes, and were found to behave as if they were individual units. Mendel also made the assumption that the body cells of his plants contained pairs of factors and that a single pair of factors controlled the expression of each plant character. Mendel believed that one of each of the pair of factors came from each of the two parents; later it was shown that chromosomes exist in pairs consisting of one member from each parent. Mendel came to the conclusion that the sex cells must be different from all the other cells of the plant and that, when they are formed, the two factors of each pair must segregate from each other. Mendel was certain that only the chance combination of these factors could account for his breeding results, and this concept is now known as the principle of independent gene assortment or Mendel's second law. About 25 years later, in 1887, van Beneden and others observed the separation of paired chromosomes during the formation of gametes. Mendel postulated that each inherited characteristic was governed by *two* factors, and that each parent contributed *one of each factor* to the formation of the new offspring. He also thought that the separation of members of different pairs of factors into the sex cells was purely a matter of chance. Thus, for each character, say round versus wrinkled seed, there were four ways in which the factors for round and wrinkled seeds could pair up at fertilization.

In order to describe the combining and segregating of the factors, Mendel labeled the dominant factor with a capital letter, say *R*, and the recessive with a small letter, *r*, as shown in Figure 8–4. A true-breeding dominant, also referred to as pure or homozygous, would be formed by the coming together of two *R*'s and would be described as *RR*; a heterozygote would be either *Rr* or *rR*; and a true-breeding or homozygous recessive would be *rr*. Mendel explained the derivation of his observed 3:1 ratio as follows:

1. If the recessive factor for wrinkled seed (*r*) in the pollen pairs with the recessive factor for wrinkled seed (*r*) in the egg, the result is a pair of recessive factors for a homozygous wrinkled seed (*rr*). The seed appears wrinkled and it contains both *rr* factors.
2. If the dominant factor for round (*R*) in the pollen pairs with the dominant factor for round (*R*) in the egg, the result is a pair of dominant factors (*RR*). The round seeds are homozygous or pure and only round-seeded plants are produced from matings using only these seeds.
3. If the recessive factor for wrinkled seed in the pollen pairs with the dominant factor for round seed in the egg, the result is a mixed or heterozygous pair of factors, wrinkled and round (*rR*). Because the round factor is dominant, only round seed are produced, even though the plant carries both *R* and *r* factors for seed shape.
4. If the dominant gene for round in the pollen pairs with the recessive factor for wrinkled in the egg, the result is another heterozygous or mixed pair of factors, round and wrinkled (*Rr*). Again, because the round (*R*) factor is dominant to the recessive (*r*) factor, the seed is round.
5. In the heterozygous situations of 3 and 4 above, although a round seed develops, the wrinkled factor *r* is not lost or diluted. The factors *R* and *r* are unaffected by their association and emerge from this association as

distinct as when they entered it. Because the recessive *r* factor *is not lost* we can explain why heterozygous *Rr* plants mated with other heterozygous *Rr* plants give rise to a 3:1 phenotype ratio, three dominant round (*RR*, *Rr*, *Rr*) to one recessive (*rr*) wrinkled (see Figure 8–4).

It is interesting that Charles Darwin also experimented with peas and noticed that the hybrid plants divided in a phenotype ratio of 3:1, but Darwin was not a mathematician and did not pursue a statistical analysis of his observations.

Many years later, when Mendel's factor was called a gene, the biologist Reginald Punnett devised a checkerboard method for arriving at all the possible gene combinations for a given characteristic. The Punnett square method, as shown in Figure 8–5, is a more direct method than Mendel's for determining the possible phenotypes and genotypes that can be produced by a given mating. For example, the Punnett square shows the behavior of factors in the cross between a homozygous or pure wrinkled-seed (*rr*) plant and a homozygous or pure round-seed (*RR*) plant.

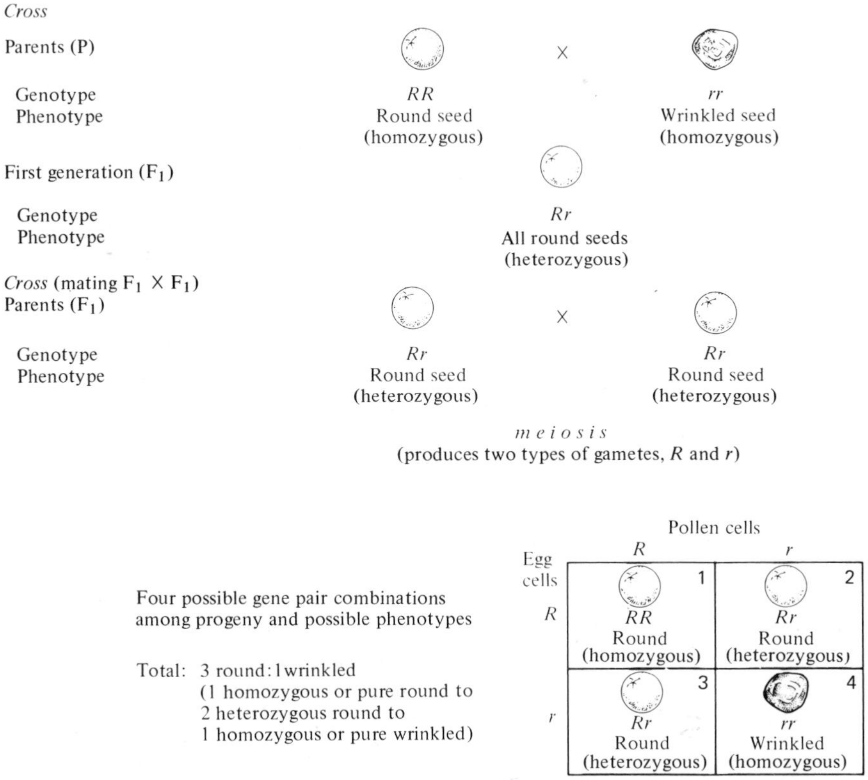

Cross

Parents (P) — × —

Genotype: *RR* | *rr*
Phenotype: Round seed (homozygous) | Wrinkled seed (homozygous)

First generation (F_1)

Genotype: *Rr*
Phenotype: All round seeds (heterozygous)

Cross (mating $F_1 \times F_1$)
Parents (F_1) — × —

Genotype: *Rr* | *Rr*
Phenotype: Round seed (heterozygous) | Round seed (heterozygous)

meiosis
(produces two types of gametes, *R* and *r*)

Four possible gene pair combinations among progeny and possible phenotypes

Total: 3 round:1wrinkled
(1 homozygous or pure round to
2 heterozygous round to
1 homozygous or pure wrinkled)

Egg cells \ Pollen cells	*R*	*r*
R	1 *RR* Round (homozygous)	2 *Rr* Round (heterozygous)
r	3 *Rr* Round (heterozygous)	4 *rr* Wrinkled (homozygous)

Figure 8–5. Punnett Square Method for Deriving the Possible Phenotypes and Genotypes in a Monohybrid Cross (Mating). The gene *R* is dominant to recessive *r*, its allele (one member of a gene pair). Therefore dominant *RR* and *Rr* will represent round seeds, recessive *rr* a wrinkled seed. *RR* and *rr* represent homozygous dominant and recessive gene pairs, respectively. *Rr*, the heterozygous gene pair, contains one of each of the allelic genes.

Gene combinations in squares 1 and 4 are homozygous: 1, a round seed; 4, a wrinkled seed. Squares 2 and 3 represent heterozygous round seeds containing the genes *R* and *r*. There is a 3:1 phenotype ratio of round-seeded (1, 2, and 3) to wrinkled seeded (4). The 3:1 ratio is, of course, a mathematical ideal that is seldom realized in a small number of offspring because of chance. As in tossing a coin, where the chance of getting a head is equal to that of getting a tail, one really does not expect always to get one head and one tail in every two tosses. The coin may come up heads twice or tails twice just by chance. In a large number of tosses, however, the result usually is close to the expected 50:50 ratio of heads to tails. Similarly, if one counts enough seeds produced by the above cross, one will find a very close fit to the expected 3:1 ratio of round seeds to wrinkled seeds.

Mendel was able to predict the probability of the appearance of any of the seven traits he studied in his pea plants, regardless of whether he studied the inheritance of a single factor or of more than one factor in a given mating. For example, in a monohybrid cross he followed one segregating pair of factors, in a dihybrid two segregating pairs of factors, in a trihybrid three segregating pairs of factors, etc. Mendel was able to predict the ratio of the phenotypes in each cross, since each pair of factors behaved independently of the other pairs of factors in a cross. In a monohybrid cross, Mendel found two phenotypes in a ratio of 3:1. In a dihybrid cross, he found four phenotypes in a ratio of 9:3:3:1.

The dihybrid cross shows how individual gene pairs segregate. We can see in Figure 8–6 that there is an assortment of the genes for seed shape and color. There are 16 possible was in which the heterozygous dihybrid gene pairs can come together at fertilization. Squares 1 and 16 are identical in genotype and phenotype to the original parents (1). Squares 4, 7, 10, and 13 (see diagonal arrow) are genotypically and phenotypically like their parents (2). Squares 2, 3, 5, 6, 8, 9, 11, 12, 14, and 15 represent new gene combinations, which in turn give rise to two new phenotype classes; round green seeds (squares 6, 8, and 14) and wrinkled yellow seeds (squares 11, 12, and 15). Thus Mendel was able to predict that, as a result of mating heterozygous F_1 parents, *9* of 16 plants would have round yellow seeds, *3* of 16 would have round green seeds, *3* of 16 would have wrinkled yellow seeds, and *1* of 16 would have wrinkled green seeds. The four phenotypic classes are present in a ratio of 9:3:3:1 (see diagonal arrow).

Although Mendel must have been delighted that he could predict the phenotypic results of a dihybrid cross, of even greater importance to him was the mounting evidence that a pair of unlike factors had been in close contact for a generation or more, yet remained pure in their later expression. In a letter to another scientist, Mendel wrote, "The course of development consists simply in this: that in each successive generation the two primal characters issue distinct and unadulterated out of the hybridized pair, there being nothing whatever to show that either of them had inherited or taken over anything from the others" (Moore, 1975). Thus when sex cells are formed in a hybrid plant and the paired factors segregate from each other and assort, each gamete receives only one member of each gene pair. The gametes of a monohybrid parent plant will contain, for example, either the round factor or the wrinkled factor for seed shape; the gamete never contains both factors for shape.

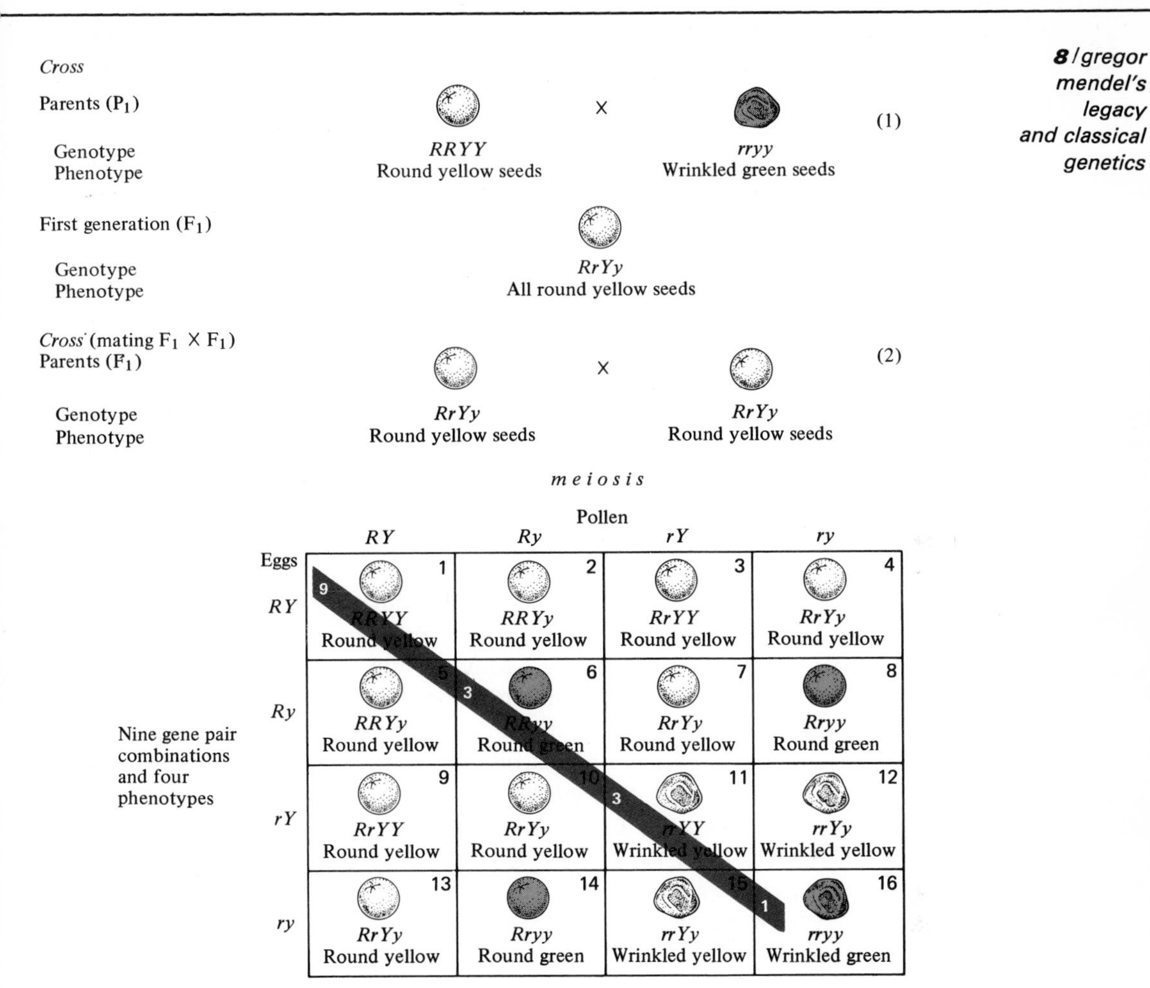

Figure 8–6. Punnett Square Method for Deriving the Possible Phenotypes and Genotypes in a Dihybrid Cross (Mating). Gene *Y* represents a yellow seed color and is dominant to its allele *y*. Gene *R* represents a round seed and is dominant to its allele *r*. Therefore *YYRR* represents a plant with homozygous dominant yellow and round seeds; *yyrr* represents a homozygous plant with green-colored and wrinkled seeds; *YyRr* represents a heterozygous plant with yellow and round seeds.

Mendel's three principles of inheritance can be summarized as follows:

1. Principle of Segregation: When sex (germ) cells are formed, the two factors (genes) that control a trait segregate or separate, each cell receiving only one factor.
2. Principle of Independent Assortment: The members of one pair of factors do not influence the way in which other pairs of factors are distributed.
3. Principle of the Purity of Gametes: Factors are unaffected by their association in the individual (organism); each factor emerges pure and distinct in the gamete.

In essence, Mendel showed that genetic information is transmitted from parent to offspring. About 100 years later Crick and Watson showed how that information is duplicated and transmitted in the form of a double-stranded mucleic acid, DNA (see Chapter 5 for discussion of DNA–RNA and gene function).

Mendel's remarkable success with his main series of experiments may be attributed largely to the plant with which he chose to work, *Pisum sativum.* It was *later* learned that each of the seven pairs of segregating factors, or genes, that Mendel studied is located on a different one of the plant's seven different pairs of chromosomes. Without having seen a chromosome, Mendel had chosen the maximum number of characteristics that could be studied in the pea without having at least two of them lie on the same chromosome. The probability of picking seven traits at random and having each gene located on each of the seven chromosomes is approximately 1 in 163. Because the seven genes that controlled the seven traits were located on separate chromosomes, Mendel never ran into the problem that later geneticists encountered—linkage. At least he did not write about it.

When genes are linked, that is, located on the same chromosome, they cannot assort independently from each other during gamete formation. Thus Mendel's second principle becomes invalid, since in a dihybrid cross one does not get a 9:3:3:1 phenotypic ratio. Mendel must have done a good deal of observing and experimenting before his formal work and gained insight into what he could expect in order to find the seven readily observable characteristics that separate into clearly distinct phenotypic ratios. As Fisher put it, "It is as though Mendel knew the answer before he started, and was merely producing a demonstration" (Sturtevant, 1965).

Mendel's phenotypic ratios also appear to be closer to expectation than statistics indicate they should be. For example, in the F_2 generation of the mating of yellow and green seeds, Mendel's count revealed numbers of 6022:2001, yellow to green seeds, a deviation of *5* from 3:1. There should have been a deviation of at least *26.*

This same statistical perfection runs through all Mendel's data. Possible explanations for this may be that Mendel had an unconscious tendency to classify somewhat doubtful individuals in such a way as to fit the expectations, or that students or assistants, who were aware of Mendel's expectations and wanted to please him, may have made some of the counts. Yet these explanations seem out of character; Mendel was a meticulous worker. Perhaps all we can say, after viewing the quality of the rest of Mendel's thinking, is that *he was right*! Investigations since the rediscovery of Mendel's work indicate that if gene segregation and gene assortment are to occur as Mendel observed,

the physical basis: transmission of units of hereditary material through the genes

1. The genes studied must be located on different chromosomes.
2. The genes studied must have a dominant–recessive relationship.
3. There must be a random assortment of chromosomes during the develoment of the gametes.
4. The genes studied must produce viable progeny that survive, so that they can be counted.
5. The separate genes must affect separate characteristics. They must not affect the *same* characteristics.

6. The separate genes must be located on the autosomes because sex-linked genes (genes located on the sex chromosomes, X and Y) do not give the predicted 3 : 1 or 9 : 3 : 3 : 1 ratios.

One can calculate the number of different gametes possible by using the formula 2^n = number of different types of gametes when n equals the number of *heterozygous* gene pairs under study. For example, if one is studying the trasmission for three heterozygous pairs, a trihybrid mating, then $n = 3$, and $2^3 = 8$ different gametes will be produced by the three heterozygous F_1 gene pairs during meiosis.

mendel's report on his research to the brünn society

There is little doubt that Mendel was aware of the significance of his results and the principles he had formulated. He had read Darwin's *Origin of Species* and noted the weak spots in Darwin's explanation of how evolution had taken place. In addition, Mendel was certain that the principles of heredity he had uncovered could help bridge the gaps in the theory of natural selection, and so after eight years of continuous study, Mendel was eager to report his findings to the world.

On February 8, 1865, at the monthly meeting of the Brünn Society for the Study of Natural Science, Mendel presented the first half of his now famous paper *Versuche über Pflanzen-Hybriden* ("Experiments in Plant Hybridization"). Many of the society's forty members had expressed interest in the project Mendel had been working on for so long and were eager to learn about the pea hybridization experiments. As Mendel read his paper, however, interest rapidly gave way to incomprehension. He spoke for the hour he had been given and then announced he would present the second half of his monograph, explaining segregation of characteristics, at the next meeting. The following month, he presented his algebraic equations, but attention was minimal. The members present failed to see the significance of what he had done. When he had finished reading, there was polite applause, but not a single person at the meeting realized he had just heard a historic presentation. Mendel's work was written up with clarity and organization, and today it is considered the "beginning of formal genetics."

Disappointed with the reception of his work but confident about its value, Mendel was certain that when it appeared in the society's *Proceedings* in 1866, the situation would soon change. Surely someone would quickly recognize the importance of the monograph. Copies of the *Proceedings* were sent to Vienna, Berlin, Rome, St. Petersburg, Uppsala, and the United States, but the paper sat unread on library shelves.

Although Mendel never doubted that his work had revealed principles of importance to the science of heredity, he soon began to feel that there was not much chance of its being recognized. His confidence in the worth of his discoveries for future research never wavered, however, and in talking with

Gustav von Niessl, secretary and active member of the Brünn society, he said, in most prophetic words, "My time will come" (Iltis, 1966).

In 1868, two years after the monograph had been published, Cyrill Franz Napp, abbot and prelate of the monastery of St. Thomas in Altbrünn, died. Gregor Mendel was elected prelate of one of the wealthiest monasteries in Austria. With the thoroughness that had characterized his research, Mendel began to perform the numerous duties of his office. As prelate, Mendel found time to carry on a little research in fruit grafting and bee crossing—he made a hybrid strain of bees that gave excellent honey, but the bees were so ferocious they stung everyone for miles around and had to be destroyed (Bronowski, 1974).

The winter of 1883 was difficult for Mendel. He was suffering from heart disease and dropsy, and was forced to remain in bed. Sick or well, however, Mendel felt compelled to continue his daily recording of the various instruments in his garden. On Sunday, January 6, 1884, Gregor Johann Mendel died at St. Thomas at the age of 62. At his funeral people of all faiths gathered at the Brünn cemetery to pay tribute to Mendel's good nature and humanity, but not one of those present was aware of his extraordinary scientific achievements in the field of heredity. The Augustinians elected a new prelate, and many of Mendel's papers were burned.

the rediscovery

Thirty-four years of silence would pass before Mendel's *time would come.* A monograph containing the important principles of Mendelian inheritance was discovered in 1900.

Mendel's misfortune was to be ahead of his time. It was years before the scientific world was ready to appreciate what he had done. First there was an intensive study of cell structure and then, from 1875 to 1885, of meiosis. Finally, scientists began to investigate the factors of inheritance as Mendel had done about four decades earlier. In many parts of the world investigators were turning to hybridization experiments, as Mendel had done, in hopes of discovering the principles of heredity that had eluded them for so long. Like Mendel, they too concentrated on a few contrasting characteristics and came to regard the numerical ratios they uncovered as extremely important. Unaware that they were not alone in their observations, three such investigators were getting ready to publish papers describing the results of their hybridization experiments.

Hugo De Vries, a Dutch botanist, was first to publish; in March 1900 two of his papers on hybridization appeared. There is a story to the effect that De Vries at first intended to suppress any reference to Mendel and thus claim that he had worked out the Mendelian scheme for himself, but in the second paper he gave Mendel full credit for the discovery. In April of the same year, a German botanist, Karl Correns, published a paper, *Gregor Mendel's Rules Concerning the Behavior of Racial Hybrids.* Correns praised Mendel's work as a beautifully organized union of experiments and interpretation, and proposed the name Mendelian rules for the principles of segregation and independent assortment. In June of 1900, Erich Tschermak, an Austrian botanist, published a paper on pea hybrids in which he recognized Mendel's work.

mendel's legacy

The simultaneous rediscovery of Mendel's principles by three botanists working independently aroused great excitement in the scientific world. Investigators began to experiment with hundreds of different plants and animals in an effort to verify Mendelian principles. In 1910 T. H. Morgan began intensive research with the fruit fly (*Drosophila melanogaster*), the result of which led to the most complete and exact understanding of heredity that mankind had achieved—the gene theory.

The first work with genetics was primarily concerned with evolutionary problems, but soon the principles Mendel had uncovered were applied to the improvement of existing agriculture and livestock. Genetic hybridization has contributed to the development of disease-resistant, higher yielding varieties of cereals, flowering plants, and vegetables as well as poultry, pigs, and cattle (Figure 8–7). Human inheritance has also been carefully studied through statistical methods, and in many cases the patterns of inherited human diseases, such as albinism, hemophilia, and cystic fibrosis have been determined.

Gregor Mendel's work established the principles of gene segregation and assortment and gave rise to the formal science of genetics. Our tribute to Mendel is our use of what he discovered.

continued investigations in classical genetics based on mendel's principles of gene segregation and assortment

Although Mendel's classical principles of gene segregation and independent gene assortment were immediately confirmed after their rediscovery in 1900, scientists were not certain that these principles were universal to the heredity of *all* organisms. Monohybrid matings did not always give a 3:1 phenotypic ratio. In the mating of chickens, for example, when a black-pigmented male mated with a white female, they produced Andalusian blue offspring (Figure 8–8). Mating of the F_1 Andalusian blue fowls resulted in three phenotypes—black, Andalusian blue, and white fowls—in a ratio of 1:2:1 instead of the expected 3:1 black to white fowls. Actually, Mendel knew that the 3:1 ratio did not always occur. After mating the red-flower variety of the four-o'clock plant to a white-flowered variety, he had observed that the new plants had *pink* flowers. It appears, however, that Mendel chose to ignore the exceptions and report only those data from his pea plant experiments that he could explain theoretically. In experiments with dihybrid matings, instead of observing four phenotypic ratios in a 9:3:3:1 ratio, as one would expect, investigators observed, in some cases, two phenotypes in ratios of 15:1, 9:7, and 13:3 or three phenotypes in a 9:4:3 ratio.

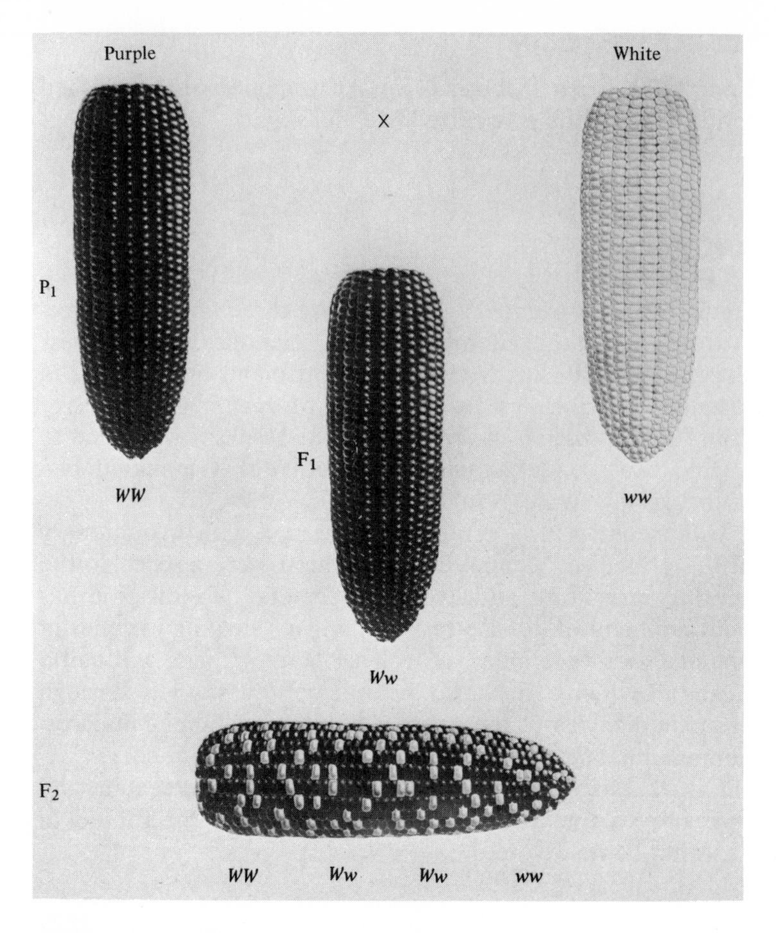

Figure 8–7. Mating in Corn. A genetic cross of a homozygous, *WW*, purple-seeded parent with a homozygous, *ww*, white-seeded parent. The F_1 or first filial generation is all purple seeded, *Ww*, showing that purple is dominant to white. Note if the F_1 were crossed to a similar F_1, in the F_2 or second filial generation the white seed would reappear. The count of kernels in the F_2 would show that three of every four seeds are purple, as one would expect based on Mendel's principle of independent gene segregation. (Courtesy of Sargent-Welch Scientific Company.)

the physical basis: transmission of units of hereditary material through the genes

It was soon found that the "strange" results from mono- and dihybrid matings were not exceptions to Mendel's principles; they were additional support of his principles. The instances of monohybrid "blending" were cases of incomplete dominance; neither allele of the gene pair dominated the other and, when the two alleles were present as a heterozygous gene pair, as in the Andalusian fowl, they gave rise to a third phenotype. In humans the phenomenon of incomplete dominance is called codominance. Examples are the MN blood genotype, where the person has both M and N antigens in his red blood cells (discussed in Chapter 12), and the sickle cell heterozygote, where the individual makes both normal and abnormal hemoglobin. In the dihybrid observations the phenotypic ratios were altered because, although the genes were on separate chromosomes, two gene pairs affected the same trait. The phenomenon of two or more genes affecting the same trait is sometimes

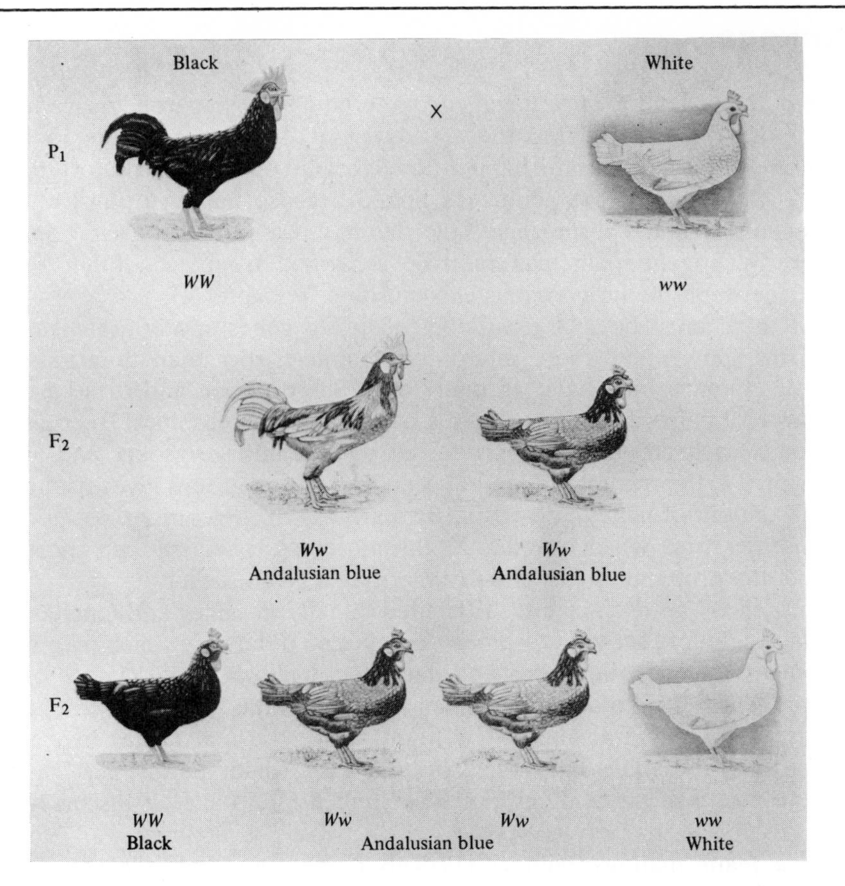

Figure 8–8. Andalusian Fowl. The mating of a black male with a white female gives rise to an "intermediate" Andalusian blue F_1. When F_1's are mated, the offspring are black, Andalusian blue, and white in a 1:2:1 ratio instead of the expected 3 black to 1 white. (Courtesy of Sargent-Welch Scientific Company.)

referred to as gene interaction. The interaction of genes is thought to occur at the phenotypic level, that is, the separate genes code for proteins that interact after their production.

It has been shown that in some cases where more than one gene is involved in the production of a phenotype, one gene may "mask" another. The masking effect is referred to as epistasis. For example, several genes are involved in the final color of your eyes. Yet if a person is an albino, the gene for albinism masks the effects of all other genes for eye color. Even though the genes for eye color are normal, they are not expressed. The eye lacks all pigment. This same gene for albinism masks the genes for pigment over the entire body, since melanin is not produced. Gene pairs located on separate chromosomes segregate independently and give rise to the expected genotypic ratio. But if the gene products interact because the genes affect the same trait, the phenotypic ratios may vary; that is, the phenotypic ratios do not always reflect the genotypic ratios in the same way. Since it is the genes that are transmitted and not the phenotype, Mendel's principle of gene segregation is as true today as it was when he did his original experiments.

gene linkage

One problem that Mendel may have encountered but disregarded is gene linkage. Suppose one is studying the transmission of two gene pairs in a dihybrid mating. If both gene pairs happen to be located on the same autosome or on the X chromosome, the genes are linked because they are physically located on or in the same chromosome, and they are passed into the same sex cell together. Such gene pairs can only be separated from each other by crossing-over between the homologous chromosomes (Chapter 4).

Logically, there must be a large number of linked genes in any organism, since organisms show many more inheritable characteristics than they have chromosomes. Humans may have as many as 100,000 genes, and these are distributed over 22 pairs of autosomes and a pair of sex chromosomes. Because of such linkage, each chromosome is referred to as a linkage group, and as stated in Chapter 4, the recombination between the linkage groups gives rise to an enormous amount of genetic diversity. By learning which genes are located on the autosomes and which on the X chromosomes, scientists can more readily predict the probability of a child's inheriting a given defect.

Genes located on autosomes are distributed equally in males and females, since both have a diploid set of autosomes. It makes no difference which parent has the "defective" gene; their sons and daughters have an equal chance of inheriting that gene. But if a defective gene is located on the X chromosome, a different pattern of gene transmission is found, since the female has two X chromosomes and the male only one. In this case which parent has the defective gene makes a major difference. (Sex-linked inheritance is discussed Chapter 11.)

In his pea plant matings Mendel checked to be sure that it made no difference which parent carried the trait. It made no difference, for example, whether the male or female plant had the wrinkled or round seeds; he still obtained the expected 3:1 phenotypic ratio. We can be certain, therefore, that the genes that Mendel studied were located on the autosomes of the pea plant.

Mendelian patterns of inheritance, dominant, recessive, and X-linked, are discussed generally below. Then specific effects of these modes of inheritance are presented in Chapters 9, 10, and 11.

consideration and application of mendelian patterns of inheritance in humans

Predicting the outcome of a genetic cross or human mating where a single gene is involved is relatively easy, thanks to our understanding of Mendel's principle of gene segregation. Using the Punnett square method, one can follow a human mating in exactly the same way that the results of mono- and dihybrid matings of pea plants are followed.

In discussing the Mendelian patterns of inheritance as they apply to humans, we are generally dealing with the inheritance of genes that are responsible for a physical defect, or disease, the occurrence of which is rare and often severe. It is difficult to classify a common normal variant because the more common traits generally do not show a *large* difference. But genes responsible for physical defects and disease are in stark contrast to the normal phenotype. Thus in the inheritance of a disease, our attention is focused on the transmission of the diseased phenotype and whether that phenotype fits the autosomal dominant, recessive, or X-linked mode of inheritance. If both phenotypes of a trait are common, it is very difficult to assess the mode of inheritance as dominant or recessive.

Genes are neither good nor bad, and the common characteristic is not necessarily the dominant or the recessive. The more common characteristic *may be* due to a gene that acts dominantly, or it *may be* due to one that acts recessively. Thus when a geneticist tries to diagram a pedigree of family inheritance for a trait, he tries to assign the genotype for one of the known modes of inheritance. If that cannot be done, the geneticist faces a major difficulty in trying to construct a genetically accurate family pedigree. The *rare* inherited physical defect or disease is rather easy to analyze, and a pedigree diagram can be constructed.

A pedigree analysis does not solve a genetic problem; it only identifies the afflicted individuals and allows for the calculation of a probability that some descendent may inherit the defect or be a carrier. A pedigree chart is particularly valuable in determining whether the inheritance of a trait in a family is consistent with the data reported in the literature. For example, with a *common* dominant disease, the dominant gene may occur alone (*Dd*) or with another dominant gene (*DD*). In the case of a *rare* dominant disease, however, the person who shows the dominant phenotype almost always has a single dominant gene (*Dd*). (A rare homozygous dominant can arise only when both parents have the same rare dominant defective gene, a very unlikely situation.) With a recessive gene for a *common* characteristic, those persons who appear normal may be either *RR* or *Rr*. But with respect to a recessive gene for a *rare* disease, those persons who appear normal are almost certainly homozygous normal (*RR*) and carry no defective genes. In fact, we are usually safe in assuming that a normal person carries a given defective recessive gene (*Rr*) only if a parent or other direct ancestor showed the disease or if the given person has previously produced a child with the defect. A similar line of argument for X-linked traits leads to a similar conclusion for women, who, in general, will not have a defective recessive X-linked gene unless an ancestor or a previously produced child shows the defect. (Men, being hemizygous, express the gene.)

Our problem in analyzing a pedigree chart for a particular family is usually that of deciding whether the mode of transmission conforms to that of a literature reference. Can we assign genotypes to each individual in the pedigree without genetic contradictions arising? If there are conflicts, we must seek the possible reasons. Are the concepts of penetrance or variable expressivity applicable? We must always consider the possibility that a different mode of transmission is operative in a particular family. If the literature gives several modes of inheritance, we must attempt to determine which pattern is consistent with the information in our pedigree, or if more than one is consistent, which is the most probable. A new pattern of inheritance may be found, even if only one has previously been published. Other problems, such as mistaken paternity, adoption, and the possibility of a new mutation, must also be taken into account, but they will not be dwelt on here. Even when a trait appears to conform to what has been reported previously, there is chance for error, since many defects can be transmitted by autosomal dominant, recessive, and X-linked modes of inheritance.

the mendelian dominant hybrid cross

In setting up a genetic cross, symbols, or a kind of shorthand or code, are used to replace words. The symbol used is usually the first letter or combination of two letters associated with the trait. When the alternative alleles of a gene are being indicated, the dominant allele is represented by the capital letter and the recessive allele by the lowercase letter. For example, when the dominantly inherited free, or unattached, earlobes and the recessively inherited attached earlobes (Figure 8–9) are being indicated, the symbol *L* represents the dominant allele and the symbol *l* the recessive allele of the gene responsible for the attachment of earlobes.

A person who is *LL* or *Ll* has free earlobes; a person with attached earlobes is *ll*. Thus a person with the genotype *LL* is homozygous dominant and has free earlobes. The genotype *Ll* is heterozygous, but because of the one dominant gene, *L*, the person also has free earlobes. The person with the genotype *ll* is homozygous recessive and has attached earlobes.

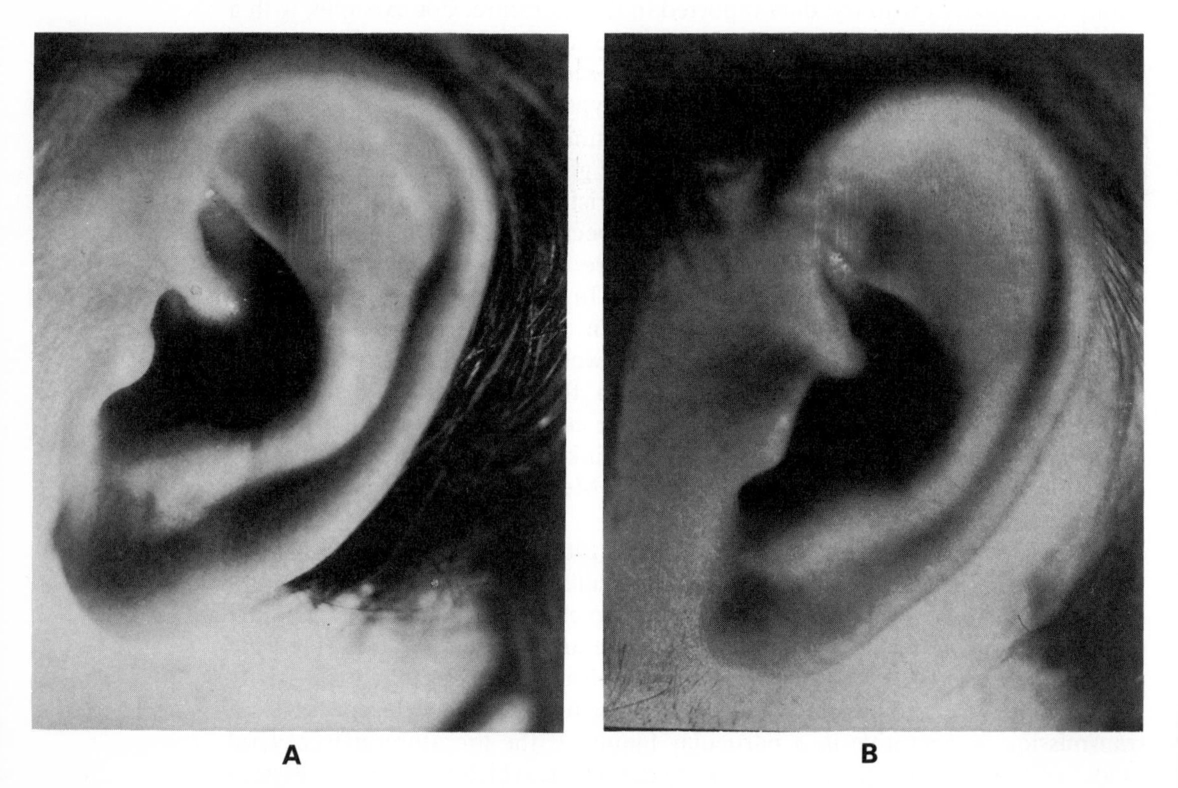

Figure 8–9. Inherited Free and Attached Earlobes. **A.** The dominant free earlobe. **B.** The recessive attached earlobe.

Example

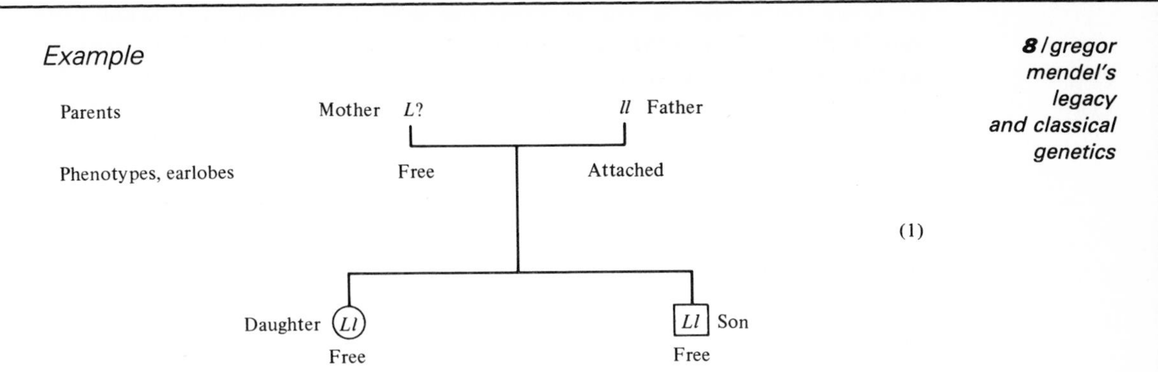

Both daughter and son have free earlobes. Since the father has attached lobes, he is *ll* and can give only *l* to his children, male or female. Since the mother has free earlobes, she must have at least one *L*. She can be heterozygous *Ll* or homozygous *LL*. Both children received an *L* from the mother, but the number of children is too small to determine the mother's zygosity. The Punnett square method for genotype and phenotype analysis is perhaps more descriptive.

The Punnett square method

1. The mother in (1) can give rise, through meiosis, to eggs containing *L* and ? (*L* or *l*) since her children both have free earlobes.
2. The father can give rise, through meiosis, only to sperm that carry a gene *l*, since he had attached lobes.

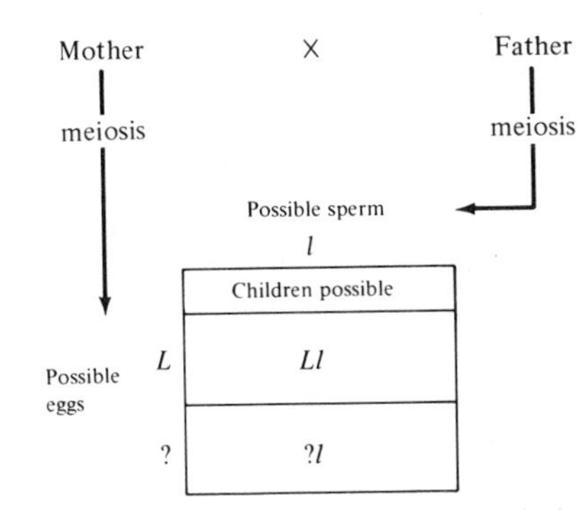

(2)

We can be certain, barring a spontaneous gene change (mutation), that on the average at least 50% of all the children will have free earlobes, since the mother must have at least one dominant gene. If her other gene is recessive, there is an equal chance of each egg carrying the *L* or the *l*; and since all the father's sperm all carry *l*, each zygote formed has a 50:50 chance of having an *L*. Since we do not have more background on the mother's parents and their parents, we cannot determine the second gene carried by the mother in Figure 8–9. If, however, the family expands, and a child with attached earlobes is born, the probability is *very* high that the mother is heterozygous *Ll*. That is the only way an *ll* child could be born to this mother, barring a fresh mutation in the egg that was fertilized.

Once it has been determined that the mother is *Ll*, each additional birth to this family will have that same one in two chance of having free earlobes. If the mother is in fact a homozygous *LL* (which cannot be proved regardless of how many children are born in this family), all the children, barring a new mutation, will have free earlobes:

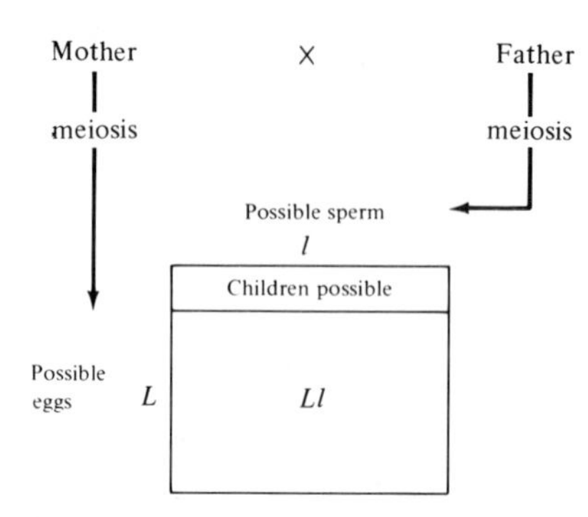

(3)

All the children are heterozygous and have free earlobes. If the mother had attached earlobes *ll*, then all the children, barring a new mutation, would have attached earlobes:

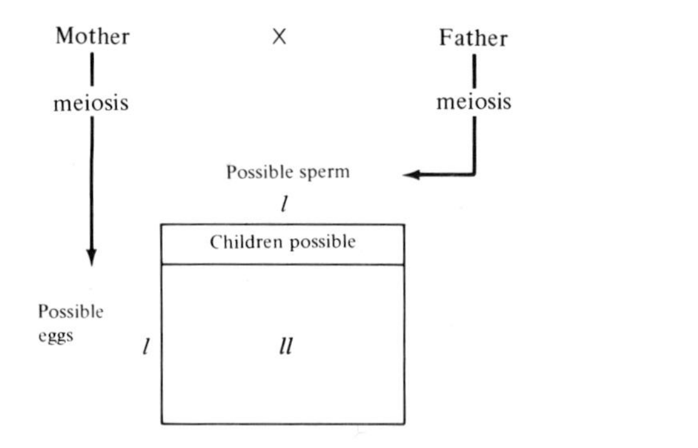

(4)

If both parents had been heterozygous with free earlobes *Ll* × *Ll*, the ratio of children with free and attached earlobes would be as follows:

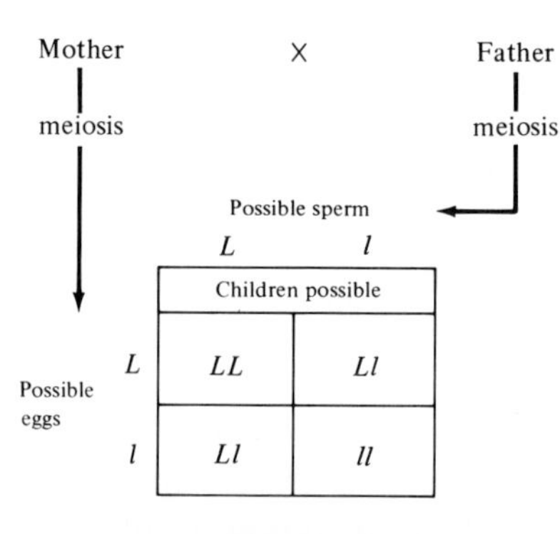

(5)

In this situation every new child has a 1 in 4 or 25% chance of having attached earlobes and a 3 in 4 or 75% chance of having free earlobes. This situation is recognized as the familiar Mendelian 3:1 ratio, which was demonstrated in the pea plant.

Some common dominantly inherited traits are shown in Figures 8–10 and 8–11. Figure 8–11 shows the dominant widow's peak and the recessive alternative, a straight horizontal hairline.

No known pathology, or disease, is associated with the dominantly inherited traits presented in Figures 8–9, 8–10, and 8–11. A number of dominantly inherited diseases are discussed in Chapter 9.

You may wish to have some fun in the classroom by determining what percentage of the class or which individuals have a dominant gene for any of the traits shown in Figures 8–9, 8–10, and 8–11. Or you might want to test for the ability to taste phenylthiocarbamide (PTC). Tasters, *TT* or *Tt*, find this chemical to be sour or bitter, while the nontaster, *tt*, does not detect a flavor.

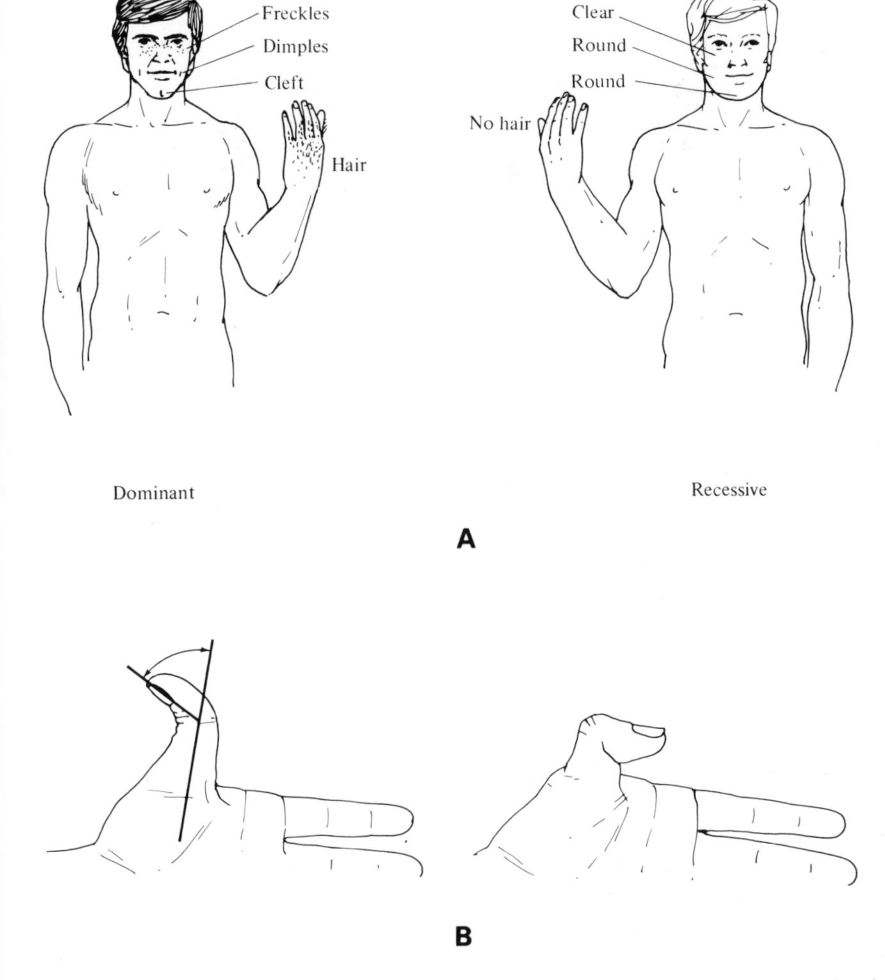

Figure 8–10. Inheritance of Some Common Traits. **A.** Contrast between dominant and recessive inheritance of common traits. **B.** Dominantly inherited hyperextensibility of the joints of the thumb. Left: Hitchhiker's thumb, the ability to bend the thumb at a sharp angle; right: pointer's thumb, the ability to "cock" the thumb at a right angle. (Courtesy of Barry McDown.)

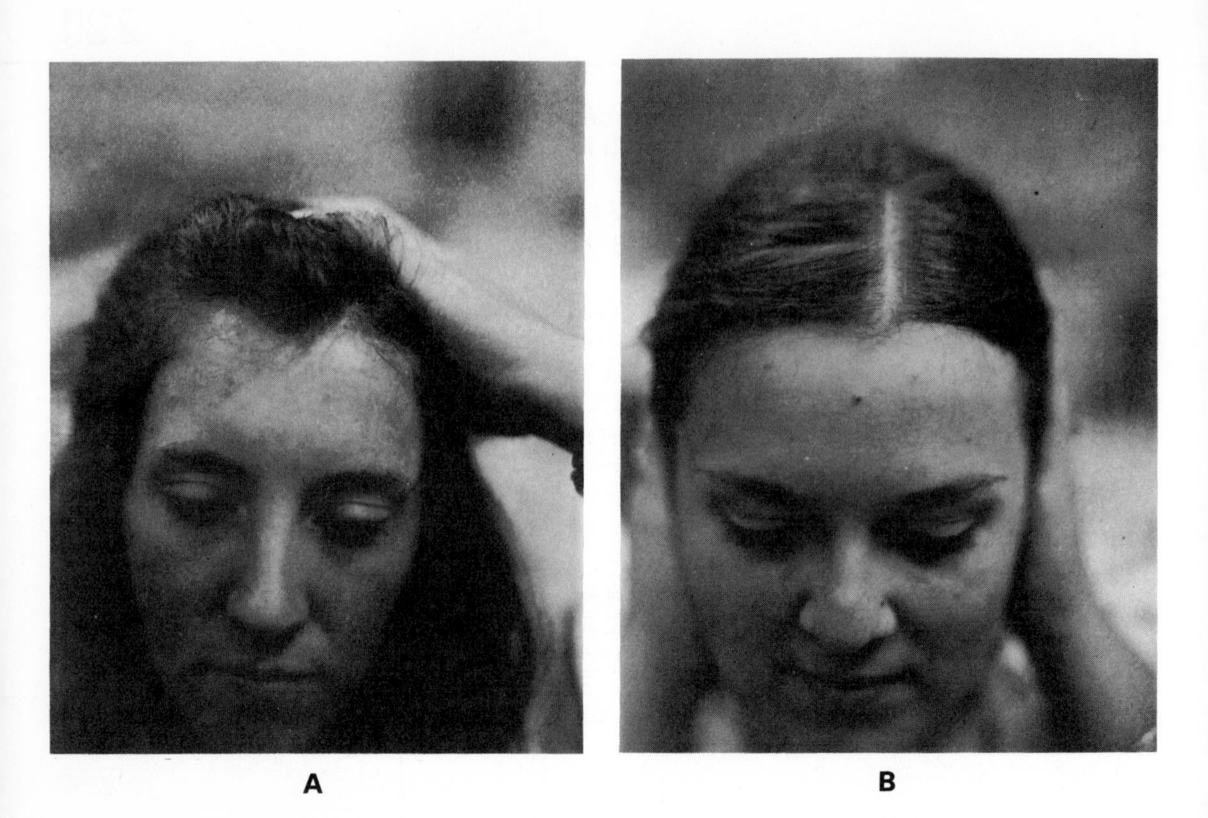

Figure 8–11. Dominantly Inherited Widow's Peak. **A.** The hairline recedes from a point, the widow's peak, directly over the forehead. **B.** The hairline is straight across above the forehead.

Paper strips with instructions for performing the test can be purchased inexpensively from scientific supply houses. As you and your classmates decide who has a widow's peak and who does not, who has free and who has attached earlobes, etc., you will quickly see that there is a gradient of gene expression. The observations will vary from those enabling easy decisions to those leading to arguments as to which trait a classmate has. Can you suggest what will have to be done in order to establish which categories the borderline cases should be in?

the mendelian recessive hybrid cross

the physical basis: transmission of units of hereditary material through the genes

The first thing the reader will probably notice from the following discussion is the similarity between the inheritance of autosomal dominant and recessive traits. Dominant and recessive traits differ only with respect to the state of the gene. In the case of the dominant, only one allele is necessary for the trait to be expressed. In the recessive case both alleles must be present for the trait to be expressed. The difference between dominantly and recessively inherited traits can be more dramatically shown in terms of an inherited disease.

Let us follow the recessively inherited cystic fibrosis through a mating and in this way see the similarities and differences between dominant and recessive traits. Persons who are homozygous dominant *CC* and heterozygous *Cc* are normal, while homozygous recessive *cc* persons have cystic fibrosis. (The incidence and social aspects of this disease are discussed in Chapter 10.)

Example

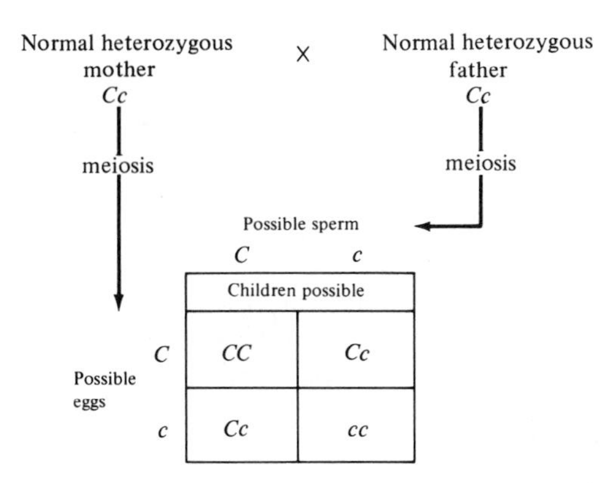

From this mating of two heterozygous *Cc* individuals, there is a 1 in 4, or 25% chance that each new birth will have cystic fibrosis, *cc*, and a 3 in 4, or 75%, chance that the child will be normal. In other words, if such a couple had four children, *on average*, one would be expected to have cystic fibrosis. But, you may ask, which child will be affected, the first, second, third, or fourth? It could be any one of the four, since the event of both recessive genes coming together in the same zygote is a matter of chance. Even more important, because such calculations give only an expected average, there is a possibility, in a given family, that none of the children—one, two, three, or all four (or more)—would be normal; that is, *all* could be affected. Further, if the first child from such a mating, is defective (*cc*), the chance of a second defective child being born into that family is still 1 in 4, or 25%, because each fertilization is an independent event. Likewise, if the first three children were born normal, and the mother is pregnant again, the chance of the fourth child being normal is still 3 in 4, or 75%.

From the Punnett square it is also obvious that there is a 67% chance that any normal child will be a carrier of the *c* gene (three out of four will be normal, and two of these three will be heterozygous *Cc*). Carriers, although generally normal themselves, serve as a source for the transmission of the defective gene to the next generation. The carrier situation is a major difference between dominant and recessive inheritance. In dominant inheritance the affected offspring obtain their gene from an affected parent, that is, the parent who expressed the defect. But in recessive inheritance the parents are usually normal; they do not show the defect. As diagrammed on the next page, in the dominant situation all heterozygous children are affected, while in the recessive situation all heterozygous children are normal. In general, in dominantly inherited diseases the affected parent is heterozygous.

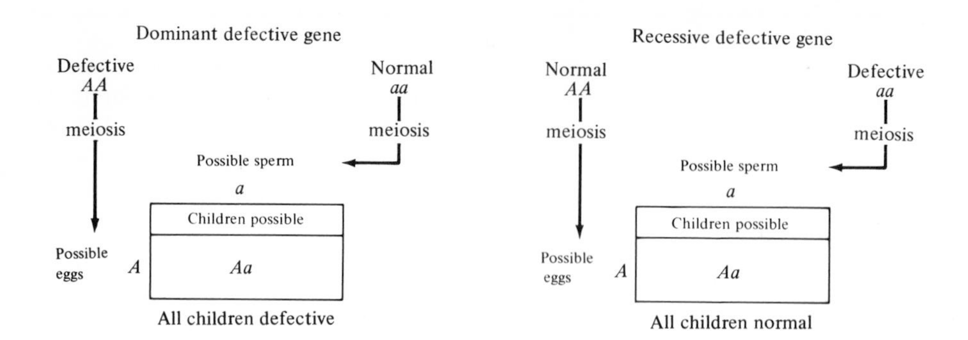

Because of the masking of the defective gene in recessive inheritance, most carriers have a defective child before they realize they are carrying a defective gene. Consequently, the genetic counselor faces more cases of distraught normal parents with one defective child than of parents who have evidence of their own dominant situation.

Another difference between dominant and recessive inheritance involves recognizing a new mutation. A new dominant mutation is not difficult to recognize if (1) neither parent has the defect, (2) the family pedigree appears to be free of the defect, and (3) the new mutation is transmitted to and observed in the next generation. However, a new mutation of a recessive gene is nearly impossible to recognize because its transfer is not immediately and directly observable. It is likely to be inherited through many generations before a normal, symptom-free heterozygote marries another heterozygote with the same recessive mutant. When that happens, they will have an affected child.

For example, Sjögren and Larsson reported a number of Swedish patients with a recessive syndrome of severe mental retardation, spasticity, and ichthyosis (scaly skin), most of whom were known to have had healthy ancestors living in northeast Sweden in the early eighteenth century. The authors suggested that an original single gene mutation probably occurred in the fourteenth century and was responsible for all the Swedish cases of this type of mental retardation occurring in the nineteenth and the twentieth centuries (Sjögren and Larsson, 1957).

Certain recessive traits have a relatively high incidence in our population. And certain recessive defects have special social overtones as they are associated with a specific ethnic group (see Chapter 15). Together, incidence and ethnic association have brought national recognition to sickle cell anemia, Tay–Sachs disease, phenylketonuria, cystic fibrosis, and the thalassemias. A number of the recognized recessive gene disorders are listed in Table 10–1. Aspects of frequency, clinical conditions, and various social implications of certain of these recessive diseases are presented in Chapter 10.

the mendelian sex-linked hybrid cross

In the early 1900's T. H. Morgan worked at Columbia University on the inheritance of various traits of the fruit fly *Drosophila*. To his surprise, while looking over his variety of stocks of mutant *Drosophila*, he observed a white-eyed male fly in the bottle that should have contained only red-eyed male and female flies! Morgan mated the white-eyed male to a red-eyed female, and *all* the offspring had red eyes. This indicated that the gene for white eyes was recessive to the gene for red eyes. Morgan then mated brother and sister F_1 red-eyed flies; in their progeny, the F_2, he found two phenotypes, flies with red eyes and flies with white eyes, just as one would expect from a single gene segregating according to Mendel's first principle. On closer examination, however, he found that *all* the *white-eyed* flies were *male*, that about half the males were red-eyed, and that *all* the females were *red-eyed*.

These observations were strange, for it appeared that eye color was associated with the sex of the fly. Morgan, however, being an experienced investigator, correctly predicted that if the genes for eye color were on the sex chromosomes, one could expect such results. He was right; further matings proved that the gene for eye color in *Drosophila* is located on the X chromosome. Morgan explained that the male has only one X chromosome, which he transmits to his daughters and *never* to his sons. The female, having XX transmits one X to each daughter as well as to each son. Thus a male can only transmit an X-linked gene to his grandchildren through his daughter. This *crisscross* pattern of inheritance only occurs when a gene is located on the X chromosome. X-linked gene transmission is distinctly different from autosomal gene transmission. The *crisscross* pattern for the inheritance of eye color in *Drosophila* is shown in Figure 8–12.

sex linkage in humans

Figure 8–13 shows the crisscross pattern of X-linked inheritance similar to that observed in *Drosophila*. Remember that the concept of a sex-linked gene being dominant or recessive applies only to the female, since only she has two homologous X chromosomes. The male is hemizygous with respect to the genes located on his single X chromosome.

To demonstrate X-linked inheritance, the gene for red color blindness (inability to see a red color) is followed through a mating sequence in Figure 8–13. The crisscross pattern of X-linked inheritance was used in the early days of *Drosophila* genetics as one way of proving that a gene was located in a chromosome. One could predict what trait would appear in the offspring by following the transmission of the parents' X chromosomes. Sex linkage also established beyond doubt that a gene existed in alternative forms. Red-color-blind parents, barring a fresh mutation, only give rise to red-color-blind children; yet a (heterozygous) normal mother and a defective father could give rise to both normal and color-blind children.

factors that can alter mendelian ratios

There are at least four factors that can alter the expected Mendelian ratios: new mutations, lethal genes, penetrance, and expressivity.

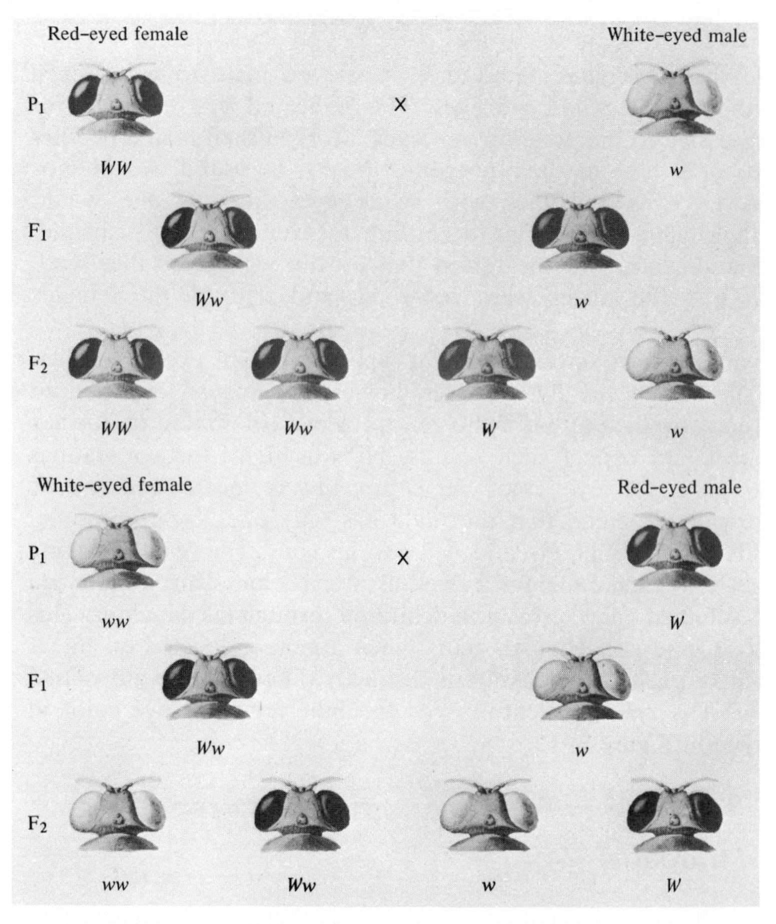

Figure 8–12. X-Linked Eye Color in *Drosophila*. When abnormal, the gene for eye color results in white eyes instead of the wild type (normal) red eyes in *Drosophila*. The gene is located on the X chromosome, and it demonstrates a crisscross pattern of father to daughter to grandson transmission. In the diagram a single *W* or *w* below the male indicates which gene is present on his X chromosome. (Males are hemizygous; that is, they have only one X chromosome.) Notice how the different gene combinations of the parents change the phenotypes observed in the F_1 and F_2 of each mating. (Courtesy of Sargent-Welch Scientific Company.)

new mutations

A gene change may occur in the sperm or egg of the parents.

lethal genes

A gene can be lethal to fetal development and can, at any time during pregnancy, cause a spontaneous abortion. In such a case, the ratio is different from what would be expected. Consider, for example, the inheritance of free versus attached earlobes; if the *LL* condition were lethal, one would find only two children with free earlobes for every one child with attached lobes.

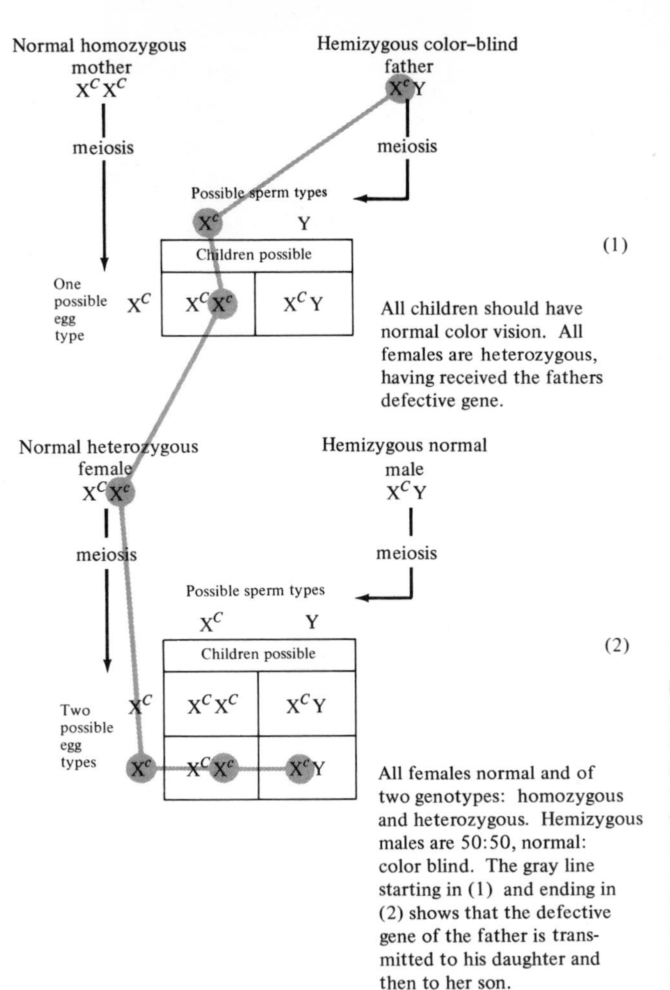

Figure 8–13. X-Linked Red Color Blindness in Humans. Homozygous *CC* and heterozygous *Cc* represent normal color vision in females; *cc* represents female red color blindness; *c*, hemizygous red color blindness in males; *C*, normal vision in hemizygous males.

Lethal genes exist in both plants and animals. In mice, for example, a single gene, *T*, is associated with a reduction in the length of the tail. If interferes with the development of the notochord (a rodlike formation of cells giving rise to the tail and other tissues) in the area of the tail close to the mouse's body. The *T* mutant gene, which restricts tail development, is referred to as brachyury. The homozygous *TT* condition is lethal (Figure 8–14). A second mutation $t°$ of the brachyury locus (the location of the gene on a chromosome) produces complete taillessness. The $Tt°$ condition allows for live birth of mice with no tail. However, the homozygous $t°t°$ is also lethal, and the ratio of expected offspring from $Tt° \times Tt°$ is completely altered. Only tailless mice $Tt°$ are live born (Figure 8–15).

In humans there are several recognized lethals that express themselves during pregnancy or within hours or days after birth. One such homozygous lethal situation occurs in the dominant disease achondroplastic dwarfism. Only the heterozygous dwarf and homozygous recessive survive (Sutton, 1902; Harris and Patton, 1971).

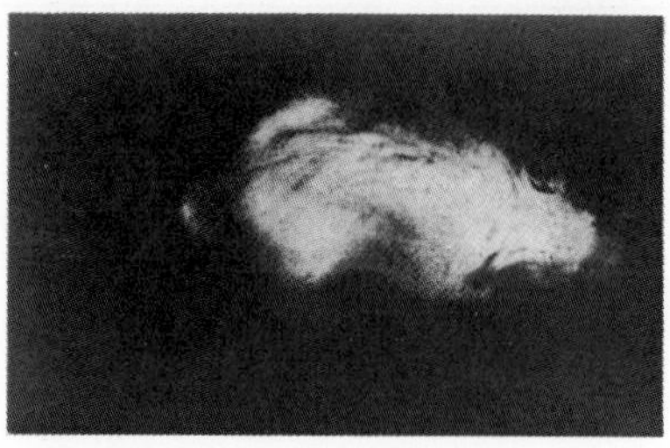

Brachy $T+$

×

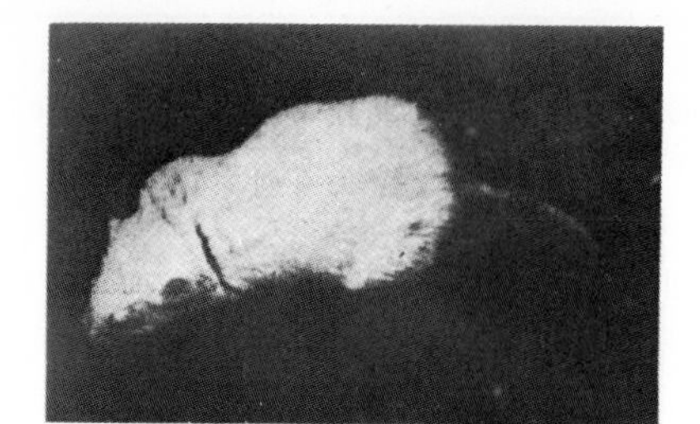

Brachy $T+$

↓

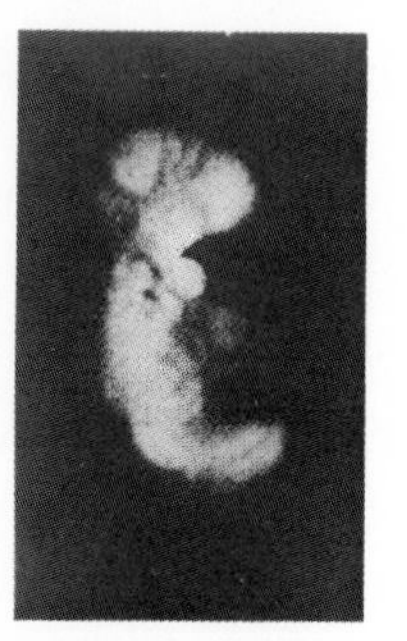

Die at 10¾ days

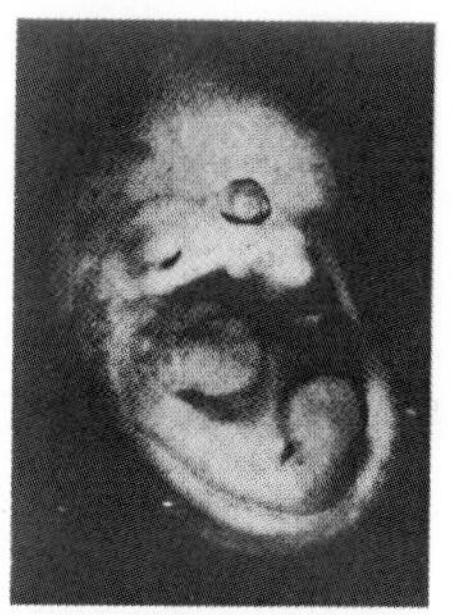

11 days

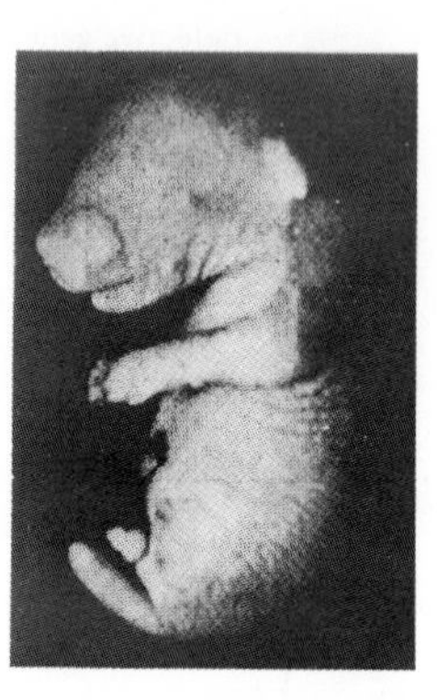

16 days

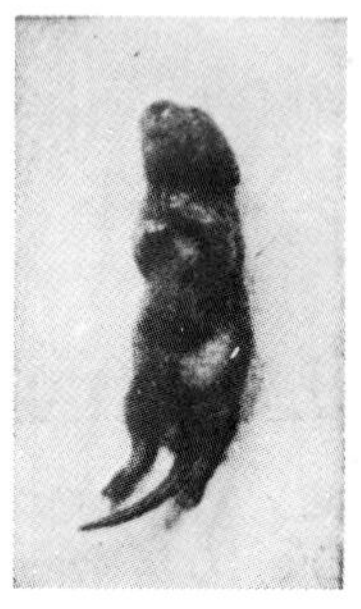

At birth

25% monster *TT*

50% brachy $T+$

25% normal ++

Figure 8–14. Inheritance and Effects on Development of the Brachyury Mutation in the House Mouse. Note the constriction that marks the end of the notochord in the 11-day heterozygous embryo. The tail beyond this point has been reabsorbed in the 16-day embryo. Homozygous *TT* condition is lethal; the heterozygous *Tt* and homozygous *tt* are viable. The lethal *TT* condition causes a 25% reduction in litter size. (From *Principles of Genetics* by Sinnott, Dunn, and Dobzhansky. Copyright ©1958 by the McGraw-Hill Book Company, Inc. Used with permission of McGraw-Hill Book Company.)

There is some discrepancy in the literature concerning the use of the terms *lethal* and *semilethal*. Here the author defines a lethal gene as one that is *expressed* between conception and the first year after birth. Thus a person with, for example, Tay–Sachs disease, whose expression is detected at six months, has inherited a lethal gene even though he may not die until age four. A semilethal gene is one that is expressed after the first birthday. The definitions assume there is no medical intervention. Examples of human lethal and semilethal genes are listed in Table 8–1. The gene responsible for the oral-facial-digital syndrome (OFD-I) is believed to be inherited as a dominant X-linked lethal in males (Stern, 1973). The OFD-I syndrome is a striking abnormality of the face, mouth, fingers, and toes (Figure 8–16). Affected females transmit the trait to about half their daughters, and the sex ratio of children born to affected females is two females to one normal male. Such data

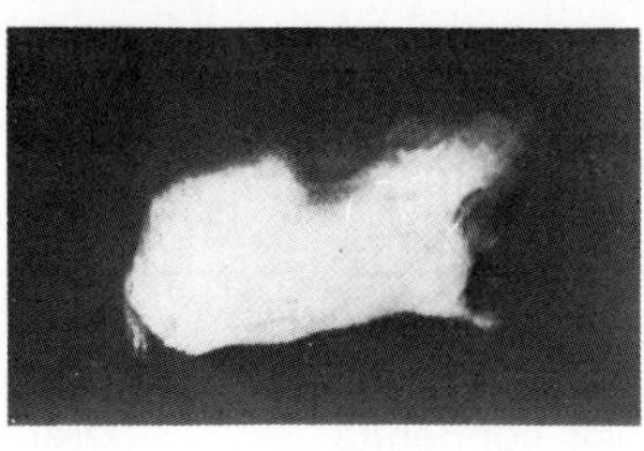

Tailless Tt°

×

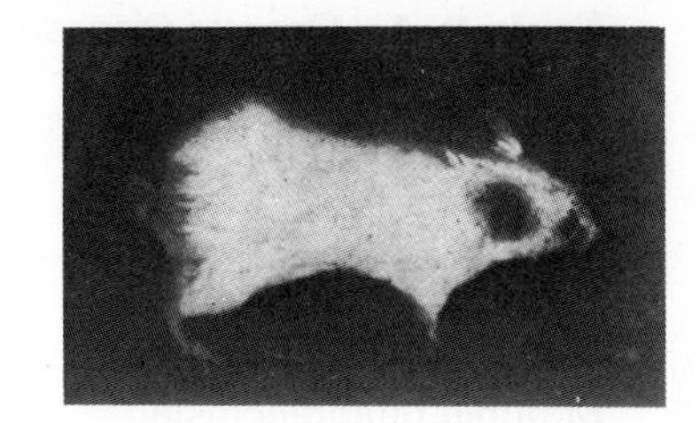

Tailless Tt°

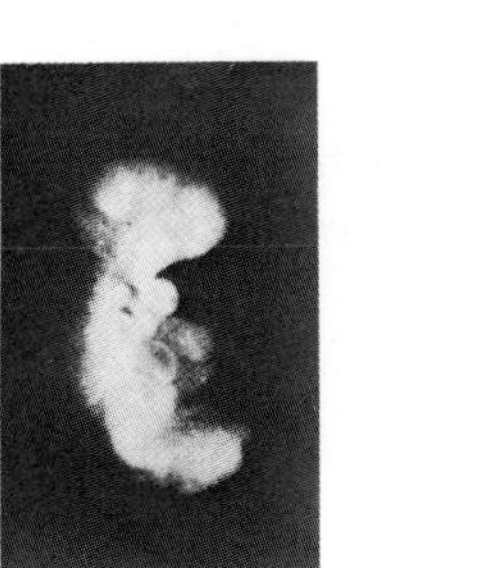

Die at 10¾ days

TT

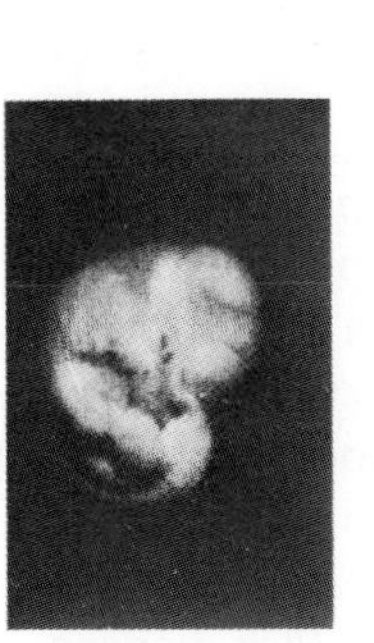

11 days

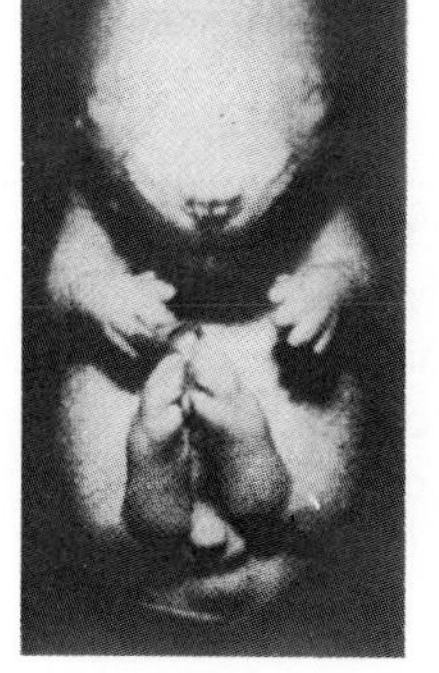

16 days

50% born tailless

Tt°

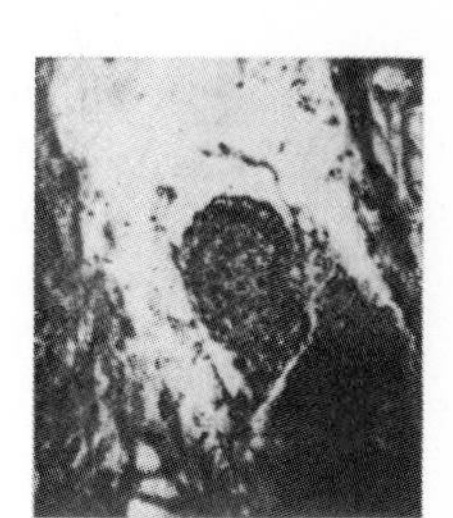

Die at 5 days (no mesoderm)

$t^{\circ}t^{\circ}$

Figure 8–15. Inheritance and Development of the Tailless House Mouse. If the mouse is homozygous *TT*, death stops development on the eleventh day, and there is partial absence of the notochord and mesodermal-derived tissue. Tt° mice survive but are tailless; $t^{\circ}t^{\circ}$ mice die after five days of incomplete development. Litter size is reduced by 50%. (From *Principles of Genetics* by Sinnott, Dunn, and Dobzhansky. Copyright ©1958 by the McGraw-Hill Book Company, Inc. Used with permission of McGraw-Hill Book Company.)

are evidence for the transmission of a dominant lethal gene. The other extreme is a semilethal gene that generally expresses itself at about age 30, with death usually following within 5 to 15 years (see Figure 9–7).

As Table 8–1 shows, lethal and semilethal diseases are transmitted with dominant, recessive, and X-linked patterns of inheritance (see Chapters 9, 10, and 11). It is very important to note that we should not associate good or bad with dominant, recessive, or X-linked. Too often students consider a dominant trait as good and a recessive trait as bad, just as they erroneously think a dominant gene is common and a recessive gene is rare. These are patterns of inheritance, and have no relationship to good or bad.

penetrance and expressivity

Expected ratios of Mendelian crosses can also be altered by lack of penetrance, that is, the failure of a gene that is present in the individual to be expressed.

Table 8–1 Genetic Diseases Associated with Lethal and Semilethal Genes

Lethal Genes (expression occurs between conception and first birthday)		
Dominant	*Recessive*	*Sex-Linked*
Hypercholesterolemia	Achondroplastic dwarfism†	Hydrocephaly
Ichthyosis vulgaris*	Cystic fibrosis	Ichthyosis vulgaris
Multiple telangiectasia	Ichthyosis congenita	Oral-facial-digital syndrome (female)
	Ichthyosis gravis	
	Tay–Sachs disease	
	Thanotophoric dwarfism	
	Werdnig–Hoffmann disease	
Semilethal Genes (expression occurs after first year of life)		
Dominant	*Recessive*	*Sex-Linked*
Huntington's chorea	Familial polycystic kidneys	Duchenne muscular dystrophy
Myotonic dystrophy	Familial polyposis	
Osteogenesis imperfecta	Fanconi's syndrome	
Retinoblastoma	Hurler's syndrome	
Tuberous sclerosis (epiloia)	Hyperlipoproteinemia type I	
Tylosis	Maroteaux–Lamy syndrome	
Xeroderma pigmentosum (mild form)	Morquio's syndrome	
	Niemann–Pick disease	
	Sanfilippo's syndrome	
	Sickle cell anemia	
	Sjögren–Larsson syndrome	
	Thalassemia major	
	Wolman's disease	
	Xeroderma pigmentosum	

*In some pedigrees the pattern of inheritance is that of an X-linked recessive gene.

†The homozygous dominant is lethal; the heterozygote is dwarf; the homozygous recessive is normal.

Expressivity is the *degree* to which a gene expresses itself, providing the gene is penetrant (shows *any* expression). Thus a gene must be penetrant before there can be expressivity. Thus of two separate persons carrying the same defective genotype:

1. One may be totally free of the disease and the other affected—penetrance.

or

2. Both may have the disease, but with different degrees of severity—expressivity.

Let's look at a number of inherited diseases and see how penetrance and expressivity are involved.

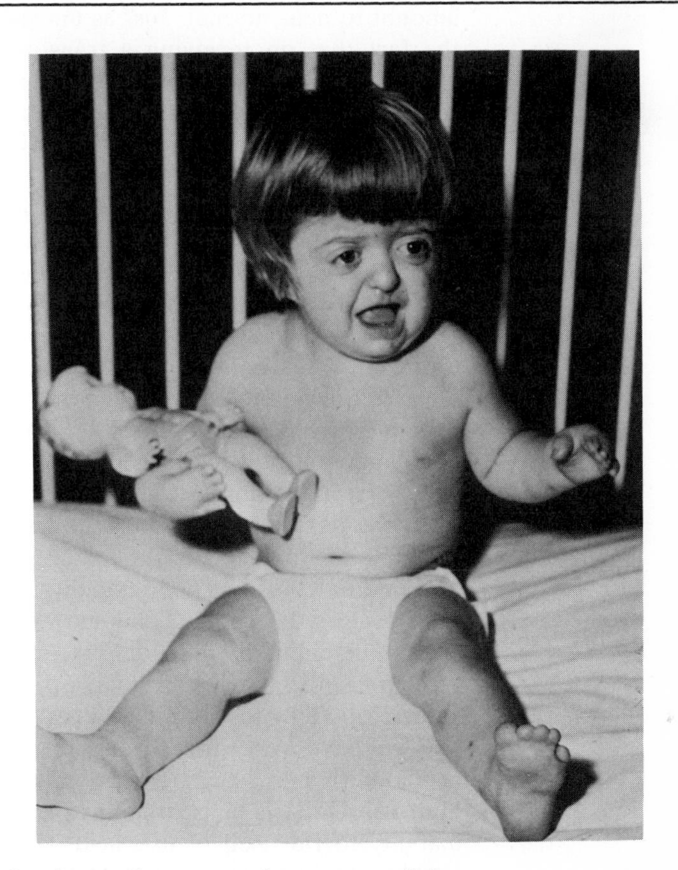

Figure 8–16. Oral-Facial-Digital Syndrome (OFD-1). This sex-linked dominant disorder interferes with mental and physical development and possibly is lethal in males. Note the fusion of fingers and toes (syndactyly), the flat appearance of the midface area, and the broad nose bridge area. (By permission of Charles H. Carter, M.D.)

Retinoblastoma is a rare autosomal dominant disease causing cancer of the eye. The dominant gene has been reported to be located on the long arm of chromosome 13 (Medical News, 1973). The gene usually expresses itself in the first six years of life. If the affected eyes are not removed surgically, the cancer spreads and leads to early death. Surgery will save the child's life, but if it involves both eyes (as in about 30% of the cases), it will leave the child blind.

The penetrance for retinoblastoma is about 80%. Thus, although children in each successive generation in a family line carry the dominant gene, 20% show no phenotypic evidence of carrying the gene. That is, in one of five of these persons, the gene does not express itself. In those children affected with retinoblastoma, the degree of expressivity is high. Penetrance is less than 100%, but the expressivity is severe, affecting one or both eyes.

Von Willebrand's disease is a dominant defect that is particularly frustrating to the clinician. The defective gene is associated with a deficiency of blood factor VIII, an antihemophilic factor needed for the normal formation of blood clots. This disease should not be confused with the sex-linked classic hemophilia A, a much more severe bleeding defect, also associated with the lack of active blood-clotting factor VIII.

Von Willebrand's disease is not a rare disorder, but it is frequently misdiagnosed because of the extreme variability of expressivity. The variability is measured by the amount of blood factor VIII present. Females with the disease show a remission during pregnancy. During routine testing the patient's level of factor VIII can change on a daily or weekly basis from a very low amount to near normal. Just as unexplainable (since we don't understand it) is the fact that on receiving a transfusion of normal plasma (plasma is blood minus the cells), the patient shows a sustained rise in factor VIII that is greater than can be accounted for by the infused plasma. The same rise in factor VIII occurs when plasma from a hemophilic patient is given to a von Willebrand's patient.

Because the von Willebrand's patient usually has more blood-clotting factor VIII than a person with hemophilia A or hemophilia B, the total clinical picture is less severe for the von Willebrand's patients. However, the variability of expressivity is such that some individuals can appear similar to hemophiliacs.

The author is reminded of a case of von Willebrand's disease that he observed as a genetic counselor. The patient had always bruised easily, but had never required treatment. Her first severe hemorrhage, of the leg, resulted from a bicycle accident at age 12. The severe pain, swelling, and hemorrhaged (purple-colored) bruises were diagnosed as a sprained knee. It took eight weeks for the bruises to clear up, and they left the knee weak and painful. Over the next six years, she required numerous casts for bruises of her joints that had begun to bleed. At age 19 her knee collapsed without apparent cause and internal bleeding was evident. She was hospitalized for 17 days. At this time doctors determined that she was low on blood factor VIII and was not a hemophiliac, but a von Willebrand's patient. Prior to tooth extractions, she was given plasma factor VIII to control bleeding, and the technique was very successful. Her orthopedist told her there is very little that can be done for her. She will have to get used to learning to live with "bone cysts and arthritis."

Gout (hyperuricemia) is another of the variations that can occur among humans. Gout is a dominantly inherited abnormality involving the body's inability to rid itself of uric acid. Until recently, hyperuricemia was a disease peculiar to humans and the Dalmatian dog. Other laboratory animals have uricase and related enzymes that degrade purines further than uric acid. However, scientists have recently learned that potassium oxonate is an excellent inhibitor of uricase. When rats are fed 5% potassium oxonate plus 1% uric acid, they become hyperuricemic in two days. Uric acid deposits in the kidney in a solid form. When the oxonate challenge is removed, the rats quickly return to a normal condition. So rats can now be used to learn more about the effects of gout and its treatment.

Humans lack uricase and therefore excrete purines as uric acid. Those who cannot excrete it fast enough develop symptoms of gout. As uric acid accumulates in the bloodstream, crystals of the weakly soluble uric acid deposit at the joints. The accumulation of uric acid in the joints causes rather severe pain. Besides the pain and discomfort, the person with gout must forgo some of the most delectable foods such as lobster. In the human form of gout, penetrance and expressivity are extremely variable. The expressivity is so variable that in some families many generations may pass without gout being detected. Heterozygous persons show an elevated level of uric acid in their bloodstream; yet only 10% of the heterozygous males and less than 10% of the heterozygous females ever develop gout. Even in those that develop gout, the symptoms and the intensity of the pain vary, as does the location of the uric acid deposits. The expressivity, or degree of expression, is dependent on diet, the metabolic output of uric acid and the rate of its breakdown, the amount of uric acid deposited at a joint, and, as is commonly known, the weather conditions. The heterozygotes who do not experience the afflictions associated with gout are referred to as dominant carriers. The dominant carrier must not be confused with the more commonly discussed heterozygous carrier of a defective gene in recessively inherited gene defects.

Ehlers–Danlos syndrome, another dominantly inherited defect, is similar in humans, dogs, and mink. This disease of the connective tissue has the same mode of inheritance and observable physical abnormalities in all three species.

Table 8–2 Examples of Penetrance and Expressivity in Human Mendelian Inheritance

	Description	*Mode of Inheritance*
High Penetrance and Expressivity		
Facioscapulohumeral muscular dystrophy	Atrophy and weakness of all facial muscles	Dominant
Nevoid basal cell carcinoma	Cysts of jaw, anomalies of ribs and spine	Dominant
Peutz–Jeghers syndrome	Blue-black spots about the mouth, eyes, elbows, fingers	Dominant
Stickler's syndrome	Nearsightedness, progressive blindness and deafness, pain and stiffness in bone joints	Dominant
Waardenburg's syndrome	Partial albinism with deafness	Dominant
Lower Penetrance and Variable Expressivity		
Blue sclerotics	Bluish outer eye, serious body defects	Dominant
Cataract	Leads to blindness (one form)	Dominant
Cleft lip	A fissure in the upper lip	Dominant*
Holt–Oram syndrome	Heart and limb defects, more severe in females	Dominant
Monilethrix	Brittleness of hair, of skin in some cases (causes sores and bleeding)	Dominant
Polydactyly	Extra fingers or toes	Dominant
Treacher–Collins syndrome†	Very unnatural facial formations	Dominant
Von Willebrand's disease	Deficiency of blood-clotting factor VIII (not to be confused with hemophilia A)	Dominant

*One form; other forms are known to be polygenic.
†The homozygous dominant condition may be lethal.

In each species penetrance is incomplete, but all those that demonstrate the defect express similar degrees of fragility of the skin (it tears easily, leaving large wounds) and hyperextensibility, or looseness of the skin, around the head and neck.

Penetrance and expressivity are not solely intrinsic properties of a gene (specifically a function of a particular gene); rather, they result from an interaction of the specific genes with all other genes and the environment. If penetrance is high, as in Huntington's disease, which is a neurological disorder leading to death, or as in Waardenburg's syndrome which leads to deafness in humans and cats, the degree of expressivity is severe. Conversely, low penetrance in general has been associated with less severe expressions of a disease and perhaps with a greater variability of the expression (Table 8–2). All the examples listed are dominantly inherited. The reason may be that recessive diseases in general do not show a family history of the disease, making calculations of penetrance most difficult.

polygenic or multifactoral inheritance

The polygenic, or *continuous,* pattern of inheritance is different from the Mendelian single-gene, or *discontinuous*, pattern in two ways: (1) the distribution of phenotypes in polygenic inheritance *is* continuous and, (2) the pattern of inheritance is difficult to establish because more than one gene is involved.

Mendel was the first to offer a simple and reasonable explanation for the way individual traits are transmitted from parent to offspring. But Mendelian inheritance involves the association of a *single* gene and a *single* phenotype, such as the gene for seed shape or the gene for seed color. The phenotypes controlled by one gene fall into clearly discrete classes. Because of the discrete phenotypic classes, such traits are called *discontinuous* traits. Discontinuous traits are either expressed or they are not. The pea seed is round *or* it is wrinkled; the seed is yellow *or* green; the seeds are not wrinkled and round or yellow and green.

As the diseases that are associated with a single gene are described in Chapters 9, 10, and 11, you will see that the affected individual carrying a single defective gene is clearly diseased and sharply different from his normal counterpart. In some cases the severity or degree of expression of a particular disease will vary from person to person. But the disease is still transmitted in a predictable pattern, and the nondiseased are sharply separated from the diseased.

Most human characteristics, however, appear to be polygenic. They are the result of the combined effects of more than one gene, each gene with a small measure of influence over the given trait. Height, weight, and skin color are normal characteristics showing the contribution of more than one gene to the final phenotype. These examples show that polygenic inheritance is usually a source of *continuous* variation, continuous in the sense that there is no sharp line that distinguishes between phenotypes. Height in man, for example, is a very subtle form of variation. Many people within a large group may look the same in height; yet if they were placed in a row, it would be easy to recognize a distribution of heights that start with the shortest person and, without major increments, grade up to the tallest person in the group.

The expression of any polygenic trait varies *continuously* over a range with few people at either extreme. Thus the distribution of a polygenic trait in a population may coincide with the standard bell-shaped normal curve. For example, if one were to measure the height of all the individuals in a large population, one would soon determine that humans without any known physical or mental defects range in height from approximately four feet to approximately eight feet. However, there are few people at either extreme, and the *number* of people at each height increases as one approaches the mean of the population, which is about five feet, ten inches.

Because height is determined by several genes, when a tall person mates with a shorter person, their offspring tend to be intermediate in height. This tendency to the intermediate when parents differ is called the *regression to the mean.* Apparently, when the genes from the taller parent and the genes from the shorter parent are brought together in the offspring, they redistribute and interact, allowing an intermediate height. This regression to the mean appears to occur in all polygenic traits: hypertension, intelligence, weight, skin color, fingerprint ridges, alcoholism (Goodwin, 1973), and a variety of human polygenic diseases.

Human skin color and fingerprint ridges have long been known to be polygenically controlled. A brief discussion of each is presented.

the inheritance of skin color

One of the most easily recognizable human traits is skin color. We easily recognize that there are major differences in skin coloration, and that these differences are transmitted from parent to offspring. Skin color is also an easily observed trait, and there is no difficulty in distinguishing between the skin of a typical white and the skin of a typical black.

The coloring of the skin, like that of your eyes and hair (discussed in Chapter 9) depends on the basic pigment melanin. In addition to melanin, however, a melanin-related compound, melanoid, is produced in a diffuse or very fine particulate form and scattered within the skin cells. Your skin color is produced by the two forms of melanin, traces of yellow-red pigments called carotenes, and the blood flowing in your veins and capillaries near the surface of your skin.

For our discussion the most important factors are those responsible for bringing about the major differences in skin color tone; these are the genes regulating melanin and melanoid. The importance of these genes is quite clear when one compares the albino, a person who cannot make the melanin pigment, with the nonalbino. The albino's skin and eyes are very sensitive to light, and his hair is white. His carotenes and hemoglobin may be normal, but the albino clearly lacks normal skin color.

The melanin pigments are very important as a protective shield against overexposure to the sun. In this respect, the melanin pigments in the skin appear to be more responsive to the external environment than are the melanins deposited in the hair and eyes. It is believed that the sun actually stimulates the synthesis and dispersion of melanin within the skin cells and that this constitutes the suntan. Light-skinned people, because of lower levels of melanin, may experience severe sunburn when exposed to the sun for a period, whereas *usually* the person with more pigment in his skin only tans.

There appears to be a genetic relationship between the genes controlling eye color, skin color, and hair color. Or is it a functional interaction correlated with natural selection? Those with darker skin usually have dark hair and dark eyes. We can understand, to some degree, the genetic and biochemical bases for this relationship through the realization that the pigment melanin is central to the formation of color in all three tissues. A mutation in a gene that completely blocks the production of this pigment should be and is reflected in the loss of pigment in all of these tissues (see Figure 10–2).

The current belief is that four to six genes are involved in the production of skin color (Nagle, 1974). Regardless of the actual number involved, we can say with some assurance that if each of the genes is in the dominant state, a maximum dark skin color is expressed. Alternatively, if all genes are in the recessive state, the minimum production of melanin results. *Perhaps* a total recessive state of all pigment gene loci would mimic the albino phenotype; however, this is difficult to determine, since we don't know for certain how many supplementary genes also act on skin color.

Based on the premise that the genes in one condition produce maximum (black) skin color and genes in the alternative condition produce a minimum (white) skin color, we can perform a mating "on paper." The various gene combinations resulting from the matings, or crosses, will demonstrate how various shades of skin color can occur. For an alternative approach to understanding the distribution of genes for skin color pigment, see Figure 8–17. Two gene pairs will be used below to demonstate a black–white mating, although it must be emphasized that more genes than that are involved in the production of human skin color.

Parents	Black		White	
Genotypes	$W_1W_1W_2W_2$	$\times$	$w_1w_1w_2w_2$	(1)
Children		$W_1W_2w_1w_2$ All will be an intermediate shade or mulatto.		
Parents	Mulatto		Mulatto	
	$W_1w_1W_2w_2$	$\times$	$W_1w_1W_2w_2$	(2)

To determine all possible gene and skin color combinations a Punnett square (3) is used. All possible gametes are derived and combined to determine the possible offspring. Note that $W_1W_1W_2W_2$ in block 1 and $w_1w_1w_2w_2$ in block 16 are like the original parents (1). Blocks number 4, 7, 10, and 13 resemble the parents (2). Blocks 2 and 5, 3 and 9, 8 and 14, 12 and 15 are identical pairs of possible skin colors, while blocks 6 and 11 represent individual possible skin colors. Of the 16 possible combinations that could result from the mulatto $\times$ mulatto marriage, 9 possible shades of black to white could be expected among the offspring.

the physical basis: transmission of units of hereditary material through the genes

Thus homozygous or pure-breeding blacks cannot mate and have a white child and homozygous or pure-breeding whites cannot mate and have a black

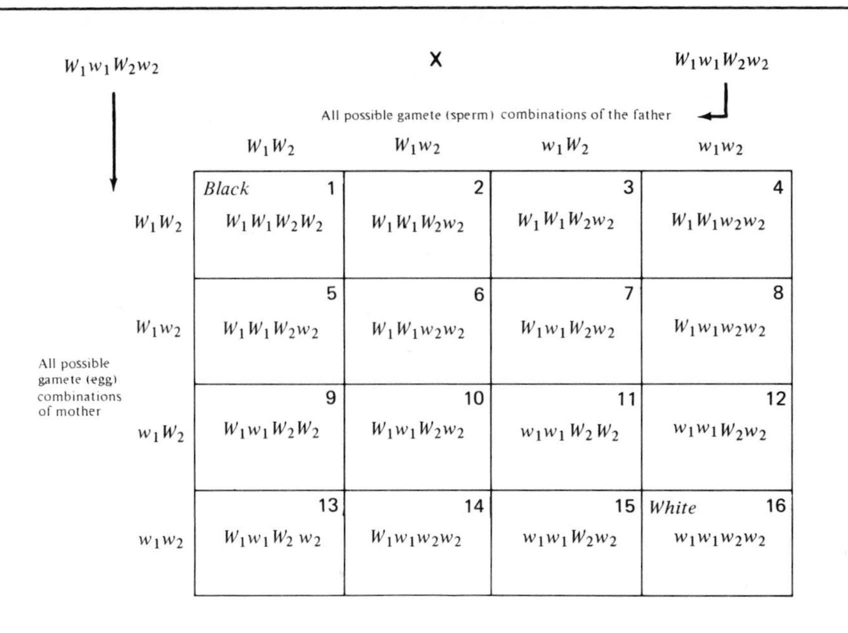

(3)

child. *If* the genes for skin color are *cumulative,* such matings will not produce the unexpected child. What one sees in the offspring of parents with different skin shades is *recombining* of pigment-producing genes into *new combinations* based on the chance processes of meiosis and random union of gametes.

The genes for skin pigmentation provide an outstanding example of gene expression that is completed after birth. White babies appear white at birth and do not exhibit noticeable darkening. Mulatto children begin to darken just after birth, and black children are light brown to red in complexion at birth and progressively complete their pigment formation.

fingerprint ridge counts

The fingerprint ridge count for all of a person's fingers provides an excellent example of polygenic inheritance that is largely unaffected by environmental factors. The count is made from the triradius to the center of the pattern, as in Figure 8–18B(4). The number of fingerprint ridges is "fixed" at about the twelfth week after conception (Carter, 1969). Look at your fingertips, and note the ridges and furrows. There are three basic patterns of ridge development: the arch, the loop, and the whorl (Figure 8–18). The simplest of the patterns is the arch. The arch is formed as the ridge arches upward near the center of the finger. Approximately 5% of all fingertip patterns are arches. Do you have the arch type? The second pattern, the loop, forms as ridges go to the center and loop back to the edge. Approximately 65 to 70% of all fingertip patterns are of the loop type. The whorl pattern is circular in appearance and spreads out from the center of the fingertip. Approximately 25 to 30% of all fingertip patterns are of the whorl type (Deaton, 1974).

To obtain the number of ridges on fingertips, Deaton suggests drawing a line from each triradius (which is the point where the ridges of the loop or

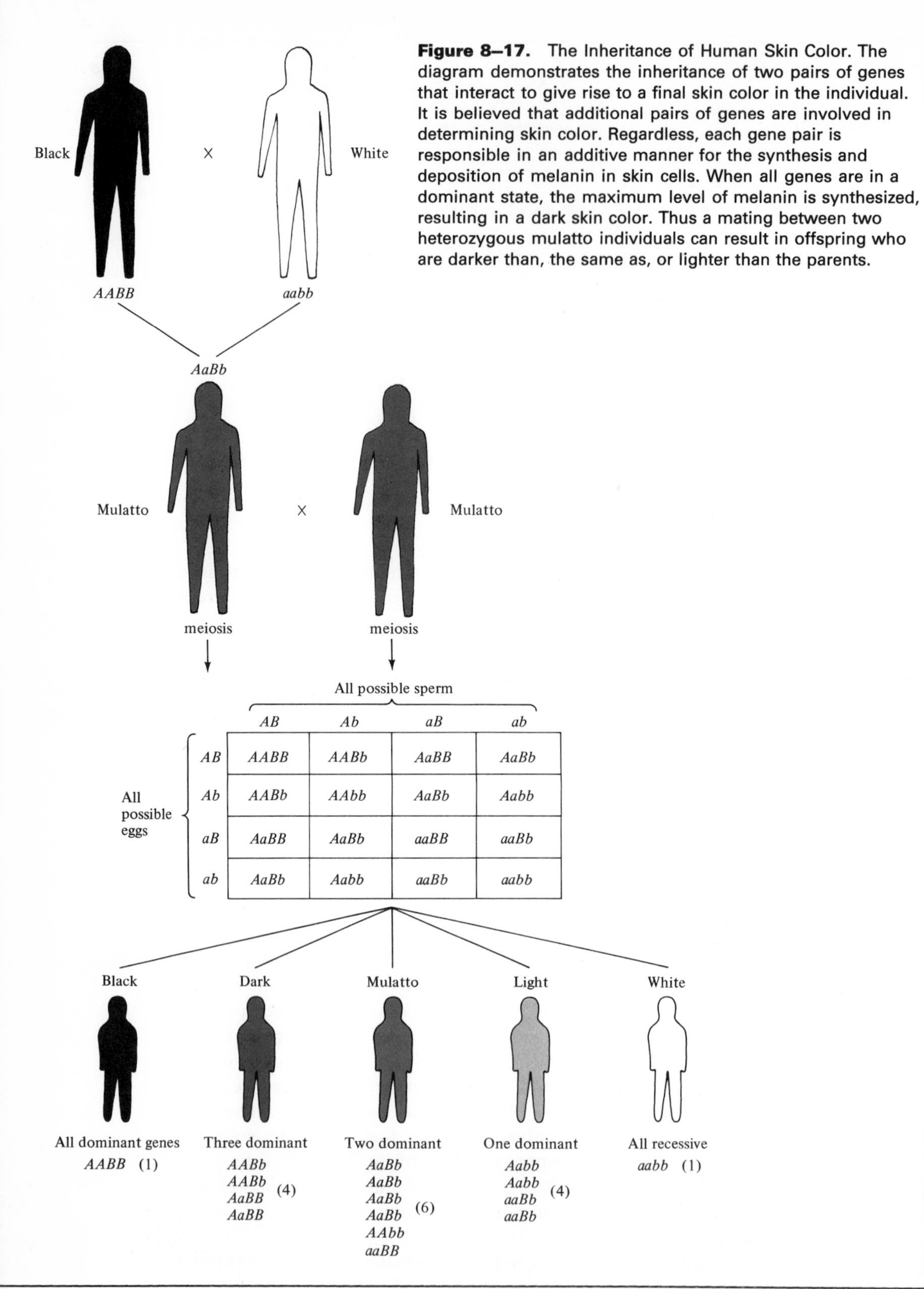

Figure 8–17. The Inheritance of Human Skin Color. The diagram demonstrates the inheritance of two pairs of genes that interact to give rise to a final skin color in the individual. It is believed that additional pairs of genes are involved in determining skin color. Regardless, each gene pair is responsible in an additive manner for the synthesis and deposition of melanin in skin cells. When all genes are in a dominant state, the maximum level of melanin is synthesized, resulting in a dark skin color. Thus a mating between two heterozygous mulatto individuals can result in offspring who are darker than, the same as, or lighter than the parents.

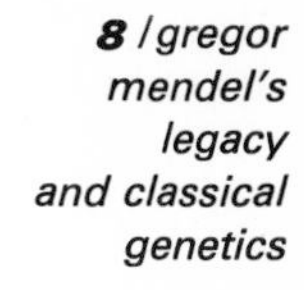

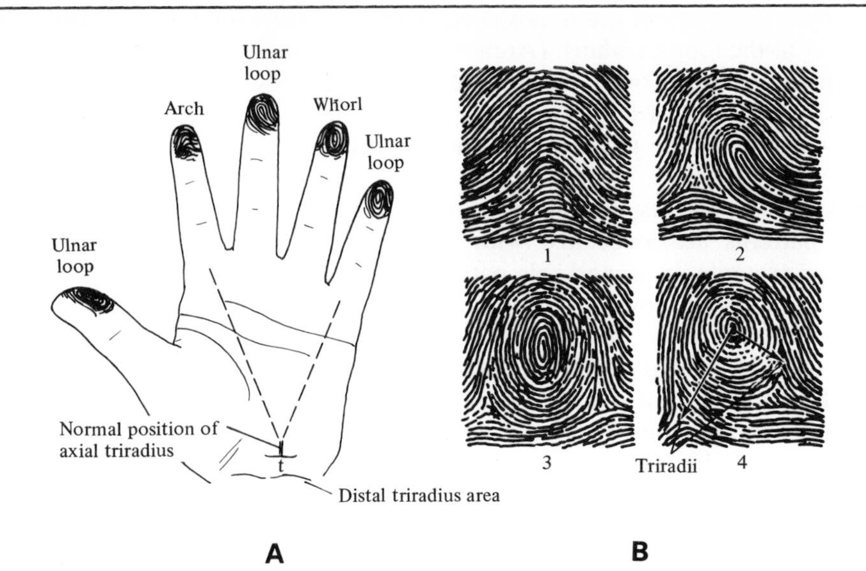

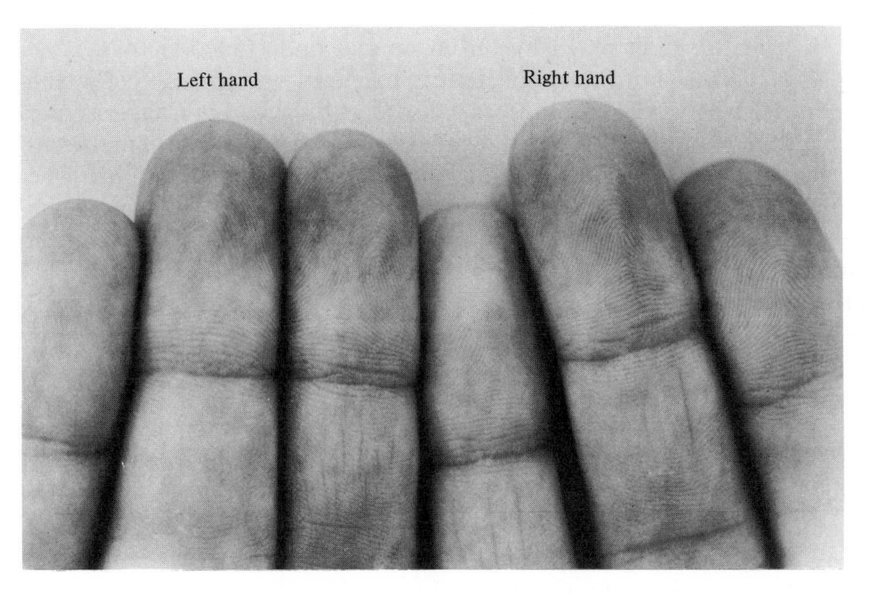

C

Figure 8–18. Dermatoglyphic Patterns. **A.** Arch, loop, and whorl fingertip ridge patterns. If the open end of the loop points toward the underside of the arm (ulnus), the loop is an ulnar loop; if toward the top part of the arm (radius), a radial loop. Normal palm creases and the axial triradius (t) are shown. **B.** Tracings of arch, loop, and whorl patterns with lines pointing out from the center of a whorl to the triradius. The highest ridge count from the center of a whorl to a triradius in 4 would be 19. **C.** The first three fingers of the left and right hands. From left to right, the fingertips of the left hand show ulnar, ulnar, and radial loops; those of the right hand, ulnar, radial, and radial loops. (**A** and **B** from John Deaton, Fingerprinting: New Way to Track Down Disease, *Consultant*, January 1974; reproduced with permission.)

whorl interrupt ridges from the opposite sides of the finger; see Figure 8–18B) to the center of the loop or whorl. (Arches do not cross ridges and therefore do not give ridge counts.) Count every ridge the line crosses. Whorls have two triradii; use the highest of the two counts. To find the total ridge count, add up the ridge count for all the fingers. In women the average ridge count is 127, in men 144.

Deaton (1974) reports that fingerprint patterns, palm print patterns, and sole print patterns are associated with the expression of various inherited diseases. For example, palm and sole patterns are associated with the chromosome disorder causing Down's syndrome. In about half the Down's children, there is the simian crease. In addition, most Down's patients have ulnar loops or an increase in whorls. But the most constant feature is a distal axial triradius. The normal and distal triradius is shown in Figure 8–18A. Other chromosome abnormalities that give particular hand and sole patterns are trisomy 13, Klinefelter's syndrome, and Turner's syndrome. A number of single-gene disorders have characteristic fingerprint pattern.

the incidence of polygenic diseases

A disease is a malfunction or change from normal body function. A genetic disease is an abnormal body function that is inherited due to a defect in the DNA or genes. Many of the chromosomal disorders discussed in Chapter 6 are associated with human diseases, but they are not inherited. They are the result of single chance events occurring during the formation of the gametes. The incidence in many cases is very low. Even in those cases of high incidence, such as trisomy 21 and the various sex chromosome abnormalities, a definite pattern of inheritance cannot be established. For example, trisomy X females would be expected to give birth to an equal number of XXY, Klinefelter's and normal sons; yet of numerous male births, no XXY males have been reported to be the offspring of trisomy X females.

A large group of inherited diseases involves the faulty expression of more than one gene. Unlike the orderly predictable transmission of the well-known single-gene-related defects, the pattern of inheritance for the polygenic, or multifactoral, diseases is not clear. We know from simple observation (the empirical approach) that if certain polygenic diseases, such as those presented in Table 8–3, have occurred in a family, the recurrence risk in that family is 4 to 8%. This is about the same risk that everyone has of receiving one of the many various genetic disorders. The recurrence risk in polygenic inherited disorders differs in the sense that one is referring to the recurrence of a specific defect.

In either event, the risk of offspring having some disorder by a fresh mutation (4 to 8%) or of a polygenetic trait recurring in a family (4 to 8%) is low compared to the risk the parents who carry a known single-gene defect take when having children. Once a dominant single-gene trait has been expressed in a family, the recurrence rate is 50 or 100%. If the gene is recessive, there is a 25% possibility that the next child will show the trait.

Table 8–3 Polygenic or Multifactoral Inheritance

Trait	*Brief Description*
Cleft lip	
Cleft palate	
Cleft lip and palate	See Figure 8–19
Pyloric stenosis	Narrowing of opening leading from the stomach into the small intestine
Hydrocephaly (one type)	Water on the brain; see Figure 8–20
Congenital hip	A dislocation of the hip at birth
Spina bifida	Open spine
Club foot	Turned-in foot, sometime abnormal in shape
Diabetes (one type)	Individual requires insulin
Hypertension	Variable, high blood pressure
Cancer	Uncontrolled reproduction of cells
Certain heart diseases	
Certain forms of mental illness	Manic-depressive, schizophrenia

In 1973 some 2 million children in the United States between the ages of 6 and 11 were afflicted with polygenetic traits such as harelip, cleft palate, clubfoot, and congenital heart defects (DHEW, 1974). In 1974 the American Heart Association reported that 23 million Americans had high blood pressure and that 60,000 persons die annually of the effects of this polygenic disorder.

multiple alleles of a gene

So far examples of one or more gene pairs that control a single trait have been presented. The dominant gene for round seeds was represented by the symbol *R*. Because there are two genes for each trait in a diploid organism, round seeds in the diploid pea plant are either *RR* or *Rr*. The wrinkled seed shape is *rr*. In this example, then, there are only two alternative genes for seed shape, *R* and *r*. Since either member of the gene pair can occupy the same physical site on the chromosome, the two members are called alleles. Thus, *R* and *r* are alleles of the gene for seed shape. A single gene occupying a given site on a chromosome, a locus, can have alternative states, or alleles, because there are changes in the DNA at that particular locus. There will be as many alleles for a given gene as there are mutations that have occurred at that locus during evolutionary time. Many genes have only two alleles, but some—for example, the gene responsible for the production of human blood antigens—have several alleles. Such genes are called multiple allelic genes.

Though a particular gene may have several alleles as with the ABO blood group alleles, only two of the alleles can occupy a person's chromosomes at any one time because we are diploids; that is, we have two of each chromosome. The series of alleles that determines the ABO blood groups is A^1, A^2, A^3; B^1,

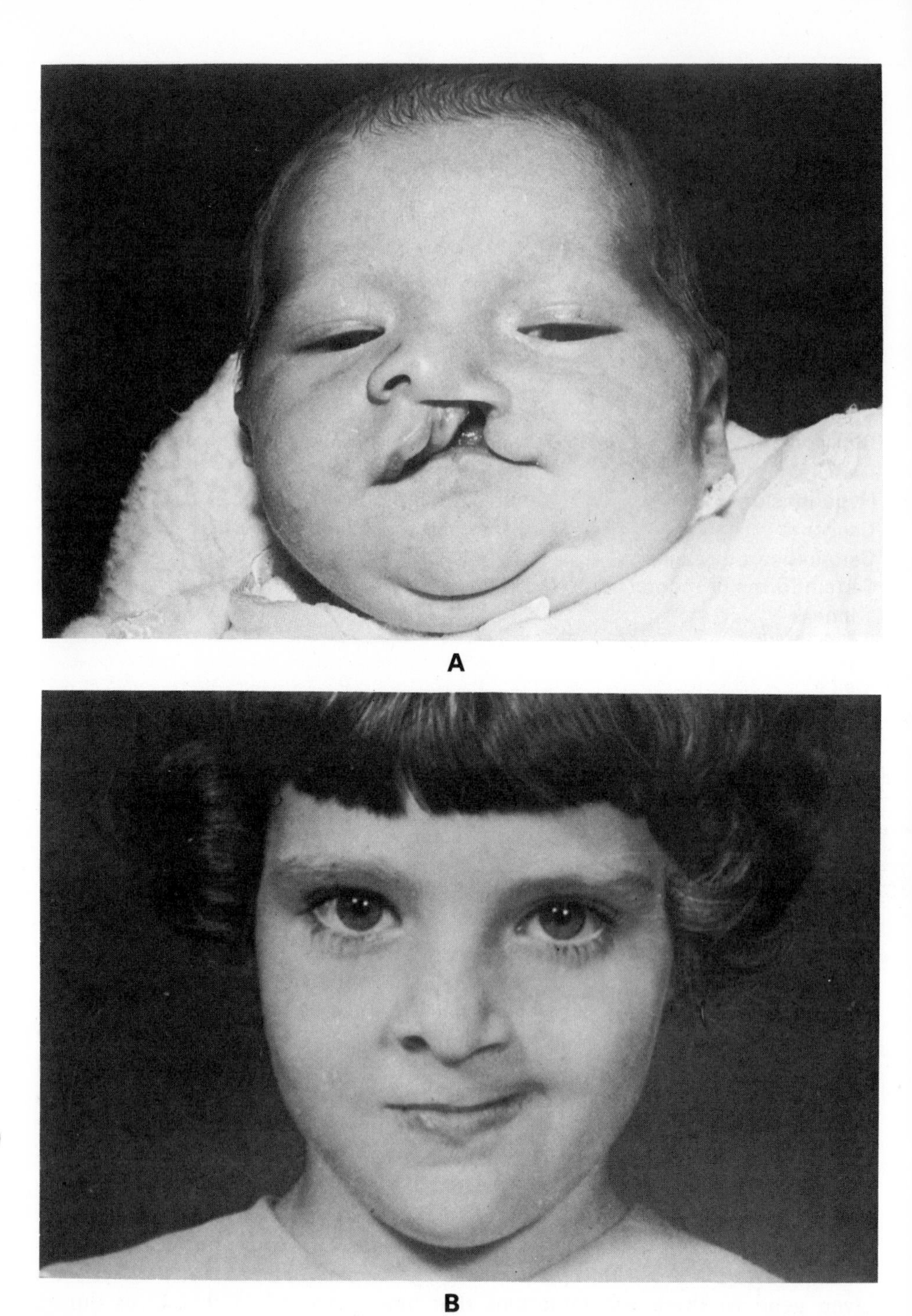

Figure 8–19. Cleft Lip and Palate. **A.** A child with cleft lip and palate at birth. **B.** At age six the same child has benefited from plastic reconstructive surgery in treatment of cleft lip and palate. Often in such cases psychological problems of social adjustment are severe. The syndrome, which is thought to be multifactoral in nature, appears in about 1 in 750 to 1000 births in the United States. Thus a total of about 6000 to 7500 affected infants are born each year. (Reprinted with permission from *Cleft Lip and Palate: Criteria for Physical Management*, Hughlett L. Morris, ed., The University of Iowa, 1963.)

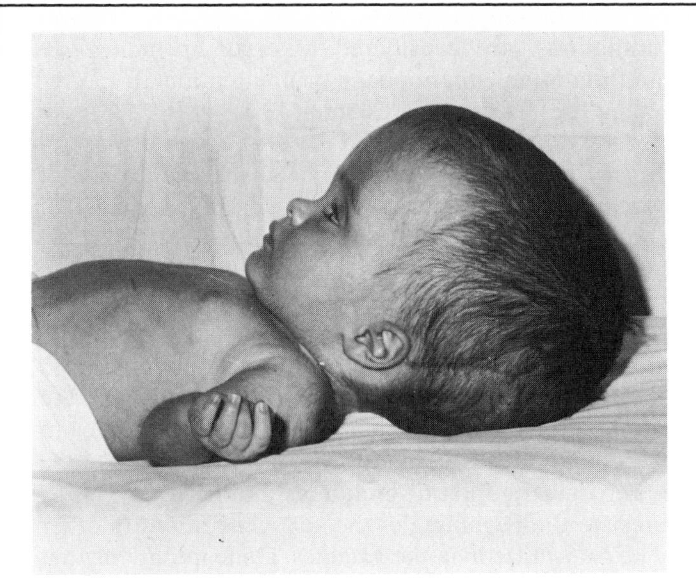

Figure 8–20. Hydrocephaly. Prior to or shortly after birth, cerebrospinal fluid collects in the skull, causing an enlargement of the head and subsequent mental retardation. Life expectancy is very short without a shunt to relieve the pressure. One rare exceptional case lived to age 68 without the shunt, although she was confined to a wheelchair at age seven and suffered mental retardation. (By permission of Charles H. Carter, M.D.)

B^2, and O. Your blood type depends on which of these alleles occupy your chromosomes. For example, if your father was blood type A^1A^1 and your mother B^1B^1, you have, barring a new mutation, received the allele A^1 from your father and the allele B^1 from your mother, and you have an A^1B^1 blood type. (Human blood-type phenotypes and genotypes are discussed in Chapter 12.)

It is not easy to determine whether a gene has multiple alleles or one is dealing with two genes at different loci. Individual families are small and do not offer a sufficient number of offspring for the necessary statistical analysis. Also, if two genes do not assort independently of one another, as Mendel's second principle states, the genes may be *closely* linked rather than allelic. To decide if one is studying closely linked gene or allelic genes, one looks to see if crossing-over occurs between the genes in questions. Should recombination occur, the genes are independent of one another, but closely linked on the same chromosome.

gene mutation, the formation of new alleles

The word *mutation* means "change." In genetics a mutation is an abrupt heritable change in a gene that cannot be attributed to gene segregation or recombination. New alleles of a gene are produced through mutation (Strickberger, 1976). Some 4 billion years ago, when molecules of DNA were formed, changes began to occur in the DNA molecule, and these changes, once incorporated into a living system, were transmitted to the succeeding generations. Under ordinary conditions of growth and reproduction, changes in the DNA, or point mutations, are considered to occur *spontaneously.* (Spontaneous mutations and mutations intentionally or unintentionally caused by changes in the environment are discussed in Chapter 13.)

As mutations occurred, they were either selected for or selected against by the environment in which they arose. (Darwin recognized the occurrence of spontaneous mutations, which he called "sports," in domesticated animals.) If the mutation was compatible with life and was reproduced, the sexual mechanism of gene recombination gained a new gene for distribution into future generations. This is why it is often said that new mutations are the raw material of evolution. At one time in our evolutionary past this was true. Today, however, it is doubtful whether a new mutation does anything more than increase or decrease the frequency of alleles that are already in the population. Most human variability now comes about through the processes of crossing-over and gene recombination (discussed in Chapter 4).

In general, geneticists are concerned about mutations that occur in human gametes because the gametes are the beginning of the next generation. In other words, mutations can only be transmitted via the gametes. Gene mutations that occur in nonsex cells, the somatic or body cells, die with the person and are of no biological consequence to future generations.

A point mutation of a gene is most likely to be very small, usually consisting of the substitution of one nucleotide for another. Nevertheless the phenotypic results can be devastating. The change may result in the complete loss of a structural or catalytic protein (enzyme), the production of only a part of a protein, or the production of a protein that has no function. In humans such effects lead to many serious inherited diseases, several of which are discussed in later chapters.

There are, however, subtle gene changes that do not appear to affect the individual's function in society in any measurable way. Such changes are sometimes referred to as *neutral* mutations. There is no way to evaluate their fitness, whether they are or will be necessary to the species' survival. In addition, there are many cosmetic types of changes that affect personal appearance. They may not appear to be of value, but the individual acceptance of such traits can affect the general population. Since physical attraction is generally a prerequisite to mating and having children, gross physical and behavioral changes restrict one's chances of finding a mate. But what of the subtle changes—dimples, attached earlobes, or certain hairlines? These changes do not appear to restrict mating patterns and the production of variation in offspring; but who really knows? As for acceptance of a subtle change affecting the species, suppose many of those with attached earlobes or dimples were not allowed to mate. Extending such mating restrictions to other subtle features would restrict the production of human variation.

pedigree charts in humans

A pedigree chart is the construction of a set of symbols that enables us to trace the incidence of a particular trait in successive generations of a family lineage.

The earliest identified pedigree chart is a clay tablet found in Iran dated about 3100 B.C. The tablet is inscribed with symbols relating to the breeding of horses (Chirshman, 1954). Pedigree analysis was used extensively by Francis Galton in the late 1800's, when he was studying family histories of famous men.

Family data, once gathered, can be summarized in a pedigree chart using the symbols shown in the Table 8–4. Brief examples of dominant, recessive, and X-linked pedigree charts are presented in Figures 8–21, 8–22, and 8–23, respectively.

The analysis of a pedigree chart, assigning genotypes and detecting the pattern of inheritance, is limited in that beyond two or three generations it is difficult to obtain reliable data on members of the family. Many times descriptions of a defect are inaccurate because of conditions of penetrance and expressivity. In addition, recessive defects may skip many generations before the phenotype appears again. In this case the defect may be mistaken for a new mutation. And in most cases genes that cause the death of the fetus via early spontaneous abortion are missed.

The distribution of traits in a family, however, can give valuable information on gene linkage and whether one is dealing with alleles of the same gene or with separate genes. Such determinations are based on Mendel's principles that alleles of a gene segregate and that nonalleles assort independently of one another. Take, for example, a family with recessive congenital deafness and deaf-mutism. In each of two generations a "recessive" deaf-mute married a "recessive" deaf-mute, and all the offspring were deaf-mutes. But in the third generation a recessive deaf-mute married a recessive deaf-mute and their

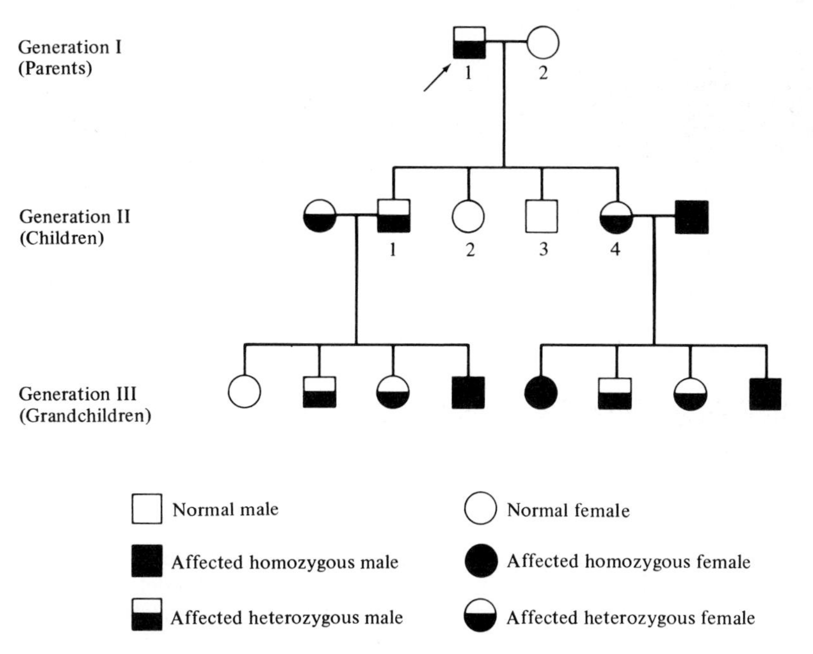

Figure 8–21. Pedigree of Dominant Inheritance. Parent I-1 is the index case; he has Marfan's syndrome. Two of his four children, II-1 and II-4, also have Marfan's syndrome. The mating of II-4 with another person affected with Marfan's syndrome produces all affected children. Other dominant defects are listed in Table 9–2.

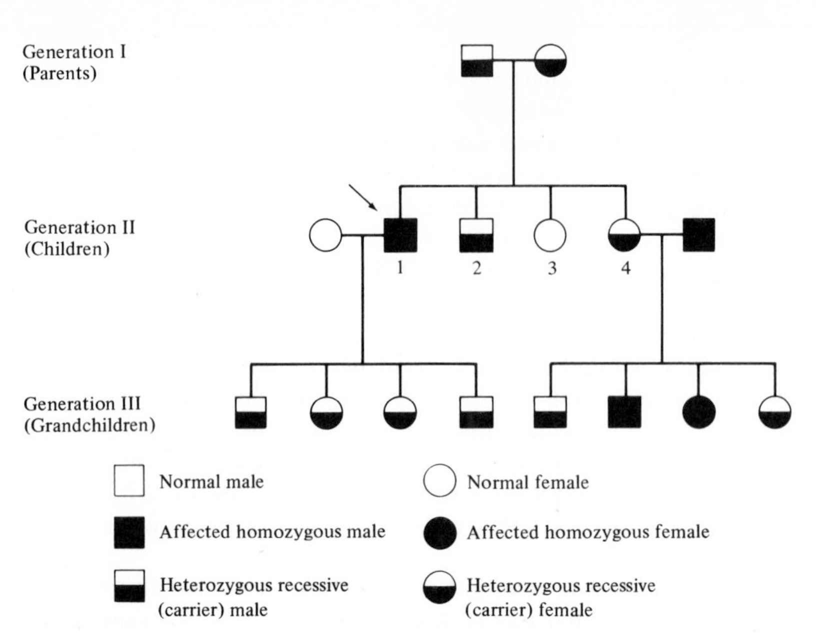

Figure 8–22. Pedigree of Recessive Inheritance. Both parents are most likely heterozygous (normal) carriers because their first child (generation II), a male, was born with alkaptonuria. Their fourth child, a girl, must have been heterozygous because mating with an alkaptonuric male produced alkaptonuric offspring. Arrow at II-1 indicates the index case. Other recessive defects are listed in Table 10–1.

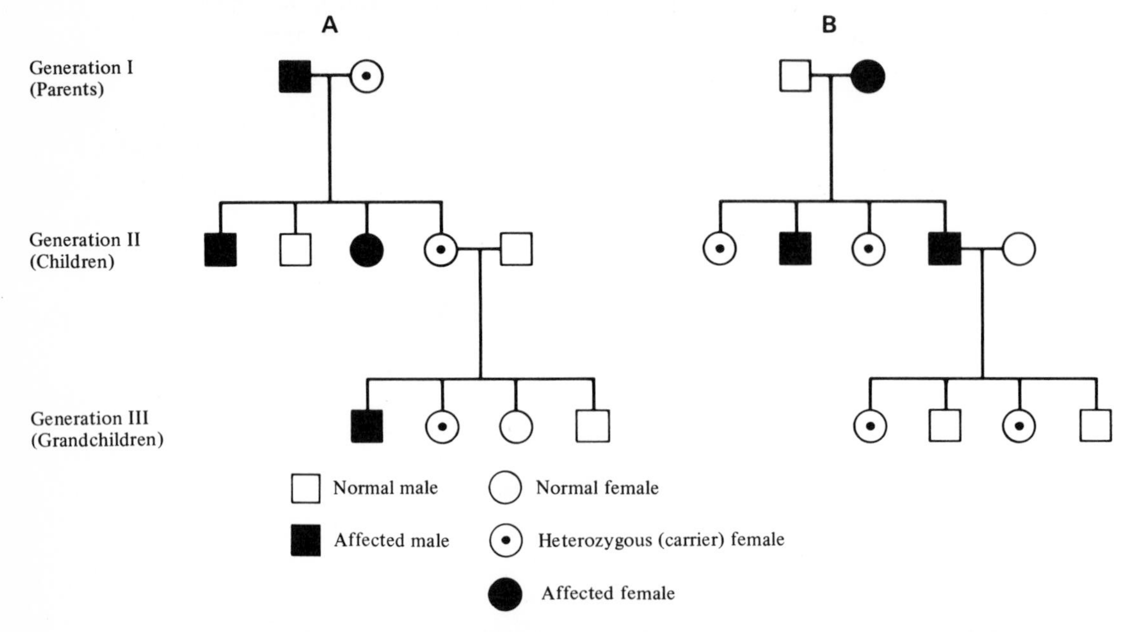

Figure 8–23. Pedigree of X-Linked Inheritance. Human pedigree for glucose-6-phosphate dehydrogenase (G-6-PD) deficiency, a sex-linked recessive trait. **A.** A G-6-PD deficient father and heterozygous (normal) mother (generation I) can give rise to normal or defective sons and daughters (generation II). **B.** Normal father and G-6-PD deficient mother (generation I) give rise to all defective sons and heterozygous daughters (generation II). A normal female and an affected male produce heterozygous normal daughters and normal sons (generation III). Other sex-linked defects are listed in Table 11–1.

Table 8–4 Symbols Most Often Used to Construct a Pedigree Chart*

Symbol	Function
○ or ♀	Normal female
□ or ♂	Normal male
○—□	Single bar indicates mating
I, II; 1 2 3	Normal parents and normal offspring, two girls and a boy in birth order indicated by the numbers; I and II indicate generations
	Single parent as presented means partner is normal or of no significance to the analysis
○═□	Double bar indicates a consaguineous mating
	Fraternal twins (not identical)
	Identical twins
2 and 6	Number of children for each sex
● and ■	Darkened square or circle means affected individual; arrow (when present) indicates the affected individual is the proband or propositus, the index case, the beginning of the analysis
and	Autosomal heterozygous recessive
⊙	X-linked carrier
and	Dead
	Aborted or stillborn

*Usually a pedigree chart contains a brief set of symbols with their explanations.

offspring were *normal*! A pedigree of this family would indicate that the parents in generation III were homozygous for deafness and mutism, but the genes were located either on separate chromosomes or at a sufficient distance from each other on the same chromosome for recombination to occur. Regardless, the point is that the deaf-mute in generation III of this family has a defective gene that was different from the defective gene of his mate. The genes were nonallelic (Figure 8–24).

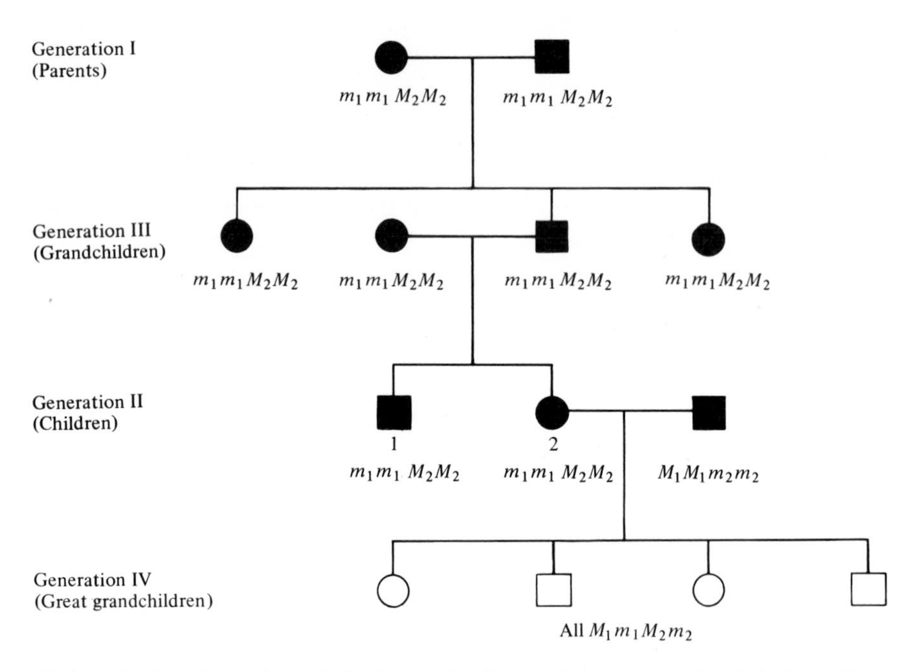

Figure 8–24. Hypothetical Pedigree Analysis of Recessive Deaf-Mutism. Parents in generation I and their children and grandchildren, generations II and III, are homozygous recessive and deaf-mute. In generation IV, all children are normal because III-2 married a deaf-mute with his defective gene for deaf-mutism at a different locus. Although recombination of the genes during meiosis produced parental gametes that carried normal and defective genes, only the gametes with normal genes were involved in the fertilization for the production of the children in IV.

references

Bronowski, Jacob. 1974. *The Ascent of Man*. Little Brown, Boston.

Carter, C. O. 1969. Polygenic Inheritance and Common Diseases. *The Lancet, 1:*1252–56.

Chirshman, R. 1954. *Iran from the Earliest Times to the Islamic Conquest.* Penguin Books, Harmondsworth (England).

Deaton, John G. 1974. Fingerprinting: New Way to Track Down Disease. *Consultant, 14* (1):87–89.

DHEW. 1974. *What are the Facts About Genetic Disease?* Publication No. NIH 74–370. Department of Health, Education and Welfare, Washington, D.C.

Goodwin, Donald W. 1973. Alcohol Problems in Adoptees Raised Apart from Alcoholic Parents. *Comprehensive Psychiatry, 14*(1):92–93.

Harris, R., and J. T. Patton, 1971. Achrondroplasia and Thanatophoric Dwarfism in the Newborn. *Clinical Genetics, 2*:61–72.
Iltis, Hugo. 1966. *Life of Mendel.* Hafner, New York.
Medical News. 1975. Bilateral Retinoblastoma Survivors Often Develop New Primary Cancers. *Journal of the American Medical Association, 234*:369–71.
Moore, Ruth. 1975. *The Ascent of Man: Sources and Interpretations.* Little Brown, Boston.
Nagle, James. 1974. *Heredity and Human Affairs.* Mosby, St. Louis.
Sjögren, Torsten, and Tage Larsson. 1957. Oligophrenia in Combination with Congenital Ichthyosis and Spastic Disorders. A Clinical and Genetic Study. *Acta Psychiatrica et Neurologica Scandinavica, Supplement 113, 32*:1–112.
Sootin, Harry. 1958. *Gregor Mendel: Father of the Science of Genetics.* Vanguard, New York.
Stern, Curt. 1973. *Principles of Human Genetics,* 3rd ed. Freeman, San Fransisco.
Strickberger, Monroe W. 1976. *Genetics,* 2nd ed. Macmillan, New York.
Sturtevant, A. H. 1965. *A History of Genetics.* Harper & Row, New York.
Sutton, W. S. 1902. The Chromosomes in Heredity. *Biological Bulletin, 4*:231–48.

We learned in earlier chapters that our hereditary material is DNA and that a specified amount of this DNA is called a gene. The gene, or more properly the chemical DNA that constitutes the gene, is the record of man's past and the biological contract for his future. We know this to be true because we have uncovered the means by which cells give rise to new cells and humans give rise to new humans. We have learned that the genes of DNA are transferred with great precision from parent to offspring generation after generation.

In general, we refer to the genes as traits, as, for example, the trait for brown eyes. The term *character* usually refers to some noticeable property of an organism, such as skin or eye color, the number of fingers or toes, shape of the skull, or the absence of an enzyme. The human species is made up of an almost infinite variety of weights, heights, voices, complexions, looks, and behavior.

We seem to know, intuitively, what is normal by the appearance of those who are abnormal. But detection of the abnormal is not always easy, and the more sophisticated the techniques we develop to measure differences among people, the more we learn about what normality is and about how serious the various abnormalities are. Further, in many cases differences among humans are noticeable, yet there can be no judgment of which trait is better, for the trait that is "best" in one environment may have no advantage in another. An example is the single defective gene for sickle cell anemia. In a country where the possibility of contracting malaria is high, persons carrying the defective gene have an advantage for survival. Such persons living in the United States, where the chance of contracting malaria is minimal, receive no known benefit;

We are the sum total of all we ever were and all we ever dreamed.

Author Unknown

human dominantly inherited diseases

in fact, there is some evidence that there is a slight biological disadvantage as well as a larger social disadvantage (discussed below and in Chapter 22). Even when inherited traits are known to be detrimental, medicine, diet, and surgery have allowed many genetically abnormal persons to survive at a level that biologically and socially simulates normality.

In Chapter 8 a few gene-controlled traits that do not appear to be detrimental to our species, such as having free or attached earlobes, dimples, or a widow's peak, were presented. In Chapters 9, 10, and 11, we will review the defective genes that cause the more severe forms of hereditary human diseases. Since there are now over 2000 known defective-gene-related diseases, only those most common, as judged by their rate of occurrence in nature or in college textbooks, will be discussed.

inherited diseases / their rate of occurrence

Changes in trends in the causes of death among humans show an increasing importance of genetic diseases. During the year 1900 the diseases tuberculosis, syphilis, typhoid fever, dysentery, whooping cough, diphtheria, influenza, pneumonia, and diarrhea together caused 647 deaths per 100,000 population; in 1920, 430; in 1940, 147; and in 1959, 41. In recent years improved programs to combat communicable diseases have reduced the numbers even further (Gowen, 1961).

Thus the infectious disease is slowly coming under control. However, genetic diseases of the heart, blood vessels, and other organs, appearing prenatally, at birth, in adolescence, or later in life, have taken the place of communicable diseases. In fact, one of the great medical advances of the current century is the clarification of the role the genes play in health, morbidity, and death. No claim can be made for the recent discovery of hereditary diseases. Egyptians of prehistoric times apparently recognized extreme inherited pathological conditions, for their god Ptah was an apparent chondrodystrophic dwarf. The Bible mentions hemophilia and epilepsy in ways indicative of some knowledge that these were inherited human defects.

The history of human defects relates that the genetic defects that are with us now are not new; many have been with us from the beginning of the human species. Clearly, this implies that it is difficult to alter the established members of a species, and that alterations may not be an improvement over the original version. Through natural selection, humans evolved to a point where they were well adjusted for survival and reproduction, and in general, continued mutations reduce this adaptability, as can be seen in the discussion of genetic diseases. Based on the number of live births in the United States in 1973 (3,141,000), some 150,000 to 200,000 genetically defective children were born. These live births with genetic defects include some 2000 to 2200 genetic diseases now recognized. Regardless of the patterns of inheritance, all forms of inherited diseases are determined at conception. Those diseases causing sterility cannot be transmitted further, since reproduction is not possible for the victim. With respect to the evolution of the species, a sterile person is genetically dead.

Table 9–1 General Age for the Expression of Selected Single-Gene and Polygenic Diseases

From Conception to 1 year	*By 10 Years*	*By 20 Years*	*By 40 years*
Achondroplastic dwarfism	Ataxia-telangiectasia	Alzheimer's disease (presenile psychosis)	Amyloidosis (primary)
Agammaglobulinemia	Batten's disease	Diabetes insipidus	Diabetes mellitus (late type)
Apert's syndrome	Duchenne muscular dystrophy	Diabetes mellitus	Familial polycystic kidneys
Cleft lip	Friedreich's ataxia	Pernicious anemia	Gout (adult; hyperuricemia)
Cleft palate	Hallervorden–Spatz disease	Xeroderma pigmentosum	Huntington's disease
Club foot	Homocystinuria		
Congenital heart defects	Hunter's syndrome		
Congenital hip	Juvenile diabetes		
Cystic fibrosis	Juvenile pernicious anemia		
Erythroblastosis fetalis (Rh disease)	Lesch–Nyhan syndrome		
Familial goiter	Myotonic dystrophy		
Fanconi's syndrome	Neurofibromatosis		
Galactosemia	Osteogenesis imperfecta tarda		
Hydrocephaly	Periodic paralysis		
Hypercholesterolemia	Tuberous sclerosis		
Ichthyosis congenita	Wilson's disease		
Ichthyosis vulgaris			
Maroteaux–Lamy syndrome			
Meningomyelocele			
Multiple telangiectasia			
Osteogenesis imperfecta			
Phenylketonuria			
Phocomelia			
Polydactyly			
Retinoblastoma			
Sanfilippo's syndrome			
Sickle cell anemia			
Syndactyly			
Tay–Sachs disease			
Thalassemia			
Tylosis			
Werdnig–Hoffmann disease			

In 1956 approximately 370 single-gene-associated diseases were known. During the next 15 years over 1500 new single-gene-associated diseases were reported and listed in McKusick's (1971) catalog of Mendelian inheritance in man. This is an average recognition rate of 100 new gene-associated diseases per year. If this rate continues, by 1980 the genetics catalog should list from 2500 to 3000 single-gene defects; by 1990, from 3000 to 4000; and by 2000, from 4000 to 5000. Add to these defective genes an equal or greater number of heritable polygenic diseases, the approximately 12% of our population that inherits a sensitivity to drugs (see Chapter 16), the defects that are chromosomal in origin, and the fact that inherited diseases are incurable, and one can readily understand why genetic diseases are the most important major health problem in the world today.

Estimates by the Department of Health, Education, and Welfare indicate that over 15 million Americans suffer from one or more types of birth defects, and that 80%, or 12 million, of the birth defects are thought to be heritable (DHEW, 1974). When measured in terms of normal life expectancy, the years of life lost because of birth defects, collectively, amount to 4.5 times the number of years lost to nongenetic heart diseases (considered the nation's number 1 health problem), 8 times the number of years lost to cancer (the nation's second leading health problem), and 10 times the number of years lost to strokes. Another way of calculating the toll of genetic diseases is to estimate the *future* life years that will be lost to them. One widely cited estimate indicates that the number of future life years to be lost in this country because of birth defects is some 36 million, with the number attributable to recognized genetic diseases being 29 million, several times as much as from heart disease, cancer, and stroke (DHEW, 1974).

As stated, genetic disease is determined, by definition, at conception. The expression of the inherited defect can occur very early during development and result in a nonviable zygote or death of the fetus. The fetus may be viable but malformed at birth, with the potential for the disease to be expressed later during life. Table 9–1 presents some variations in the times when a defect is first observed, or when it becomes significant. Of these diseases the dominantly inherited are treated in this chapter; Chapters 10 and 11 deal with the recessive and X-linked inherited diseases.

Inherited diseases are caused by a single gene, polygenes, and chromosomal disorders. Chromosomal diseases are presented in Chapter 6, and polygenic diseases in Chapter 8. The largest number of inherited human disorders currently cataloged are the result of single-gene defects. The basic patterns of transmission of single-gene defects are presented in Chapter 8. *Dominantly inherited diseases often involve a change in a structural protein* (*noncatalytic*). *Recessively inherited diseases* (Chapter 10) *often show an altered catalytic protein, or enzyme* (Thompson and Thompson, 1973).

Some of the more than 1000 dominantly inherited conditions are listed in Table 9–2. Several disorders from this list will be presented in detail with reference to the particular demands on society. A brief review of the pattern of dominant inheritance is shown in Figure 9–1.

Of the many dominantly inherited traits, several specific traits are easily observed, have significant social overtones, and are usually the subject of various questions in discussions of human inheritance. They are eye diseases

Table 9–2 Examples of Common Autosomal Dominant Conditions*

Condition	*Brief Description*
Achondroplasia	One form of dwarfism
Acute intermittent porphyria	Inability to metabolize hemoglobin, nausea, spells of violent behavior, pigmented stools
Amyotrophic lateral sclerosis	Degeneration of nerve cells
Aniridia†	Absence of the iris
Apert's syndrome	Fusion of fingers and toes, numerous skeletal abnormalities
Atopic allergic disease	Ragweed hay fever
Basal nervous cell carcinoma	Aberrant behavior, prone to skin cancer, calcified ovaries
Blue sclerotics	Sclera of eye blue instead of white, serious body defects
Brachydactyly	Hand malformation, shortened fingers
Camptodactyly	Permanently stiff, brittle, and bent fingers
Chondrodystrophic dwarfism	Normal size head and trunk, short appendages
Chronic pyelonephritis	Nerve deafness and kidney infection
Cleidocranial dysostosis	Lack of bone hardening, collarbone defects
Congenital cataract	Leads to blindness
Congenital ptosis	Drooping upper eyelid due to paralysis
Congenital stationary night blindness	Twilight or night blindness
Craniofacial dysostosis (Crouzon's syndrome)	Hearing impaired, beak nose, short top lip
Dimples	Indentations in the cheeks and chin
Distal myopathy	Muscle degeneration
Dubin–Johnson disease (hyperbilirubinemia II)	Jaundice, leads to kidney malfunction and later treated by dialysis
Earlobes	Free are dominant over attached
Ehlers–Danlos syndrome	Connective tissue disorder, fragility of the skin and peripheral blood vessels
Elliptocytosis	Oval-shaped red blood cells
Epiphyseal dysostosis	Growths due to defective bone formation
Eye color	Dark dominant over blue (many modifying genes)
Facioscapulohumeral muscular dystrophy	Weakness of all facial muscles
Familial idiopathic cardiomyopathy	Diseased heart muscles

Table 9–2 continued

Condition	*Brief Description*
Familial polycystic kidneys	Cysts develop in kidneys about age 40, produce hypertension, uremia
Familial xanthomatosis	Increased levels of cholesterol, skin eruptions
Freckles	Pigmental spots on the skin
Gardner's syndrome	Polyps or finger-like projections growing in large intestine (can become cancerous)
Gilbert's disease (hyperbilirubinemia I)	Overproduction of bilirubin
Gynecomastia	Breast development in males
Hair	Curly dominant over straight
Hair, white forelock	White tuft of hair in dark background of hair, associated with additional spotting
Hairline	Widow's peak dominant over straight hairline across forehead
Hair texture	Fuzzy hair that breaks easily
Hair texture	Wooly hair that resembles sheep's hair
Hapsburg lip	Lower lip and jaw protrude beyond upper lip and jaw
Hirschsprung's disease	Congenital intestinal destruction
Holt–Oram syndrome	Heart problems, narrow shoulders
Huntington's disease	Progressive degeneration of nervous system
Hypercholesterolemia	Raised levels of cholesterol in blood
Hyperlipidemia	Raised levels of fats involved in early age heart attacks
Hyperuricemia (gout)	Raised levels of uric acid in blood
Hypospadias	Opening on bottom of penis; intersexual development
Infantile polycystic kidney‡	Early cyst development in kidneys
Intestinal polyposis (Gardner's syndrome)	Nodules or polyps in the intestines, can become cancerous
Juvenile pernicious anemia‡	Insufficient red blood cells in the young
Lobster claw (cleft hand)	Severe abormalities of the hands and feet
Loss of nipples and breast	No breast or nipple development

*The list includes those dominant diseases and traits most often encountered in college texts.
†Other inherited eye diseases will be presented in the section on inherited eye diseases.
‡These diseases have shown both autosomal dominant and recessive transmission.

(continued)

Table 9–2 continued

Condition	*Brief Description*
Marfan's syndrome (arachnodactyly)	Affects bone, muscle, and connective tissue; appendages are long and thin
Methanethiol	Given off in urine after eating asparagus
Microphthalmus	Very small eyes
Migraine (headache)	Often described as allergenic disease
Multiple exostosis	Bony growth on the surface of a bone
Myoclonic epilepsy	Progressive epilepsy
Myotonic dystrophy (Steinert's disease)	Severe atrophy of temporalis muscle, weak eyelid muscles
Nail-patella syndrome	Congenital ankle and knee abnormalities
Niemann–Pick disease‡	Accumulation of lipids in liver and spleen
Neurofibromatosis (non-Recklinghausen's disease)	Tumorlike formations on the skin and in nervous system
Osteogenesis imperfecta	Fragility of the bones, often dwarfism, deafness
Pelger–Huet anomaly	Abnormal white blood cells
Phenylthiocarbamide	Ability to taste PTC as bitter or sour
Pheochromocytoma	A particular cell causing hypertension by releasing epinephrine
Phocomelia	Seal- or flipper-like limbs, absent digits, some mental retardation
Piebaldness	Variable spotting of white areas on dark skin
Porokeratosis	Hardening and scaling of skin around the sweat glands
Porphyria (four types; see Chapter 5)	Inability to metabolize hemoglobin, variable emotional and physical problems
Polydactyly	Extra fingers or toes
Rendu–Osler–Weber syndrome (hereditary hemorrhagic telangiectasia)	Facial features similar to Bloom's syndrome, but diseased blood vessels more prone to rupture; cirrhosis of liver common
Retinitis pigmentosa‡	Common cause of blindness, deposit of pigment on the retina
Retinoblastoma	Cancer of one or both eyes
Rh-positive blood type	Person has the Rh antigen
Right eye preference	Use of right eye over left
Sexual precocity (some forms)	Early development and puberty-type changes

Table 9–2 continued

Condition	*Brief Description*
Sipple syndrome	Medullary carcinoma of the thyroid and pheochromocytoma (a chromaffin cell tumor producing hypertension)
Smoking respiratory syndrome	Absence of the alpha$_1$-antitrypsin enzyme
Stickler's syndrome	Painful joints, nearsightedness, progressive blindness and deafness
Syndactyly	Malformation of hands and fingers—crooked, fused, or shortened
Treacher–Collins syndrome	Jaw and facial malformation, deafness
Tuberous sclerosis (epiloia, Bourneville's disease)	Tumorlike formation in various organs
Tylosis	Callous formation on hands and feet, hardening and scaling of skin
Von Hippel–Lindau syndrome	Affects retina and cerebellum, cysts in kidneys and pancreas
Waardenburg's syndrome	Progressive deafness in humans and cats
Werdnig–Hoffmann infantile muscular dystrophy‡	Early death due to muscle degeneration
Zollinger–Ellison syndrome	Tumors of endocrine glands

*The list includes those dominant diseases and traits most often encountered in college texts.
†Other inherited eye diseases will be presented in the section on inherited eye diseases.
‡These diseases have shown both autosomal dominant and recessive transmission.

and hair and skin coloration. Although skin color (discussed in Chapter 8, see Figure 8–17) may have the greater social implications, eye and hair colors have significant cosmetic value and are socially important as opposite-sex attractants.

eye coloring

Studies in animals such as the fruit fly *Drosophila* have confirmed that eye color and eye shape are inherited characters. As we have learned from the *Drosophila* studies, the genes controlling eye color and shape can be located in the autosomes and in the sex chromosomes (Figure 9–2). Although the location of a gene on an autosome rather than a sex chromosome may result in

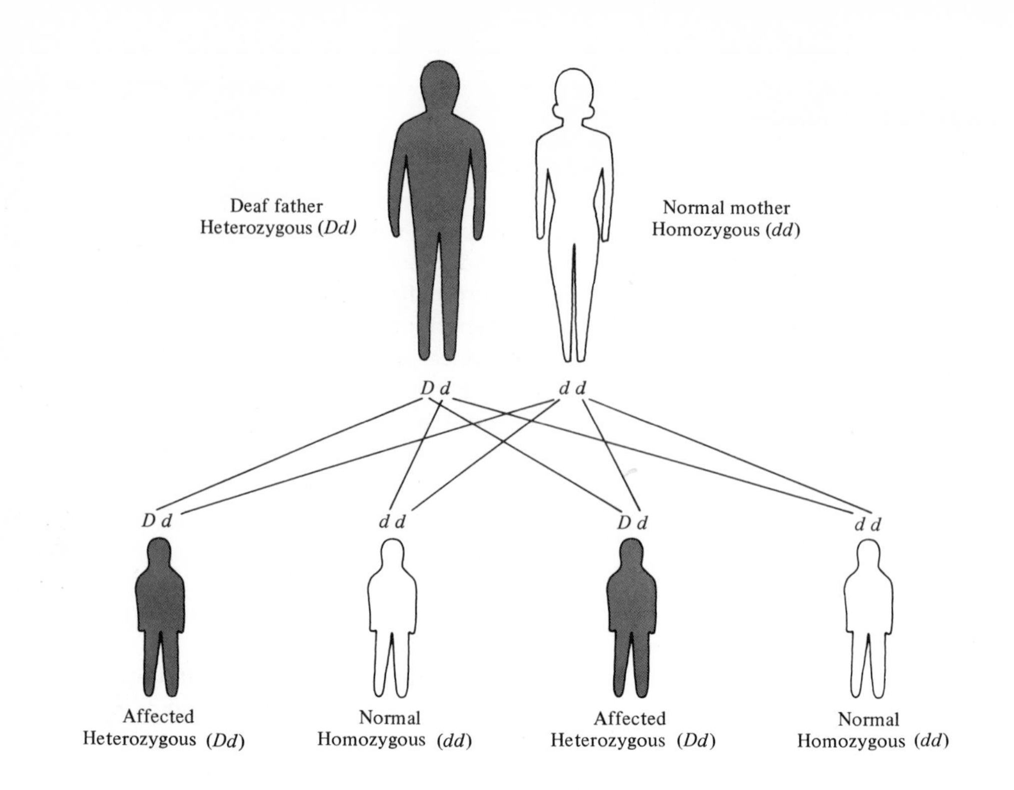

Figure 9–1. A Review of Autosomal Dominant Inheritance. One affected parent has a single faulty gene *D* that dominates its allele or normal counterpart *d*. Each child's chance of inheriting either the *D* or the *d* from the affected parent is 50%. A number of other autosomal dominant diseases are listed in Table 9–2.

A

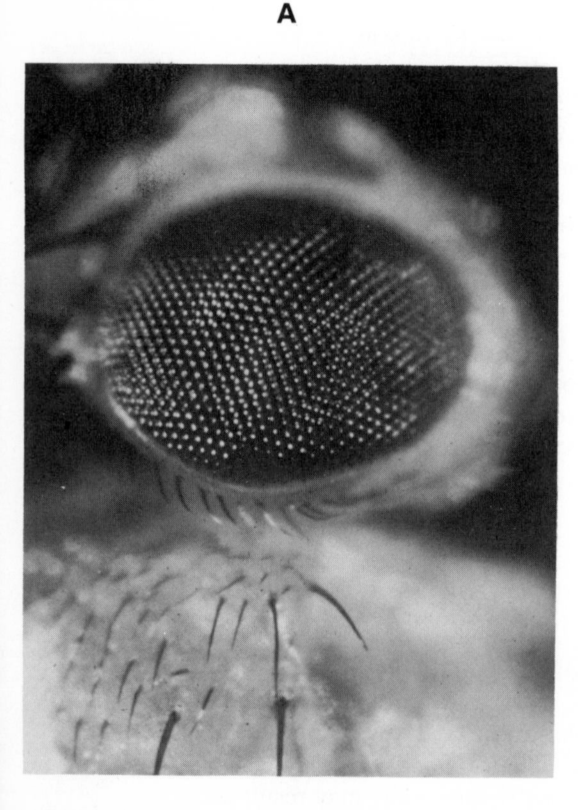

Figure 9–2. **A**. The wild type *Drosophila* eye. It would appear red in life color. **B.** The mutant white eye devoid of pigment. This mutation shows only a color mutation. **C.** Eyeless mutation shows a reduced number of eye facets. **D.** Lobe eyes show fewer remaining eye facets. **E.** Bar eye shows a reduced number of facets and an eye shape different from **C** and **D.** Note that **C, D,** and **E** are structural mutants that alter the shape of the eye, the remaining facets still have wild type red color. (Courtesy Carolina Biological Supply Company.)

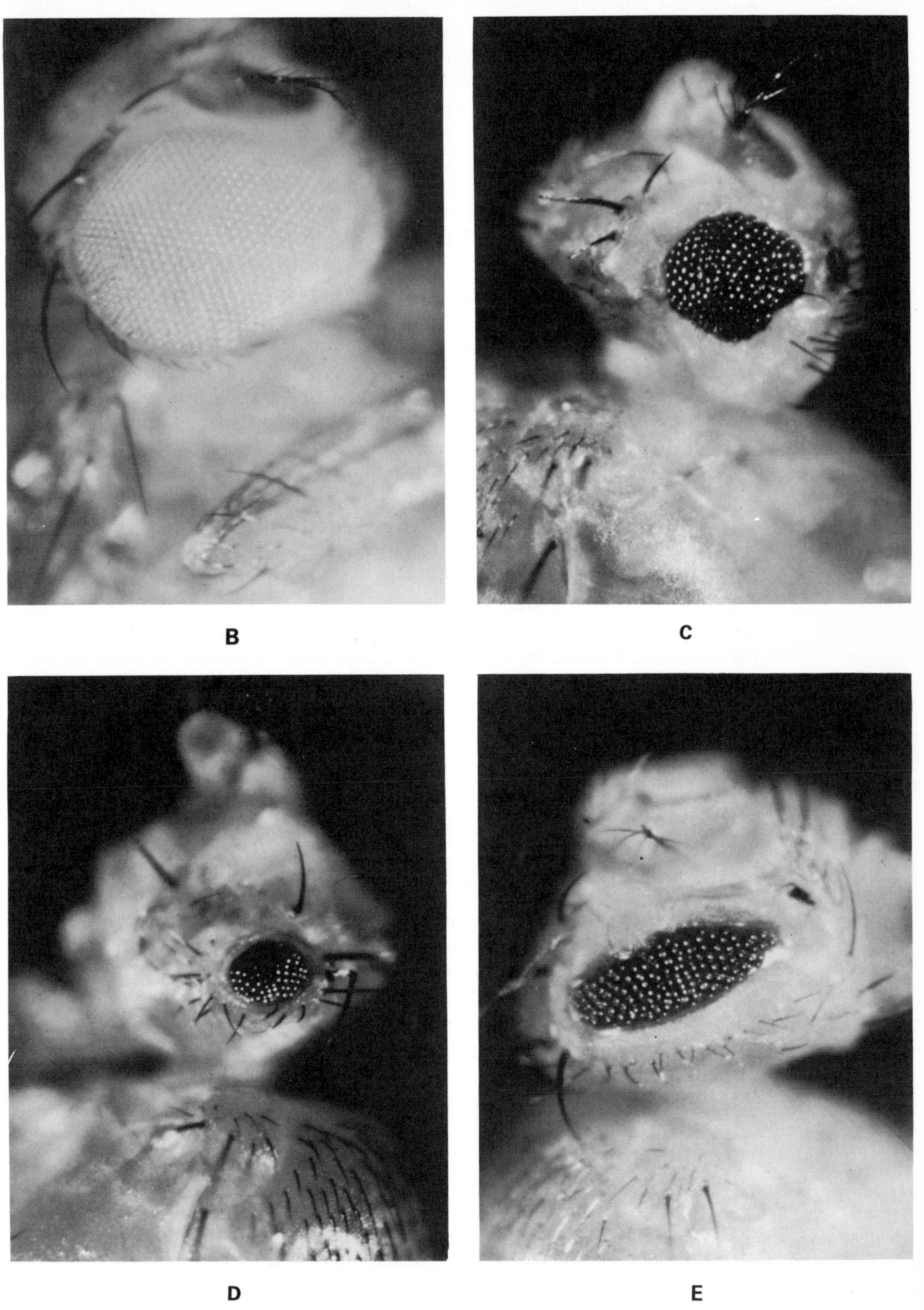
B
C
D
E

a different inheritance pattern, the degree of expression of a gene generally is not affected by chromosome location.

Eye color was perhaps the first trait for which a pattern of inheritance was established in man (Davenport, 1907; Brues, 1946). In order to understand the role of the gene for eye color in humans, we must first accept that eye color is an observable effect produced by the reflection of light from a single pigment whose synthesis and distribution are gene-controlled. Blue-, gray-, green-, and pink-appearing eyes are optical phenomena. Eyes do not contain those individual pigments. An eye merely looks blue, green, or a shade of gray. The gene responsible for the variously colored eyes apparent in the population controls the production of the pigment melanin (see Figure 10–2). Melanin varies in color from black or dark brown to shades of light yellow. The variety of eye colors results from the reflection of light, which is affected by the different size, shape, and number of melanin particles present on the iris of the eye (Figure 9–3). There are a few other auxiliary pigments that supplement melanin in producing special color effects in the eye, but in general the basic eye colors are the result of an interplay between reflected light and various amounts of melanin granules localized in or on the iris.

The iris is a colorful ring of tissue immediately surrounding the pupil, the very center of the eye. The iris can be thought of as having a front and rear section, as shown in Figure 9–3. The front of the iris is that layer of tissue closest to the front of the eye. The manner in which melanin is deposited on either or both the front and the rear of the iris determines the observable eye color. It is assumed that a separate gene (not shown on Figure 9–3) is responsible for the deposition of the melanin on the iris. It has been stated that the first human would have had to have dark eyes for protection from the harsh sun and high levels of ultraviolet radiation.

brown, blue, green, gray, and albino eye color

When dark-eyed people mate with dark- or blue-eyed people, they generally produce dark-eyed children. Brown-eyed people mating with brown-or blue-eyed people generally produce brown-eyed children. And matings between two blue-eyed people produce blue-eyed children. Such data suggest that the dark or brown trait, *B*, is dominant over blue, *b*.

In dark eyes a black to dark brown melanin is heavily deposited on the front of the iris. In blue eyes there is a light deposit of tiny melanin granules in the rear of the iris. Incoming light, which meets no reflective surface until it meets these tiny granules at the rear of the iris, is reflected back to the surface of the iris as blue. The pigment is not blue, nor is the iris. The phenomenon is an optical illusion. It is similar to the way scattered light reflected off dust particles gives the sky such beautiful shades of blue.

In green eyes the pigment on the rear of the iris is similar to that of the blue eyes; however, a lighter-colored or yellowish brown melanin is deposited on the front of the iris. The reflected light now shows the combined effects of blue integrated with brown-to-yellow pigment and appears green. Additional pig-

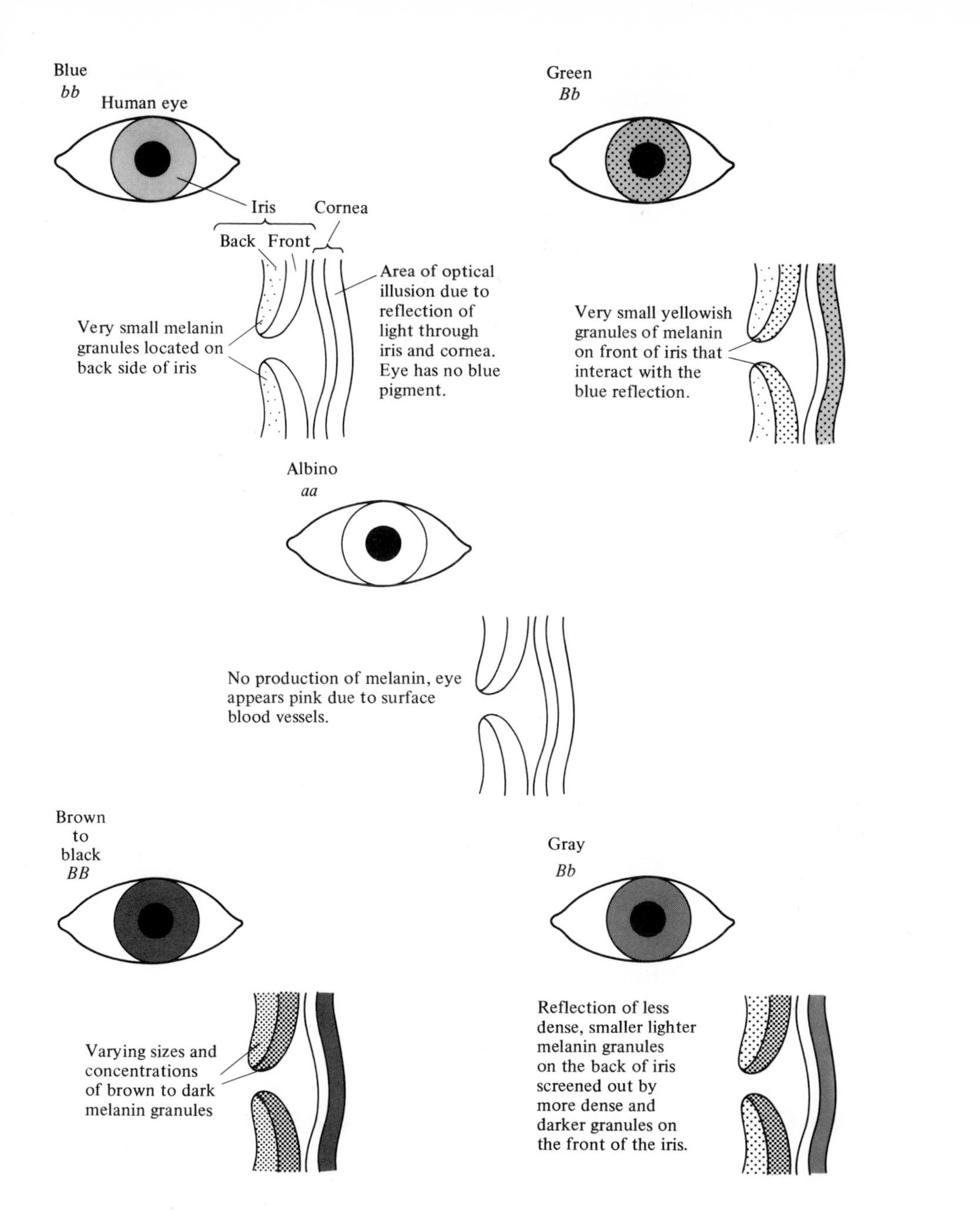

Figure 9-3. The Inheritance of Human Eye Color. Eye color is the result of the reflection of light from various shades of melanin pigment granules. The pigment granules are located on one or both sides of the iris, depending on one's genetic endowment for the production of melanin and for the location of that pigment. As the diagram of the eye indicates, the eye color observed is an optical phenomenon based on light scattering off the pigment granules. The genotypes as presented indicate that one dominant gene must be present for melanin synthesis. Various genetic and environmental dispositions lead to light and yellowish melanin granules, the interaction of which produces the more common eye colors: black, brown, gray, and blue. The albino produces very little or no melanin pigment because of the recessive *aa* condition, which is independent of other eye color genes *B* and *b*; if one is homozygous *aa*, the *B* genotype will not be expressed (epistasis). The pink coloring of the albino eye is due to a reflection of the blood vessels nourishing the iris of the eye.

ment in the front of the iris results in gray-yellow or gray-brown eyes (Scheinfeld, 1965; Rufer, Bauer and Soukup, 1970).

In the albino eye pigment is absent from the iris. In reviewing the biochemical pathway of Figure 10–2, we can see that the albino mutation blocks the production of the melanin pigment. The pink effect is the result of light reflected off the tiny blood vessels that nourish the eye. The gene for albinism is clearly different from the gene that regulates the production of melanin for brown- and blue-eyed phenotype, *B_* and *bb,* respectively.

As pigment is made and deposited on the iris, many exceptions to dark, brown, green, gray, and blue eyes can and do occur (for example, tortoise, yellow, black, and ruby colors). The rarer iris colors, such as tortoise or ruby, occur as a result of the interaction of gene product and the environment during development.

The various beautiful patterns of iris coloring are also inherited. Patterns of flecks, strips, and patches can be uniformly distributed over the iris or restricted to certain areas. Such patterns may indicate the involvement of other genes.

A particularly interesting phenomenon is that of someone with two differently colored eyes, a condition called heterochromia iridis. Usually, the person has one blue eye and one brown. The incidence of heterochromia iridis, or unmatched eyes, is 1 in 600 live births, so it is rather common; nevertheless, the first time you encounter this phenomenon, you may be slightly startled. The condition does not appear to have any other associated abnormalities. It is an unexplainable sporadic event that occurs during development. It can be speculated that a heterozygote *Bb*, for reasons not understood, expresses both genes unilaterally, one on each side of the body. There are at least two inherited defects that can give rise to this situation. In people who have brown eyes with a cataract or glaucoma in one eye, the diseased eye may appear to be blue. There is also a particular inherited nervous disorder that affects only one side of the body and apparently results in one blue eye.

Now that the readers of this text understand that chemistry is involved in the giving forth of beautiful eyes, perhaps they will push for color changes via chemical injections. This may one day happen.

inherited conditions leading to blindness

In the United States approximately 500,000 people (an incidence of 2 per 1000) are legally blind. To qualify as legally blind, you would, even with vision correction, have distant vision of less than 20/200 in the better eye. The 20/200 means that you can see at 20 feet what a normal person sees at 200 feet. Normal vision is designated 20/20. This definition means that few of the legally blind are totally blind or unable to distinguish between light and dark, but they nevertheless have serious visual handicaps. Using the National Health Survey's definition of blindness as the "inability to read ordinary newsprint even with the aid of glasses" (*Estimated Statistics*, 1966), there are about 1 million people in the United States afflicted with blindness.

America's blind have achieved varying degrees of social status. They range from the beggar with the tin cup to people such as the late Helen Keller, whose social achievements were outstanding. Blindness pushes most, however, into rather menial jobs or into social dependency. In 1969 direct financial assistance to the blind totaled approximately $268 million. In addition, over 800 agencies offering professional services and rehabilitation programs to the blind were spending about $470 million yearly (Scott, 1969). Together, the costs reached at least $738 million in 1969. Today the costs are over $1 billion per year.

The general causes of blindness are separated into two major categories: accidents 5% and disease 95%. Of the cases of blindness caused by disease, 75% occur before age 15 and approximately 50% are due to hereditary causes.

how is blindness inherited?

A number of inherited diseases that can lead to blindness, if not treated, are listed in Table 9–3 according to inheritance patterns. The list is not complete and is intended only to show that eye defects can be inherited in all manners. Smith (1970) lists some 250 inherited conditions that show ocular involvement.

In one Canadian study 180 members of one family lineage expressed the

Table 9–3 Hereditary Diseases of the Eye That Frequently Affect Visual Function*

Dominant	*Recessive*	*X-Linked*	*Undetermined*
Aniridia	Adult optic atrophy	Choroideremia	Absence of cones
Anophthalmia	Albinism	Diabetes insipidus	Absence of rods
Arachnodactyly (Marfan's syndrome)	Buphthalmos	Leber's optic atrophy	Cerebromacular degeneration
Blue sclerotics	Cataract	Lowe's optic atrophy	Craniofacial anomalies
Buphthalmos	Cryptophthalmos	Megalocornea	Degeneration of retina (pigment)
Cataract	Day blindness	Microphthalmia	Epithelial-endothelial dystrophy (Fuchs's syndrome)
Coloboma	Glaucoma	Night blindness	Glaucoma
Concomitant strabismus	Hydrophthalmos	Nystagmus	Hereditary ataxia
Corneal dystrophy	Laurence–Moon–Biedl syndrome	Pseudoglioma	Leber's tapeto-retinal degeneration
Corneal erosion	Macular corneal dystrophy	Red or green color blindness	Macular corneal dystrophy
Dacryocystitis	Microphakia	Retinitis pigmentosa	Pseudoxanthoma elasticum
Glaucoma	Microphthalmia	Yellow-blue color blindness	
Heterochromia iridis	Retinitis pigmentosa		
Juvenile optic atrophy	Tay–Sachs disease		
Nystagmus			
Phakomatoses			
Ptosis			
Retinitis pigmentosa	*Polygenic*		
Retinoblastoma	Diabetes mellitus		

* Certain eye defects have been found to be transmitted by more than one Mendelian mode of inheritance. Retinitis pigmentosa, for example, has been described as dominant, recessive, and X-linked.

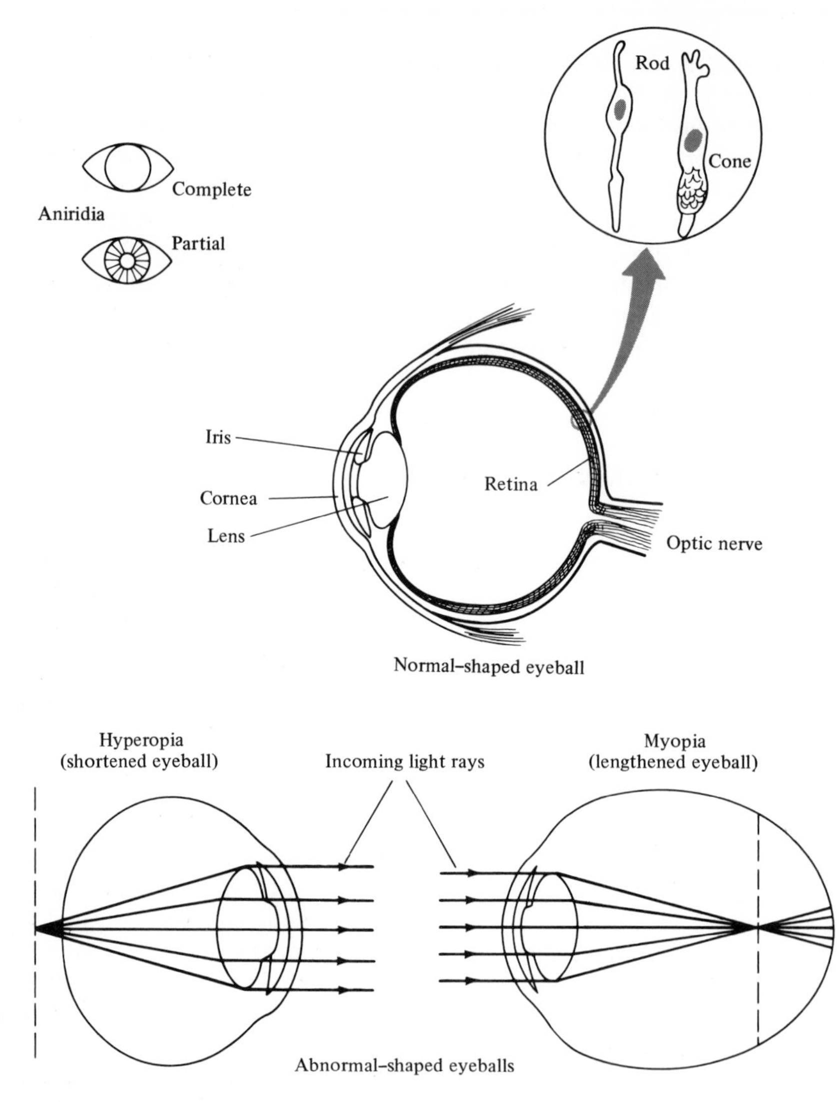

Figure 9–4. Inherited Disorders of the Eye. The most common inherited eye diseases are represented in the diagram. Poor vision can be inherited as single-gene and polygenic defects of the eye and as a part of variously inherited muscle, nerve, and skeleton syndromes. Inherited defects in the cones lead to various forms of color blindness. The retina or inner lining of the eye is the site for many forms of inherited disorders, for example, retinoblastoma, retinitis pigmentosa, and detached retina. The inherited shape of the eyeball can also present certain common vision problems: the recessive myopia (nearsightedness), the dominant hyperopia (farsightedness), and the X-linked recessive microphthalmia, where the eyeball has normal shape but is very small.

dominant gene for aniridia, partial or complete absence of the iris. Very little can be done for this condition, since the loss or nondevelopment of the iris occurs before birth. However, the National Society for the Prevention of Blindness asserts that half of all blindness can be prevented. This 50% must include most of the accidentally caused blindness, which accounts for only 5% of all blindness. One hereditary disease that contributes to the number of blind persons is diabetes. It is estimated that some 50% of all diabetics develop cataracts and disorders of the retina (retinopathy) that lead to permanent blindness. The National Society predicts that by the year 2000 there will be more cases of blindness from diabetic retinopathy than from all other causes combined.

descriptions of some inherited eye diseases

Some of the diseases listed in Table 9–3 are illustrated in Figure 9–4. A normal eye is presented to serve as a reference in locating the various parts of the eye that can be affected through the inheritance of a defective gene.

Buphthalmos involves defects in the light-filtration angle of the anterior chamber of the eyeball and related defects in the ciliary muscle. Both autosomal recessive and dominant patterns of inheritance are observed.

Cataract is an opaque condition of the lens or the cornea. Development of a cataract and the age at which the cataract is likely to appear are inherited (Scott, 1969). A dominant gene is usually involved, but cataract may be inherited in an autosomal recessive manner. The most severe forms occur only in males; this suggests that expression is modified by the sex of the individual. Cataract may also be inherited as part of a number of inherited genetic syndromes.

Coloboma and *anophthalmia* involve lesions of the iris. Inheritance is dominant, but expressivity varies. New mutations, chromosomal trisomies, and gene defects that are expressed earlier in development are involved in some cases of these eye defects (Fraser, 1964).

Cryptophthalmos is caused by a recessive autosomal gene that produces a condition in which the eyelids fail to separate (Figure 9–5).

Leber's tapeto-retinal degeneration, including retinal aplasia, was found to be the cause of 9% of the blindness in English schoolchildren. It is often associated with other inherited abnormalities. The pattern of inheritance is not understood. Retinal coloboma and aniridia alternate in families with pedigrees of this disease, suggesting the possibility that the three diseases represent different expressions of the same gene.

Microphthalmia is inherited in an autosomal recessive and recessive sex-linked manner. The eyes are small and nonfunctional; the individual is blind.

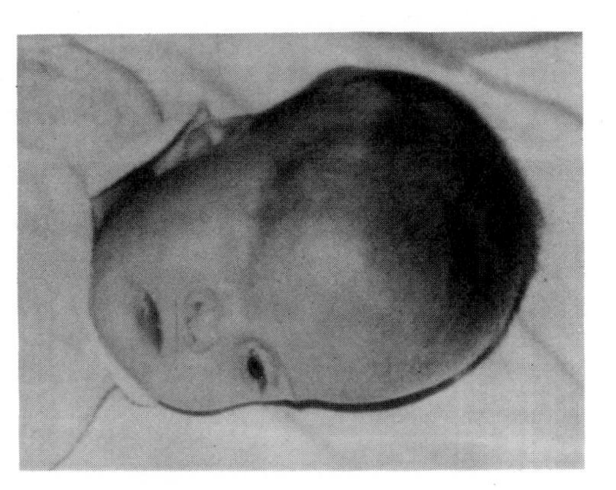

Figure 9–5. Cryptophthalmos (Unilateral). In this recessively inherited disorder, the eyelid and cornea fail to develop on one or both eyes, and the skin of the forehead continues to grow down to the cheeks. Many other body organs may also be affected when cryptophthalmos is present. (By permission of Charles H. Carter, M.D.)

Optic atrophy is the degeneration of the optic nerve and is caused by a recessive sex-linked gene or by a dominant or recessive mutation. It occurs as part of neurological syndromes, and in some cases there is no effect on the central nervous system (Fraser, 1964).

Pseudoglioma with infantile retinal detachment is an eye defect that appears to be almost like retinoblastoma except that it is a nonmalignant tumor of the retina. With time, the retina becomes detached, and several other eye defects result as a consequence. Often it is associated with mental retardation and deafness. It is inherited as a sex-linked recessive gene.

Retinitis pigmentosa is the production of a pigment beside the blood vessels of the retina. It is inherited in some families as a dominant, in others as a recessive, and in still other families as an X-linked disease. It has also arisen as a mutation of at least eight different gene loci (McKusick, 1960).

Retinoblastoma is cancer of the retina. It often appears to be controlled by a dominant gene with incomplete penetrance. It affects 1 in 34,000 live births. There is some evidence to indicate that inherited retinoblastoma is a member of a set of mutations that appear with increasing frequency as mother's age increases. This increase is likely to be attributable to such causes as a longer period of exposure to radiation (Fraser, 1964).

From their study of blindness in English schoolchildren, Fraser and Friedman (1967) concluded that not less than 19 autosomal dominant, 30 autosomal recessive, and 10 sex-linked genes function directly in causing blindness.

Genes are also involved in various eye defects that usually do not result in blindness as legally defined. Among these is myopia, or nearsightedness, which is usually inherited as an autosomal recessive. Hyperopia, or farsightedness, and astigmatism, or irregular conformation of the cornea, are inherited in a dominant manner (Figure 9–4). Excessive curvature of the cornea is also caused by a dominant gene. Nystagmus, a condition in which the eyes undergo uncontrollable rolling movements, is inherited through a sex-linked recessive gene in some families and through an autosomal dominant in others. Congenital night blindness results from a sex-linked recessive. Day blindness, failing eyesight in bright daylight, is inherited through a recessive pattern. Red and green color blindness is caused by a sex-linked recessive gene. Total color blindness, clouded cornea, displaced lenses, displaced pupils, irises with missing segments, crossed eyes, drooping eyelids, mirror reading, albino eyes, and dry eyes are also inherited defects.

the inheritance of hair color

the physical basis: transmission of units of hereditary material through the genes

Is it true that gentlemen prefer blonds or that redheads have bad tempers? The cosmetic industry has thrived on coloring agents, rinses, and shampoos for the various types of hair. Even the natural blond uses conditioning products made especially for fine, wavy, oily, or dry hair. And what of those people who wish to switch hair color, those not satisfied with the genetic expression of their inherited genes for hair color, form, and texture? Persons involved in the masking of their true genetic inheritance of hair color only mimic the acts nature performs on others. Of the easily observable traits of social significance, hair and skin colors are the most easily seen, and hair color is the most easily changed by the environment (a person with chemicals).

hair pigment and hair color

Many children are born with hair color that is different from that which they will have as adults. As many parents can testify, some dark-haired children slowly lose the dark hair and become brown- or blond-haired later on. The reverse is also true. With time and exposure to the environment, a fixed pigment system for hair color appears to become established. This fixation of individual pigmentation is so permanent in the normal person that, regardless of environmental assault through exposure to sun, bleach, or drugs, close observation will reveal many unchanged pigment granules in the hair. This is not to say that these granules are never lost or bleached; they do undergo change with age. This change is a natural decolorization process that affects the hair follicle. The timing of the natural graying process appears to be under genetic control, since children of parents who gray at certain ages appear to gray at approximately the same age. The process of graying is generally a slow one; it is never completed as quickly as suggested by the saying "He turned gray overnight." Even tension, fright, and sickness, although capable of bringing on premature graying of the hair, do not decolorize hair already on the scalp. In general, gray hair comes about with the growth of new decolorized hair, which is mixed in varying amounts with the normal hair color.

dominance in hair coloring

Within the strands of hair, granules of melanin are deposited in varying concentrations. Melanin from dark brown to yellowish brown color is similar to that deposited on the iris of the eye. A heavy deposit of black to dark brown melanin within the hair strand yields very dark or black hair; decreased quantities of dark pigment and increased quantities of yellowish pigment yield brown, light brown, and dilute blond. The arrangement of the granules, their hue, and the reflected light from them account for the various hair colors one sees. If you have ever been in a place with flashing colored lights, you have noticed that your hair appears to be a different color depending on the light reflected off the pigment granules. If there are no granules of melanin pigment, because of a block in the pathways (see Figure 10–2) for the production of melanin, the individual is an albino (*aa*), as shown in Figure 9–6. In the aged person gray to gray-white hair results from a reduced quantity of dark pigment and in an increased, but small, amount of the yellowish pigment.

The pigmentation for the production of hair involves, in addition to melanin, the synthesis of a red pigment. The production of the red pigment indicates that there is at least a second gene locus for hair coloring, a gene for the production of red pigment. The evidence for this is that persons with black hair can have strands of red hair at the same time, giving a chestnut hue if enough red hair strands are present. Also, red hair is found on people with any

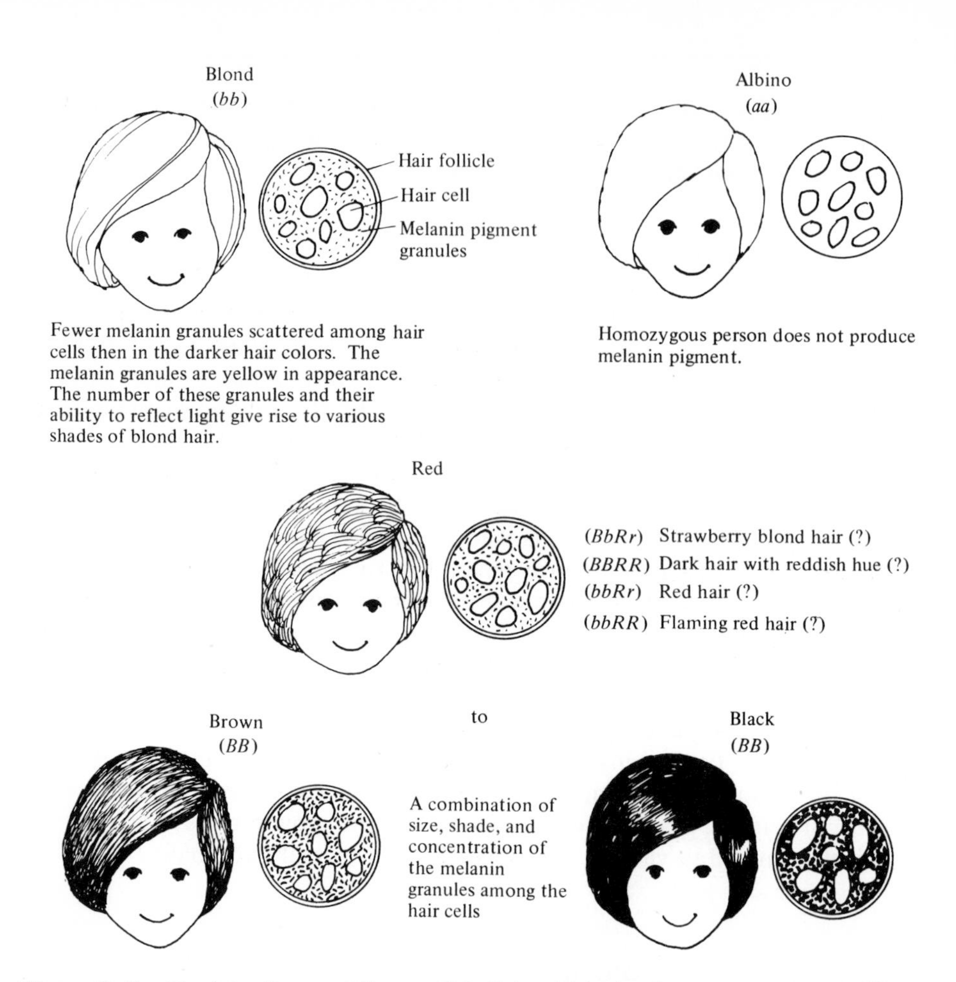

Figure 9–6. The Inheritance of Human Hair Color. Melanin pigment granules of different sizes, shapes, and color (black to brown to yellow-brown), scattered in varying numbers among the hair cells, reflect light in such a manner that hair appears black, brown, brownish blond, or blond. For black or brown hair at least one dominant gene is required. Albinos have a separate gene in a recessive state *aa.* Blond hair is a result of fewer and lighter-colored (yellow-brown) melanin granules. The more diluted the melanin, the lighter or blonder the hair. Red hair presents a separate problem in that at least one additional pair of genes is involved, the dominant allele being responsible for the synthesis of a red pigment. Various amounts of red pigment interacting with varying concentrations of melanin pigment result in dark hair with a reddish hue, light red hair, reddish blond hair, or the flaming redhead. Because the exact gene complement and interaction of gene products with the environment are unknown, the genotypes and phenotypes for the different shades of red hair are in doubt (?).

eye color, whereas black- or dark-brown-haired people usually have dark or brown eyes, and blonds usually have blue, gray, or green eyes.

Probably the strangest phenomenon, well known in the copper-mining industry, is that persons with red hair may find their hair turning green if they remain long in contact with the mining operation. The reaction is not completely understood, but it does represent another case of environmental change in one's phenotype because of a chemical compound. Persons who do not have the red pigment in their hair strands do not develop green hair, and not all redheads experience the hair color change. So, as has been stated previously, the interaction of the genotype with the environment is complex and usually not understood.

the transmission of genes controlling hair color

Melanin, the same pigment that affects eye colors, also influences the color of human and animal hair. It follows, then, that a mutation, or change, in the gene or genes responsible for the synthesis of melanin will affect the color of one's hair as well as his eyes. It is well established, however, that the genes controlling the deposit of melanin in the hair are different from those that control the deposits of melanin in the eyes. We can be sure of this because some dark-haired people have blue eyes and some blonds have brown eyes. Eye color and hair color, then, appear to be independently inherited traits. If we deal with one trait at a time, we can predict the probability of a child's having a specific hair color or eye color.

Parents often are puzzled by a child's hair color (or eye color): "Both my husband and I have black hair, but our child has blond hair. How can that be?" Let's explore this question using a pedigree chart.

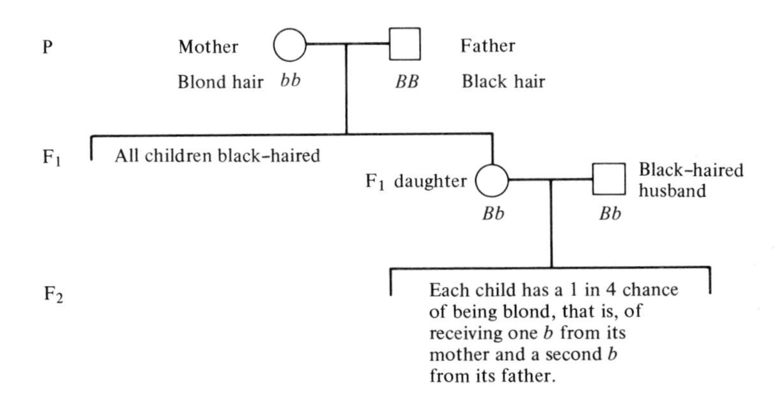

Black and brown hair are dominant to blond hair. Since blond hair is recessive, it will be represented by *bb*. Persons of *BB* or *Bb* genotype will have black or brown hair. The pedigree chart shows how two black-haired parents could produce a blond-haired child. Both the father and mother must be heterozygotes. Each could have black hair, yet contribute a recessive gene *b*. A homozygous *bb* child would have blond hair. You should now be able to work out the probabilities in your family for hair color, eye color, and other dominantly inherited traits, such as dimples and free earlobes.

huntington's disease

Huntington's chorea is usually characterized by mental and physical deterioration (Figure 9–7). The term *chorea* is the Greek term for "dance," and Huntington's patients display a most unusual "dance." They have jerky,

Figure 9–7.(*Opposite*) Huntington's Disease. A severe dominantly inherited neurological disorder that leads to a slow agonizing death. The expression of the disease involves loss of memory, choreic-like movement, loss of physical movement, and decrease in mental ability. **A.** Photograph of Woody Guthrie, renowned folk singer before the expression of Huntington's disease. **B.** Woody Guthrie sometime after the expression of Huntington's disease. **C.** The last photograph ever taken of Woody at Creedmoor Hospital, October 1966. Included with Woody are his son Arlo (seated at center) and Mrs. Majorie Guthrie. Woody died at age 55 after 13 years of illness. (Courtesy of Mrs. Marjorie Guthrie, President, Committee to Combat Huntington's Disease, Inc.)

twisting, uncontrollable muscle spasms, which are often mistaken for the willful throwing of themselves or of objects they may be holding. Often progressive mental changes accompany the disease because of the degeneration of the central nervous system and the loss of cells in the frontal lobes of the cerebral cortex of the brain. The loss of these brain cells sometimes leads to fits of depression, insanity, or suicide (Heathfield, 1967). The progressive change in personality is one of the striking features in diagnosing a Huntington's patient. The personality change in many cases is believed to be caused by the patient's fear of what the disease will do to him.

Medical dictionaries list about 50 varieties of chorea, for example, Saint Vitus's dance. Huntington's chorea, now preferably called Huntington's disease (HD), causes a slow, agonizing, humanly demeaning deterioration of the patient. From the onset of symptoms the patient steadily deteriorates for 10 to 15 years until he dies.

incidence of huntington's disease in the united states

It is difficult to provide an accurate estimate of frequency because this disease is difficult to diagnose and it usually occurs late in life. (Many persons die before they show the symptoms and are consequently never counted as HD.) The Committee to Combat Huntington's Disease (CCHD) believes that current estimates are inaccurate and that there are numerous unreported cases of HD. CCHD predicts that when all cases are found, the number of HD patients in the United States will reach 100,000.

early recognition of huntington's disease

Although Huntington's disease was described in America by Waters as early as 1848 and again in 1863 by Lyon (Myrianthopoulos, 1966), it was considered rare and particularly difficult to diagnose. In 1872 George Huntington, in a paper about various kinds of chorea, presented what is now considered the classic description of the HD patient. A portion of Huntington's paper *On Chorea* is presented below in order that the reader can appreciate the clarity and foresight that Huntington displayed at a time when the Mendelian principles of heredity had not yet been rediscovered (Huntington, 1872).

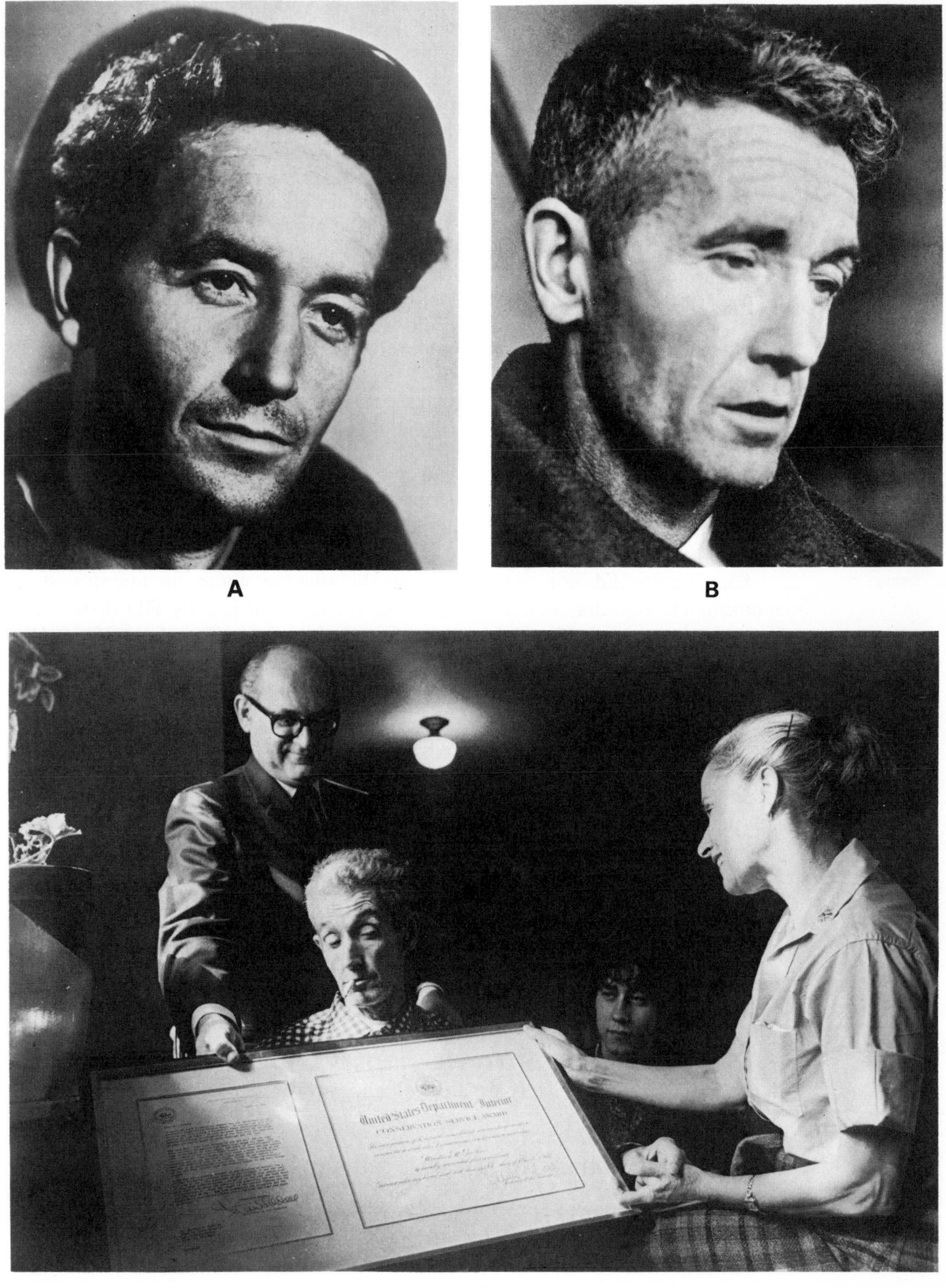

A

B

C

> And now I wish to draw your attention more particularly to a form of the disease which exists, so far as I know, almost exclusively on the east end of Long Island. It is peculiar in itself and seems to obey certain fixed laws. In the first place, let me remark that chorea, as it is commonly known to the profession, and a description of which I have already given, is of exceedingly rare occurrence there. I do not remember a single instance occurring in my father's practice, and I have often heard him say that it was a rare disease and seldom met with by him.
>
> The hereditary chorea, as I shall call it, is confined to certain and fortunately a few families, and has been transmitted to them, an heirloom from generations away back in the dim past. It is spoken of by those in whose veins the seeds of the disease are known to exist, with a kind of horror, and not at all alluded to except through dire necessity, when it is mentioned as "that disorder." It is attended generally by all the symptoms of common chorea, only in an aggravated degree, hardly ever manifesting itself until adult or middle life, and then coming on gradually but surely, increasing by degrees and often occupying years in its development, until the hapless sufferer is but a quivering wreck of his former self.
>
> . . . .
>
> . . . I know nothing of its pathology. I have drawn your attention to this form of chorea, gentlemen, not that I considered it of any great practical importance to you, but merely as a medical curiosity, and as such it may have some interest.

Because of Huntington's description the disease carries his name. His conclusion—"I know nothing of its pathology"—remains generally true today, over 100 years later.

Huntington did not recognize that the disease could be manifested in children. He felt that it was a disease of adults. And usually HD does occur between the ages of 25 and 55. But recent data indicate that about 2% of HD patients are under 12 and 5% over 60 when the disease begins. The vast majority, however, show their first symptoms between ages 30 and 45. In juvenile HD the physical symptoms are in general distinctly different from those for the adult form. The child experiences seizures of muscle rigidity with very limited use of his muscles—just the opposite of the jerky, spastic muscle seizures experienced in the adult forms. It is estimated that about 12% of all HD patients show the rigid form (Myrianthopoulos, 1966).

Huntington suggested that the disease occurs more often in males than in females, and although a number of studies deny this sex-influenced tendency, a report by Brackenridge (1971) indicates that there is statistical evidence showing that more HD patients are male. This is consistent with a more recent report, given at the 1972 Centennial Symposium on HD, that in 75% of the cases of juvenile HD, the affected parent, the person who transmitted the dominant gene to the child, was the father. Another report at this symposium indicated that in one case of identical twins one twin had the rigid form of HD, and the other had the choreic form. This indicates that the form of expression depends on what nerve cells are affected and where the nerve cells are located.

the immigration of huntington's disease to the united states

There are geographical hot spots (areas of highest incidence) of HD cases that can be accounted for historically through the migration of persons carrying the HD gene from Europe to other parts of the world. Stevens and Parsonage

(1969) speculate that Huntington's disease began as a fresh mutation in England several hundred years ago. Vassie (1932) stated that HD came to the United States in 1630 via two afflicted brothers and their unaffected wives, as they migrated from Bures, a small village of Suffolk, England, to Boston. These immigrants subsequently produced 11 generations of HD-afflicted offspring. It is believed that a daughter of one of the original settlers and her husband went to East Hampton, Long Island. The immigrants and their descendants, often referred to as the Bures family group, were accused of witchcraft, being possessed by the devil, theft, quackery, prostitution, family feuds, use of profanity, lack of religious attitude, and drunkenness. They were punished by brandings, beatings, and death (Vassie, 1932). Vassie suggested that the "witches" whose trials and deaths occurred at Salem were persons afflicted with HD. Hedges (1973) stated that seven of the women of the Bures family group were accused of being witches. One, named Elizabeth, left her husband and daughter to return to England, only to be hanged there for witchcraft. The daughter Elizabeth, left Boston, later became known as the Witch of Groton. In 1691 Cotton Mather declared from his Boston pulpit that she should be hanged. Her fate is unknown. The disease has now been diagnosed all over the world, with Japan appearing to have the fewest number of cases.

the committee to combat huntington's disease / case histories

In 1967, the year folk singer Woody Guthrie died of Huntington's disease, his widow, Marjorie, and six other persons from families with the disease formed the Committee to Combat Huntington's Disease (CCHD). The committee now has 50 chapters across the country. Huntington's disease groups also meet in England, Scotland, Australia, Canada, and Belgium. A principal aim of a Huntington's disease committee is to provide *all* families having an afflicted member with information and counseling to help them find whatever medical and financial assistance is available.

The biggest problem faced by families with HD, says Marjorie Guthrie, is finding a proper care facility that will take patients with the long-term terminal disease. If the reader wishes to update his knowledge of HD, he can receive excellent free material by writing to The Committee to Combat Huntington's Disease, 250 West 57th Street, New York, N.Y. 10019.

Marjorie Guthrie has overcome the depression of losing her husband to HD. She has three children: Arlo (he appeared in the movie *Alice's Restaurant*), Joady, and Nora-Lee. The children, especially Arlo, the oldest, witnessed their father's deterioration; his death came 13 years after the onset of the symptoms. Woody wrote a penetrating account of his progressive disease in his autobiography, *Bound for Glory*, and also described the gradual deterioration of his mother, Nora Belle Guthrie, due to HD. Woody died in a mental institution.

The Guthrie children realize there is a 50% chance that each of them, like their father, will come down with HD. Arlo decided to have children, knowing that if he should come down with HD, his children will also have a 50% chance of having HD. It is to be hoped that, with an additional 30 to 40 years of research time, a control or cure may have been found if either of his two children should inherit HD. Mrs. Guthrie, when interviewed about Arlo's choice to have children, said, "It's the quality of life that counts, not the quantity, and if you have 40 to 50 years of a good life that's good too!"

Arlo Guthrie had children. Tony Navarro has not. The Navarro family was written up in the November 1971 issue of *Today's Health* (Harmetz, 1971). Tony Navarro is married and loves children, but he is afraid that he may have HD and transmit it to his offspring. In 1971 he was 33 years old. His two brothers and a sister had HD. Brother Ed at 46 was in the terminal stages of HD; he had no bowel control, could not walk, talk, or feed himself. He had to be strapped in bed to prevent him from injuring himself by his wild involuntary movements. Brother Rudy was 38 and very bright. Four years earlier the symptoms started; in his chemistry classroom, he kept dropping his pencils and tripping. By 1971 he had trouble speaking, his left eye squinted, and his face was slightly twisted. He had already undergone behavioral changes. Once an avid reader, he no longer read; a conversationalist, he now had little to say. His sister said that he spoke of suicide. Sister Eva, at age 42, had had HD for 10 years. She had not yet been hospitalized, but she was having great difficulty with her speech and with eating, and could not walk.

Mrs. Navarro, Tony's mother, lost her husband, sister-in-law, and mother-in-law to HD (Harmetz, 1971). So much tragedy for one family. With a fully penetrant gene like HD, the 1 in 2 chance for each child of a defective parent presents the potential parent with a very sobering decision when it comes to having a child.

Stories like those of the Guthries and Navarros are not uncommon. Below is an excerpt from a letter a woman wrote to her congressman in 1974 asking his support for House bill HR 12215 (the National Huntington's Disease Control Act).

> Dear Congressman
>
> . . . I am writing to you because I am the wife of one of the victims of Huntington's Disease, and I am earnestly and urgently requesting your help to get the House bill passed.
>
> For the past 14 years, this insidious inherited disease has caused us anguish I do not have the space to describe. My husband is 59 but looks more like 75. Once a brilliant auditor, he is now mentally like a five year old, and cannot even attend to personal needs. The dying brain cells first destroy the personality, then gradually destroy the body, but it drags on for so long; 24 years for his father.
>
> Thus, medical expenses are catastrophic. I cannot nurse him myself as I must work to support him, a full time and a part time job. The state mental hospital does not want cases of Huntington's, and of course I do not want to place him at ——. The only answer at present is a nursing home, where he has been for almost a year.
>
> Monthly medical expenses run from $350 to $375, and insurance covers only about $15 of this. This financial drain does not allow me to save much, and I will be forced to retire at 65, in seven years. Then we must depend on our two sons, but they must save for the same probable fate. Each son has a 50–50 chance to escape or inherit the disease. If the older son does not escape, then his own little son must face the same odds. All family decisions are affected by this heavy cloud that hangs over us.

If these bills could be passed, perhaps there could be some financial aid for our family and other affected families. If research is funded, perhaps my sons and my grandson can be saved.

Please use your influence to help in any way in this matter. It will be greatly appreciated.

Sincerely,
Mrs. ——

familial intestinal polyposis (gardner's syndrome)

Familial intestinal polyposis is also referred to as adenomatosis, hereditary multiple polyposis, and multiple papillomas. The first complete description of the disease was provided by Gardner and Richards (1953). The polyps, fingerlike projections growing out of the lining of the large intestine, first appear as nonmalignant growths. Then with time, generally 1 to 15 years, they undergo a change and become cancerous. The only cure at present is surgical removal of that part of the large intensine, and this can only be done if the disease is diagnosed early. The disorder is difficult to diagnose, since there is no single outstanding symptom. Approximately 2 per 100,000 in the general population have this particular disease as a result of spontaneous gene mutation. The frequency of cases in the population must be at least twice that. The disease is dominantly inherited with a high degree of penetrance, and it is expressed during the late 20's or early 30's. The polyps may be found over the entire length of the large intestine, but in 95% of the cases they are found within 1 to 4 inches of the anal opening.

marfan's syndrome (arachnodactyly)

Marfan's syndrome was described in 1896 by A. B. Marfan. It is transmitted as a dominant trait with variable expression and affects the bone, muscle, and connective tissue. The patient tends to have long, thin appendages—arms, legs, and fingers—and is often stoop-shouldered and awkward. The disease is sometimes referred to as arachnodactyly (spider-fingeredness) because of the excessive length of the fingers and toes (Figure 9–8). The majority of Marfan's patients have myopia and blue sclera of the eyes. Many of these patients have severe loss of vision or blindness resulting from retinal detachment and glaucoma. Marfan's patients also appear to have a tendency to aortic disease, which leads to the rupture of the aorta and death. The mutation rate is estimated at 5 per 1 million gametes. Many texts state that Abraham Lincoln may have been afflicted with Marfan's syndrome.

hypercholesterolemia

Hypercholesterolemia is inherited through a dominant gene and is associated with high levels of blood cholesterol. Tests for abnormal amounts of fat in the bloodstream may make it possible for the victims to avoid potential heart

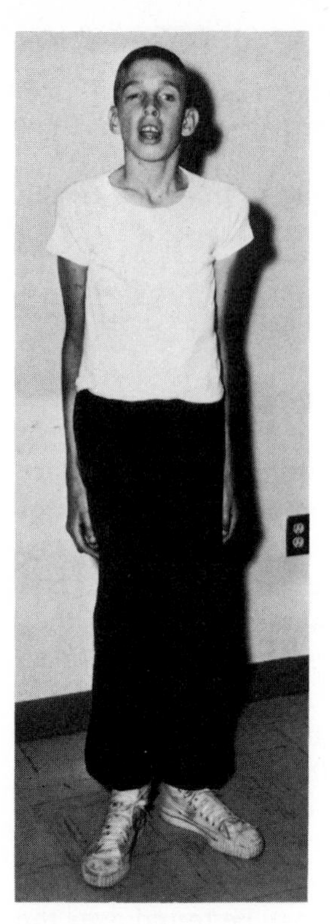

Figure 9–8. Marfan's Syndrome. The main manifestations are due to a single-gene-determined defect of connective tissue. A defect in the amino acid sequence of collagen may be responsible. The patient demonstrates long arms, legs, and digits and is generally awkward in his walking and running. Eye, heart, and spinal column defects also are commonly part of the syndrome. (By permission of Charles H. Carter, M.D.)

attacks. The incidence of the disorder is very high, approximately 1 in every 200 live births (*Bio Medical News*, 1972). Research has shown that a restricted diet and the use of certain drugs can lower the cholesterol levels in some patients to near normal.

familial polycystic kidneys

Virchow suggested in 1856 that a deposit of salts in the renal tubules would later lead to cyst formation. Some 70 years later Cairns established that the formation of cysts in the kidney is an inherited disorder. Since then it has been established that the disease is dominantly inherited, and that it is completely penetrant if the person who has the defective gene lives to the age of 80 (*New England Journal of Medicine*, 1969). The disorder rarely occurs before the age of 20 and is usually diagnosed between ages 45 and 55. This, like Huntington's disease, is an example of delayed gene expression. Once the cysts begin their development, however, they continue to enlarge and uremic poisoning occurs. Treatment at this point is kidney dialysis.

polydactyly

Polydactyly is a dominantly inherited disorder, where the patient has extra digits on one or both hands and/or on his feet. The usual case is an extra finger

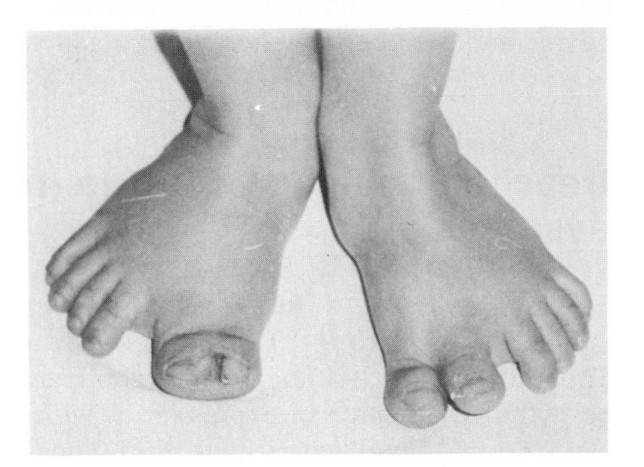

Figure 9–9. Polydactyly. This inherited dominant trait involves extra digits, fingers and/or toes. Also note the syndactyly or fused large toes on the left foot. (By permission of Charles H. Carter, M.D.)

on one or both hands. The location of the extra digit is variable. Extra fingers have been located next to the thumb and next to the small finger. The person with the normal number of digits is homozygous recessive *dd*. Heterozygous *Dd* individuals have variable penetrance and different degrees of expression. One hand may have six fingers, both hands may have an extra digit, one hand and one foot may be involved, etc. (Figure 9–9).

There are at least 14 additional diseases that sometimes involve extra digits: basal cell nervous syndrome, Carpenter's syndrome, trisomies 13 and 18, Conradi's disease, Ellis–van Creveld syndrome, focal dermal hypoplasia, Laurence–Moon–Biedl syndrome, oral-facial-digital syndrome, Smith–Lemi–Opitz syndrome, lissencephaly syndrome, dyschondroplasia, facial anomalies, and polysyndactyly syndrome. Should otherwise normal persons have extra digits, surgery, which is effective, may be performed for cosmetic reasons (social acceptance).

ragweed hay fever (an atopic allergic disease)

The term *atopic* refers to a group of allergic diseases that have a hereditary basis. Classical studies in clinical allergies indicate that there are a number of allergic diseases that appear to be transmitted within family lines. Statistics on these families support the idea that there is a genetic predisposition to certain allergic diseases. As most geneticists are aware, it is difficult to measure a predisposition or susceptibility for the development of a disease (see Chapter 16). Levine and Stember (1972) studied the genetic control of the ragweed hay fever allergy and concluded that an immune response (IR) gene is responsible for the production of an antibody that reacts specifically with the protein of the ragweed pollen. When the antigen (ragweed protein) enters the body, it elicits production by certain cells of a specific protein, the antibody, to neutralize that ragweed protein or antigen. The protein antigen of ragweed pollen has been named antigen E (see Chapter 12). Therefore, because an antibody is one of the forms of immunoglobin made by the body, we say that antigen E stimulates immunoglobin E (IgE) synthesis for body protection. Thus when certain persons are exposed to ragweed pollen, the IR gene allows for the synthesis of

IgE to neutralize the antigen E (ragweed pollen). The IR gene, according to Levine and Stember, is inherited as a genetic dominant.

treacher–collins syndrome (mandibulofacial dysostosis)

The defective gene of the Treacher–Collins syndrome is most often expressed in the face (Figure 9–10). The first case of this syndrome was reported by Thomson in 1846. Although it is a dominant trait and penetrance is believed to be 100%, the degree of expression is widely variable. In general, the patients have normal intelligence, but they can display an absence of the lower eyelashes, deafness, defective jaw formation, cleft palate, congenital heart defects, and vertebral anomalies. All experience some malformation of the face. This is a rare dominant trait, and it is believed that 60% of the cases are due to new mutations. The 60% means that in 6 out of 10 new cases the defective child has normal parents. Almost all the children who receive the defective gene from a defective parent receive it from their mother. The cause is not understood.

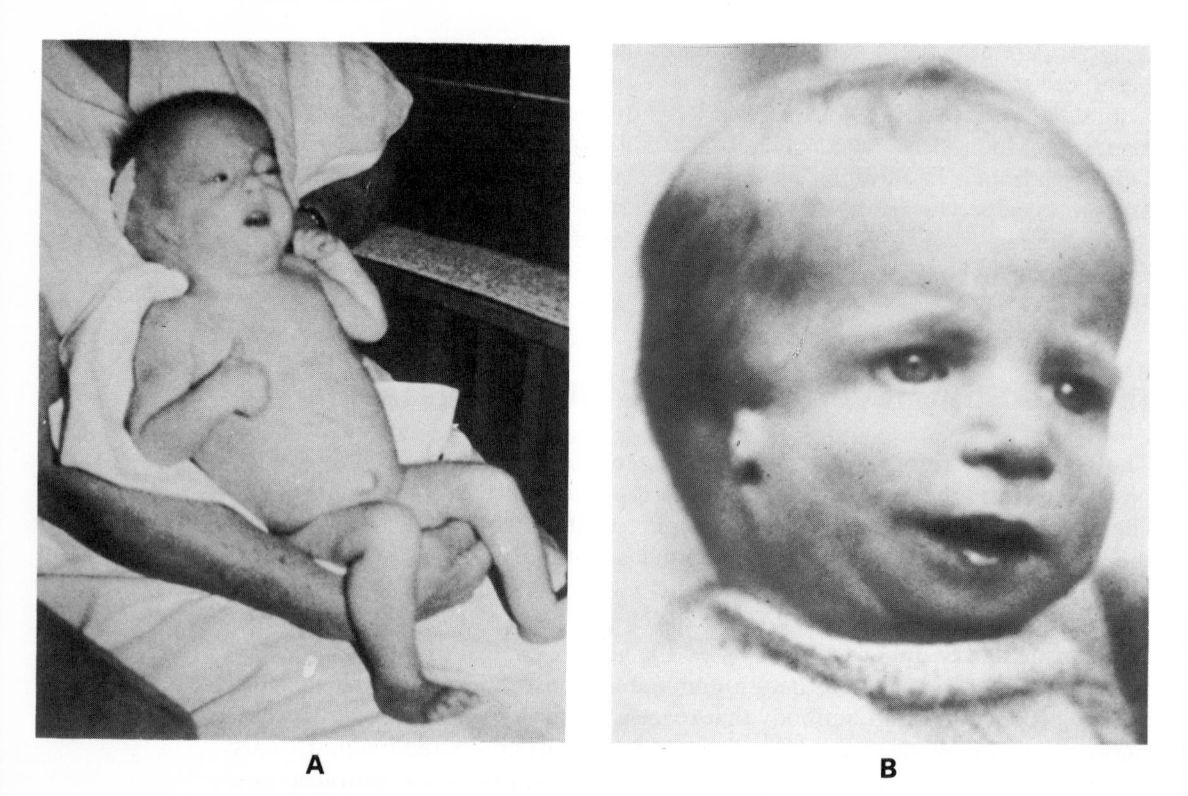

A B

Figure 9–10. Treacher–Collins Syndrome (Mandibulofacial Dysostosis). This dominantly inherited disease mainly affects the skull and face. Patients generally have conductive hearing loss, but most have normal intelligence. (**A** by permission of Charles H. Carter, M.D.; **B** by permission of G. R. Fraser, M.D., and the *Journal of Medical Genetics*.)

hereditary deafness in humans (dominant, recessive, and sex-linked)

Rather than separate the various forms of hereditary deafness, the author feels it is better to discuss them together, even though the patterns of recessive and sex-linked inheritance are presented in Chapters 10 and 11.

Both Hippocrates and Aristotle thought that deaf-mutes were unable to speak because they had an irreparable lesion of the body. This belief held back attempts to teach the deaf to speak until sometime in the sixteenth century (Fraser, 1964). At this time a Benedictine monk, Pedro de Ponce, taught several profoundly deaf children of Spanish nobility to speak. By the nineteenth century, the church, government, private institutions, and scientists were deeply involved in the care and understanding of the deaf. In 1846 charges were leveled that marriages between cousins gave rise to deaf children. Recall that this was well before Mendel announced his laws of inheritance. Mygge supported the idea that consanguineous marriages caused increased numbers of deaf children by showing that, in fact, marriages within the same family line were more likely to produce a deaf child than marriages among unrelated individuals. During this time in history, heredity was understood only in cases where a parent's trait was expressed directly in the child (now known as complete dominant inheritance). When a normal child was born to deaf parents or normal parents had a deaf child, these events were not understood. They were not considered to be a direct contribution of the parents to their child; rather, the child acquired a disposition for being deaf.

With the disclosure of Mendel's principle of gene segregation, the occurrence of unexpected cases of deafness suddenly became clear. These were cases of deafness due to the transmission of a recessive gene, and only when the two recessive genes were inherited by the same zygote was the child deaf. The normal parents in these cases must have been heterozygotes carrying one normal and one defective gene for deafness. Mendel's law also helped clarify the role of consanguinity because if relatives carrying recessive genes for hereditary deafness mated, the chance of the recessive genes coming together in a zygote was greater than it would be if each married at random. (An explanation of the increase in chance is given in Chapter 14.)

In the early part of the nineteenth century, the deaf were extremely handicapped economically. Government agencies, private institutions, and social welfare agencies were not sufficient to provide an economic base for the deaf, and their employment was out of the question. (Even today, although the picture has been drastically altered, a continuous drive to hire the handicapped is still necessary.) Then the poor deaf, which included the vast majority, did not procreate in significant numbers. As the deaf slowly overcame their handicaps in developed countries like the United States, they began to reproduce in larger numbers to the extent that Alexander Graham Bell (1883) wrote a monograph, *Upon the Formation of a Deaf Variety of the Human Race.* Because Bell concluded that a high percentage of the deaf marriages gave rise to deaf children, we assume that many of the deaf cases that Bell and others were describing in the early literature were cases of dominant inheritance.

Table 9–4 Hereditary Deafness Without Associated Physical Defects

Type of Deafness	*Brief Description*
Dominant	
Congenital profound	Deaf-mute, requires special speech training
Progressive nerve	Hearing loss of high tones begins in childhood, moderate low-tone loss in adults
Unilateral	Deafness in one ear, the other normal
Low frequency	Low-sound deafness begins at about age 30, progressive moderate loss of all frequencies
Mid-frequency	Loss of mid-sound frequency in childhood, progressive moderate loss of all frequencies
Otosclerosis	Slow progressive hardening of ear bones
Recessive	
Congenital profound	Deaf-mute, requires a special speech training
Early-onset neural	Progressive hearing loss from birth, deaf by age six
Congenital moderate	Moderate hearing loss at birth, nonprogressive
Sex-Linked	
Congenital neural	Severe congenital hearing loss, deaf-mutism
Early-onset neural	Early childhood deterioration of hearing
Moderate	Childhood to puberty beginning of progressive moderate hearing loss

Today in the United States there are no less than 6 million partially or totally deaf people. Of these about half of the children and a third of the adults have a hereditary form of deafness (Konigsmark, 1969).

As with blindness, a number of genes are associated with complete (profound) or partial deafness and other hereditary defects that include deafness as a part of the syndrome. The pattern of inherited deafness can be dominant, recessive, or sex-linked. Table 9–4 lists cases of inherited deafness not associated with other physical defects (Konigsmark, 1969b). As indicated, deaf-mutism and various types of progressive hearing loss can be inherited via all three modes of inheritance. The more difficult cases of deaf-mutism are the profound, since these victims have not had an opportunity to hear sound and most never will, even though they may learn to speak.

A recent technological advance may change the life of soundlessness for the congenitally deaf if the auditory nerve is intact. The Mid-American Hearing Research Foundation has stated that if means can be found to stimulate the auditory nerve, the deaf may be able to hear (*Bio Medical News*, 1973). A new technique of electrically stimulating the acoustic nerve via an implant is enabling the "stone deaf" to hear the ring of the doorbell or the telephone; in the most successful case, one man carried on limited telephone conversations with his wife.

the physical basis: transmission of units of hereditary material through the genes

A terminal is implanted under the skin behind the ear and then stimulated with a light electric current. The chief obstacle reported at the First International Conference on Electrical Stimulation of the Acoustic Nerve As a

Treatment for Profound Sensorineural Deafness in Man at San Francisco involves the isolating of the implant electrode from the outside to prevent infection. It was agreed that there are a number of problems yet to be resolved. However, the implant holds promise. Implants can enable patients to communicate much better than they could with lip reading alone.

The Phonics Corporation of Silver Springs, Maryland, has devised a TV phone service for the deaf and those unable to use a phone booster. The company has produced an eight-pound unit that includes a standard typewriter keyboard, which is clipped to the antenna of any TV set. A telephone conversation is then possible between two persons with TV phones. The typewritten conversation is read on the TV screen.

One condition, Waardenburg's syndrome (Figure 9–11), deserves discussion because the characteristics (white hair, lack of pigment in the iris, and deafness) of this syndrome have been found to occur in cats (Figure 9–12), mice, and dogs (Bergsma and Brown, 1971). Waardenburg's syndrome accounts for approximately 5% of human congenital profound deafness (deaf-mute). Animals with the analogous inherited defect are under study in hopes that they will serve as model systems from which we can learn more

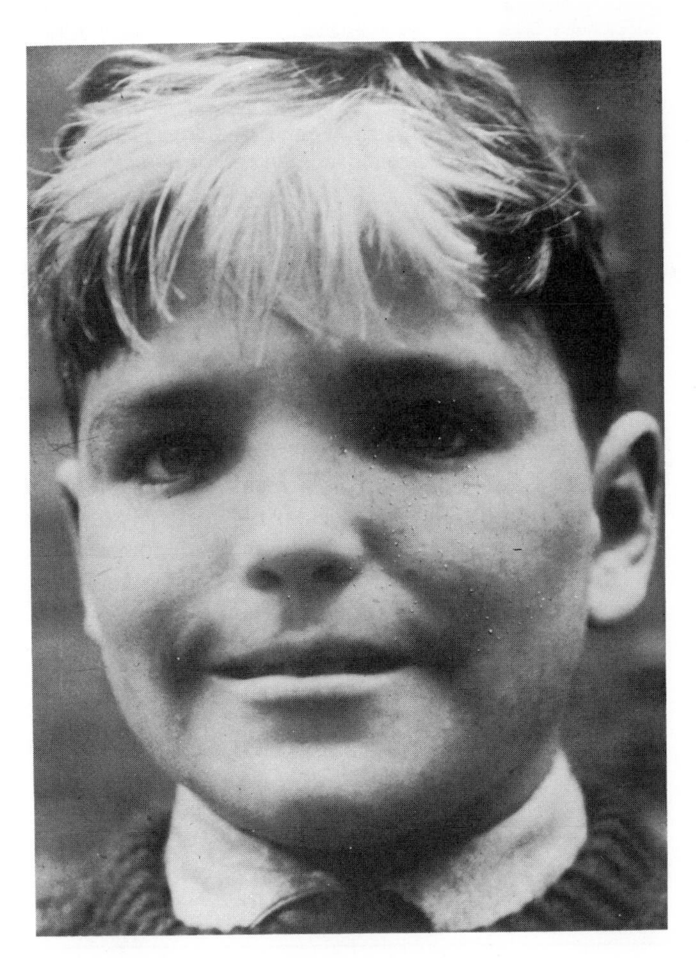

Figure 9–11. Waardenburg's Syndrome. A boy with white forelock and deafness, an example of this dominantly inherited syndrome. Deafness in this case is due to lesions of the membranous labyrinth. Waardenburg's syndrome is responsible for 5% of the cases of deafness in humans. (By permission of G. R. Fraser, M.D., and the *Journal of Medical Genetics*.)

Figure 9–12. Variation of Coat Color of White Deaf Cats. The cat hair–deafness syndrome is analogous to Waardenburg's syndrome in man, where white forelock, reduced pigment, and congenital deafness occur. (From D. R. Bergsma and K. S. Brown, White Fur, Blue Eyes, and Deafness in the Domestic Cat, *Journal of Heredity,* May–June 1971; reproduced with permission.)

about the development of hearing. It has been determined that in cats a dominant gene, *W*, is associated with the development of white fur, blue eyes, and deafness. The *W* gene is completely penetrant in producing white hair and shows incomplete penetrance with respect to deafness and the production of a blue iris. Thus various aspects of this syndrome occur in all cats carrying the *W* gene, just as the expression of Waardenburg's syndrome is variable among humans.

Hearing loss can occur as a result of many inherited diseases involving all the major body systems. Hearing loss can involve low-, medium-, and high-pitch sound. The time of the gene's expression in the different diseases is variable, as is the degree of expression. Deafness can be inherited as a dominant, recessive, or sex-linked condition.

references

Bell, A. G. 1883. Upon the Formation of a Deaf Variety of the Human Race. *Memoirs of the National Academy of Sciences, 2*:177–82.

Bergsma, D. R., and K. S. Brown. 1971. White Fur, Blue Eyes, and Deafness in the Domestic Cat. *Journal of Human Heredity, 62*:171–85.

Bio Medical News. 1972. Offspring Tested for Parents' Ills. *3* (May): 1.

Bio Medical News. 1973. Electrical Implants for Totally Deaf. *4* (Nov.): 7.

Brackenridge, C. J. 1971. A Genetic and Statistical Study of Some Sex-Related Factors in Huntington's Disease. *Clinical Genetics, 2*: 267–86.

Bures, A. M. 1946. A Genetic Analysis of Eye Color. *American Journal of Physical Anthropology, 4*: 1–36.

Davenport, C. B. 1907. Heredity of Eye Color in Man. *Science, 26*: 589–92.

DHEW. 1974. *What Are the Facts About Genetic Disease?* Publication No. NIH 74-370, pp. 3–5, 6, 20. Department of Health, Education and Welfare, Washington, D.C.

Estimated Statistics on Blindness and Vision Problems. 1966. The National Society for the Prevention of Blindness, New York.

Fraser, G. R. 1964. Review Article: Profound Childhood Deafness. *Journal of Medical Genetics, 1*: 118–51.

Fraser, G. R., and A. I. Friedmann. 1967. *The Causes of Blindness in Children*. Johns Hopkins Press, Baltimore.

Gardner, E. J., and R. C. Richards. 1953. Multiple Cutaneous and Subcutaneous Lesions Occurring Simultaneously with Hereditary Polyposis and Osteomatosis. *American Journal of Human Genetics, 5*: 139–46.

Harmetz, Aljean. 1971. Must They Sacrifice Today Because of Threatened Tomorrows? *Today's Health*, November, p. 44.

Heathfield, K. W. G. 1967. Huntington's Chorea. *Brain*, 90: 203–32.

Huntington, George, M.D. 1872. On Chorea. *The Medical and Surgical Reporter, XXVI*, No. 15 (April).

Konigsmark, Bruce W. 1969. Hereditary Deafness in Man. *New England Journal of Medicine, 281*: 713–78 (part 1), 774–78 (part 2), 827–32 (part 3).

Levine, B. B., and R. H. Stember. 1972. Ragweed Hay Fever: Genetic Control and Linkage to HL-A Haplotypes. *Science, 178*: 1201–1203.

McKusick, Victor A. 1960. *Medical Genetics*. Mosby, St. Louis.

McKusick, Victor A. 1971. *Mendelian Inheritance in Man: Catalogs of Autosomal Dominant, Autosomal Recessive and X-Linked Phenotypes*, 3rd ed. Johns Hopkins Press, Baltimore.

Myrianthopoulos, Ntinos C. 1966. Huntington's Chorea. *Journal of Medical Genetics, 3*: 298–314.

New England Journal of Medicine. 1969. The Enigma of Familial Polycystic Kidneys. *281*: 1013–14.

Rufer, V., Jan Bauer, and F. Soukup. 1970. On the Heredity of Eye Color. *Acta Universitatis Carolinae Medica, 16*: 429–34.

Scheinfeld, Amram. 1965. *Your Heredity and Environment*. Lippincott, Philadelphia.

Scott, Robert A. 1969. *The Making of Blind Men*. Russell Sage Foundation, New York.

Smith, David W. 1970. *Recognizable Patterns of Human Malformation*. Saunders, Philadelphia.

Stevens, David, and Maurice Parsonage. 1969. Mutation in Huntington's Chorea. *Journal of Neurology, Neurosurgery and Psychiatry, 32*: 140–43.

Thompson, James S., and Margaret W. Thompson. 1973. *Genetics in Medicine*. Saunders, Philadelphia.

Vassie, P. R. 1932. On the Transmission of Huntington's Chorea for 300 Years—The Bures Family Group. *Journal of Nervous and Mental Disease, 76*: 553–73.

The three patterns of Mendelian single-gene inheritance and the terms used to describe them were introduced in Chapter 8, and we have considered in Chapter 9 a number of human dominantly inherited diseases. We now turn to human recessively inherited diseases. A review of the recessive mode of inheritance is presented in Figure 10–1.

An autosomal recessive trait is expressed only if the individual receives the defective gene from both parents. Usually, the parents are phenotypically normal. Table 10–1 lists 57 recessively inherited human diseases. This is only a small fraction (1/16) of the 800 recessive defects listed in McKusick's (1971) catalog of human inheritance. A few of these recessive disorders are discussed to acquaint the reader with the variability and range of diseases that recessive genes can cause and the particular demands on society that result.

Unlike dominant inheritance, for every person recognized with a recessive trait, there are many more persons in the population who are heterozygous carriers for that trait. For example, among blacks about 1 in every 400 live births has the homozygous recessive genotype $Hb^{S}Hb^{S}$ and will demonstrate sickle cell anemia. Yet about 1 in every 10 adult blacks is a heterozygous carrier, $Hb^{A}Hb^{S}$. (Hb is short for hemoglobin, the protein affected by a defective sickle gene. Hb^{A} represents the gene for normal hemoglobin A, and Hb^{S} the gene for defective hemoglobin S).

Cystic fibrosis occurs in 1 in 1000 to 1500 live births; yet about 1 in 22 persons carries the gene. Of the Ashkenazic Hebrews, 1 in 6000 have Tay–Sachs, but 1 in 25 is a carrier. In non-Hebrews 1 in 60,000 have Tay–Sachs disease, and 1 in 400 carries the gene. The incidence for phenylketonuria is 1 in 10,000 to 20,000 live births; yet about 1 in every 100 persons is a carrier. These data should make it clear that marriage within one's own

Appearances are deceptive.

Aesop: The Wolf in Sheep's Clothing

human recessively inherited diseases

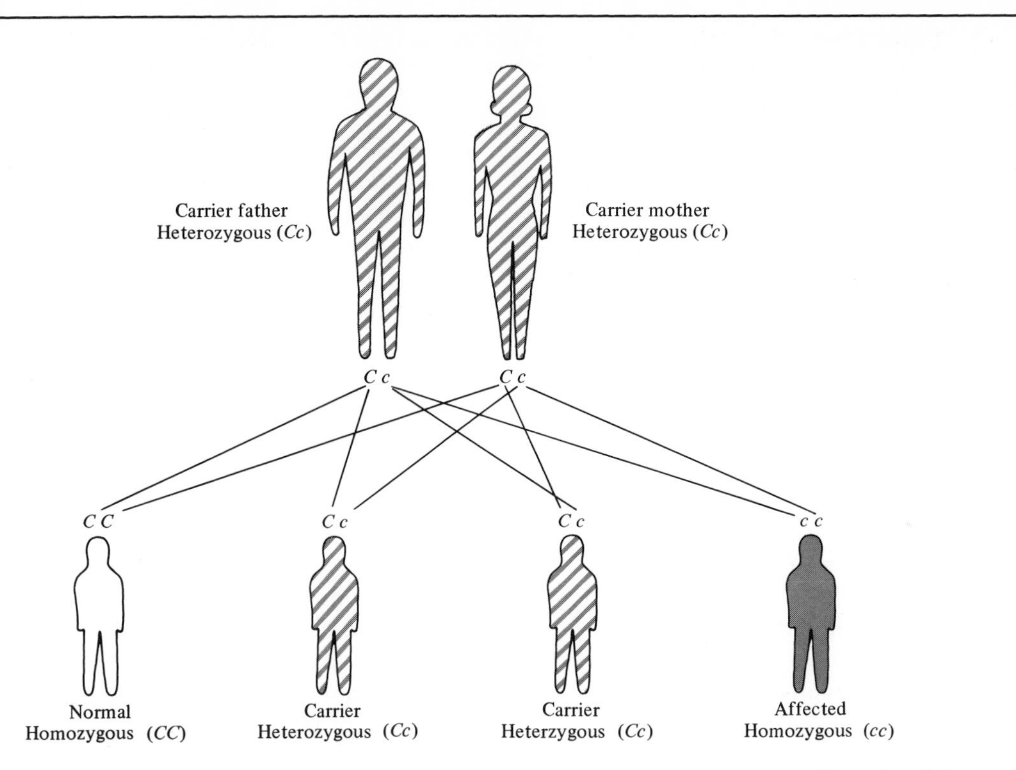

Figure 10–1. A Review of Autosomal Recessive Inheritance. Both parents, if carriers (of cystic fibrosis in this example), are normal because they also carry a dominant normal gene *C* that masks the faulty allele, the recessive counterpart *c*. The odds for each child are (1) a 25% risk of inheriting both recessive genes, which can cause a serious birth defect, (2) a 25% chance of inheriting *CC* and being unaffected, and (3) a 50% chance of being a carrier, *Cc*, and normal, as are both parents. Over 800 autosomal recessively inherited diseases have been identified. For other examples see Table 10–1.

ethnic group, or between cousins, will increase the chance of bringing two recessive genes together in the offspring. Consanguinity is discussed in Chapter 14.

As stated in Chapter 5, there are about 100 genetic diseases associated with the loss or alteration of a specific enzyme. Most of these enzyme-associated disorders have a recessive mode of inheritance. Testing the potential parent for the presence of these particular enzymes can sometimes reveal the carrier state. In other cases a particular physical appearance may indicate a carrier. For example, persons heterozygous for the recessive (severe) form of xeroderma pigmentosum, a disease that in many cases leads to skin cancer, may be excessively freckled (see Figure 13–3). Such identification has value in giving one the knowledge to choose the route best suited to his individual needs, but it also opens up a Pandora's box of moral and ethical issues involving abortion and the right to life.

The majority of heterozygous carriers do not *appear* to express many symptoms. However, new and more sophisticated techniques of sampling and measuring now reveal that some of the heterozygous carriers *do* show symptoms. If a person does not suspect he is carrying a trait, however, the symptoms generally go unrecognized. For example, sickle cell carriers Hb^AHb^S

Table 10–1 Examples of the Common Autosomal Recessive Conditions	
Condition	*Brief Description*
Albinism	No melanin pigment in skin, eyes, or hair; no synthesis of melanin; totally light-sensitive skin and eyes; white hair; shortened life expectancy
Alkaptonuria	Pigmentation of cartilage, dark eye spot, urine darkens on exposure to air
Ataxia-telangiectasia (Louis–Bar syndrome)	Progressive central nervous system degeneration
Bloom's syndrome	Extensive red blotches over face, increased incidence of leukemia, sensitivity to light, narrow head, diminished immunoglobulin
Carpenter's syndrome	Acrocephaly, syndactyly, polydactyly, mentally deficient
Cartilage-hair hypoplasia	Fine sparse hair, loose-jointed hands and feet, short stature
Craniometaphyseal dysplasia (Pyle's disease)	Tall in height, thick bony wedge over nose, poor teeth, abnormal spine curvature
Cretinism (nongoitrous)	Lack of physical and mental development, large head, lack of thyroid tissue
Cystic fibrosis	Mucus congestion of lungs; affects all exocrine glands, especially pancreas; treatment with antibodies, etc., can extend life to about 18 years
Diastrophic dwarfism	Severe dwarfism, rigid deformed ears, calcification of cartilage
Ectromelia	Congenital absence of one or more limbs
Ellis–van Creveld syndrome	Disproportionate dwarfism; appendages short, body normal; polydactyly; about half have congenital heart defects
Familial dysautonomia (Riley–Day syndrome)	Drooling; excessive sweating; subnormal intelligence; tastebud-less tongue; deterioration of autonomic nervous system; insensitivity to corneal pain, burns, and pain caused by sharp objects
Fanconi's syndrome	Retarded growth and mental development, tending toward leukemia, deafness, heart defects

Table 10–1 continued

Condition	*Brief Description*
Friedreich's ataxia (hereditary ataxia)	Degeneration of spinal cord, foot deformity, death generally before 30
Galactosemia	Excess of galactose in liver, enlarged liver, cataracts, mental retardation
Gaucher's disease	Enlargement of liver, lymph nodes and spleen; severe psychomotor impairment
Generalized gangliosidosis	Similiar to Hurler's syndrome but occurs before birth; low birth weight, rapid deterioration from birth
Gierke's disease	Enlarged liver, excess glycogen in liver, retarded growth, mental retarded, mimics diabetes
Growth hormone deficiency	Midget with normal body proportions due to deficiency of growth hormone
Hartnup disease	Pellagra (scurvy)-like symptoms, skin disorders.
Hereditary microcephaly	Small head, sloping forehead, severe mental retardation, short stature
Homocystinuria	Error in methionine metabolism, fine hair, coarse skin, crowded teeth, nervous, narrow skull, homocystine in urine, 50% retarded
Hurler's syndrome (mucopolysaccharidosis I)	Severe growth retardation, heart problems, stiff joints, coarse face, mental deterioration, death in childhood
Hyalinosis cutis et mucosae	Face covered with yellow-white nodules, abnormal tooth shape, yellow-white nodules on tongue and lips
Krabbe's disease	Severe atrophy of the brain, severe motor deterioration
Larsen's syndrome	Flat face, wide-spaced eyes, cleft palate, congenital heart defects
Laurence–Moon–Biedl syndrome	Obesity, polydactyly, retinal pigmentration, mental deficiency
Leprechaunism (Donahue's syndrome)	Elflike face, small chin, thick lips, abnormal metabolism in major organs and testes or ovaries
Leroy's syndrome (I-cell disease)	Marked slow growth, thick tight skin, limited joint movement, enlarged cell lysosomes

Table 10–1 continued

Condition	*Brief Description*
Maroteaux–Lamy syndrome (mucopolysaccharidosis VI)	Coarse face, stiff joints, varying
Maple syrup urine disease	Maple syrup odor of urine, progressive degeneration, convulsions, deafness, early death
Mohr's syndrome (oral-facial-digital II)	Cleft upper lip, conductive hearing loss, lobbed tongue, broad tongue
Morquio's syndrome (mucopolysacchardosis IV)	Dwarfism, abnormal bone shape in hands and chest, death before age 20
Niemann–Pick disease	Defect in lipid metabolism; enlarged liver and spleen
Oculocutaneous albinism	Subnormal vision; skin and eyes sensitive to sun; tendency to develop skin cancer
Phenylketonuria	Excess phenylalanine in blood; severe mental deterioration; eczema in 33% of patients
Phocomelia	Flipper limbs, some mental retardation; absent digits
Pompe's disease (glycogenesis II)	Enlarged tongue and heart; muscle weakness
Progeria (Hutchinson–Gilford syndrome)	Small face, dwarfism, heart disease, wrinkled skin
Pseudoxanthoma elasticum	Thick and coarse skin, poor vision, mottled pigments in eyes, intestinal bleeding
Pyknodysostosis	Underdeveloped facial bones, parrot-like nose, blue sclera, wrinkled skin on hands, short stature
Sanfilippo's syndrome (mucopolysaccharidosis III)	Progressive mental deterioration, stiff joints, long-term survival possible
Scheie's syndrome (mucopolysaccharidosis V)	Limited joint movement, cloudy cornea, hearing loss, excessive body hair, chondrotin sulfate B in urine
Seckel's bird-headed dwarfism	Congenital small head, sparse hair, birdlike face, mental retardation, small brain
Sexual ateleiotic dwarfism I	Excessive premature wrinkling of facial skin, dwarfism, shrill voice
Sickle cell anemia	Abnormal hemoglobin, blockage of small blood vessels, acute pain, death about age 20

Table 10–1 continued

Condition	*Brief Description*
Sjögren–Larsson syndrome	Spastic arm and leg movement, seizures, profound mental retardation, scaling of body skin, absence of sweating
Smith–Lemli–Opita syndrome	Mental retardation, microcephaly, undescended testes, facial anomalies
Tay–Sachs disease (amaurotic idiocy)	Cherry red spot in eye macula; after 6–9 months rapid deterioration of vision and motor skills; death at about 2–4 years
Thalassemias	Lack of sufficient hemoglobin, premature death in many cases
Tyrosinosis	High levels of tyrosine in blood, cirrhosis of liver, tremors, muscle spasms, seizures
Werdnig–Hoffmann disease (congenital amyotonia)	Juvenile muscular dystrophy
Werner's syndrome	Premature graying of hair, juvenile cataract, short stature, beak nose
Wilson's disease	Progressive tremors; drooling; cirrhosis of liver; copper in urine; greenish ring around cornea; progressive dementia-paranoid; early adolescent death
Wolman's disease (familial xanthomatosis)	Vomiting, diarrhea, enlarged liver and spleen, severe anemia, late childhood death
Xeroderma pigmentosum	Lack of the DNA repair enzyme stunted growth, epilepsy, extreme sensitivity to ultraviolet radiation that appears to cause cancer growths on skin, shortened life span

make a normal and an abnormal hemoglobin, but there is sufficient normal hemoglobin unless the heterozygote comes under unusual oxygen stress. In screening programs, the sickle cell carrier can be easily identified through the separation of normal and abnormal hemoglobin in his red blood cells (see Chapter 18).

The carrier can also be detected through measurement of the amount of enzyme present in certain tissues. In measuring the amount of any given enzyme, one makes a well-supported assumption that

1. In a given dominant homozygous condition (for example, *AA*) two *A* genes make *x* amount of enzyme.

2. In a heterozygous condition (for example, *Aa*), one *A* gene makes $x/2$ or half the amount of the enzyme.
3. In the homozygous recessive condition (*aa*), the genes make little or no enzyme.

The assumption does hold true for a number of diseases. For example, in phenylketonuria, Tay–Sachs, acatalasia, galactosemia, and argininosuccinicaciduria, heterozygous carriers demonstrate about half the enzyme of the normal homozygote, while the recessive homozygote makes little or no measurable enzyme. The complete absence of the enzyme may mean the actual loss of that gene from the chromosome through a deletion. But most often the absence is due to an alteration of the base-pair structure of the gene. This alteration results in an abnormal polypeptide such that no enzyme is made or a protein is made but it lacks enzymatic function.

From the molecular consideration of the gene, presented in Chapter 5, the existence of genes with different degrees of activity, including total absence, is fully expected. If a certain sequence of base pairs in the gene leads to the production of a polypeptide chain with normal activity, a sequence that differs from it in a single base pair can lead to the production of a polypeptide having hardly any difference from the activity of the normal, a minor difference, or a profound difference, depending on the type of base-pair change and its position in the sequence of DNA. If, instead of the substitution of one base pair for another, a pair is lost, the transcription and translation of the base-pair sequence into an amino acid chain is likely to result in severely defective protein. The reading of the genetic message begins at one end of the gene; the protein enzyme would be normal up to the point where the base pair was lost. It would be abnormal for the rest of its length and, therefore, less active or inactive.

We can readily see that if substance A must be converted to substance B because B is an essential metabolic material for the organism, then the enzyme that effects the conversion must be made. If an organism is of the recessive genotype and little or no enzyme is made, the consequences could be lethal. When the conversion of a substance does not occur, the situation is usually referred to as a metabolic block. We say that the organism is mutant for the substance not produced or deficient for the enzyme.

In certain cases we need only supply the substance that the body cannot make, and no ill effects will be noted. All of us, for example, are mutant in that we lack the necessary enzymes to synthesize 8 to 10 of the amino acids our body requires for building different proteins (Table 5–3). We circumvent this problem by taking in these amino acids through our diet, and the individual amino acids are plugged into the correct pathway for the synthesis of a necessary protein.

When we know the order of substances converted one into another by the action of different enzymes, we have recognized a biochemical pathway. At each step a specific enzyme is required for the conversion of one product into another, and when that enzyme is defective, that pathway is shut down. Thus in many homozygous recessive diseases, pathways are known to be altered because of the lack of an enzyme. This is represented diagramatically in Figure 10–2, which shows various homozygous recessive genotypes and the resulting inherited disorder. Albinism, alkaptonuria, cretinism, phenylketonuria, and tyrosinosis, all caused by recessive genes, are closely related metabolically.

The concert-like interaction of the various biochemical pathways serviced by the different enzymes can be extraordinarily complex. The various metabolic reactions will not be presented here, since they are not essential to an appreciation of the inheritance of human disease. The easily defined pathway for the conversion of phenylalanine, as presented in Figure 10–2, is sufficient to make the point that at each step in the process, an enzyme is required to convert one substance into another. In Figure 10–2 phenylalanine, which is one of our essential amino acids and must be taken in through our diet, enters the body and travels to the liver for enzymatic conversion into tyrosine. As the amount of tyrosine reaches certain levels, separate enzymes in separate pathways convert tyrosine into thyroxine, melanin, maleylacetoacetic acid, and water. If the various enzymes are not available at the appropriate steps, certain metabolic diseases will occur. Several of them will be discussed briefly.

phenylketonuria

It took twenty years, from 1934 to 1954, to establish the relationship between body chemistry and genetics in phenylketonuria (PKU). Since 1954, the PKU program has become a model for the screening of inherited defects in the newborn. Problems associated with the screening programs are discussed in Chapter 22.

Around 1934 a Norwegian mother brought her two mentally retarded children to Dr. Ashborn Følling to determine the cause of the musty odor of their urine. Their urine contained a small amount of phenylpyruvic acid, which turns ferric chloride bright green. Since these children were mentally retarded, Følling next tested children in institutions for the retarded, using the ferric chloride urine test. He found that urine from other mentally retarded children showed the same chemical reaction. This was the beginning of a new approach in establishing the cause of mental retardation (Hsia, 1971).

By 1954 scientists knew the homozygous recessive *pp* did not synthesize the enzyme phenylalanine hydroxylase. Without the enzyme, phenylalanine is not converted into tyrosine, and as a result, the level of phenylalanine increases in the liver. It overflows into the bloodstream and is carried about the body.

The heterozygote, or carrier, can generally be detected with the use of the phenylalanine tolerance test. Persons are given a sizable amount of phenylalanine. The normal homozygote breaks down the phenylalanine rather quickly; the heterozygote demonstrates a delay in breaking down the compound. As far as is known, the heterozygote does not show mental impairment. However, how does one assess a loss in IQ that perhaps could have been 5 to 10 points higher if the person were not a heterozygote? Although the mode of action is still not understood, high levels of phenylalanine (phenylpyruvic acid) are associated with severe mental retardation.

For example, three children born to one phenylketonuric mother, each having a different father, were retarded, even though they were not genetically

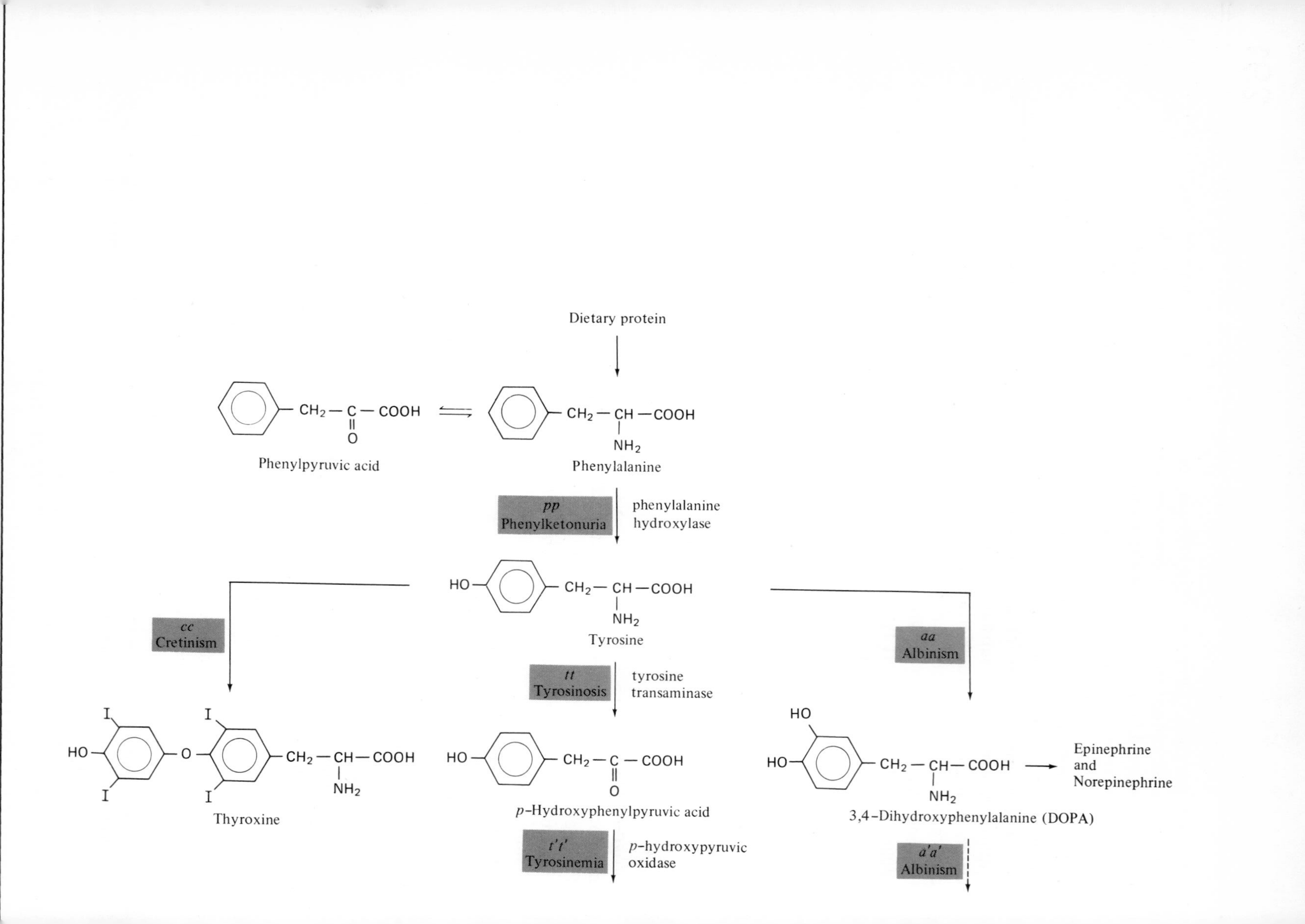
Dietary protein
Phenylpyruvic acid
Phenylalanine
pp
Phenylketonuria
phenylalanine
hydroxylase
Tyrosine
cc
Cretinism
aa
Albinism
tt
Tyrosinosis
tyrosine
transaminase
Thyroxine
p-Hydroxyphenylpyruvic acid
Epinephrine
and
Norepinephrine
3,4-Dihydroxyphenylalanine (DOPA)
t′t′
Tyrosinemia
p-hydroxypyruvic
oxidase
a′a′
Albinism

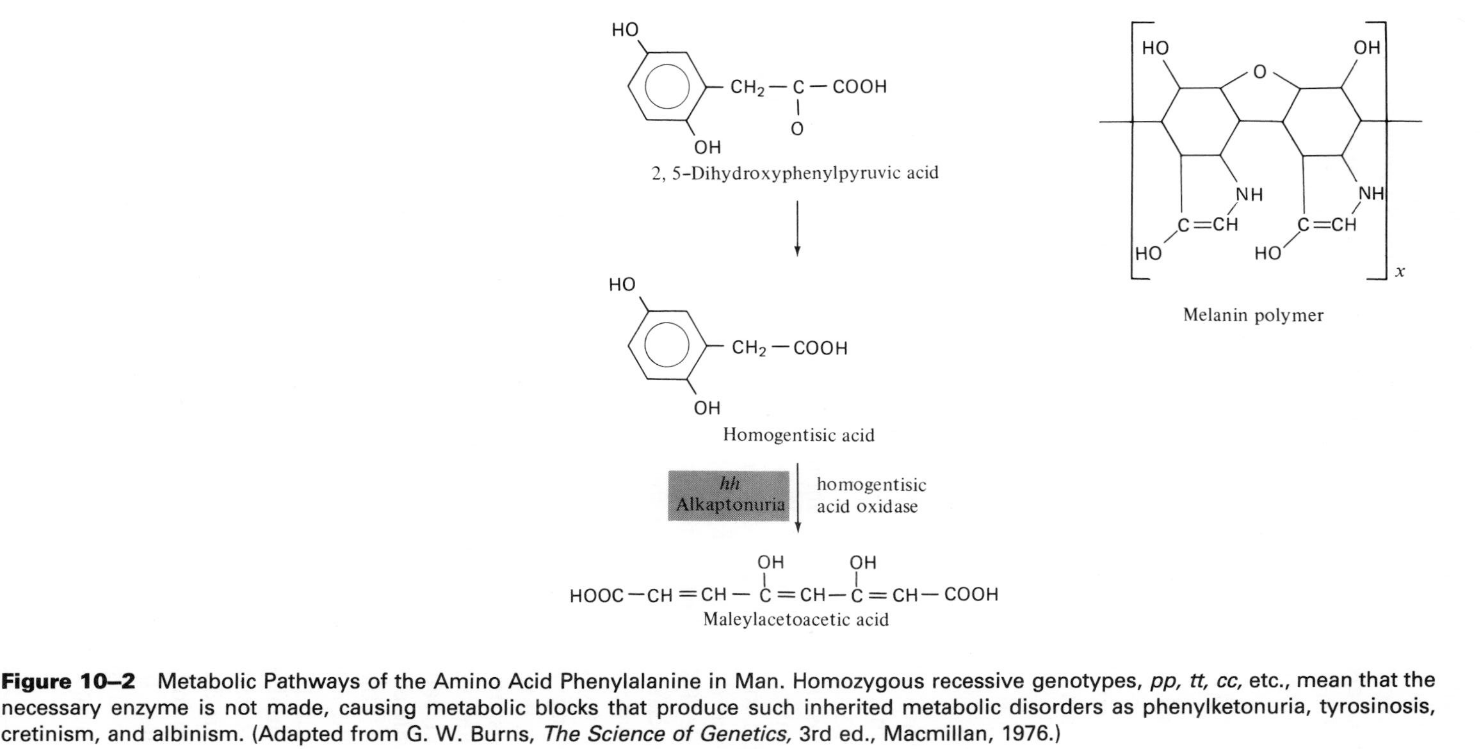

Figure 10–2 Metabolic Pathways of the Amino Acid Phenylalanine in Man. Homozygous recessive genotypes, *pp, tt, cc,* etc., mean that the necessary enzyme is not made, causing metabolic blocks that produce such inherited metabolic disorders as phenylketonuria, tyrosinosis, cretinism, and albinism. (Adapted from G. W. Burns, *The Science of Genetics,* 3rd ed., Macmillan, 1976.)

defective. Such observations suggest that the exposure of the fetus in utero to the mother's high levels of phenylalanine may cause irreversible brain damage. However, PKU children born to normal parents experience gradual mental retardation with time *after birth*. There is an average loss of 2 IQ points per week in the untreated PKU child. This indicates that under normal conditions the gene for the production of phenylalanine hydroxylase is "turned on" at about the time of birth (another example of gene control as discussed Chapter 5.) Data collected from PKU studies also substantiate reports that our nervous system is not complete at birth. Data from studies on treating PKU children indicate that brain differentiation is completed at about age six. It is *believed* that after age six a child can be taken off the low-phenylalanine diet (described below) because high levels of phenylalanine may no longer affect mental development (National Academy of Science, 1975).

Affected persons are generally pleasing to look at. They are usually blond, blue-eyed, and fair-skinned because of decreased melanin production (see Figure 10–2). About a third, the more severely retarded PKU patients, have eczema as well as a musty odor of the urine. For other associated problems, see Figure 10–3.

the PKU diet

Terry is a happy five-year-old. He has never known the gastronomic delight of eating a steak, drinking a malted milk, or eating a simple everyday sandwich. Children like Terry, who are phenylketonuric or PKU children, live on a very restricted diet. Most protein has the amino acid phenylalanine in its makeup, so PKU children have low-protein diets in order to keep their intake of phenylalanine at a low level. The diet treatment for recognized PKU children has received a great deal of attention in the literature. About half the critics do not feel that the PKU low-protein diet helps these children. An equal proportion of the critics believe that the diet is of benefit in preventing mental retardation and, later, in helping moderate certain behavior. This particular issue is dealt with more directly in Chapter 22.

Although having your child tested for PKU is mandatory in nearly all states, whether or not the PKU child is placed on the diet is up to the child's parents. A diet is usually worked out for each individual case, since every individual is unique, even in the expression of a disease. As the diet is planned and tested, blood samples are monitored to determine what foods in each particular case will elevate phenylalanine levels in the blood. In general, the PKU child is not able to eat anything rich in protein, such as milk or milk products, fish, most meats, peanut butter, or eggs. In place of milk, infants are given a prescription powder which resembles milk when mixed with water. It contains the essential amino acids minus phenylalanine. This powder, which is sold under various trade names, costs about $9 a can with six cans per case. A can lasts about four days. The cost, based on 1975 figures, runs about $60 per month just for the milt substitute. A substitute wheat starch flour for making bread and cookies, imported from Canada, costs about $1.95 per pound. Together, the special diet and medical costs can become a burden on a family. Estimates place the costs of care and treatment of each PKU child at $100,000 per lifetime.

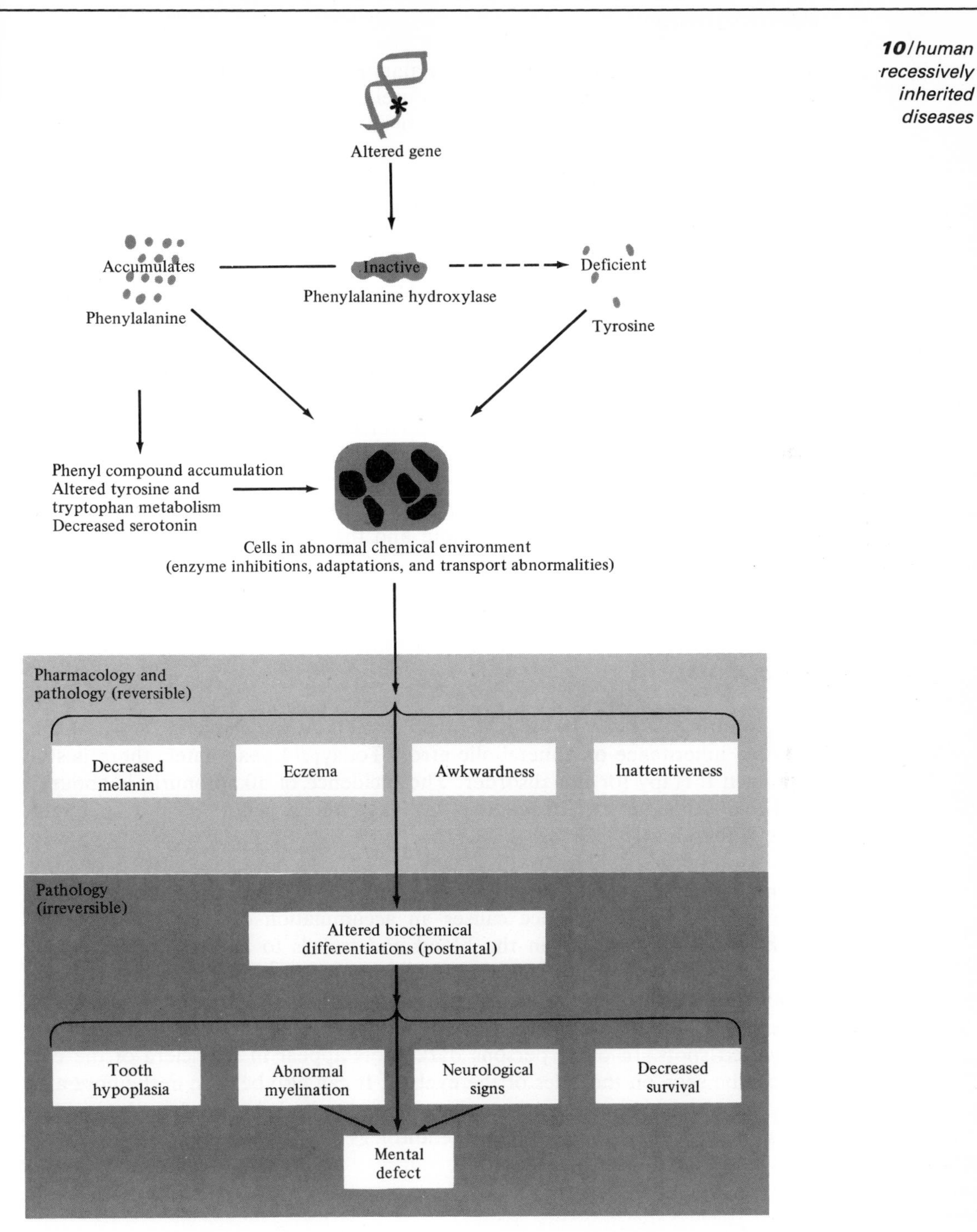

Figure 10–3. Metabolism of Phenylalanine and Associated Pathology. The metabolic defect causing phenylketonuria is in the gene that codes for phenylalanine hydroxylase. The direct result is failure of the enzyme's substrate, phenylalanine, to be metabolized to tyrosine. Depicted are the varied results of the accumulation of phenylalanine, the alterations in its metabolic pathways, and the decreased production of tyrosine.

tyrosinosis

Approximately 5% of ingested phenylalanine is utilized directly as an amino acid for the making of new body protein. However, the majority of the ingested phenylalanine, in the presence of the enzyme phenylalanine hydroxylase, is converted into tyrosine. Since tyrosine is made within the body from the conversion of phenylalanine, tyrosine is not considered an essential amino acid. Tyrosine is also incorporated directly as an amino acid into protein. Through the action of specific enzymes it is converted into the pigment melanin, epinephrine, norepinephrine, dopamine, and the hormone thyroxine. Still other molecules of tyrosine are utilized in pathways that create biological energy for the cell (Knox, 1969). If the enzyme tyrosinase is absent, as it is in the affected homozygous recessive person, then blood levels of tyrosine increase. The increase is ultimately associated with cirrhosis of the liver, agitated behavior, muscle spasms, tremors, and seizures, as well as a form of tyrosine albinism. Partial correction of tyrosinosis also involves a diet with reduced phenylalanine; when there is less phenylalanine to be processed into tyrosine, less tyrosine will accumulate in the blood. In general, tyrosinase negative people have normal intelligence, which clearly distinguishes between the effect of high levels of the amino acid phenylalanine and that of high levels of tyrosine on mental development.

alkaptonuria

Alkaptonuria was cited by A. E. Garrod in 1902 as the classic example of the recessive inheritance of a metabolic error. Today, 74 years later, there is still no known therapy for this disorder. The incidence of alkaptonuria is about 5 per 100,000 live births. Thus, about 160 alkaptonuric people are born per year in the United States.

Alkaptonuria, similar to PKU and tyrosinosis, is caused by a missing enzyme, in this case an absence of homogentisic acid oxidase, an enzyme of the liver and kidney. Its absence causes an accumulation of homogentisic acid, which reaches high levels in the blood and travels to all parts of the body. When the urine of an alkaptonuric person is left standing in the light, the homogentisic acid in the urine turns very dark. The accumulated product also becomes deposited in the body, cartilage, and tendons. It appears as darkly pigmented spots. In older persons dark spots appear in the sclera of the eye and can be seen on the sides of the eyeball. It can also be seen in the nose and ears. The spots appear blue due to overlying tissue. It is also deposited at the joints and causes arthritis of the spine and legs as the person ages.

albinism

In contrast to PKU, tyrosinosis, and alkaptonuria, albinism results from the *lack* of a substrate rather than an accumulation (Figure 10–4). As seen in

Figure 10–4. Albino Male. "Unzie" became a well-known albino as he traveled with the Forepaugh–Sells Circus. He came to the United States from Australia. He had pallid white skin, pink eyes, and white hair. (From Bernard Kobel, Choice Circus Photographs.)

Figure 10–2, albinism can occur when one inherits either of two separate recessive genetic defects. In each case a separate enzyme converts one compound into another. In oculocutaneous albinism, the form most readily recognized in the population, the body is incapable of synthesizing the pigment melanin. Recall the association of melanin with eye color (Figure 9–3), hair color (Figure 9–6), and skin color (Figure 8–17).

In oculocutaneous albinism the defect is believed to occur because of a failure in the conversion of tyrosine into the next compound (beta-3,4-dihydroxyphenylalanine), which is further converted into melanin (Figure 10–2). Because tyrosine is utilized to produce a number of other body proteins, there is no accumulation of tyrosine in the albino. We know, therefore, that the missing enzyme must be one that is utilized along the path from tyrosine to melanin.

cystic fibrosis (CF), the great impersonator

If you were to meet Mark, a very handsome, smiling child of six, you would not immediately suspect he is seriously ill, nor would you suppose he had a life expectancy of another 6 to 12 years. Mark is burdened with one of the most common semilethal recessive hereditary disorders known, cystic fibrosis, or, as it was originally called, fibrocystic disease of the pancreas. The symptoms vary widely. They are often unclear and mislead the physician toward a diagnosis of one of the more familiar diseases like pneumonia, celiac disease (intestinal malabsorption leading to diarrhea and malnutrition), asthma, and malnutrition.

Mark is one of many thousands of CF children in the United States. Approximately 1 in 1000 to 1500 live births have cystic fibrosis. An incidence of 1 in every 1000 live births means some 3141 cystic fibrosis children are born every year in the United States. It is estimated that there are 5 million heterozygous carriers of the cystic fibrosis gene in the white population.

The disease was previously thought to be found predominantly among the white population. The incidence of CF in the black population was unknown until a recent study in Washington, D.C., showed it to be 1 in 17,000 live births (Kulczycki and Schauf, 1974). From this figure it appears that CF is about 17 times more frequent in the white population.

clinical conditions

The various symptoms of CF can be expressed at different ages and with different degrees of severity (Black and Mendoza, 1976). Although there is great difficulty in diagnosis, suspected persons can now be identified as CF through the measurement of the amount of salt in their sweat. Cystic fibrosis patients produce an excessive amount of salt in their sweat, fingernails, and hair. Other tests are available and used on the very young. The salt test is accurate and easily performed and is the preferred method on older patients. About 15% of the cases diagnosed do not even involve the pancreas, and many cases do not involve the liver. Shortly after birth 10% of all CF patients undergo surgery for the relief of intestinal obstruction. As the child ages, various symptoms of bronchitis, malnutrition despite a good diet, large foul-smelling bowel movements, abnormally salty sweat, an increased sensitivity to taste and smell, and an unusually large appetite develop.

Eighty-five percent of the CF patients have a pancreatic problem. Because all the body's mucus-secreting glands function abnormally, the ducts of the pancreas become plugged with thick, gluelike mucus. This mucus block restricts the flow of the digestive enzymes necessary for the breakdown of food in the intestines. Without the pancreatic enzymes, the digestion of food is greatly impaired; 50% of all fats and proteins and 10% of all carbohydrates are lost. A deficiency of vitamins A, D, K, and E also occurs (Anderson, 1960). The loss of digestive enzymes accounts for the malnourished look of infants and children with cystic fibrosis. The diet for these cases is supplemented with capsules of enzymes.

The thick mucus secretions also collect in the lungs, creating severe breathing difficulty (emphysema). The lungs are affected in all CF patients. "Sooner than later" cystic fibrosis patients develop a chronic lung disease as a result of the thick mucus clogging the small air passages of the lungs. Approximately 90% of CF deaths are the result of lung disease and associated problems (di Sant'Agnese, 1969). Changes in lung pressure restrict blood flow; this increases the blood pressure in the lungs, which subsequently leads to chronic heart strain and eventual heart failure, the second leading cause of CF deaths. The third major cause of death in CF patients is infection.

screening tests to identify cystic fibrosis carriers

Although effective screening tests have been developed for many genetic disorders (See Chapter 18), a simple and reliable method of detecting carriers of the cystic fibrosis trait has not yet been found. Researchers are working to develop tests to identify carriers. For example, Barbara Bowman at the University of Texas, Galveston, isolated a substance in the blood of both cystic fibrosis patients and carriers that inhibits ciliary action on oyster gills. It is believed that this factor impairs the wavelike motion of the tiny hairs (cilia) in our windpipes (trachea), which would move the mucus up the trachea, clearing the lungs. Although the presence of this factor was noted in 1967, investigators have been unable to isolate it from the blood serum until now. The significance of developing a test for carrier identification lies in our search for a medical cure and the application of the knowledge to genetic counseling.

recognition and a case history of cystic fibrosis (CF)

The disease was first recognized in 1936, and in 1965 D. K. Danks reported that it is an inherited disorder of children and young adults (Danks, Allen, and Anderson, 1965). In 1939, 80% of CF patients died within the first two years of life. Between 1949 and 1954 the mean age at death was about four years. By 1969 the mean age was 15 years, and now it is near 18.

The increase in life expectancy is a result of improved therapy. A brief description of Chris's treatment for CF will enable the reader to understand the marvels of modern medicine and appreciate that genetic afflictions like CF present social as well as medical problems.

Chris is a robust young child of four who was diagnosed as having CF early in his first year of life. Early diagnosis was prompted by the loss of an earlier child to CF. The family recognizes that CF is a terminal illness and provides all that is possible for the life of their child. Chris's balance between life and death is made possible by the use of expensive medicines, physical therapy, and frequent hospitalizations. In fact, Chris spent the first six months of his life in a hospital. He sleeps under a plastic tent, while a $600 electric moisturizer sprays a steady stream of medicated mist into the tent. The medication loosens the mucus in his lungs, and in the morning Chris is able to cough it up. At times the tent is not sufficient, however, and Chris breathes directly from a mask inhalator, which provides a medicated mist (Figure 10–5). In conjunction with the inhalation therapy, Chris's parents—like most, if not all, CF parents—use a technique of postural drainage. This technique helps the mucus to drain from the air passages of the lungs. The child is placed in a proper position over the mother's lap, and she claps his chest. The vibrations of the chest clapping help drain the lung mucus. Chris receives the chest-clapping treatment two or three times a day. The procedure requires a good deal of time. He takes from 20 to 40 pills every day. For example, he takes eight digestive enzyme pills, three antibiotics, and at least one vitamin tablet before *each* meal. In addition, Chris, in an ordinary day, takes in calcium, cough syrup, liquid antibiotics, other vitamins, and a very special high-protein low-low-fat diet. These are the daily items Chris requires routinely when he is well.

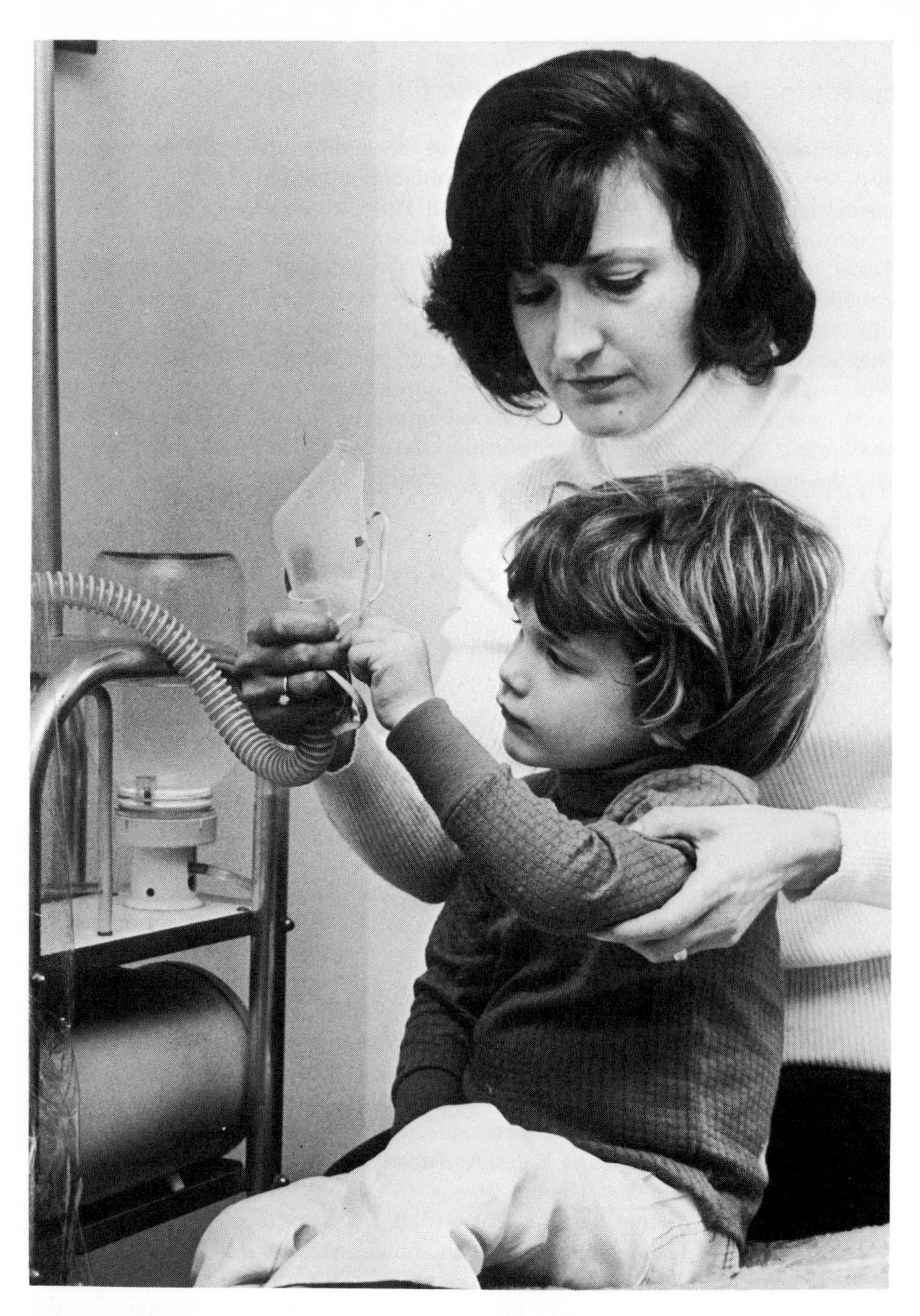

Figure 10–5. Cystic Fibrosis. Chris sits in front of his mist inhalator as his mother prepares to help him inhale a medicated mist. The machine provides an antibiotic-medicated mist to loosen the mucus that must be coughed up to open his air passages.

If he has an infection, additional special care is required. For example, during Chris's second year of life, his mother had to give him three injections of the antibiotic garamycin per day, and at times he received up to 10 injections per day. His direct medical costs in his first two years of life were $17,000. Similar to other CF children, Chris also has a continuous dental problem. Cystic fibrosis children in general have weak, cavity-prone teeth that require constant attention.

In Chris's case the home also had to be air-conditioned. The total financial bill, which would have to include a considerable expense for electricity, was not calculated, but the national average is $10,000 per year per CF patient.

the effect of cystic fibrosis on the family

As indicated by di Sant'Agnese (1969), the financial and emotional stress on a family can be devastating. The illness can lead to family misunderstandings, tension, and guilt feelings among parents and brothers and sisters. The daily and nightly care of the ill child puts a severe strain on the marriage and on the individuals' goals and values. It has been said that three out of every four marriages involving a CF child end in divorce. The causes are many and varied: feelings of guilt coupled with the inability to accept the physical manifestation of that guilt, one partner blaming the other (or one set of grandparents blaming the other grandparents), a desire to remarry and have normal children, and a feeling of complete hopelessness. Nervous disorders also occur at a higher rate among the parents of CF children than in the general population, and they tend to coincide with either the birth or the death of a CF child.

The sense of loss following the death of a CF child is compounded by the fact that most CF children are of average or above intelligence, are particularly cooperative and agreeable on the whole, and present an attractive picture physically with no gross abnormalities (Anderson, 1960). The parents establish a particularly close working and loving relationship with their child. The death of the child, although expected at any time from the moment CF is diagnosed, often completely immobilizes a parent who has been unable to face the possibility.

For details on research in cystic fibrosis, write to the National Institute of Arthritis and Metabolic Disease, National Institutes of Health, Bethesda, Md. 20014. To receive help for a child who has cystic fibrosis, consult state or county departments of health and welfare. At the community level, contact local rehabilitation services. You may also consult the social service department of a community hospital or a local chapter of the National Cystic Fibrosis Research Foundation. Visiting nurse and homemaker services, available for chronically ill patients in many communities, can greatly ease the burden of CF patients.

the inherited mucopolysaccharide diseases

Six diseases have been identified that result from the body's inability to break down large molecules of mucopolysaccharides, a mucuslike sugar–protein complex. The mucopolysaccharides are found in a variety of connective tissue in all parts of the body, including the cornea of the eye and the walls of the blood vessels. Thus a defect in the utilization of these large molecules would affect the entire body. This is exactly what happens in these six particular disorders. Most of the body is involved in the diseases, and the mucopolysaccharidoses are, therefore, considered to be among the most debilitating of all the inherited diseases.

Current belief is that six separate genes are involved in the synthesis of six specific enzymes, each required to break down a particular mucopolysaccharide that accumulates in certain body cells (Table 10–2). The individual mucopolysaccharidoses, designated I to VI, are also named for the persons most associated with their description and recognition.

Hurler's syndrome has a higher incidence than any of the others. It occurs in about 1 in 10,000 live births. Death usually occurs before the age of 10 (McKusick et al., 1965). Recently, Hurler's and Scheie's patients have been shown to be deficient in the enzyme alpha-L-iduronidase. Normal persons have an excess of this enzyme, and it was thought that a transfusion of this enzyme might help the affected person. Several attempts were made, by various methods, but they did not work (Booth and Nadler, 1973). Enzyme replacement has also been tried with other inherited diseases, but so far no success has been reported. Hurler's syndrome has its counterpart in the animal world in the form of snorter dwarfism in cattle. "Snorter" dwarf cattle (so called because of their difficulty in breathing) inherit a mucopolysaccharide disorder similar to that of humans.

Hunter's syndrome is also referred to as gargoylism and, unlike all other mucopolysaccharidoses, is transmitted as a sex-linked recessive. Hunter's patients in general live longer than Hurler's patients, as this defect is much less severe than Hurler's syndrome. Sanfilippo's patients are in general the most severely deficient mentally. Morquio's patients are the most severely dwarfed, and they experience respiratory paralysis in later life. Scheie's syndrome has been described as a variant of Hurler's with stiff joints, claw hands, excessive body hair, and an additional heart-valve abnormality.

The mucopolysaccharidoses, perhaps more than any other group of inherited diseases, demonstrate classic pleiotropy, the expression of many abnor-

Table 10–2 The Inherited Mucopolysaccharide Storage Diseases

Mucopolysaccharidosis (Syndrome)	*Pattern of Transmission*	*Mucopolysaccharide Found in Urine*
I (Hurler's)	Autosomal recessive	Chondroitin sulfate B, heparitin sulfate
II (Hunter's)	Sex-linked recessive	Chondroitin sulfate B, heparitin sulfate
III (Sanfilippo's)	Autosomal recessive	Heparitin sulfate
IV (Morquio's)	Autosomal recessive	Keratosulfate
V (Scheie's)	Autosomal recessive	Chondroitin sulfate B
VI (Maroteaux–Lamy)	Autosomal recessive	Chondroitin sulfate B

malities as a result of a single gene defect. In these disorders the whole body is affected; very little is normal. What is looked for, one could say, is the degree of severity—the less severe the disease, the more normal the patient.

diabetes mellitus

A 12-year-old boy asked his mother how long he could live. She said, "Until you're about 32." The young boy had juvenile diabetes. Somehow he had learned he could not live as long as his normal classmates. The boy's father and three younger brothers are also diabetic. The adult form of diabetes, which the father had, is not as severe as the form his children have, juvenile diabetes. Every morning they receive insulin injections, and they keep a strict diet.

In a separate case a man was picked up by police and jailed on a charge of driving while intoxicated. The man was actually suffering from insulin reaction, and he needed medication while he was being held in a cell for a period of four hours. Cases like this one are socially difficult. The officer stopped the car as it weaved from one lane to another, an action that most likely prevented an accident and injury to the ill man and perhaps to others. To help in such situations, persons who are diabetic often wear identification bracelets or other types of identification. The type of identification shown in Figure 10–6 deals specifically with the case in point, the apparently drunk driver.

what is diabetes and when was it recognized?

Diabetes is a syndrome resulting from the body's inability correctly to metabolize and utilize the different forms of carbohydrates—sugars and starches—in the diet. Normally the body converts carbohydrates into body heat and into glycogen, which is stored in the liver. Stored glycogen is ready for rapid conversion to blood sugar as the level of blood sugar decreases.

Under normal circumstances the body regulates the use of carbohydrates through the action of the hormone insulin, which is secreted by certain (beta) cells located in the pancreas. In the diabetic there is an impairment of insulin activity. Either insulin production is insufficient, as in juvenile diabetes, or the

I AM A DIABETIC

If I am found unconscious or behaving abnormally, my condition probably is the result of an overdose of insulin. I am not intoxicated. If I can swallow, give me sugar, candy, fruit juice, or a sweetened drink. If I am unable to swallow, or if recovery does not take place promptly, call my physician or send me to a hospital immediately.

Name ________________ Phone ________

Address ________________________

Physician ________________ Phone ________

Address ________________________

Insulin type and dosage ________________

Figure 10–6. A wallet card comes in the pamphlet *Diabetes and You* (Department of Health, Education and Welfare, Diabetes and Arthritis Control Program, pamphlet No. 567). It can be obtained from the Superintendent of Documents, U. S. Government Printing Office, Washington, D.C., for 25¢

body produces both insulin and a blocking (anti-insulin) molecule that prevents the insulin from properly controlling the entry of blood sugar into the body cells. This abnormality, the presence of increased inhibitor, is inherited as a Mendelian codominant. Its frequency in the population is as high as 24%, indicating a gene frequency as high as 13%. Not all individuals having the abnormality will have diabetes, but under the appropriate environmental circumstances, inhibition of insulin will result in the disease. There is some evidence that the later the onset of diabetes, the greater the chance that it is due to a blockage of insulin action rather than to a lack of insulin.

In either case, the diabetic is unable to use all of the sugars taken in through the diet, and the level of sugar in the blood increases to a point where it is secreted in the urine. The condition of high levels of sugar in the urine is referred to as hyperglycemia and is common in any diabetic who is out of control, that is, who fails to regulate his blood sugar level. In this situation the blood becomes acid owing to the increase in certain molecules called ketones. As the level of ketones rises, the blood becomes more acid, and the patient eventually loses consciousness, goes into a coma, and dies unless he is given insulin. The patient can avoid such problems by eating the proper diet and following his doctor's advice.

Diabetes was first referred to in an Egyptian document of 1500 B.C. It was also described by Aurelius Celsus between the years 30 B.C. and A.D. 50. In the seventeenth century Thomas Willis, a physician to Charles II, tasted urine of a diabetic patient. He described the diabetic urine as "wonderfully sweet as if it were imbued with honey or sugar." Mellitus in Latin means honey-sweet and so the disease of diabetes became diabetes mellitus. The medical use of insulin came about much later, in about 1922. Before the advent of synthetic insulin, the production of animal insulin was coupled with the production of food. Insulin-containing organs were gathered at slaughterhouses for processing into insulin for human consumption. The average diabetic over a 30-year period required the production and destruction of about 10000 head of cattle (Ornstein, 1967).

Although insulin has been available for over 50 years, and the oral antidiabetic agents for almost 20, control of diabetes remains imperfect. According to the Public Health Service, the mean life expectancy of the diabetic is at least 25% shorter than that of the nondiabetic. And during this shortened life span, the diabetic runs a high risk of heart disease, stroke, blindness, kidney failure, and gangrene. The ultimate answer still eludes medicine, and even the primary question—What is diabetes?—has not been completely answered. Nevertheless, over a period of 50 to 60 years, insulin has allowed more diabetics to live longer. But this extended life of the diabetic has created certain problems; as the diabetic grows older, there may be new expressions of the disease. And we are hard pressed to handle certain middle-aged diabetics.

how common is diabetes?

Diabetes affects either sex and is ten times more prevalent after age 45. Taking the diagnosed and the suspected but undiagnosed diabetics, it is estimated that there are 8 to 12 million diabetics in the United States (Ellenberg, 1974). In short, diabetes occurs in approximately 1 person in every 20 to 25.

The incidence of diabetes increases rapidly with age. The number of diabetics under age 25 is about 2 per 1000. In the 45 to 54 age group, the rate is 20 per 1000, and in the 65 to 74 group it is 65 per 1000. Although the incidence appears to be the same in both sexes under age 45, it is much higher among females after age 45. Below age 45, nonwhite females have a higher incidence than white, and nonwhite females over 45 have twice the frequency of the white female. Nonwhite males and white males are about the same.

The distribution of diabetic cases with respect to age is 10% under age 25, 20% at 25 to 44, 50% at 45 to 64, and 20% over 64. Thus 90% of all diabetics are 25 or older. Data from the Joslin Diabetes Foundation indicate that only 1 child in 2500 expresses diabetes under age 15 (Joslin et al., 1959).

the inheritance of diabetes

Diabetes is associated with family history, obesity, and advancing age. With respect to inheritance, no definite pattern has yet been established. Many reports indicate both dominant and recessive modes of inheritance with reduced penetrance. Because of reduced penetrance we are not able to predict when or where diabetes will be expressed, even if a person carries the defective gene. Available evidence, from several scientific investigations on the inheritance of diabetes, suggests that no single, fully penetrant gene controls predisposition to diabetes. To summarize a great number of papers on this subject, we can say only that diabetes mellitus is determined primarily by inheritance, probably through a polygenic mechanism, and secondarily by such environmental stresses as dietary imbalance, overnutrition, obesity, and in some cases viral infection.

There is general agreement among geneticists that a genetic component exists. This becomes quite clear when we look at family histories. For example, when both parents are diabetic, abnormal glucose tolerance curves may appear in the children long before the advent of clinical diabetes. Glucose tolerance curves are established by determining the time necessary for the person to reduce blood sugar levels after the ingestion of a large dose of sugar. The risk for genetically related individuals when diabetes has been diagnosed in the family is 2.5 times that for persons in the general population. Twin studies also lend support to the theory that diabetes is inherited. In fact, certain twin studies indicate that there is a strong tendency for the inheritance of non-insulin-requiring diabetes (the form that is treated by controlling diet). Reported twin data vary, but they suggest that inheritance is involved. It is difficult to separate the types of inheritance because of the variety of clinical symptoms, the extreme differences in time of expression, and the lack of a basic definition of diabetes.

There is some evidence for multifactoral or polygenic inheritance. The frequency of diabetes in the offspring of two diabetic parents is three times that of children having only one diabetic parent. If a parent with diabetes also has an affected brother or sister, the risk to that parent's offspring is doubled. Conversely, the risk is doubled for a brother or sister of a diabetic if that

Table 10–3 Gene and Chromosome Diseases That Include the Expression of Diabetes Mellitus

Gene Syndromes	*Chromosome Abnormalities*
Ataxia-telangiectasia	X0 (Turner's)
Cockayne's	XXY (Klinefelter's)
Cystic fibrosis	Trisomy 21
Flynn–Aird	Chromosome 18
Friedreich's ataxia	short arm deletion
Laurence–Moon–Biedl	
Lipodystrophy	
Prader–Labhart–Willi	
Testicular deficiency	

diabetic has a diabetic child. Lending support to the multifactoral hypothesis is the variety of chromosome and gene disorders in which diabetes is expressed as a part of the pathological syndrome (Table 10–3).

There are two outstanding conditions associated with the onset of diabetes.

1. A genetic predisposition. Persons with diabetic relatives have a greater tendency to become diabetic. For example, if both parents are diabetic, there is about a 100% probability that their children will express diabetes if the children live to *age 80.*
2. Obesity. A review of the numerous reports on diabetes and whether or not it is an inherited disorder shows that all investigators agree that excessive weight triggers the onset of diabetes. Eighty percent of all diabetics in the United States are overweight. Interestingly, overweight diabetics rarely require insulin; that is, they can be maintained by diet restriction (Simpson, 1969).

There is substantial evidence that a second very different form of diabetes, diabetes insipidus, is inherited as a sex-linked condition. Diabetes insipidus results from a disorder of the pituitary gland and is characterized by intense thirst and the excretion of large quantities of urine. Sex-linked disorders are discussed in Chapter 11.

socioeconomic aspects of diabetes mellitus

The National Health Interview Survey of July 1964 through June 1965 revealed that over a half million diabetics were limited in their ability to work, keep house, or engage in scholastic activity. Of these, about 40% were completely unable to perform a major activity. During the 12 months covered by this study, employed diabetics lost an average of 15 working days compared to 5 days for the nondiabetic. Diabetes is responsible for 90% of the foot and leg amputations not arising from accidents or war wounds, and it is the second leading cause of blindness.

The annual cost of diabetes to the United States economy in 1964–65 was estimated at $2 billion. Today it must be considerably higher, since there are many more diabetics and cost figures across the board are higher.

hope beyond insulin?

The question is asked many times: "Is there something that can be done besides controlling diet or taking insulin?" At the present the answer is no. But what of the future? It would appear that several options are available, only one of which is genetically oriented, that is, to change the DNA in such a manner that the disease no longer occurs. This approach is not yet feasible.

Another approach is that of transplanting a normal pancreas into a diseased patient. A number of pancreas transplants have been performed, but of the 31 patients who had had a pancreas transplant through 1973, two survived and they lived for only a year. These two patients, who had been very ill and insulin-dependent, no longer had to take insulin. This is a remarkable development, but organ transplants involve a number of problems. Because of recipient rejection of the transplanted organ, the physician has to administer all sorts of chemotherapeutic agents—anti-immunogenic agents and corticosteroids—to combat rejection. The drugs have one side effect or another on the patient. Further, the anti-immune suppressing drugs leave the patient open to infection. Even if science could overcome *all* the transplant problems, where would one obtain the millions of pancreases should everyone want the transplant?

Another approach is that of transplanting a normal pancreas into a diseased cells and the beta cells from the pancreas and grow them in culture. Rats and mice that were made diabetic with injections of alloxan have literally been cured with injections of islet cells that had been grown outside the body in culture. But, here, too, there is the same problem of rejection that occurs with the whole gland.

A machine that can automatically perform certain pancreatic functions for diabetics was developed recently by researchers from Toronto's Hospital for Sick Children and the University of Toronto's Institute of Biomedical Engineering. The device, which is about the size of a television set, can continually measure a diabetic's blood sugar levels, compute insulin requirements, and inject needed insulin. If a patient's blood sugar is low, the machine can inject dextrose into the bloodstream. The scientists say that a smaller, portable version of their device may be technologically feasible.

Scientists at the Joslin Diabetes Foundation, Boston, are attempting to develop a mechanical pancreas that could be implanted in the diabetic. A tiny computer would detect glucose concentration in tissue fluid and release insulin from a small reservoir into the patient.

In fiscal year 1974 about $12.5 million was spent for research on diabetes. Years of criticism of Congressional neglect of this serious and common disease led in 1974 to passage of the National Diabetes Mellitus Research and Education Act, which will funnel $40 million into programs of research and training (*Laboratory Management*, 1974). The law calls for regional diabetes research and training centers, and establishes a 17-member National Commission on Diabetes to formulate long-range plans to combat the disease.

the inherited lipid (fat) storage diseases

The lipids are fatty substances found in all the body tissues including the bloodstream. They can be in the form of neutral fats, stored by the body as an energy reserve, or as a part of more complex molecules found in cell membranes. As cells die, the fats within the cells, as well as the fatty products of various body reactions, must be degraded and reorganized into new cell-building materials. In genetically capable people, individual genes promote the synthesis of specific enzymes to break down cell lipids. If such enzymes are missing, as in persons with particular genetic deficiencies, the lipids accumulate in certain tissues. In many cases the lipid accumulation results in mental retardation, enlargement of the liver and spleen, seizures, blindness, skeletal deformities, kidney failure, and nerve damage. In humans most of the known lipid storage diseases bring about premature death (Brady, 1973).

Studies on the lipid storage diseases began in 1881 with an observation made by an ophthalmologist, Warren Tay. Tay observed a reddish brown spot on the retina of a one-year-child who was progressively degenerating with time. Six years later, Bernard Sachs, a neurologist, reported the same characteristics in a young patient. Sachs, through continued studies, found that the disease was inherited and involved an accumulation of lipid in nerve cells of the brain. Today we know the lipid to be a particular ganglioside, GM_2, and the cause to be lack of the enzyme hexosaminidase A. The disorder became known as Tay–Sachs disease and was later found to occur more frequently in children of the Ashkenazic Hebrews than in other people (see Chapter 15).

A second lipid storage disease was reported by C. E. Gaucher in 1882. In Gaucher's disease the lipid glucocerebroside accumulates in cells of the spleen, resulting in mental retardation, mild anemia, decrease in blood platelets, and enlarged spleen and liver. Sixteen years later a third lipid disorder was reported by Johannes Fabry and William Anderson. Patients with Fabry's disease have a particular type of skin rash and kidney damage.

As new lipid disorders became known, it became clear that at least 9 of the 10 accumulated-lipid diseases (Table 10–4) had a common ceramide (*N*-acylsphingosine) to which additional molecules were added. In a normal person 10 separate enzymes are involved in breaking down the particular sphingolipid to the common ceramide sphingosine. Should any one of these enzymes be missing because of an altered gene, a particular lipid storage disease results.

From Table 10–4 it can be seen that 9 of the 10 listed lipid storage diseases are inherited as recessive disorders, and one, Fabry's, is inherited as an X-linked recessive. In every case the enzyme deficiency has been identified. We can now determine whether the developing fetus has any of the lipid storage diseases by analyzing fluid and cultured cells taken by amniocentesis (presented in Chapter 6). It has also been shown that almost all body tissue will reveal a lipid storage disorder. Measurements made on parents of children with a lipid storage disease have revealed that the heterozygous parents have less than the

Table 10–4 Inherited Lipid Storage Diseases in Humans

Disease	*Accumulated Sphingolipid*	*Deficient enzyme*	*Inheritance**
Ceramide lactoside lipidosis	Ceramide lactoside	β-Galactosidase	AR
Fabry's	Ceramide trihexoside	Galactosidase	X-LR
Fucosidosis	H-isoantigen	Fucosidase	AR
Gaucher's†	Ceramide glucoside (glucocerebroside)	β-Glucosidase	AR
Generalized gangliosidosis	Ganglioside GM_1	β-Galactosidase	AR
Krabbe's	Galactocerebroside	β-Galactosidase	AR
Metachromatic leukodystrophy	Ceramide galactose-3-sulfate (sulfatide)	Sulfatidase	AR
Niemann–Pick†	Sphingomyelin	Sphingomyelinase	AR
Tay–Sachs†	Ganglioside GM_2	Hexosaminidase A	AR
Tay–Sachs variant (Sandhoff's disease)	Ganglioside GM_{2+}	Total hexosaminidase A and B	AR

* AR, autosomal recessive; X-LR, sex-linked recessive.

†Highest frequencies of Gaucher's, Neimann–Pick, and Tay–Sachs diseases are found in the Ashkenazic Hebrews.

normal amount of the enzyme. Thus a system of detecting the recessive carrier is now possible.

Because the lack of an enzyme causes the accumulation of lipids, it was thought that adding the missing enzyme to a young patient might offset the disease. Enzyme therapy was tried, but it failed. Today there has been little if any success with this approach, for the body rapidly clears itself of the injected enzyme. Aspects of various forms of therapy in inherited disorders are presented in Chapter 19.

references

Anderson, D. H. 1960. Cystic Fibrosis and Family Stress. *Children, 7*:9–12.

Blanck, R. R., and E. M. Mendoza. 1976. Fertility in a Man with Cystic Fibrosis. *Journal of the American Medical Association, 235*:1364.

Brady, R. O. 1973. Hereditary Fat-Metabolism Diseases. *Scientific American, 229*:88–97.

Booth, C. W. and H. L. Nadler. 1973. Plasma Infusions in an Infant with Hurler's Syndrome. *Journal of Pediatrics, 82*:273–78.

Danks, D. K., J. Allen, and C. M. Anderson. 1965. Genetic Study of Fibrocystic Disease of Pancreas. *Annals of Human Genetics, 28*:323–56.

di Sant'Agnese, P. A. 1969. Unmasking the Great Imposter—Cystic Fibrosis. *Today's Health, 47*(Feb.):38–41.

Ellenberg, Max. 1974. Beyond Insulin: Intensifying the Search for a Cure. *Laboratory Management, 12*(Aug.):14.

Hsia, David Y-Yang. 1971. A Critical Evaluation of PKU Screening. *Hospital Practice, 6*: 101–12.

Joslin, E. P., et al. 1959. *The Treatment of Diabetes mellitus.* Lea and Febiger, Philadelphia.

Knox, E. W. 1969. Inherited Enzyme Defects. *Hospital Practice, 14*: 33–41.

Kulczycki, L. L., and Victoria Schauf. 1974. Cystic Fibrosis in Blacks in Washington, D.C. *American Journal of Diseases of Children, 127*: 64–67.

Laboratory Management. 1974. Diabetes Bill Becomes Law. *12*(Sept.): 20.

McKusick, Victor A. 1971. *Mendelian Inheritance in Man: Catalogs of Autosomal Dominant, Autosomal Recessive and X-Linked Phenotypes,* 3rd ed. Johns Hopkins Press, Baltimore.

McKusick, Victor A., et al. 1965. The Genetic Mucopolysaccharidosis. *Medicine, 44*: 1–21.

National Academy of Science, 1975. *Genetic Screening: Programs, Principles and Research.* National Research Council, 2101 Constitution Ave. NW, Washington, D.C. 20418.

Ornstein, Leonard. 1967. The Population Explosion, Conservative Eugenics and Human Evolution. *Bioscience, 17*: 461–64.

Simpson, N. F. 1969. Diabetes in the Families of Diabetics. *Canadian Medical Association Journal, 98*: 427–32.

It has been established that mortality rates are higher for males than for females in every age category. One of the major reasons for higher death rates in males is their greater suseptibility to bacterial infections leading to death (Gadpaille, 1972). Even the discovery and prescription of broad-spectrum antibiotics did not change the ratio of male to female deaths due to bacterial infections. This suggests that the antibiotics work more efficiently in a system that itself is better able to cope with infection. This theory supports the idea that antibiotics do not cure a patient; they help bring the infection into balance with the normal body defense systems, but in the final analysis the body cures itself.

Many authors have asserted that the presence of the female's second X chromosome (XX) offers an advantage in coping with stress and disease over the single X of the male (XY). In terms of sex-linked inherited diseases, the female does have an outstanding advantage. The female can be heterozygous, since she carries two homologous X chromosomes, and she displays the usual dominant and recessive patterns of expression. That is, she carries two genes for each trait, one on each X chromosome, just as she carries two genes for each trait on the autosomes discussed in Chapters 8, 9, and 10. The male, having only one X chromosome, is in a vastly different situation. He carries only one gene for any sex-linked trait. In this case the male is referred to as hemizygous. His sex-linked genes do not show dominance or recessiveness; he has only one gene for a trait and it is expressed. He does not have a second gene to mask the effect of a particular defective gene. Sex-linked inheritance is discussed in Chapter 8; a brief review of X-linked inheritance is presented in Figure 11–1.

We are truly heirs of all the ages; but as honest men it behoves us to learn the extent of our inheritance, and as brave ones not to whimper if it should prove less than we had supposed.

Author Unknown

human sex-linked inherited diseases

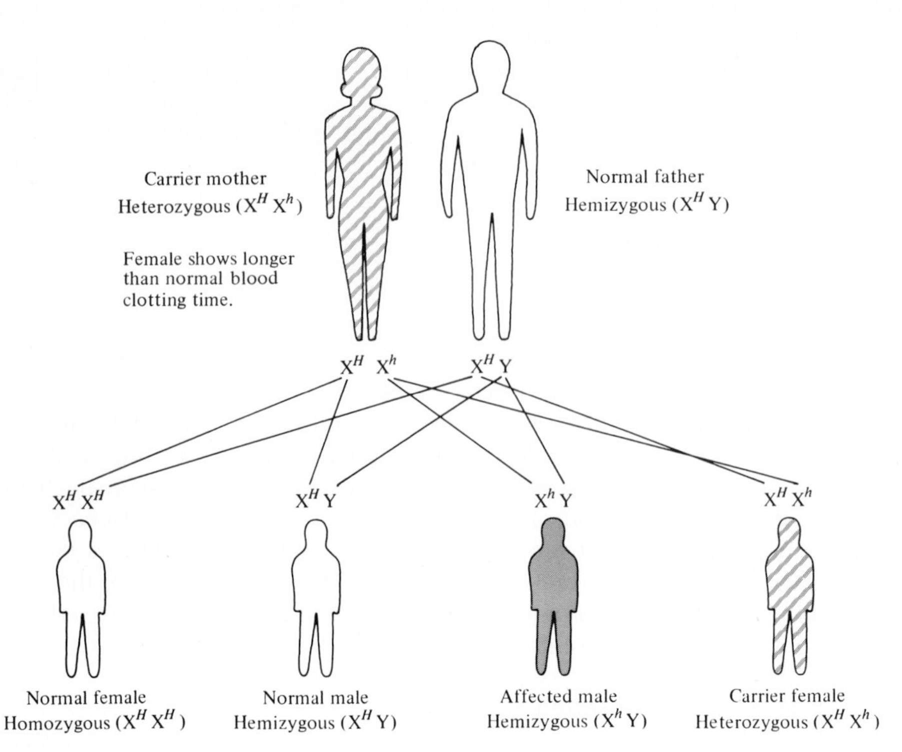

Figure 11–1. Review of X-Linked (Sex-Linked) Inheritance. Most X-linked disorders occur when the mother is a carrier. She carries a faulty gene on one X chromosome; the other X is normal. As shown here for hemophilia, each female child has a 50% risk of inheriting the one faulty X and being a carrier like her mother and a 50% chance of inheriting no faulty gene. Each male child has a 50% risk of inheriting the faulty X and the disorder and a 50% chance of inheriting normal X and Y chromosomes. Over 150 X-linked inherited diseases have been identified; for other examples, see Table 11–1.

Figure 11–1 makes it apparent that the two X chromosomes of females and the one X of males form a pattern of inheritance distinctly different from that seen in non-sex-linked, or autosomal, gene transmission. The X-linked pattern is based on the fact the X-bearing sperm can fertilize only an X-bearing egg. Thus, under normal conditions, a male's X chromosome is found only in his daughters, and his Y chromosome is found only in his sons. Thus in each daughter there is an X chromosome from each parent. In the sons the X chromosome comes from the mother.

For a recessive disease the female may be a homozygous dominant, heterozygous, or a defective homozygous recessive. As a heterozygote, she is a carrier, just as with autosomal recessive disorders. The heterozygous female has a one in two chance (50:50) of giving one particular X chromosome to her son. If the son receives the X chromosome that carries a defective gene, he will pass it on to every daughter he fathers.

the physical basis: transmission of units of hereditary material through the genes

Because the male has only one X chromosome and the female has two, a particular gene that has nothing to do with the individual's sex but happens to be located on the X chromosome is transmitted in a crisscross pattern of inheritance, mother to son to granddaughter. Never, under ordinary circumstances, does it go from father to son, as an autosomal gene can. In general, X-linked disorders in the hemizygous male show complete penetrance and high expressivity. But X-linked defective genes usually show a less severe expression in the female.

As with recessive autosomal diseases, it was believed until recently that a recessive sex-linked disease did not affect the heterozygous person. However, as was pointed out in Chapter 8, there are a number of instances where the autosomal heterozygote does show symptoms, and this is also true for the heterozygous female with a gene for an X-linked disease. For example, as a group females who are heterozygous for the sex-linked disorder hemophilia show a variety of blood-clotting "times." That is, if we were to measure the amount of time it takes a group of normal persons' blood to clot and compare this to the times collected for a group of heterozygous carrier females, we would see that some females are borderline hemophiliacs while others are almost normal. So although the carrier does not appear to have a bleeding disease, certain symptoms are present. The condition is simply not severe enough to be recognized or, in most cases, require medical treatment. In this particular case the longer it takes for the carrier to form a clot, the less active blood-clotting factor VIII she has.

Some of the more common of the 150 sex-linked inherited diseases are listed in Table 11–1. Certain of these X-linked disorders are discussed with respect to their social demands on society.

genes on the Y chromosome

Very little is known with certainty about the Y chromosome—only that it is about half the size of an X chromosome (see Chapter 6) and that under normal conditions it confers maleness during an early period of embryonic development (see Chapter 7). The Y chromosome is transmitted directly from father to son, and all genes located on the Y are passed from father to son. This mode of gene transmission is called *holandric inheritance.* Holandric genes are by definition exclusive to the male sex.

Although many sex-linked traits are known to be associated with the X chromosome, only three traits have been reported to be associated with the Y chromosome: hairy ear rims, webbed toes, and porcupine-type body hair. However, recent studies indicate that none of these three holandric traits is fully documented, although logically there is no reason why the human Y chromosome should not carry Y-linked genes. Perhaps we do not have the necessary means for appraising holandric genes. In organisms for which we have extensive pedigrees, however, like *Drosophila,* the Y chromosome contains very few Y-linked genes. In mice the Y chromosome appears to contain a gene for the production of a tissue-compatibility antigen.

what is color blindness?

Color blindness is the inability to see one of the primary colors, red, yellow, or blue, and has nothing to do with the appearance of color in the eye. There are

Table 11–1 Examples of the More Common X-linked Inherited Diseases

Condition	*Brief Description*
Agammaglobulinemia	Failure to make sufficient gamma globulin
Anhydrotic ectodermal dysplasia	Lack of sweat, uneven sized breasts
Anophthalmia	Congenital absence of the eyes
Borjeson's syndrome	Mental deficiency, epilepsy, endocrine disorders
Brown tooth syndrome	Defective enamel, possible corncob appearance of teeth
Cerebellar ataxia	Atrophy of the cerebellum, clumsy, explosive speech
Charcot–Marie–Tooth muscular dystrophy	Progressive weakening of the muscles of the hands and feet
Choroidoretinal degeneration	Sex-linked retinitis pigmentosa
Christmas disease	Hemophilia B, a milder form of hemophilia, lack of blood-clotting factor IX
Combined immunodeficiency disease	Severe lack of immune defense mechanisms, antibodies, etc.
Congenital cataract	Cataract present at birth, requires surgery
Congenital deafness	Profound deafness, deafmutism
Congenital ichthyosis	Hard scaly skin, hyperpigmentation, hypogonadism
Congenital night blindness	Poor vision at twilight
Deutan color blindness	Inability to see green color
Diffuse cerebral sclerosis	Hardening of the brain
Duchenne muscular dystrophy	Weakness of trunk and hypertrophy of muscles
Fabry's disease	Inborn error in glycolipid metabolism, later renal disease
Fatal granulomatosis (children)	Development of numerous tumors or granulomas
Glucose-6-phosphate dehydrogenase deficiency	Deficiency of G-6-PD, causes favism, drug sensitivity
Gout	High levels of uric acid in blood, arthritis
Hemophilia A	Lack of active blood-clotting factor VIII
Hereditary bullous dystrophy, macular	Fluid-filled macula blisters
Hunter's syndrome (mucopolysaccharidosis II)	Growth retardation, secretion of chondroitin sulfate B, cartilage affected
Hydrocephaly	Blockage in the aqueduct of Sylvius, fluid buildup enlarges skull
Hyperparathyroidism	Oversecretion of the parathyroid glands
Hypochromic anemia	Reduced hemoglobin in red blood cells

Table 11–1 continued

Condition	*Brief Description*
Hypophosphatemia	Vitamin-D-resistant rickets, elevated alkaline phosphatase
Ichthyosis vulgaris	Fish skin disease; dry, hard, scaly skin
Keratosis follicularis spinulosa	Mild hardening of hair follicles
Late spondyloepiphyseal dysplasia	Fusion of vertebrae epiphysis
Lesch–Nyhan syndrome	Self-mutilation associated with a defect in purine metabolism
Lowe's syndrome	Mental retardation, congenital glaucoma
Manic-depressive	Certain states of manic-depression segregate with the Xg blood locus
Menke's syndrome	Kinky hair disease, growth retardation, small jaws
Microphthalmia	Abnormally small eyes
Nephrogenic diabetes insipidis	Deficiency in decarboxylation of keto acids, ketonuria
Neurohypophyseal diabetes insipidis	Malfunction of pituitary gland, thirst, hunger
Norrie's disease	Congenital blindness, retinal malfunction
Nystagmus, hereditary	Constant involuntary cyclical movement of the eyeball
Ocular albinism	Eye problems resulting from the lack of eye pigments
Oral-facial-digital syndrome (OFD-I)	Cleft tongue and palate, digit anomalies, mental retardation
Protan color blindness	Inability to see red color
Spastic paraplegia	Spasms or convulsions due to lesions of the spinal cord
Spatial visualization	Ability to visualize (spatially) abstract objects
Spinal ataxia	Poor motor function, clumsy, awkward, resulting from spinal problems
Total color blindness	Sees only shades of black and white
Wiskott–Aldrich syndrome	Antibody deficiency, eczema, infections, early death

at least two separate genes that determine the amount of light sensitive pigment present in the cones of the eye. Other genes are responsible for the electrochemical conversion of cone-pigment excitation into color vision.

By strict definition color blindness is considered a disease, even though the defect does not appear to have other biologically noticable implications. The social importance of seeing color in our industrial society can be found everywhere—in advertising media, display rooms, automobiles, airplanes, and virtually everything from human hair to colored socks. Almost everything is dyed, painted, or impregnated with a color. Our pleasure at perceiving color has evolved into a need for countless technologists, engineers, designers, artists, dyers, lighting specialists, television experts, decorators, and commercial art photographers, all to attract the consumer. Of greater social significance, think of the red or green color-blind driver confronted with traffic lights. Certainly red and green color-blind people must learn to make the proper adjustment. And color vision provides esthetic pleasure. We enjoy seeing, in color, various art forms and the unending array of color in nature. But perhaps the color blind appreciate the world of textures better than those with color vision.

color perception

The eye functions like a TV camera. it has the ability to transmit nerve signals to the brain for conversion into both color and images. The retina contains two kinds of organelles that function in vision, the rods and cones. The two have separate pigment systems. The rods contain rhodopsin, which bleaches from pink to white when exposed to light. The bleaching phenomenon of rhodopsin somehow converts nerve signals into a picture in varying shades of gray; thus the rods are responsible for twilight vision, which is colorless (Rushton, 1975). These black and white receptors constitute our most sensitive light detection system. Only recently has their sensitivity been matched by a specially built electronic photocell (Rushton, 1969). The rods are not involved in color vision.

Color perception is a function of the cones. The retina houses three types of cones, each containing a single pigment sensitive to one particular color—red, green or blue (MacNichol, 1964). It is believed that a separate gene controls the formation of each of the three cone pigments. The colors we see depend on the way the three pigment receptor systems are stimulated. For example, if the yellow and blue cones are stimulated equally, we see the color green. When stimulated, the cone's pigments transmit the excitation in sets of two to the optic fibers; the fibers integrate the visual excitation into a color pattern that is sent on to the brain. Such combined functioning of the rods and cones and the brain allows for the perception of about 200 shades of black to white images and some 7.5 million shades of color! We see the colors in the so-called visual spectrum from 400 to 700 nanometers (a measurement of light wavelength). Light waves shorter than 400 or longer than 700 nanometers are not seen by the human eye, but special machines have been devised to study these wavelengths. Other organisms can see in the shorter and longer wavelengths. Insects, for example, can see in the ultraviolet range.

the types of color blindness

the physical basis: transmission of units of hereditary material through the genes

To identify the types of color blindness, we can project red, green, and blue lights on a screen. A normal trichromat, a person with normal vision, sees white in the area where red, green, and blue overlap. An abnormal trichromat has to have more of one of the three colors in the overlap area in order to see white. One who sees white in an area where just two of the three colors overlap is a dichromat, or someone who is color blind. Persons without the red-sensitive pigment in their retinal cones are called protanopes, and those without green-sensitive pigment are deuteranopes. Occasionally, a male will lack both pigments or will be unable to distinguish yellow from blue; each of these conditions is rare. The rarest of all is the individual without any color sensation, the monochromat, who sees only black and white. About 8% of the males and less than 1% of the females in the United States are color blind. About 60% of the color-blind males are green color blind.

classic hemophilia or bleeder's disease

Classic hemophilia, referred to as hemophilia A, is a defect wherein the patient lacks active blood-clotting factor VIII, one of at least 12 factors in the blood that in concert bring about blood-clot formation. Previously it was assumed that classical hemophilia was the result of insufficient production of AHG (antihemophilic globulin or blood-clotting factor VIII). Recently Zimmerman, Ratnoff, and Powell (1971) reported that the plasma of persons with hemophilia contains normal quantities of a protein that is similar to AHG but lacks activity. Thus, as suggested by Kisker (1976), hemophilia may be caused not by a lack of gene product (lack of production of AH) but by a structural alteration in the gene product that results in the loss of clotting activity. In other words, there is no quantitative loss of factor VIII production, but a qualitative loss in its activity. Obviously, if clotting does not occur, one bleeds to death. A closely related inherited deficiency in factor VIII, von Willebrand's disease, was presented in Chapter 8. Von Willebrand's disease involves a dominant autosomal gene. In contrast, hemophilia is an X-linked recessive disorder, meaning that the female with hemophilia has a defective gene on each of her X chromosomes. A second form of X-linked hemophilia is also known. In this milder form, referred to in the literature as Christmas disease, or hemophilia B, the patient lacks blood-clotting factor IX.

It is estimated there are 100,000 hemophiliacs in the United States. About 30% of the new cases of hemophilia A and B are the results of new mutations (Schneider, 1974). A new mutation means there is no history on either side of the family, but suddenly a child is born with the disease. Such a high percentage of mutations for such a very severe disorder is unusual. The other 70% of the cases result when a female carrier gives rise to a hemophilic son (Figure 11–1). The son, should he survive and should his wife be totally free of the defective gene, can sire only heterozygous carrier daughters, like his mother, or normal sons. This pattern of X-linked recessive inheritance can be more easily understood if the reader will review Figure 8–23.

Homozygous *hh* (hemophilic) females are most likely selected against either early in life because of accidents or at their first menstrual period. It is possible that the homozygous recessive female expresses a less serious hemophilia than the hemizygous male, but this is unlikely since the heterozygous female *Hh* demonstrates increased blood-clotting times. These heterozygous females vary in the amount of active blood factor VIII they carry in their bloodstream. Today, both male and the rare female hemophiliacs can be sustained by the administration of the missing active blood-clotting factor VIII.

hemophilia does not belong to royalty

Hemophilia has been described in the Hebrew Talmud, in the Old Testament, in accounts of the royal families of Europe, and even in ancient Egyptian scriptures. In early Egypt, if the first child bled to death from a minor wound, no further children were permitted in the family. The Talmud prohibited circumcision if two successive children bled to death. And, of course, there is the famous pedigree of hemophilia that started as a new mutation with Queen Victoria (Figures 11–2 and 11–3). Additional information concerning hemophilia of the royal families is presented in Chapter 20.

Hemophilia is distributed in a very democratic manner, if we exclude the fact that many more males than females are affected. The disease is inherited without regard to race, color, or creed. In fact, it is not limited to humans; hemophilia is known to occur in dogs and mice, and it is reasonable to assume that this type of disorder can be found among many other animals.

Figure 11–2. Queen Victoria. She was the first known carrier for hemophilia in her family line and is believed to be responsible for beginning the transmission of the gene for hemophilia in her descendants. The transmission of hemophilia can be seen in Figure 11–3, a pedigree of the Royal family. (From Radio Times Hulton Picture Library, British Broadcasting Corporation.)

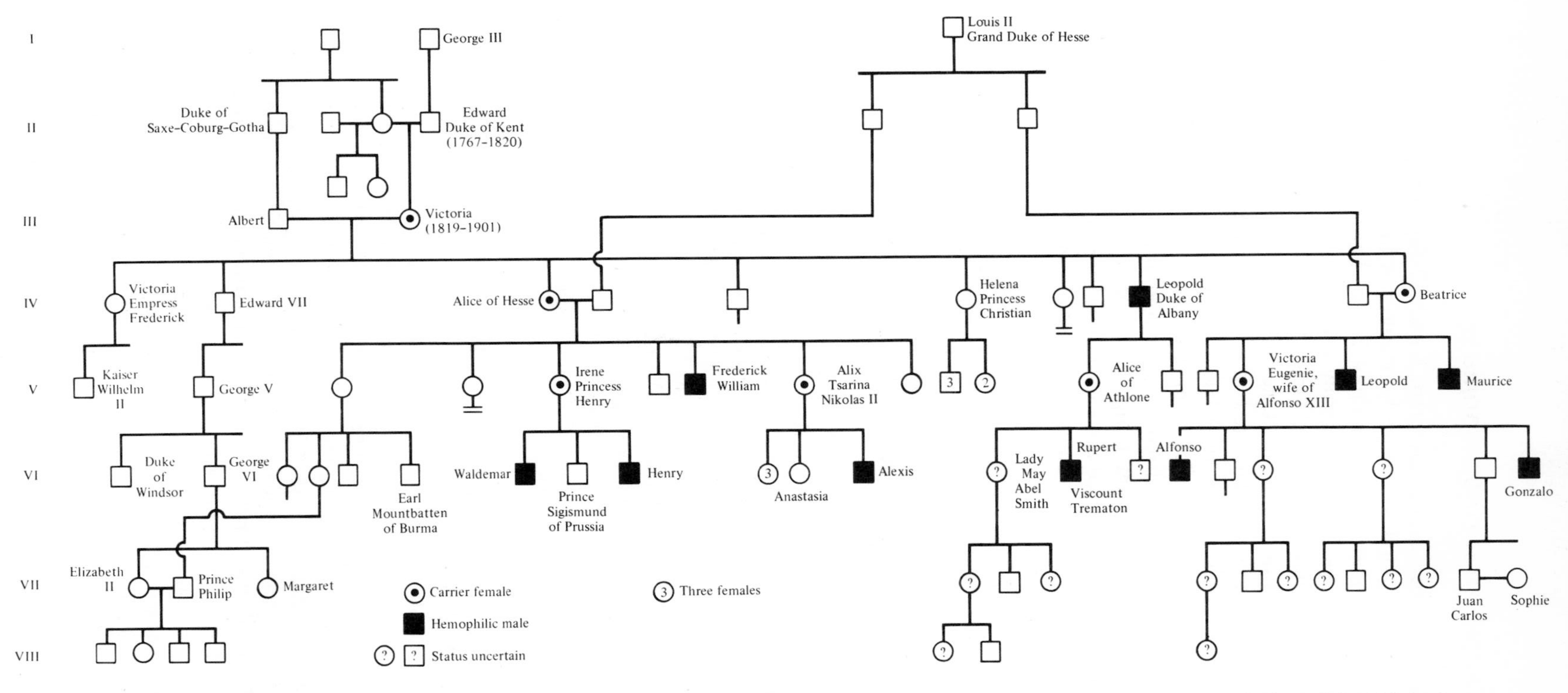

Figure 11–3. Pedigree of Queen Victoria and Her descendants. A pedigree showing Queen Victoria (1819–1901) and the transmission of the X-linked hemophilia through her descendants. (From Victor McKusick, *Human Genetics,* 2nd ed., © 1969, Prentice-Hall, Inc., Englewood Cliffs, N.J.)

therapy and expense for hemophiliacs

Denny needed to have two teeth extracted. So what? So Denny needed 156 pints or 19.5 gallons of blood to ensure he would not bleed to death once a tooth was removed. Leonard had a kidney hemorrhage. He needed four gallons of blood. The boys did not actually receive these large volumes of blood. They received the harvested blood factor VIII from whole blood. Factor VIII can now be obtained in two forms. One form is a cryoprecipitate; that is, factor VIII of *one* donor is concentrated at a blood bank and frozen until needed. The other form is via factor VIII concentrates; the concentrate is prepared by pharmaceutical companies from large *pools* of plasma (Kasper and Southgate, 1974). The plasma is dried to a powder form, and the resulting concentrate can be stored at room temperature or in a home refrigerator. There does not appear to be a price advantage, but for self-injection or home use, the concentrate may have advantages in terms of ease of application, shelf life, and known dosage.

In either case, the average cost for blood factor VIII is about $4000 to $8000 per year per hemophiliac. The routine medical care costs are very high. In one case, for example, a boy had 130 transfusions in one year. The doctor's fee was $15 per transfusion or a round figure of $2000. Figures such as these apply to the patient who requires factor VIII to control a particular bleeding situation.

The best therapy, as suggested in Schneider's (1974) report, would be a daily injection of factor VIII similar to the way the diabetic takes insulin. Most people, however, cannot afford it. In one such case daily injections cost the patient $22,000 per year. The father is very wealthy and can afford it for his son. Even if everyone who needed it could afford it, where would the blood factor come from? Today, collectively, 5000 blood banks take in 15 million units of blood per year. This must cover all routine or major surgery in the United States. If each hemophiliac were given the required amount of factor VIII, it would take all the 15 million units of blood per year. Obviously, there is a dilemma over a very scarce resource of blood. One answer to this problem can come from science—the synthesis of a functional factor VIII type of blood-clotting molecule.

Hypnosis is one form of therapy for this inherited disease. The therapy is based on the fact that a relaxed patient bleeds less often than one under tension. This has been shown to be the case, and in some areas of the country, such as Denver, Colorado, self-hypnosis is being successfully used by some patients to lessen the severity of hemorrhage and the number of times they have attacks. In general, hypnosis cannot stop bleeding once it has started, but hypnosis can help by relaxing a person, so that he may not have as many spontaneous (accident-free internal) hemorrhages. A relaxed person produces less adrenalin. Perhaps this lower production of adrenalin keeps the blood pressure lower and thus accounts for the decrease in spontaneous internal bleeding. Hypnosis is not a new concept. It is believed that Rasputin the Terrible used hypnosis to save the life of Alexis, the heir to the Russian throne (see Figure 11–3, generation VI).

In general, however, like many genetic diseases, hemophilia creates a sense of helplessness in the victim. The hemophiliac must face the fact that he is dependent on the lifesaving factor VIII forever. For a great many, this dependency has led to a life of despair or even to suicide. Hemophilic patients have a normal range of intelligence, but they require a great deal of encouragement and, in many cases, special arrangements for their education and employment. Their unemployment rate is five times as great as that of nonhemophiliacs, and 15% have never worked because of the severity of their illness or a poor self-image.

The recently acquired ability to recognize the genetic carrier of hemophilia offers prospective parents a choice as to whether or not a hemophilic child should be born. This, of course, is a very delicate and sensitive issue, and it arises in relation to any disorder where the parents can be identified as carriers or the fetus can be identified as diseased. Yet the nation's ability to support additional cases of such expensive diseases as hemophilia is questionable. There are about 8000 new cases of hemophilia A every year in the United States. If the nation can afford additional cases, how many more can it afford and for how long? Perhaps you can discuss in class what can be done and by whom to avoid or administrate social problems of a genetic nature. The national cost for hemophilia today, based on $20,000 per year for each of the 100,000 patients, would be $2 billion. Current legislation now provides a guarantee that all hemophilic patients will receive free treatment at their local health centers. A major financial strain has been lifted from the individual family. Perhaps a series of new clippings will serve to illustrate how particular legislation of this type occurs (Figure 11–4).

the inherited muscular dystrophies

A mother watched the third of her four afflicted sons die with muscular dystrophy. He was young, only 20, and the third son to die in a period of 15 months. The father is disabled, and the mother's time is spent caring for the rest of the children, the house, and the ailing husband.

There are at least six types of muscular dystrophy (Table 11–2), but the Duchenne type, the X-linked form, is the most severe and accounts for about 90% of all muscular dystrophy cases (Zundel and Taylor, 1965). The family referred to had the misfortune to be afflicted with the X-linked type. Among the children of a Duchenne carrier woman, there is a 50% chance that each son will be dystrophic and a 50% chance that each daughter will be a carrier.

Table 11–2 Inheritance Patterns for the Muscular Dystrophies

Dystrophy Classification	*Mode of Inheritance*
Duchenne	X-linked recessive
Facioscapulohumeral	Autosomal dominant
Limb-girdle	Autosomal dominant
Distal myopathy	Autosomal dominant
Ocular myopathy	Unclear (X-linked?)
Myotonic	Autosomal dominant
Werdnig–Hoffman	Autosomal dominant

It's Blood Donors Or Welfare

HARRISBURG, Pa. (UPI) — Bonnie and Francis Marshall have found only two ways to keep their hemophiliac son Kevin alive — recruit 36 blood donors every month or go on welfare.

The Marshalls, who appealed to the state to keep them off welfare, were disappointed yesterday when they learned they must supply donors in exchange for free treatments for Kevin, 12.

"I know that's impossible," Mrs. Marshall said. "I've been begging blood for 11½ years."

The donors also must travel 100 miles to Philadelphia Children's Hospital, which has agreed to supply the $1,000-a-month treatment.

"That means a $3 turnpike toll, lunch and gasoline," Mrs. Marshall said. "I can't get people to do that."

A spokesman for Gov. Milton Shapp said this was the only way the governor could find to fulfill his promise to help the boy. He said the governor's office already had lined up 11 "sure donors."

Shapp telephoned the Marshalls Wednesday when he learned of the problem. He reversed a decesion by other state officials that Marshall would have to quit his $7,800-a-year job and accept welfare to get help for Kevin.

The governor enrolled Kevin in a special state-federal program for hemophiliacs at Philadelphia Childern's Hospital, which waived its residency requirement.

But Mrs. Marshall did not learn about the donors until yesterday.

"Everything was so beautiful yesterday and now this," she said. "But I can't complain or they will say I'm not grateful."

Terry Dellmuth, Shapp's special assistant for human services, said the state will supply a case worker or public nurse to help the Marshalls recruit donors.

—DAD WON'T HAVE TO GO ON WELFARE—

Kevin May Get Free Blood

HARRISBURG, Pa. (UPI) — The parents of a 12-year-old hemophiliac boy reached a "compromise" with state officials, enabling them to obtain free $1,000-a-month treatments.

A spokesman for Gov. Milton J. Shapp said the tentative agreement would keep Bonnie and Francis Marshall of suburban Camp Hill off welfare.

"The details of the compromise still must be worked out," he said.

The spokesman said Kevin Marshall probably would be able to obtain treatments at Hershey Medical Center.

"In exchange," he said, "we will organize blood donors for Hershey."

The hospital normally requires cash payments for the expensive hemophiliac treatments.

"Gov. Shapp came through," Mrs. Marshall said. "I knew he would. It was only a matter of time."

The compromise ended four days of anguish for the Marshalls, who were told by the state last Tuesday they would have to go on welfare to get help for Kevin. The father would have been forced to quit his $7,800-a-year job to qualify for welfare.

Kevin's plight prompted hundreds of telephone calls from persons all over the nation offering to donate blood.

A sailor in San Diego said he and his friends would give a six-month supply of blood.

Shapp called the family Wednesday when he learned about the case. He promised to find a way to keep them off welfare.

But the Marshalls were disappointed by the governor's first solution. It would have required them to recruit 36 donors a month and drive 100 miles to Philadelphia to give blood in exchange for Kevin's treatments.

Mrs. Marshall, who has been lining up donors since Kevin was five months old said that was "impossible."

"The compromise was suggested by Mrs. Marshall herself," the governor's spokesman said. "We hope we can work it out."

John Gamaldi, who was assigned to supervise the case by the health department, said the agreement with Hershey is "not official yet. He said he has a "grace period" because the boy's last supply of medicine, purchased from private funds, will last about 10 days.

Free Blood Due Hemophiliac, 12

HARRISBURG, Pa. (UPI) — Kevin Marshall, a hemophiliac whose family had been faced with going on welfare to provided treatment for him, finally is assured of free treatments.

Gov. Milton J. Shapp announced yesterday that the 12-year-old's parents need not pay for the $1,000-a-month treatments nor find donors to provide the necessary blood. In addition, the boy will be treated at the nearby Hershey Medical Center instead of 100 miles away.

The Marshalls appealed to the state after they had exhausted all private donations. They were told they must go on welfare to be eligible for help. But this meant Francis Marshall would have to quit his $7,800-a-year job and support the five members of his family on the welfare check.

Told of the Marshalls' plight, Shapp promised aid. At first he proposed enrolling Kevin in a state-federal program at Philadelphia Children's Hospital. But this meant the Marshalls would have to recruit nine people each week to give blood and drive 100 miles to Philadelphia.

Mrs. Marshall, who had been lining up blood donors since Kevin was 5 months old, said that proposal was "impossible."

Shapp then changed the order. Under the final agreement, Kevin need travel only 15 miles from his home in suburban Camp Hill to the Hershey facility for treatments.

Shapp said blood donations will be accepted by Red Cross chapters across the country and sent to Hershey.

A

Bill Seeks Financial Aid for Hemophiliacs

WASHINGTON (UPI) — Legislation to provide financial assistance for the estimated 100,000 hemophiliacs in the nation will be introduced in the next Congress by Sen. Harrison A. Williams.

Williams

The New Jersey Democrat said in making the announcement Sunday his bill would assure that every hemophiliac would be able to afford new treatment techniques.

"Recent breakthroughs by health researchers have made it possible for hemophiliacs to live relatively normal lives if they are able to avail themselves of new theraphy methods," said Williams.

Hemophiliacs have inherited a congenital deficiency of a plasma factor needed for blood to clot.

Williams, chairman of the Senate Labor and Public Welfare Committee which has jurisdiction over health matters, noted that the recently discovered clotting factor can cost between $10,000 and $20,000 a year per patient, depending on the severity of the case and the availability of healthy blood from donors.

Only two states — New Jersey and Pennsylvania — have financial-aid programs for hemophiliacs. But, Williams said, even those states cannot possibly fund the treatment needed and the federal government must step in.

B

Figure 11–4. Toward Federal Legislation. **A.** The case of Kevin Marshall as released in the press in 1972. A number of articles on this particular case appeared across the country. Notice that a governor of a large state became personally involved. **B.** This article, indicating that states could not carry the large financial responsibility and that the federal government should step in, appeared in 1973. In 1974 federal legislation was passed providing free treatment for all hemophiliacs.

clinical symptoms of duchenne muscular dystrophy

Duchenne muscular dystrophy (MD), with rare exceptions, is a childhood or adolescent affliction. In the majority of cases the first symptoms of falling and clumsiness appear within the first six years of life. The most striking feature is the enlarged, weak, and doughy-feeling calf muscle. The muscular dystrophy progresses at a rate inversely proportional to the age of onset. The earlier the first symptoms appear, the more rapid the progression; the later, the slower the progress of the muscle deterioration. Early-onset patients are usually confined to wheelchairs by age 9 or 12, as muscle and skeletal deformities develop rapidly in young patients. About 75% of these patients die by age 20, with about 5% living past age 50.

treatment of muscular dystrophy

To date there is no meaningful treatment known for any of the muscular dystrophies. Research in this area is being funded by the federal government and private foundations. Perhaps you recall the annual Labor Day Muscular Dystrophy Telethons produced by comedian Jerry Lewis. Recently, a selenoprotein that may be involved in supplying energy for the growth of heart and skeletal muscles was found to be missing in lambs that displayed a form of muscular dystrophy. This particular selenoprotein can be obtained from certain bacteria in large quantities, and this will aid in the investigation of the involvement of selenoprotein in human muscular dystrophies. A recent report (*Chemical and Engineering News,* 1976) suggests that muscular dystrophy may not be a muscle disease. A chemist at the University of Kentucky says evidence gained from studying alterations in protein and membranes of human red blood cells of dystrophic versus normal suggests that MD is a systemic disease (involves many body tissues, not just muscle tissue). The reader is referred to the article "The Muscular Dystrophies" (Zundel and Taylor, 1965) for detailed discussions of the non-X-linked dystrophies and to the publications on muscular dystrophies of the National Foundation/March of Dimes, 622 Third Avenue, New York, N.Y. 10017.

severe combined immune deficiency diseases

A brand new playroom complete with a jungle jim, swing, and scooter is a new permanent addition to the isolator system that has housed a child named David since twenty seconds after his birth on September 21, 1971 (Figure 11–5). David suffers from a sex-linked hereditary disorder known as severe combined immune deficiency disease. David has no thymus gland, lymph nodes, or tonsils. His supply of white blood cells is inadequate, and he lacks the ability to make antibodies. David is incapable of resisting disease-causing germs.

David was delivered germ-free by Baylor College of Medicine physicians who anticipated his disease through tests made on his mother. He was placed immediately in a transparent sterile isolator. Since then David has been in

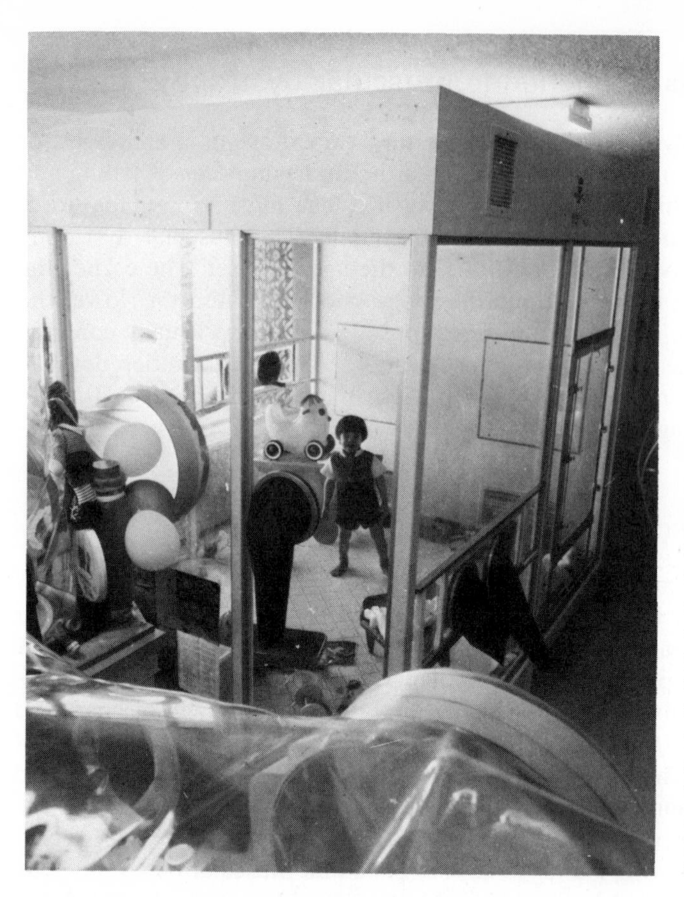

Figure 11–5. David in His Playroom. This third edition of David's isolation room measures 9 by 7 by 6 feet and has filtered air. All food, clothing, toys, etc., are sterilized before they are passed in to David. It is not known how long he will have to live in such isolation. Scientists are still looking for a means to restore his immune response system. (Photo by Bob Simmons, Baylor College of Medicine.)

three flexible plastic isolators—each one larger than its predecessor to accommodate David's increased growth and development. The plastic film of the flexible isolators is no obstacle to hearing or seeing in either direction. The total isolator system is housed at the Clinical Research Center of the Baylor-affiliated Texas Children's Hospital (*Inside Baylor Medicine,* 1974).

Air passed into the germ-free playroom is sterilized by circulation through high-efficiency filters in the ceiling of the room. At different levels in the clear playroom walls, there are two pairs of rubber gloves which the family, nurses, and doctors can use to assist and play with David (Figure 11–6). There is also a glove on either side of the porthole that connects David's smaller isolator with the playroom. These gloves are used to assist him in moving from one unit to the other. He eats and sleeps in his smaller, flexible isolator, but spends most of his walking time in the playroom. At age three David said to his father, whom he saw working in the room, "You let me out of this bubble, and I'll help you," and to his mother, "When I get out of this bubble, I'll go with you to the kitchen."

the physical basis: transmission of units of hereditary material through the genes

At an earlier age David was able to spend much of his time at home in his smaller germ-free isolator. But as a consequence of continued development, he will have to spend much more time in the playroom at the Clinical Research Center. At the present his physical and psychological development are described as normal.

Figure 11–6. David and a Physician Holding Hands. David has never been touched by an ungloved hand. His total experience of the outside world consists of viewing people through the walls of his plastic isolator and the exchange of feeling given through rubber-glove-covered hands. (Photo by Bob Simmons Baylor College of Medicine.)

David is no longer germ-free. He has accidentally acquired at least 11 different microorganisms none of which has yet to produce an infection. As we are well aware from animal studies, a completely germ-free state is not desirable. Microorganisms in the intestines are essential for digestion and production of certain vitamins. The problem is, a normally noninfective bacteria in one environment, such as the intestine, may behave quite differently outside the intestine. In general, people with normal defense mechanisms need not be concerned. David, however, will have to be continuously monitored to ensure that these or other organisms do not establish themselves throughout his system. Recently, David was able to leave the shelter in a special space suit similar to those worn by our astronauts.

X-linked immune deficiency

Severe combined immune deficiency may occur spontaneously, or, as in David's case, it can be inherited. The disorder can be inherited as an autosomal or X-linked recessive type of the disease. In David's case the disease is sex-linked; this form has a frequency of 1 in 10,000 live births. His mother is a heterozygous carrier. Thus only males in David's family are likely to have the deficiency, and with the X-linked pattern of inheritance, each male has a 50:50 chance of having the disease. Since David's mother had previously delivered an immune-deficient boy, who died when he was seven months old, Baylor physicians performed an amniocentesis (analysis of amniotic fluid; see Chapter 6) when she was five months pregnant to determine the sex of the child. To date the disease itself cannot be diagnosed prenatally. The fetus was male, and preparations were made for a sterile delivery in case the male lost the 50:50 probability of being normal. This child, of course, was David. David has a normal five-year-old sister, who has a 50:50 chance of being a carrier like her mother. When old enough, she should receive genetic counseling. (Point for class discussion: How old is old enough for counseling?)

possibility of therapy for david

As stated by the Clinical Research Center group (*Inside Baylor Medicine,* 1974):

> David's case is part of an extensive, ongoing research project to find cures for immune deficiency. The isolation is a holding pattern until effective means of treatment are available. It was at first thought David's deficiency was time-related. That is, Baylor physicians and scientists thought his immune system might develop naturally. This has not happened and David's doctors say they cannot assume it will.

One possibility is that David's immunological defect is caused by the absence of a hormone that stimulates cells to fight infection. The study of one such hormone, thymosin, is still in the experimental stages at Baylor College of Medicine's immunology laboratories.

Another possibility is that the child's defect is due to the absence of cells responsible for immunity. In that case, a transplant of these cells, which are contained in the bone marrow, might correct the defect. The problem here would be the same as in any transplant—rejection. But in this case the rejection reaction would come from the graft, not from the recipient, since only the graft would have the immunological capability for rejection.

A successful transplant may be accomplished through two possible approaches. The first requires the use of tissue from a donor who has the same tissue type as the recipient. David as yet has no donor with his same tissue type. The second requires a means of blocking the graft rejection reaction, and methods for accomplishing this are not yet available (*Inside Baylor Medicine,* 1974). In terms of the cell bone marrow transplant, it is hoped that a transplant from a normal person would provide the immune-competent cells missing in David to correct his deficiency. The statistical odds of finding a compatible donor for David are 1 in 32,000. To date a suitable donor has not been found. Even if a donor is found, there is no assurance the technique will work.

economic support

When David is in the hospital, which is one week out of every seven, he is maintained and cared for at the Clinical Research Center (CRC) in the

Baylor-affiliated Texas Children's Hospital. This research center is supported by a grant from the National Institutes of Health. Because David is cared for under a grant for the CRC, his parents are responsible financially for only his clothing, toys, and transportation to and from the hospital. The large number of well-trained research physicians and scientists collaborating on David's case are supported by individual government and private research grants.

There are a number of genetically related social questions to be asked in this particular case. Perhaps the one that is the most soul-searching is: How will this experiment end? A young child is involved. Certainly, the benefits of science are shared by all and the information gathered from this case will be very useful. But what about David? How will it end for him?

the lesch–nyhan syndrome, an X-linked genetic abnormality of purine metabolism

The Lesch–Nyhan syndrome is not being presented here because it is an X-linked recessive trait of high incidence. Rather, it is a rare disorder. It is being presented because it has some unique features, and it should provide the reader with some idea of the complexity of inherited genetic defects. This disease involves an overproduction of purines, especially guanine. This *overproduction* is associated with the *absence* of an enzyme, hypoxanthine-guanine phosphoribosyl transferase (HGPRT), which is accompanied by increased activity of a very similar enzyme, adenine phosphoribosyl transferase (APRT). As a result of this abnormal purine metabolism the patient displays certain abnormal behavior. Thus this disease is the first example of a well-defined molecular abnormality known to be directly associated with a reproducible pattern of abnormal behavior.

Clinically, all patients have been males under 10 years of age. An affected female has never been found, and the heterozygous female carrier demonstrates the level of enzyme found in the homozygous normal female.

The patients show involuntary muscle spasms, muscle distortions primarily of the limbs and digits. They are mentally retarded and show compulsive self-mutilation, biting and destroying fingers and lip tissue (Figure 11–7). Later they may show signs of uremia with damage to the kidney. The patient has difficulty swallowing and vomits frequently. According to Nyhan, none of his patients has ever been able to walk, and none has ever been successfully toilet trained (Nyhan, 1972).

The child appears to be normal during the first few months after birth, with symptoms beginning in the latter part of the first year. With the appearance of the first teeth, a dramatic behavior change takes place. The patient chews away at his lips and fingers until the primary teeth are extracted. The unusual aspect of the self-mutilation is that the patients scream in pain while they bite themselves. They are happy only when securely protected from themselves by physical restraint. Many of these children scream all night until their parents or guardians are taught how to restrain them securely in bed.

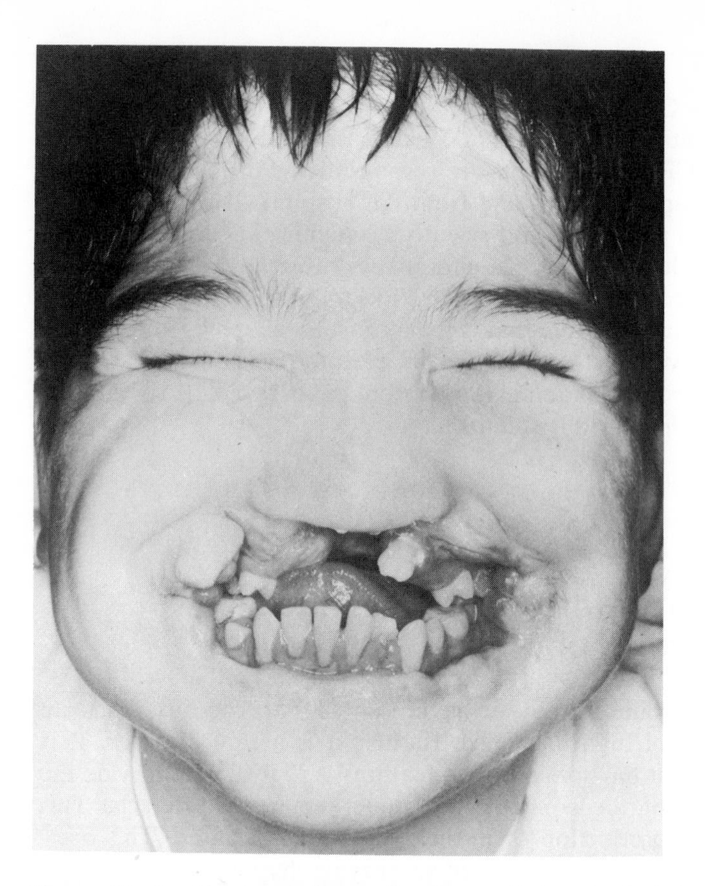

Figure 11–7. The Lesch–Nyhan Syndrome. Notice the lip-biting damage suffered by a young boy with Lesch–Nyhan disease. This rare disorder is carried and transmitted to male offspring as an X-linked recessive gene. Its clinical symptoms are mental retardation with compulsive self-mutilation behavior and life-threatening uremia with kidney damage. Scientists have now developed tests to tell whether prospective mothers carry the abnormal gene and to assess, with considerable accuracy, the genetic status of unborn children. (Courtesy of Dr. William L. Nyhan and the National Institute of General Medical Science.)

the physical basis: transmission of units of hereditary material through the genes

In contrast, they are unusually engaging children when they are restrained. They have all had good senses of humor. They smile and laugh easily. However, when protective coverings or restraints are removed, their personality changes immediately. They appear terrified. As they get older, they learn to call for help. Often while they are screaming or calling, they are already tearing at their flesh. These patients would harm others, according to Nyhan (1972), but they are restricted by their limited ability to move about. Physicians or nurses working with them are often kicked or hit, and broken eyeglasses are common around these patients. They often laugh uproariously when successful in hitting someone or knocking his glasses off. Nyhan states that the behavior of these patients is a very striking feature of the disease. He states this is the first instance in which a stereotyped pattern of human behavior and a distinct biochemical abnormality have been linked. An understanding of the mechanisms involved could contribute to an understanding of a biochemical basis of behavior and to behavioral genetics.

J. Edwin Seegmiller (1971) reported that the neurological symptoms of muscular distortion, spasticity, self-mutilation, and mental retardation can be related to the absence of HGPRT in the nervous tissue of patients with the disease. Persons with a lowered level of HGPRT sometimes demonstrate a second disease, X-linked juvenile gout, rather than the Lesch–Nyhan syndrome. The HGPRT enzyme is responsible for the formation of the purines guanine, hypoxanthine, and lesser amounts of xanthine.

A detailed study of the mutant enzyme in cell strains developed from 11 different patients with the Lesch–Nyhan syndrome has revealed at least three different phenotypes with respect to levels of enzyme activity. This study indicated that a number of different mutations can occur in the structural gene coding for the HGPRT enzyme. The different mutations result in reduced enzyme activities. At least 30 patients have been described who have reduced levels of hypoxanthine-guanine phosphoribosyl transferase activity in their erythrocytes, but have gout rather than the Lesch–Nyhan syndrome. The reasons are as yet unclear.

X-linked hydrocephaly

Hydrocephaly, or the excessive accumulation of cerebrospinal fluid within the ventricles (brain cavities), occurs in at least nine genetic syndromes. Only one of these is known to be transmitted in an X-linked manner—stenosis (blockage) of the aqueduct of Sylvius.

About 2 in every 1000 live births are in some state of hydrocephaly. As the cerebrospinal fluid accumulates, the resulting pressure leads to mental retardation, loss of motor abilities, and in most cases early death. About 30% of all cases of hydrocephaly are X-linked, with the accumulation of fluid resulting from congenital narrowing of the aqueduct or a membrane growth across the aqueduct.

D. S. Bickers and R. D. Adams (1949) reported a family in which seven male infants in two generations demonstrated massive head enlargement at birth. Subsequently, many additional patients have been reported. The head enlargement is generalized and is often present at birth. In many cases the head is so large that a drainage shunt is necessary before vaginal delivery can be completed. A few patients have had less severe hydrocephaly at birth and have lived for many years, but have been mentally retarded. In one case the author is personally aware of, a patient who lived to age 58, the patient was confined to a wheelchair at about age six and had an IQ of about 70.

possible therapy

Normally the cerebrospinal fluid helps to cushion the brain against shocks. The fluid is produced in the four hollow chambers, or ventricles, of the brain, circulates through the ventricles, around the brain and spinal cord, and is then absorbed into the bloodstream. If a passage between any of the ventricles is blocked or severely narrowed, the fluid accumulates, increasing the pressure in ventricles and damaging the surrounding brain tissue. Since the bones of an infant's skull remain soft and pliable for at least six months after birth, one of the first signs of hydrocephaly is enlargement of the baby's head. If untreated, the condition is fatal in two thirds of the cases, and in the rest retardation almost always occurs. Until recently, treatment involved inserting a flexible silicone tube (from the outside) into the largest ventricle of the brain and placing the other end of the tube in the chest cavity. The excess fluid drained off through the tube and was absorbed in the body cavity. The trouble with this procedure is that the child is totally dependent for years on the reliability of the silicone shunt. If the tube clogs, the child can become ill and/or die.

Recently, Epstein and Hochwald at New York Medical Center demonstrated a way in which hydrocephaly could be prevented (*Newsweek*, 1973). From research on cats, it was postulated that if the human infant's head is wrapped in elastic bandages, the resulting pressure will promote the absorption of the excess fluid by the capillaries, the small blood vessels of the brain. They instructed the parents of 11 infants on the wrapping procedure, and the children have been periodically checked. Treatment in each case was discontinued after six months, by which time the skull had become firm. Now at ages two to three all the children show normal neurological development and are free of symptoms.

Although this treatment appears to be quite effective when the head enlargement occurs after birth, those with heads that enlarged before birth most likely will not benefit from this technique, and so research goes on.

genes whose expression is modified by sex

This and the preceding chapters described genes with known patterns of inheritance. These genes are located in either an autosome or a sex chromosome. If they are in an autosome, males and females can expect to inherit the gene with equal frequency. That is, either sex has an equal chance of demonstrating the inheritance of the particular gene. In sex linkage a different pattern of inheritance is revealed. Certain genes, those located on the X chromosome, are inherited in a crisscross pattern. A father cannot, under ordinary circumstances, directly transmit one of his genes to his son if the gene is located on his X chromosome.

There is an additional dimension to autosomal inheritance. The expression of some genes that are known to be located on an autosome is limited or influenced by the person's sex. The expression of sex-limited genes, which are located in autosomes and present in both sexes, is limited to one sex. In sex-influenced traits the genes are also located in the autosome and found in both sexes, but one of the sexes shows a higher incidence of the trait. In either case, the genes in question are a part of every male's and every female's autosomal chromosomes; yet for reasons that have to do with the sex of the individual, their expression can be limited and influenced. Sex-limited and sex-influenced traits, by definition, do not involve genes carried on the sex chromosomes (although they could; for example, if an X-linked gene were expressed in males, but never in females, even though a female was homozygous recessive, then the expression would be limited to a single sex).

sex-limited gene expression

Examples of sex-limited genes or traits are the formation of breasts and ovaries in women, prostate in men, hypospadias, testicular failure, distribution of body hair, and the ability to form an egg versus a sperm. There are many possible sex-limited traits. However, they are difficult to assess in areas outside of those associated with sex itself. It is the expression of many sex-limited traits that probably accounts physiologically and morphologically for the formation of two sexes. The same idea holds true in other animals and some plants.

sex-influenced gene expression

There are many examples of sex-influenced traits—traits expressed in both sexes, with one sex having a higher incidence than the other. Females, for example, suffer from autoimmune disease more often than males, and males more often express baldness. Some examples of sex-influenced gene expression are given in Table 11–3. Sex-influenced genes are also known to exist in other species of animals.

In sex-limited and sex-influenced traits, both males and females carry the same autosomal genes; yet their biological sexes are different, and the different

Table 11–3 Expression of Sex-Influenced Traits

Trait	*Ratio If Known*
More common in males	*M:F*
Gout (autosomal)	8:1
Congenital pyloric stenosis	4:1
Clubfoot	2:1
Fanconi's syndrome	2:1
Phenylketonuria	2:1
Harelip	3:2
Alkaptonuria	
Baldness	
Cleft lip and palate	
Down's syndrome	
Hirschsprung's disease	
Huntington's disease (late)	
Mental retardation (genetic causes)	
More common in females	*F:M*
Congenital hip dysplasia	4:1
Rothmund–Thomson syndrome	7:3
Albright's osteodystrophy	2:1
Cretinism (nongoitrous)	2:1
Anencephaly	
Scleroderma	

sexes modify the frequency of expression of the genes. The development of a particular sex in a given environment has much to do with the expression of a given gene. Sex-limited and sex-influenced traits are excellent examples of environmental control of gene expression.

references

Bickers, D. S., and R. D. Adams, 1949. Hereditary Stenosis of the Aqueduct of Sylvius as a Cause of Congenital Hydrocephalus. *Brain*, *72*:246–62.

Chemical and Engineering News. 1976. Muscular Dystrophy May Not Be a Muscle Disease. *54*:23.

Gadpaille, J. 1972. Physiology of Maleness and Femaleness. *Archives of General Psychiatry*, *26*:193–206.

Inside Baylor Medicine. 1974. Playroom Addition Enhances Child's Germ-Free Environment, *5*(4):1, 7. Texas Medical Center, Houston, Texas.

Kasper, C. K., and M. T. Southgate. 1974. The Many Facets of Hemophilia. *Journal of the American Medical Association*, *228*:85–92.

Kisker, T. C. 1976. Modern Concepts of Hemophilia. *Southern Medical Journal*, *69*:230–32.

Newsweek, 1973. Ending the Brain Drain. July 2, p. 75.

Nyhan, W. L. 1972. Clinical Features of the Lesch–Nyhan Syndrome. *Archives of Internal Medicine*, *130*:186–92.

Rushton, W. A. H. 1969. O Say Can You See? *Psychology Today*, *3*:46–53.

Rushton, W. A. H. 1975. Visual Pigments and Color Blindness. *Scientific American*, *232*(3):64–74.

Schneider, Mary Jane. 1974. Halting Hemophilia. *The Sciences*, *14*:21–26.

Seegmiller, J. E. 1971. Genetic Abnormalities of Purine Metabolism. In Jean de Grouchy, F. J. G. Ebling, and I. W. Henderson (eds.) *4th International Congress of Human Genetics*. Excerpta Medica, Amsterdam.

Zimmerman, T. S., O. D. Ratnoff, and A. E. Powell. 1971. The Immunologic Differentiation of Classic Hemophilia (Factor VIII deficiency) and von Willebrand's Disease with Observations on Combined Deficiencies of Antihemophilic Factor and Proaccelerin (Factor Y) and on an Acquired Anticoagulant Against Antihemophilic Factor. *Journal of Clinical Investigation*, *50*:244–54.

Zundel, W. S., and F. H. Taylor. 1965. The Muscular Dystrophies. *New England Journal of Medicine*, *273*(10):537–43, (11):596–601.

Blood chemists discovered as early as the eighteenth century that blood is composed of two major components: cells and a liquid, the plasma. The cells are of three kinds: red blood cells, white blood cells, and platelets. Only gradually, however, did it become apparent that whole blood is a very complex fluid that cannot be transfused or exchanged indiscriminately between humans.

plasma, serum and immunoglobins

In a person with normal clotting ability, the blood begins to coagulate, or clot, soon after it becomes exposed to air. If freshly drawn blood is placed in a test tube, you can readily see a colorless to slightly yellow fluid called serum exude from the blood clot. The serum contains dissolved mineral solutes and approximately 7% of the blood protein (Raffel, 1961). If clotting is prevented and all the blood cells are centrifuged out, a liquid portion called plasma remains. Plasma contains the serum, various proteins, and at least 12 clotting factors such as thrombin and fibrinogen (see Chapter 11).

From plasma, albumin and globulin can be isolated. Although there is twice as much albumin as globulin, the plasma globulins contain the vast array of antibodies produced for protection against diseases. The antibodies collectively are often referred to as the humoral globulin. An antibody is made by specialized plasma cells and interacts with a specific agent, called an antigen,

Blood will tell, but often it tells too much.

Don Marquis

human blood types as defined by antigens and antibodies

which stimulates its production. An antigen is any protein the body does not recognize as its own. The protein may come from infectious organisms such as a toxin, or a carbohydrate–protein complex that challenges the body's immunological system.

Any substance, whether from a virus, bacterium, or a body cell, that acts as an antigen has a number of reactive sites, or antigenic determinants, on its surface. Some whole cells have as many as 200 different kinds of antigenic reactive sites on their surface, and others have as few as two or three.

It was once thought that for a substance to serve as an antigen these reactive sites must be foreign to a particular animal. Luckily, this belief is generally true. We now recognize, however, that under certain circumstances humans do make antibodies that react to antigenic sites on their own cells. Such autoantibodies, made against oneself, often produce tissue injury and cause autoimmune diseases. The antibody produced in response to stimulation by an antigen is very specific and interacts only with a specific antigen or one of very similar chemical composition. Each plasma cell specializes in the synthesis of one or perhaps several different antibodies. In fact, if an antigen is altered and then reintroduced into the blood system, a second type of antibody will be made to respond to the subtle change made in the antigen.

what do the different antibodies do?

As a result of the specific antibody–antigen complex, the antibody can cause antigenic molecules to clump, and in some cases it can cause whole cells containing the antigen (for example, a red blood cell) to lyze, or break open. Antibodies coat whole cells or molecules, making them more digestible to phagocytes (cells capable of engulfing and digesting invading organisms such as bacteria), and if the antigen is a toxin (a poisonous molecule), the antibody may neutralize it. Thus a number of self-preserving steps can be promoted by the production of antibodies specific to various antigens. In a discussion of antigens and antibodies, we are reminded of the old cliché about which came first, the chicken or the egg; antigens are defined as substances that induce the formation of antibodies, and antibodies are defined as substances induced in the body by antigens!

the makeup of the antibody

Although there is a seemingly endless variety of antigens, they stimulate only five general classes or kinds of antibodies: immunoglobulin G (IgG), immunoglobulin M (IgM), immunoglobulin A (IgA), immunoglobulin D (IgD), and immunoglobulin E (IgE) (Table 12–1). Of the five immunoglobulin molecules found in humans the best known, immunoglobulin G (IgG), is present in the greatest quantity.

The genes responsible for producing the five immunoglobulins are unique in that they represent a protein interaction otherwise unknown to our human

Table 12–1 Classification of Human Immunoglobulin Antibodies

Immunoglobulin	*Light Chain**	*Heavy Chain**	*Complete Antibody**	*Percent of Total*
IgG	$\kappa\lambda$	γ	$\kappa\lambda\gamma_2$	80
IgM	$\kappa\lambda$	μ	$\kappa\lambda\mu_2$	10
IgA	$\kappa\lambda$	α	$\kappa\lambda\alpha_2$	10
IgD	$\kappa\lambda$	δ	$\kappa\lambda\delta_2$	Trace amount
IgE	$\kappa\lambda$	ε	$\kappa\lambda\varepsilon_2$	Trace amount (increases slightly with age)

*κ, kappa; λ, lambda; γ, gamma; μ, mu; α, alpha; δ, delta; ε, epsilon.

gene pool. Some ten years ago a British scientist, R. R. Porter, and an American scientist, G. M. Edelman, discovered that when immunoglobulins, or antibodies, were treated with certain chemicals, they fell apart into a number of small components. For this work Porter and Edelman recently shared a Nobel prize. The separate immunoglobulin molecules are each made up of four polypeptide chains: two light (L) chains, molecular weight 20,000, and two heavy (H) chains, molecular weight between 55,000 and 75,000.

Immunoglobulin G is the only one of the five immunoglobulins that crosses the placenta of an immunized mother in significant quantity to provide the early forms of antibody protection for the newborn. That is, the newborn has in its blood IgG that was produced in the mother. By crossing the placenta into the fetal bloodstream, IgG affords the newborn a certain amount of immunity to infectious agents. But if the mother has been previously sensitized to one of five Rh antigens (discussed in detail later) and the fetal cells of her new pregnancy carry the appropriate Rh antigen, the IgG that crosses the placenta may lyze the red blood cells of the fetus, causing the disease known as erythroblastosis fetalis. If a mother is given purified anti-Rh IgG before she becomes initially sensitized by the fetal cells that are normally transferred during placental separation of the fetus, Rh sensitization does not occur. Treatment of Rh-negative mothers at each birth with anti-Rh IgG is now common practice. The current cost of a single injection of anti-Rh IgG can be from $75 to $175—a very small price for the value received in protecting future children.

production of the immunoglobulins

A condition known as hypergammaglobulinemia causes an overproduction of a given immunoglobulin. Diseases of the liver and connective tissue and certain forms of cancer (leukemia and lymphoma) cause a general increase in all five immunoglobulins.

In Waldenström's macroglobulinemia the level of IgM is raised; in the cancers myeloma G and myeloma A immunoglobulins IgG and IgA are

Table 12–2 Inherited Immunological Diseases

Disease	*Mode of Inheritance**	*Incidence of Malignancy, %*	*Type of Malignancy*
Ataxia-telangiectasia	AR	10	Lymphoma, sarcomas, Hodgkin's disease
Burton's agammaglobulinemia	X-LR	10	Lymphatic leukemia
Combined immune deficiency disease	X-LR		
DiGeorge's syndrome	F		
Immunologic amnesia	AR		
Swiss agammaglobulinemia	AR		
Thymic dysplasia (Nezelof's syndrome)	F		
Wiskott-Aldrich syndrome	X-LR	15	Lymphoreticular

*AR, autosomal recessive; X-LR, sex-linked recessive; F, familial, tendency inherited, but mode of transmission undetermined.

increased, respectively (Porter, 1968). Immunoglobulin IgE is of great clinical importance because antibodies of this class, when combined with antigens, are responsible for allergic reactions. Although the concentration of IgE in normal blood serum increases gradually during childhood, it generally remains low, except in persons with various allergies.

Conditions of hypogammaglobulinemia (insufficient globulin production) are known to be associated with certain inherited conditions such as ataxia-telangiectasia (Table 12–2).

immunological disorders

autosomal recessive

There are two known autosomal recessively inherited immunological anemias: ataxia-telangiectasia and Swiss agammaglobulinemia. Persons with immunologic anemia have normal levels of IgG, but fail to produce antibodies against various disease-causing organisms. Such people are under frequent threat of infection, eczema, and a lack of white blood cells (lymphopenia). Patients with recessively inherited ataxia-telangiectasia produce lower than normal levels of IgA. These people do not reject skin grafts in the normal time period; they suffer from progressive ataxia or brain (cerebellar) abnormalities, pulmonary infections, and cancer. Those with Swiss agammaglobulinemia suffer from an extreme lack of circulating cells and are unable to produce high levels of any particular antibody. Generally, children born with this inherited defect contract viral infections at birth, and they rarely survive one year (Guttmann, 1972).

sex-linked

There are three known X-linked inherited immunological diseases: Wiskott–Aldrich, Burton's, and the combined immune deficiency agammaglobulinemias. The Wiskott–Aldrich syndrome begins in infancy with recurring expressions of otitis media (inflammation of the middle ear), thrombocytopenia (decrease in the number of blood platelets necessary for clot formation), eczema, and recurrent infections. IgA is elevated and IgG is normal; but levels of IgM are lower than expected. In Burton's and combined immune deficiency disorders, patients lack sufficient IgG, IgA, and IgM. The lymph nodes are poorly defined or absent, and plasma cells are absent. These persons cannot be immunized against bacterial infections of any type. These defects are congenital, and the patients must receive repeated immunoglobulin injections to survive.

If depression of the immunoglobulins and lymph nodes is severe, a child may not live outside a sterile compartment, as is the case of David (see Chapter 11). David, who was born with X-linked combined immune deficiency disease, has an inadequate supply of white blood cells to fight off infection. He has no plasma cells for the production of immunoglobulins (antibodies), no lymph nodes, and no thymus gland.

hypersensitivities, the allergic diseases and immunoglobulin E (IgE)

Perhaps the best known of the immunological diseases are allergies, referred to as hypersensitivities by physicians. When an individual has been immunologically sensitized to an antigen, as in the use of a vaccine or through natural occurrence, future exposure at a later date can lead to a booster effect of the immune response resulting in tissue damage.

Some allergic or hypersensitivity reactions are immediate. For example, sensitivity to insect stings or to a drug, such as penicillin, may bring on a severe reaction that, if not promptly treated, can lead to death. Less serious or delayed allergic reactions occur in persons sensitized to grass pollens, animal danders, house dust, and so on.

Immunoglobulin E (IgE) is most important with respect to allergic reaction to various summertime antigens such as ragweed. This immunoglobulin is difficult to distinguish chemically from IgA. IgE, however, has a high content of carbohydrate and is maintained at much lower levels in the serum. In normal persons 1 in 5000 immunoglobulin molecules is IgE (Guttmann, 1972).

If one injects ragweed pollen into the skin, the person becomes locally sensitized. The reaction of antibody with ragweed pollen (antigen) results in a chain of acute physiological and pharmacological reactions. For example, as the ragweed antigen combines with IgE, a special cell, a mast cell, releases histamine. The histamine immediately dilates the blood capillaries, alters permeability of the cells of the various tissues, which in turn, causes a localized skin reaction (wheal-and-flare). If the IgE should be beneath the bronchial mucosa, adjacent to the smooth muscles in the chest area, the antigen–antibody complex with its subsequent release of histamine can cause severe and immediate bronchial constriction, stimulating an attack of asthma. General symptoms of ragweed allergy are sneezing, running nose, and tearing eyes.

The best way to escape these types of allergenic responses is to avoid the antigen, in this case ragweed pollen, or prevent its contact with immunocompetent cells, those cells capable of synthesizing IgE. Another way is through desensitization therapy. Small injections of antigens are used to immunize the patient. The allergen (antigen) elicits the synthesis of IgG antibodies that block the interaction of the allergen and the tissue-fixed IgE.

autoimmune disease in man

Certain pathological conditions in humans result in the body's immune system not being able to recognize itself or products of the body. Thus in some people, for reasons unknown, plasma cells produce antibodies that attack their own tissues. This condition is referred to as autoimmune disease. A number of the autoimmune diseases result from the formation of antibodies against various portions of one's own kidneys. There is evidence that antibodies react with self-antigens residing in the glomerulus of the kidney, causing glomerulonephritis.

There are a number of disorders suspected of being caused by autoimmune reactions. Addison's (pernicious) anemia patients often have serum antibodies that react to gastric mucosal cells. In Hashimoto's thyroiditis there are circulating antibodies specific to the thyroid tissue. Other suspect autoimmune diseases are diseases of the central nervous system, hemolytic anemia, thrombocytopenia, systemic lupus erythematosus, rheumatoid arthritis, cirrhosis of the liver, and ulcerative and granulomatous enterocolitis. The high incidence of Hashimoto's thyroiditis and of circulating thyroid antibodies in Down's and Turner's (X0) patients suggests an association between autoimmunity and chromosome aneuploidy. Autoimmune diseases are more common in females than in males, and incidence increases with age (Porter, 1968; Guttmann, 1972).

immunoglobulin M and the ABO blood groups

Blood groups A and B are considered to be carbohydrate antigens that elicit a prolonged synthesis of immunoglobulin M (IgM) antibodies. It was once believed that the A and B blood groups' antibodies were unique in that their genetic control for synthesis was established in humans without prior processes of immunization. It is now believed that we produce antibodies specific to A and B blood antigens because chemical groups in our foods (plant and animal), which closely resemble our blood groups' antigens, continually stimulate our immunological system. If this theory is correct, we are indeed fortunate that the IgG locus is not involved in the production of our anti-A and anti-B antibodies because the IgG antibodies have the ability to cross the placenta, and ABO blood type incompatibility between mother and fetus occurs more frequently than the better known Rh incompatibility. The number of pregnancies where the mother is of one blood type, say A, and her fetus is of a different blood type, AB, B, or O, is much higher than the number of pregnancies where the mother is Rh negative and the fetus is Rh positive. If the material antibodies to fetal blood type AB, B, or O were of the IgG type, the antibodies could cross the placenta and many ABO incompatible pregnancies would be in the same danger as the incompatible Rh pregnancies.

reviewing human blood types

Early beliefs and early attempts at blood transfusions were extremely dangerous for the patient. In 1874, for example, Hamlin reported that the "white corpuscles of cow's milk were converted into red blood corpuscles when milk was used in transfusing humans" (Schmidt, 1968). In 1890 Jenkins reported a human transfused with lamb's blood as treatment for typhoid fever (Powell, 1970). Aside from these curious attempts at transfusions, many deaths occurred as a result of transfusing the blood of one human into another. In fact, it took some 275 years and countless tragedies before Karl Landsteiner, in 1900, discovered and classified the "agglutinating" and "agglutinable" blood factors that accounted for transfusion problems. For this achievement Landsteiner received the 1930 Nobel prize in medicine.

Landsteiner showed that red blood cells vary with respect to specific proteins, now termed antigens, found on the cell's surface. He classified the two different kinds of antigens: antigen A on the red blood cells in type A blood and antigen B on the red blood cells in type B blood. In addition, he found that certain people have both antigens A and B (type AB blood), and others have neither antigen A nor antigen B. Blood cells with no A or B antigens were designated type O. Those persons with type A blood were shown to have a factor in the serum portion of their blood causing the agglutination or clumping of type B cells when the two types were mixed, and those with type B blood were shown to have a factor that caused the clumping of type A cells (Figure 12–1). The serum of type A blood contains antibody B, and the serum of type B blood carries antibody A. Type O blood does not contain either antigen A or antigen B, but does contain both A and B IgM antibodies (Table 12–3).

For safe administration of blood from donor to patient, it is important to determine at least the ABO blood group of each individual prior to a transfusion. It is generally accepted that the effect of the donor's antibodies, especially those of a type O donor, on the cells of the recipient is not important because the antibodies of the donor are immediately diluted during limited transfusion. The problem is that the recipient's antibodies may attack the donor cells if the blood groups are different.

Table 12–3 Blood Types and Generally Safe Transfusion Possibilities

Blood Group						*Distribution in United States, %*	
Phenotype	*Genotype*	*Has Antigens:*	*Has IgM Antibodies Against:*	*Can Give Blood to:*	*Can Receive Blood from:*	*White*	*Black*
A	I^AI^A, I^Ai	A	B	A, AB	A, O	41	27
B	I^BI^B, I^Bi	B	A	B, AB	B, O	10	21
AB	I^AI^B	A, B	None	AB	All	4	4
O	ii	None	A, B	All	O	45	48

Donors of group O can, in theory, give blood to each of the three other types, as well as to their own type, because red blood cells of blood type O do not contain A or B antigens. Type O blood is referred to as the universal donor. However, when a *large* amount of type O blood is transfused, the large number of antibodies of the type O donor might cause antigen–antibody problems. In practice is has been found best to use a donor of the same blood group as the recipient, and the universal donor is used only in an emergency. A blood antigen–antibody coagulation test is simple to run and is readily available to determine whether a given donor's blood will be compatible with the recipient's (Figure 12–1).

For Landsteiner's ABO blood types one can postulate four separate phenotypes (A, B, AB, and O) and six genotypes (I^AI^A, etc., see Table 12–3).

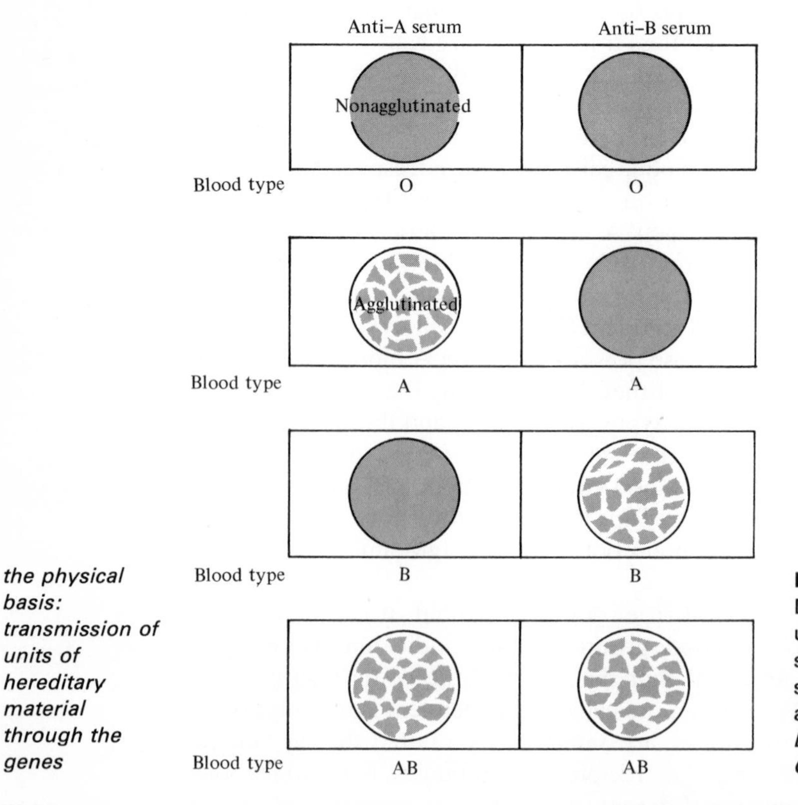

the physical basis: transmission of units of hereditary material through the genes

Figure 12–1. Blood Typing. Demonstration of blood typing using specific anti-A and anti-B serums on whole blood samples containing the AB antigens. (From G. J. Stine, *Laboratory Exercises in Genetics*, Macmillan, 1973.)

The genetics of the ABO system was worked out by von Dungren and Hirsfield in 1910, and elaborated upon by Bernstein in 1925. The ABO blood groups' antigens are controlled by multiple alleles at a single locus (multiple alleles are discussed in Chapter 8). The A and B blood group alleles are inherited without apparent dominance between the A or B genes (codominance); however, both A and B alleles are dominant to the gene for O. It should be pointed out that, on purely technical grounds, the common reference to ABO antigenic blood groups of man is incorrect, but the designated ABO blood group system is permissible. The expression *antigen blood groups* implies that blood groups are recognized by the antigens a person possesses. For example, a person of blood type A has antigen A on his red blood cells. Persons of blood type O, by definition, have neither antigen A nor antigen B on their red blood cells. Thus the system should be referred to as the AB antigen blood system.

There are other antigens on the red blood cells. One in particular, antigen H, to be discussed later, should be added to the system, making the common human blood system known as ABH blood antigen system.

An uncommon subgroup of A was discovered in 1911 and group A was divided into A_1 and A_2. More recently, other subgroups of A, A_3 and A_4, have been found. Three slightly different variants of B have also been reported. Although only the A, B, AB, and O groups are important in transfusions, subgroups, such as those of A, are relevant to certain legal problems.

the MN blood group system

The MN blood group system was discovered in 1927. This system differs from the ABO system in that no anti-MN antibodies are found in human serum. Landsteiner, who discovered the system, suggested that two alleles, *M* and *N*, determine the presence of either or both of the corresponding antigens M and N, thus yielding three genotypes and three phenotypes. Like ABO substances, M and N antigens are inherited in ordinary Mendelian fashion, neither M nor N being dominant (Carpenter, 1975).

In 1947 another antigen-specific antibody, S, was found, which ultimately proved to be associated with the M and N locus. In 1951 an allelic antigen, s, was discovered. Thus we have the following possible combinations of *M*, *N*, and *S* alleles: *Ms*, *Ns*, *MS*, and *NS*. It is possible to distinguish six phenotypes and ten genotypes of the MN system using the alleles of *M*, *N*, and *S* (Table 12–4).

The ABO and MN blood group systems are only two of at least 15 different blood group systems known in man, each one determined by a separate genetic locus. Examples of others are P, Rh (CDE), Lutheran, Kell, Duffy, Kidd, Lewis, Diego, Bg, Auberger, Xg, Dombrock, and Stoltzfus.

social utilization of blood group analysis data

Through modern analysis of the ABO groups, a variety of legal and social problems can be dealt with more accurately. Some of the problems clarified are

Table 12–4 Genotypes and Phenotypes of the MN System

Allelic Combination (Genotype)	*Antigenic Phenotype*
MS/MS, MS/Ms	M/S
Ms/Ms	M/s
MS/NS, MS/Ns, Ms/NS	MN/S
Ms/Ns	MN/s
NS/NS, NS/Ns	N/S
Ns/Ns	N/s

cases of disputed parentage when children are born out of wedlock, when there is a charge of rape, or when there is a maternity mix-up. In addition, blood analysis is helpful for identification in kidnapping cases, in cases of hit-and-run accidents, and in homicide. In these cases geneticists and medical doctors are brought before the court to give testimony on the genetic basis of the findings in question. The expert can, in most cases, present genetic blood group evidence for the exclusion of a father in a paternity suit if the "father" is of a blood group incompatible with evidence presented. The expert can reach his decision because of his knowledge of the Mendelian mode of inheritance for the ABO blood group alleles. Blood group testing can be used, however, only in the exclusion of paternity. When the three well-known blood systems (ABO, MNs, and Rh-Hr) are used together, there is better than a 62% chance of excluding a falsely accused man in a paternity suit. The use of all the 15 known blood groups would eliminate a falsely accused man more than 90% of the time. But blood group analysis can be used only to exonerate; it cannot prove fatherhood.

An interesting case was described by Gradwohl (1948) in illustrating the use of the ABO and MN blood groups in a case of disputed paternity. The father wanted blood tests carried out because one of his friends had jokingly suggested he might not be the actual father of his child. Although he stated he had no reason to doubt the faithfulness of his wife, he decided to have the tests made to ease his mind. The mother, child, and husband were tested in the usual manner for ABO and MN blood types. The results of the tests were as follows: the mother was OMN, the child was A_2N (a subtype of the A antigen), and the father was A_2BN. With this rather unusual combination of blood groups, it was very easy to state that this man could have been the father of the child, although actual paternity could not be definitely proved by these tests. Because the child possessed only the antigen N for the MN pair (which was present in the mother's blood), the father would have had to have antigen N in his blood. With a group O mother and a group A_2B father, any child would be either A_2 or B but not A_2B. Additional court cases where blood typing was used in disproving paternity are presented in Chapter 22.

With the rapid advances and discoveries in the field of blood grouping, we may look forward to a time, predicted by Landsteiner, when blood grouping will be as specifically identifying as fingerprints, but the time has not yet arrived. Medically, certain diseases are known to be correlated with certain blood groups. These are discussed in Chapter 16.

In a second example of the medicolegal aspects of human blood typing, Burns (1976) cites a court case involving an error in child identification. The identification bracelet of family 1's child was found to be missing, and family 2 discovered that their child had family 1's name tag. Confronted with the mistaken identity problem, family 1 refused to switch children. Fortunately, blood tests quickly demonstrated that neither child could have belonged to the family that had received it, and that each could have belonged to the other parents.

	Parental Blood Groups	*Blood Group of Child Taken Home*
Family 1	A×AB	O
Family 2	O×O	B

An exchange of the children then satisfied both families. Obviously, the tests did not prove that the child received by family 1 belonged to family 2, only that it could have. If the families had been A×B and O×B, and the children had been O and B, it would have been impossible to make a supportable decision, since either family could have produced either child. In this case blood would have been typed for other antigens.

In cases of illegitimacy the courts vary a great deal concerning the degree to which blood group genetics is utilized in the formation of a verdict. One of the more celebrated cases of illegitimacy occurred in California. A well-known movie actor was accused of fathering the child of a former female friend. Three qualified physicians analyzed blood tests performed on the child, its mother, and the alleged father. The results presented to the court were as follows:

1. Alleged father, O
2. Mother, A.
3. Daughter, B.

An examination of the blood group findings shows that the man was not the father of the child, since a new mutation is highly improbable. Yet the first jury became deadlocked, and a second trial resulted in his being found guilty and sentenced to pay child support for the next 21 years.

Laws of various states differ considerably with regard to the weight attached to blood tests in paternity cases. Frequently, the importance of such evidence is up to the court. In New York results from blood group tests are definite or conclusive evidence only if they establish nonpaternity. In other states blood tests constitute admissible or introduced evidence but are accorded no greater weight than other evidence in the trial. Still other states will not even permit the introduction of blood tests as evidence. According to Burns (1976):

> Some of the more realistic laws and practices concerning blood tests include those of
>
> *Alabama:* admissible only if definite exclusion is established; courts are permitted to order blood tests (statute).
> *California:* conclusive evidence (statute).

Colorado: defendant entitled to have blood tests received in evidence when exclusion is indicated (statute).
Connecticut: admissible only if definite exclusion is established (statute).
Kentucky: court authorized, at its discretion, to admit tests that show a possibility of paternity (statute).
Maine: admissible in combination with expert testimony when nonpaternity indicated (statute).
Massachusetts: admissible only if exclusion is established (statute).
Mississippi: conclusive evidence (statute).
New Hampshire: admissible if exclusion is established (statute).
New York: blood tests that demonstrate nonpaternity constitute conclusive evidence (statute).
North Carolina: admissible when presented by expert testimony (statute).
Ohio: admissible in combination with expert testimony when nonpaternity indicated (statute).
Pennsylvania: conclusive evidence if nonpaternity is established (statute).
Rhode Island: admissible if definite exclusion is established (statute).
Wisconsin: admissible if exclusion is established (statute).

Happily, the number of states in which blood tests may not even be introduced as evidence is decreasing.

the rhesus antigenic factors and hemolytic disease (anemia of the newborn)

In hemolytic disease destruction of the red blood cells generally begins in the fetus and continues in the newborn. The two important clinical features, jaundice and anemia, are both the direct results of such hemolysis. Jaundice, a yellowish discoloration of the skin and whites of the eyes, occurs as blood cells are destroyed, causing the release of an excessive amount of brown-yellow bile pigment into the blood. Anemia is the result of red cell destruction. Each year in the United States an estimated 2 to 3 million pregnant women face the hazard of becoming sensitized to the Rh blood factor, a particular antigen found in the red blood cells. Those who are sensitized are exposed to the Rh antigen in the blood of an Rh-positive fetus so that their immune system is able to produce more quickly the antibody against the Rh factor when the next exposure occurs. At current prices the cost to prevent this sensitization for all Rh-negative women would be about $150 million to $500 million. Not all Rh-negative women become sensitized by such pregnancies, but thousands do. The consequence is serious and often tragic. Once a woman becomes sensitized to the Rh factor, future pregnancies will bring about more rapid antibody formation in response to the fetal Rh+ cell. These antibodies then cross the placenta into the fetal bloodstream and in many cases result in infants with hemolytic disease. Some are stillborn; some die shortly after birth with neurological defects; and others suffer permanent brain damage. Hemolytic disease of the fetus and newborn has been occurring in 1 in every 150 to 200 pregnancies in the United States (Lin-Fu, 1969).

Hemolytic disease, a disease causing lysis of the red blood cells, is commonly referred to as erythroblastosis fetalis because the condition starts in utero and affects the erythropoietic system of the fetus, often causing the appearance of circulating erythroblasts. The clinical symptoms are jaundice, anemia, an enlargement of the liver and spleen, and massive edema or accumulation of fluids in the tissue, a condition referred to as hydrops fetalis (massive edema after birth commonly leads to death owing to heart failure).

If the infant is severely jaundiced shortly after birth, the condition is referred to as icterus gravis neonatorum. Between 5 and 10% of the Rh incompatible neonatals develop kernicterus, a form of brain damage caused by an excessive amount of bile pigment in the blood. About 70% of the infants with kernicterus die within seven days of birth. Among those who survive, many have neurological handicaps. At one time kernicterus was responsible for 10% of all cases of cerebral palsy. In less severely affected infants, congenital anemia may dominate the clinical picture. Before treatment became available, neonatal mortality was 50% among infants born alive with hemolytic desease (Lin-Fu, 1969).

Rh incompatibility

The symbol Rh is taken from the word *rhesus*, a species of monkey, *Macaca rhesus*. The antigenic factor Rh is found in the blood of the rhesus monkey as well as in humans. The Rh antigen, like the A and B antigens, is found on the red blood cells. Red blood cells carrying this antigen are Rh positive (Rh+), and red blood cells that do not carry the Rh antigen are Rh negative (Rh−) (Figure 12–2). In the United States 85% of the white and 92% of the black population is Rh positive. The Japanese, Chinese, and "pure" American Indians are 99% Rh+. As with the ABO blood system, problems occur only if a blood transfusion occurs between patients differing in Rh sensitivity. The Rh antigen of an Rh+ person will elicit the production of antibodies in an Rh− transfusion recipient. The first transfusion of Rh+ blood into an Rh− recipient usually sensitizes the individual who produces low levels of antibody without a severe reaction. Once sensitized, however, the recipient can experience a severe reaction or even death on subsequent Rh+ blood transfusions.

The same events may occur during pregnancy when the fetus inherits the gene for Rh+ blood from the father and the mother is Rh− (homozygous recessive). The Rh+ red blood cells of the fetus may cross the placenta (primarily during the process of birth) into the Rh− blood of the mother. The mother is sensitized, but is not usually severely affected by the production of the antibodies formed in her blood system against the Rh+ antigen. There may be severe effects on the Rh+ fetus, however, if not during the first pregnancy, then in subsequent Rh+ pregnancies. During the first or later pregnancies, the antibody level may increase to a point where it crosses the placental barrier in significant quantity and destroys the red blood cells of the Rh+ fetus.

Marriage between an Rh− woman and an Rh+ man is often described as Rh incompatibility, the same term that is used to describe the pregnancy of an Rh− woman with an Rh+ fetus. In an Rh incompatible mating, if the husband is homozygous for the Rh factor, that is, if he carries two Rh+ genes, all the offspring will be Rh+, but if the husband is heterozygous, or carries only one Rh+ gene, the chances are that 50% of the offspring will be Rh+ and 50% Rh−. An Rh− woman married to an Rh+ man can therefore have either Rh+

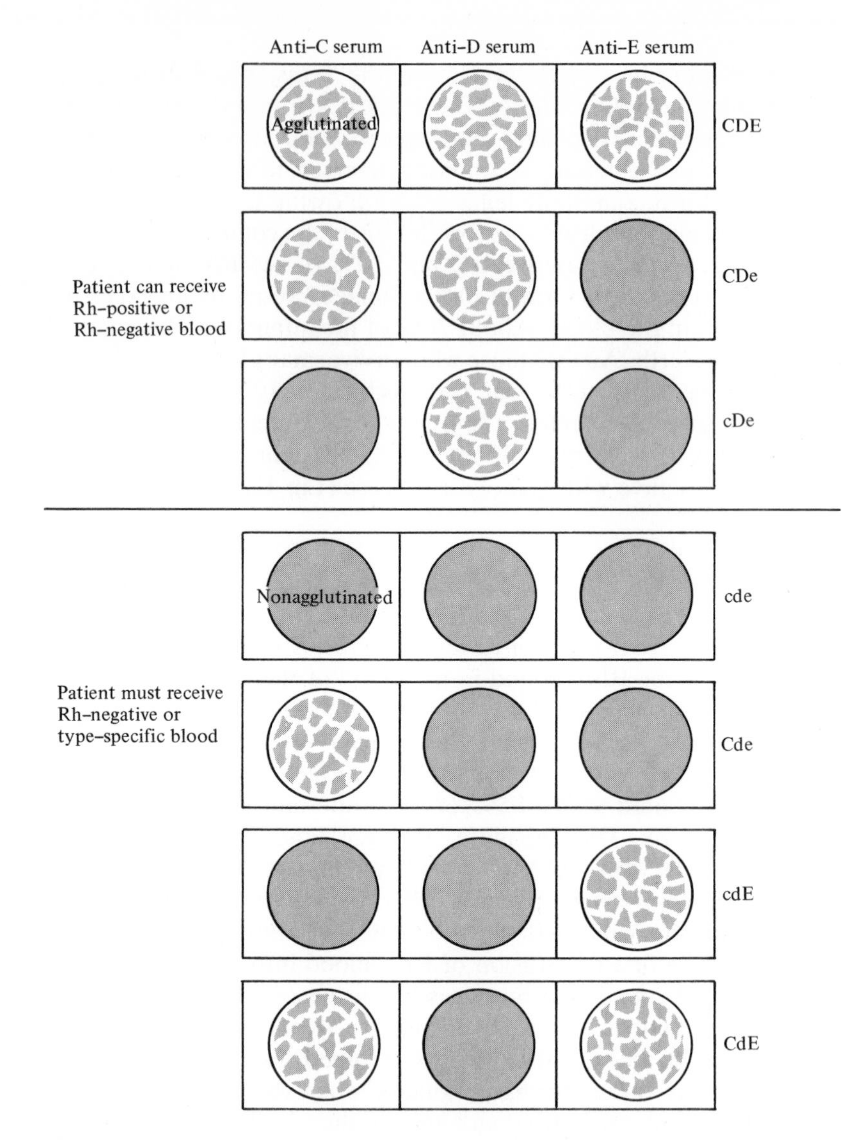

Figure 12–2. Rh Blood Typing. Rh antigens using specific Rh and anti-C, anti-D, and anti-E serums. (From G. J. Stine, *Laboratory Exercises in Genetics,* Macmillan, 1973.)

or Rh− children, depending on the zygosity of the husband. Only on becoming pregnant with an Rh+ fetus is she faced with the potential danger of the Rh sensitization. However, not all Rh− women with Rh incompatible pregnancies become sensitized. Rh incompatibility is estimated to occur in 13% of Caucasian marriages and 5% of black marriages.

the physical basis: transmission of units of hereditary material through the genes

Our knowledge of the Rh system has enabled us to solve the problem of erythroblastosis; however, the genetic and clinical terminology still remains confusing. Presently, three Rh factors are known, and they are produced from a very restricted region of the chromosome. Weiner postulates that a single gene is responsible for the production of all three Rh+ factors, R^0, R^1, R^2; that is, there is a multiple allele system, as is known to exist in the ABO system.

Table 12–5 Comparison of the Weiner and the Fisher Rh Symbolisms

Genotype Symbols		
Weiner	*Fisher*	*Simple Phenotype*
rr	*cde/cde*	Rh−
R^0r	*cDe/cde*	
R^2r	*cDE/cde*	
R^1r	*CDe/cde*	Rh+
R^1R^2	*CDe/cDE*	
R^1R^1	*CDe/CDe*	

Fisher, on the other hand, believes there are three closely linked genes, each involved in the production of a separate antigen (a multiple factor theory). He labels the three factors C, D, and E (Table 12–5 and Figure 12–2).

primary Rh sensitization during pregnancy / the placental barrier

In the Rh sensitization of the mother (often referred to as isoimmunization), the placenta plays an important role. The placenta forms at the site where the fertilized ovum is implanted and becomes a highly vascularized network of blood vessels. Depending on the length of the pregnancy, the placenta can cover one third to one half of the internal surface of the uterus under the developing embryo. An organ of both fetal and maternal origin, the placenta consists of blood vessels, vascular spaces, and small amounts of supportive tissue. The blood circulation systems of mother and fetus are entirely separate, but there is an extensive surface of contact, since the fetal vessels extend into the chorionic villi, which in turn extend into the intervillus spaces that are filled with maternal blood (Figure 12–3). The placenta's main function is to provide for the exchange of substances between mother and child. This is brought about either by simple diffusion of gases, water, and various electrolytes or by active transport of nutrients out of the maternal system into the fetal system. The possibility that red blood cells might cross this placental barrier was scarcely considered until the early 1940's.

In 1942 Levine suggested that fetal red blood cells make their way into the maternal circulation in sufficient amounts to stimulate isoimmunization of the mother (Levine, 1942). Later, Chown (1954) demonstrated the presence of fetal erythrocytes in the maternal circulation. It has now been demonstrated that 71% of all women with an Rh incompatibility have fetal red blood cells in their circulatory system after delivery (Figure 12–4). However, it is also well known that not all of these mothers are sensitized. An additional problem resulting from an incompatible Rh pregnancy is the development of jaundice and mental retardation in the newborn. During pregnancy excessive bilirubin, resulting from the lyzed red blood cells of the Rh+ fetus, is handled by the

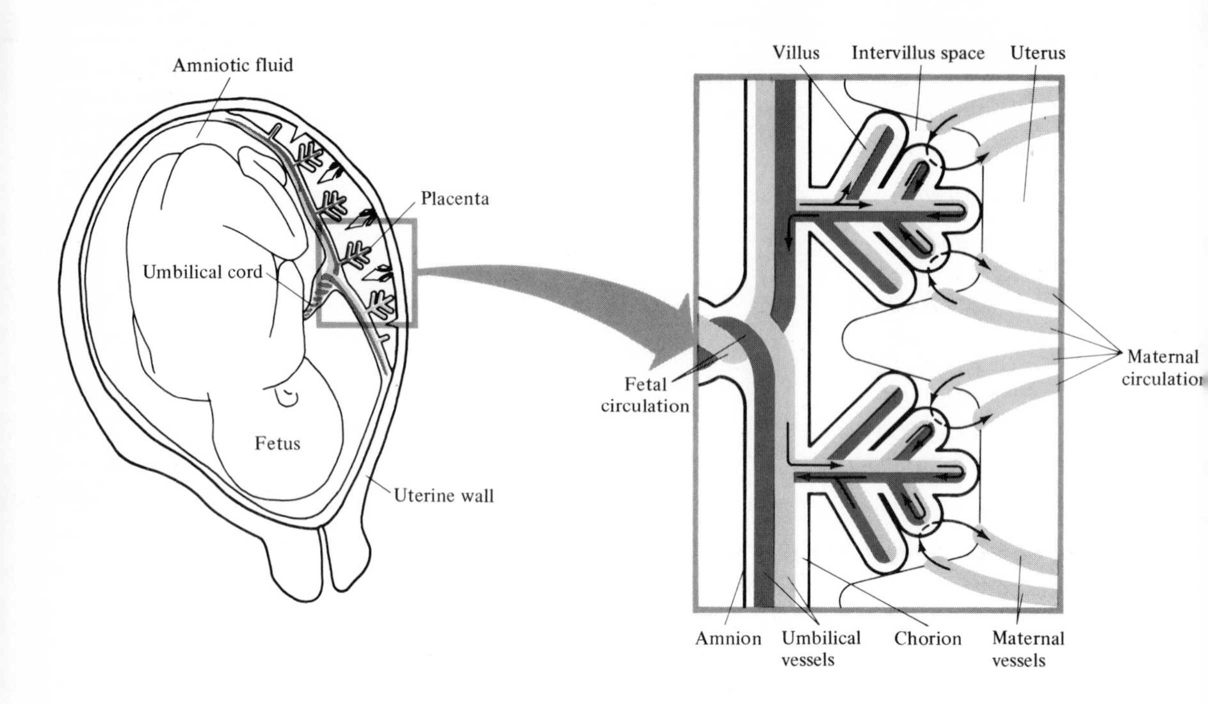

Figure 12–3. Scheme of Placental Circulation. Arrows indicate the routes of fetal and maternal circulations within the placenta. Dashed lines represent oxygen, nutrient, and waste exchange through the placental barrier. (From *Blood Groups, Antigens, and Antibodies as Applied to Hemolytic Disease of the Newborn,* © Ortho Diagnostics Inc. 1968.)

maternal circulation. At birth, however, the child's system cannot tolerate the volume of bilirubin due to cell lysis; if it is not removed, bilirubin accumulates and can cause jaundice and brain damage (Figure 12–5).

ABO blood group protection in cases of Rh incompatibility

It has been estimated that 13 of every 100 Caucasian marriages are Rh incompatible. Yet of these 13 only one or two will be plagued with the Rh hemolytic disease. The question is: Why so few? The large difference between the number of children expected to have the disease and those actually found indicates that there is a protective mechanism operating in the Rh-negative mother that, somehow, inhibits the formation of Rh antibodies.

the physical basis: transmission of units of hereditary material through the genes

One form of protection, called ABO incompatibility, is offered by the ABO blood system. In cases of ABO incompatibility, the mother and father are of different blood types, a situation similar to that of Rh incompatibility. For example, if the mother is of blood type O and the father is of type B, a natural antibody–antigen reaction is possible if the type B red blood cells of a fetus get into the maternal bloodstream. If the red blood cells carrying the antigenic B enter the maternal bloodstream, the anti-B antibodies she has will complex with and lyze the type B red blood cells. Thus, should the father be blood type

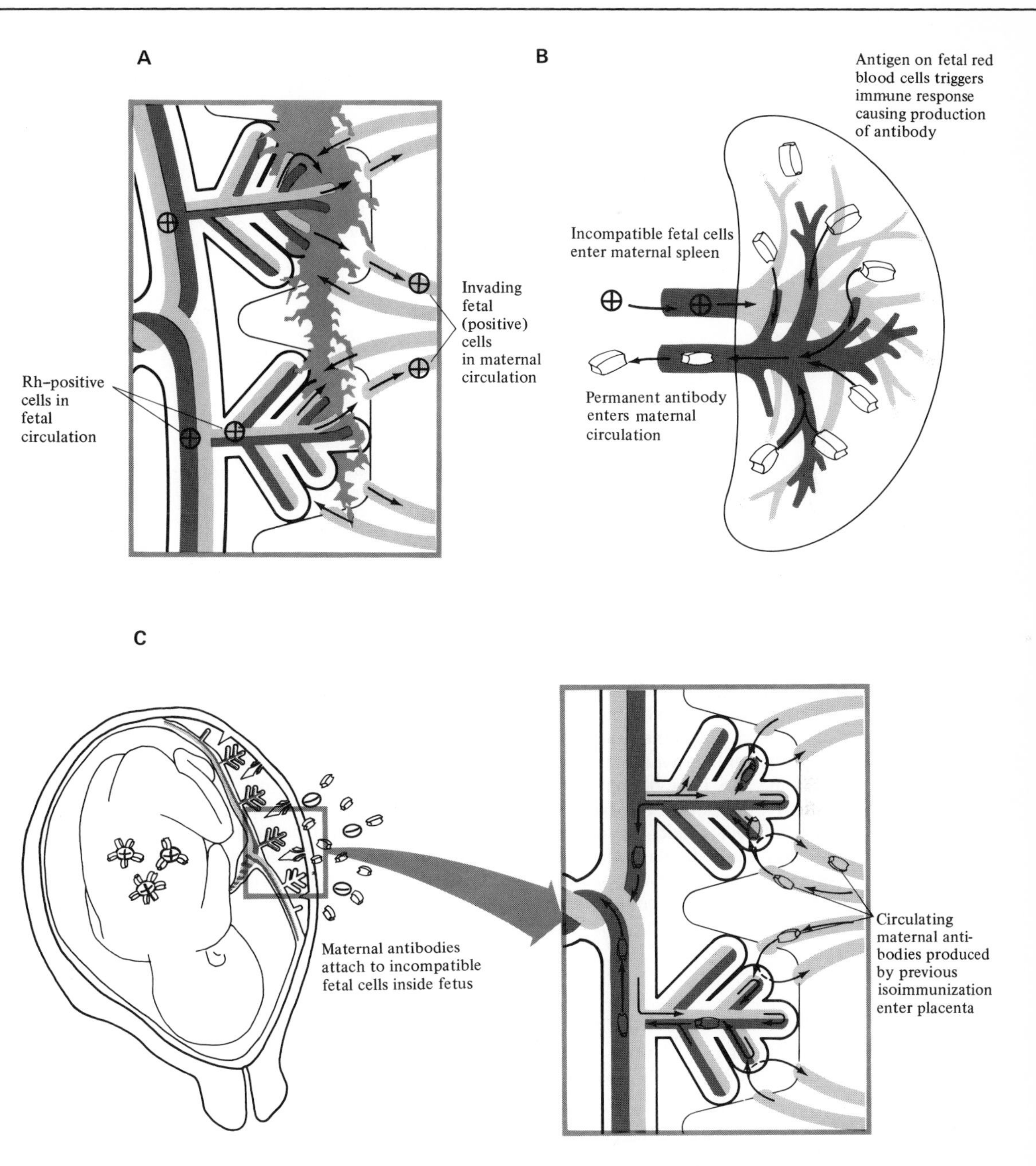

Figure 12–4. Rh Incompatibility. **A.** Separation of placenta following delivery. The placental vessels (villi) and connective tissue rupture, allowing escape of fetal blood cells. Prior to complete constriction of the open-ended maternal blood vessels, Rh+ fetal red blood cells may enter the mother's bloodstream. **B.** The mother's spleen after delivery of the Rh incompatible child. **C.** Subsequent incompatible pregnancy. Residual antibodies produced in response to red blood cells of previous incompatible fetus are transported through the placental barrier. They attach to the specific red cell antigen sites of the incompatible fetus of the current pregnancy. Sensitized cells do not have a normal life span; the baby suffers from anemia and its consequences. (From *Blood Groups, Antigens, and Antibodies as Applied to Hemolytic Disease of the Newborn,* © Ortho Diagnostics Inc. 1968.)

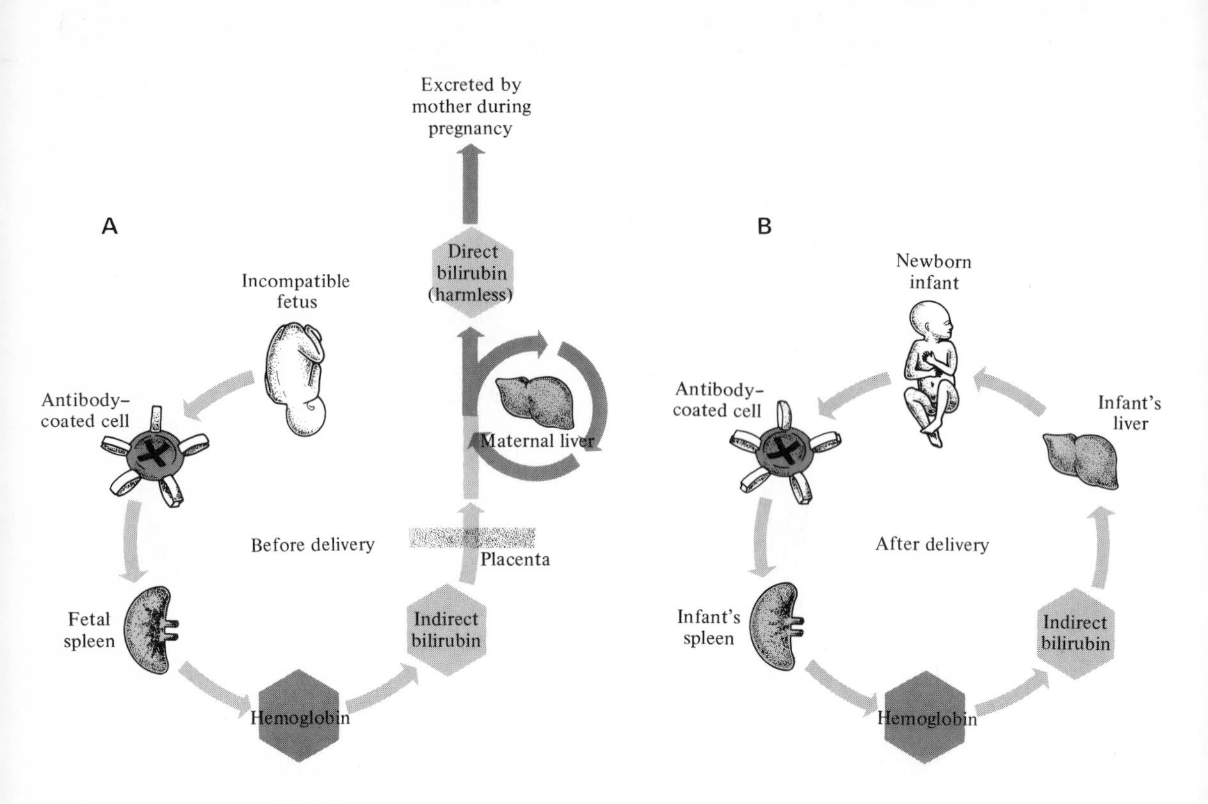

Figure 12–5. The Metabolism of Bilirubin. **A.** Before delivery jaundice in the fetus does not occur because bilirubin produced by the breakdown of cells in the fetal spleen passes via the placenta to the maternal circulation. Serum albumin transports the fetal bilirubin to the maternal liver where an enzyme (glucuronyl transferase) converts it to excretable direct bilirubin. The liver of the neonate does not produce glucuronyl transferase and cannot convert bilirubin to an excretable form. Consequently, bilirubin accumulates and if not removed will collect in tissues and cause jaundice and brain tissue damage. **B.** After delivery, the infant's liver cannot handle the amount of bilirubin and the child stays jaundiced with yellow-appearing skin and eye color. (From *Blood Groups, Antigens, and Antibodies as Applied to Hemolytic Disease of the Newborn,* © Ortho Diagnostics Inc. 1968.)

B, Rh positive, the lysis of the B cells as they enter the maternal bloodstream prevents the mother's immune system from making any anti-Rh antibodies. Children with Rh hemolytic disease usually have ABO compatible parents.

the physical basis: transmission of units of hereditary material through the genes

This form of self-protection became known when anti-A and anti-B antibodies were observed in a mother of type O blood, and the observation immediately suggested that there might be a way of copying the natural protection provided by ABO incompatibility. The mother could be given a measured amount of anti-Rh immunoglobulin after birth. This would be done with the intent of destroying the Rh incompatible fetal cells before they sensitized the mother. If the fetal cells were lyzed quickly enough, her immune system would not react to the presence of the Rh+ incompatible cells. The mother's blood system would function in each successive pregnancy as though it were still the first pregnancy with respect to the Rh incompatibility phenomena (Figure 12–6).

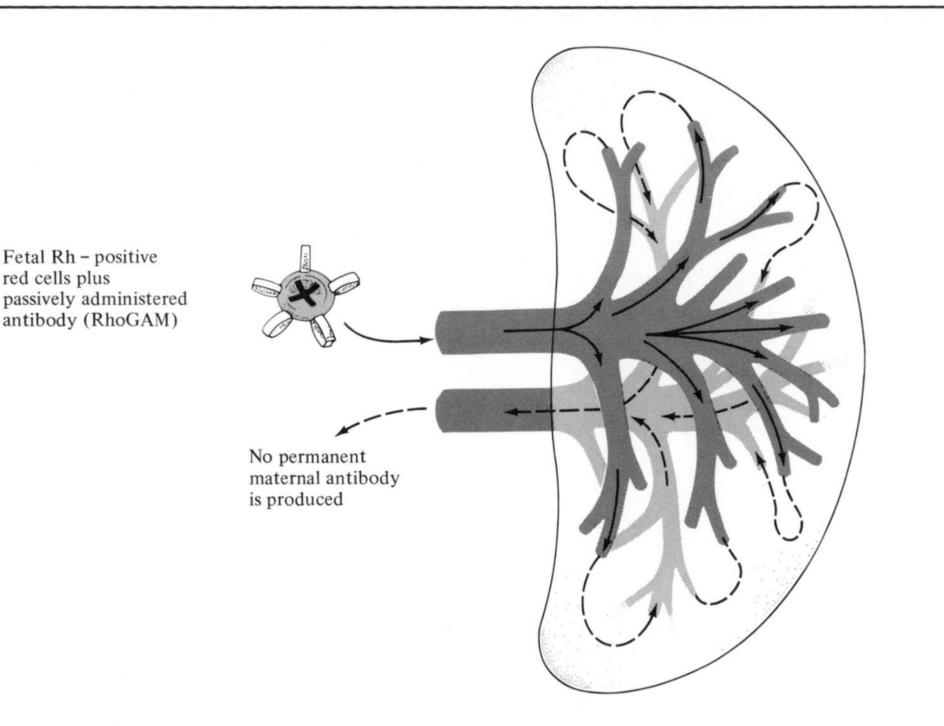

Figure 12–6. Prevention of Primary Immune Response to Rh (D) at Delivery of an Incompatible Fetus. RhoGAM Rho(D) immune globulin (human) is injected into the mother within 72 hours of delivery, and the immune globulin enters the spleen and lymph nodes. Incompatible Rh (D) positive fetal cells are not capable of initiating a primary response in the presence of adequate antibody of the same specificity. This ensures that the mother does not become sensitized to the Rh+ antigen of her child at birth. Thus her second pregnancy will appear to be her first with respect to Rh+ antigen exposure. RhoGAM should be administered after each Rh incompatible pregnancy. (From *Blood Groups, Antigens, and Antibodies as Applied to Hemolytic Disease of the Newborn,* © Ortho Diagnostics Inc. 1968.)

An immunoglobulin containing a high concentration of anti-Rh antibodies has been developed, and when it is given to the mother, it keeps her from making her own anti-Rh antibodies. In a sense the administration of anti-Rh antibody fools the maternal system. The mother does not make an antibody and therefore is not sensitized if she carried an Rh incompatible fetus. All unsensitized Rh-negative women who give birth to Rh-positive infants should receive anti-Rh immunoglobulin within 72 hours after delivery.

references

Burns, George W. 1976. *The Science of Genetics*, 3rd ed. Macmillan, New York.

Carpenter, Philip L. 1975. *Immunology and Serology*, 3rd ed. Saunders, Philadelphia.

Chown, B. 1954. Anemia from Bleeding of the Fetus into the Maternal Circulation. *The Lancet, 1*:1213.

Gradwohl, R. B. H. 1948. *Clinical Laboratory Methods and Diagnosis*, 4th ed. Mosby, St. Louis.

Guttmann, Ronald D. (ed.). 1972. *Scope Monograph on Immunology*. Upjohn Company, Kalamazoo, Mich.
Levine, P. 1942. Erythroblastosis Fetalis and Other Manifestations of Isoimmunization: Symb. in Erythroblastosis fetalis, 2nd American Congress in Obstetrics and Gynecology. St. Louis.
Lin-Fu, J. S. 1969. Rh-Negative Mothers. *Children, 16*: 23–27.
Porter, Ian H. 1968. *Heredity and Disease*. McGraw-Hill, New York.
Powell, J. B. 1970. Human Blood Antigens and Antibodies. *Carolina Tips 33*(14): 53–55.
Raffel, Sidney. 1961. *Immunity*, 2nd ed. Appleton-Century-Crofts, New York.
Schmidt, P. J. 1968. Transfusion in America in the Eighteenth and Nineteenth Centuries. *New England Journal of Medicine, 279*: 1319–20.

part 5

mutation / a source of change during the continuation of life

A genetic mutation is a sudden *change* in a genotype having no relation to one's ancestry. In other words, the change leads to a new DNA that was not present in the DNA of one's parents. Because the mutation occurs in DNA, there is the possibility that this mutant gene may be transmitted to offspring. Our concern with a genetic mutation, then, is for the effect on future generations, on what is called the *genetic load*, the number of defective genes in the human gene pool. As stated previously, each of us carries from five to eight recessive lethal genes and perhaps a much larger number of recessive genes that, when homozygous, cause mental retardation, physical malformation, and other inherited defects. The only way to avoid an increase in our genetic load is to stop mutations from occurring, but mutation continues to occur in *all* living species because *all* living species have DNA as their hereditary material, and one property of DNA is the ability to mutate spontaneously (Figure 13–1). In humans the majority of mutations are not useful and appear to cause distress. So humans as individuals would benefit most by having the number of new mutations reduced rather than increased. A point can be made for the continued benefit of mutation in that it is a source of variability. But in a world that has over 4 billion people, sufficient human variations may already be available in the human gene pool.

As humans attempt to engineer their biological evolution, they could do with fewer new mutations and the use of greater care in selecting human variants for the propagation of future generations. At the moment, however, positive eugenics, as put forth by Aristotle, Galton, and others (see Chapter 3)—the mating of *only* the "superior" individuals of society—is not feasible, for we cannot define "superior" individuals and, even more important, we do

And a truth that has lasted a million years
Is good for a million more.

Ted Olson

genetic mutations caused by environmental mutagens

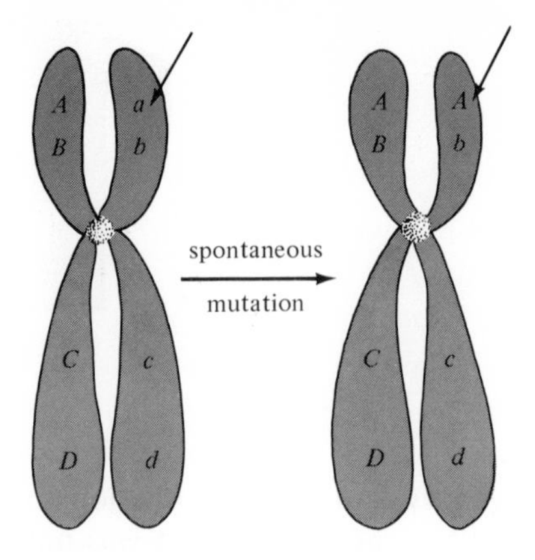

Figure 13–1. Spontaneous Mutation. Note that gene *a* in the left chromosome changed to gene *A* on the right. In this case, the original heterozygous pair of alleles (genes) *Aa* became, through spontaneous mutation, homozygous *AA*. All genes have a certain low probability of spontaneously changing from one state to another in time. This source of gene change over time is thought to be the mechanism through which evolution of various species occurs.

not have the knowledge to control gene mutation. As noted earlier, it is theorized that spontaneous mutations of DNA gave rise, over a period of billions of years, to the variety of species that have inhabited the earth. And we know that spontaneous mutations occur in the DNA of humans and other plants and animals now. Even in the mating of superior individuals, there is a *chance* that a spontaneous mutation might occur in one of the parents' gametes, and their child could be born defective.

Each of us has many thousands of genes located in the DNA of our 46 chromosomes. During a lifetime, every gene in every cell of our body has a chance of spontaneously undergoing mutation. If the mutation in the DNA occurs in body or somatic cells, the mutation affects only the individual, and when he dies that particular mutation is lost from the human gene pool. However, should there be a spontaneous mutation in the DNA of the gametes, the changed DNA can be transmitted to the next generation, and the next, and so on. What is spontaneous mutation?

spontaneous mutations

With a biological complexity unmatched by any other organism, we rely on the faithful reproduction and proper functioning of many thousands of genes to maintain our species. However, out of the many genes required for the production of a new human, just by chance, one or more genes may mutate (Figure 13–1). Such gene changes may arise out of the very nature of the complicated mechanism of gene replication. Copies of genes are formed out of a large number of nucleotides that must be lined up in just the right pattern to form a particular gene. Ideally the fidelity of DNA replication is such that a specific pattern of nucleotides forming a gene has the maximum probability of being reproduced. Sometimes, however, perhaps through the accidental jostling of nucleotides, a wrong nucleotide may be incorporated as a strand of DNA is being synthesized. The result is a spontaneous mutation—that is, the wrong nucleotide is incorporated without any intentional environmental intervention. Such mutations are thought to be the mechanism by which slow evolution

mutation

Table 13-1 Spontaneous Mutation Rates for Some Human Inherited Diseases

Inherited Trait	*Description*	*Mutations per Million Sex Cells*
Achondroplasia	Dwarfism	4–14
Aniridia	Partial or complete loss of the iris	5
Chronic anemia	Lowered levels of circulating hemoglobin	500
Huntington's disease	Degeneration of central nervous system	5
Microphthalmos	Unusually small eyes	5
Neurofibromatosis	Fibroid nodules of the skin	130–140
Retinoblastoma	Cancer of the eye	4–8
Tuberous sclerosis	Tumor-like formations in various organs and on the skin	0.1–0.3
Waardenburg's syndrome	Deafness	4

occurs. Each spontaneous mutation has, through the history of life, been subjected to the rigors of the environment. Those mutations that survived were passed on. In general, a beneficial change was one that enhanced the ability to survive in order to reproduce.

The rate of spontaneous mutation has been shown to vary from gene to gene in every organism tested, but each gene in a particular environment has its own characteristic rate of mutation. Neel (1962) analyzed data from various sources on nine different inherited diseases of humans and estimated an average mutation rate of approximately 4 mutations per 100,000 sex cells produced. This figure may appear low, but consider for the moment that each male ejaculation contains 200 to 500 million sperm. If an ejaculation contained 300 million sperm at a mutation rate of 4 per 100,000, each ejaculation would contain 12,000 mutant sperm. Of those mutant sperm (1 in every 25,000), the majority would be harmful. Examples of the kinds of spontaneous mutations and their frequency in human sperm are found in Table 13–1.

inducible genetic mutations

When a mutation is caused by the application of a mutagen, the mutation is said to have been *induced.* Radiation and chemicals are environmental mutagens used to induce mutations.

If a mutagen acts on the gametes of an organism, some of the offspring may inherit a mutant gene. The mutant gene may be so disadvantageous that death occurs before birth, and if death occurs at a very early stage of fetal growth, the pregnancy may not even be detected. If pregnancy goes to term, however, an abnormal offspring may be born, but the appearance of such an offspring is not in itself evidence that a mutation has occurred, since it may also be due to teratogenesis. Teratogenesis is the production of malformations by an environmental agent during embryonic development.

Like a spontaneous mutation, a mutagen may affect somatic or sex cells. Mutation in somatic cells may cause diseases such as cancer but such mutations, as previously stated, do not become a part of the human gene pool. One very important point to keep in mind is that although science is now able to induce mutations by using mutagens, the nature of the mutation that occurs is still a matter of chance. What the scientist does, by using a mutagen, is to increase the rate above that at which a particular mutation would have occurred spontaneously.

radiation-induced mutations

In 1927, H. J. Muller showed for the first time that the number of mutant *Drosophila* could be increased by subjecting the gametes of flies to X-radiation (X-rays). His early work with X-rays on a biological system was important for two reasons: he showed (1) that radiation could cause a mutation and (2) that he could produce a number of mutations in excess of the number of mutations that occur spontaneously. In other words, he was able to increase the frequency of mutations in a population. In 1946 Muller received the Nobel prize in medicine for his work in the field of radiation genetics.

Since the discovery of X-ray-induced mutation in *Drosophila* in 1927, it has been learned that all high ehergy or penetrating radiation is mutagenic. That is, such radiation increases the frequency of mutation in the gametes of radiation-exposed individuals. High energy radiation is radiation that has the potential to ionize atoms and molecules, the constituents of all forms of matter. Ionizing radiation converts electrically neutral atoms and molecules into particles (ions) that carry an electrical charge. When such ionization occurs the number of positive charges produced is always equal to the number of negative charges formed. High energy radiation passing through water, for example, may cause ionization in several ways, among which are

$$H_2O + \text{radiation} \rightarrow H^+ + OH^-$$

$$H_2O + \text{radiation} \rightarrow H_2O^+ + e^-$$

where the symbol e^- stands for an electron. An X-ray or gamma ray passing through a cell may cause ionization of atoms and molecules within the cell.

In genetic terms the greatest danger is ionization of atoms making up the DNA because this may result in genetic mutation. A *direct hit* on DNA has the potential of causing immediate damage. Such is the case if ionized components of the DNA do not recombine exactly as they were prior to the direct hit. There is no protection from direct hit damage. Mutation of the DNA can also be caused *indirectly* by the ionization of atoms or molecules close to but not physically connected to DNA. For example, the reactive particles formed from the ionization of a water molecule near a DNA molecule could react with an atom of DNA resulting in damage to DNA as though the DNA had received a direct hit. Some protection against indirect action is possible because diffusion of the radicals or electrons must occur before they interact with the DNA. Chemical protective agents can be used to absorb the radicals and electrons before they reach the DNA (Figure 13–2).

Ionizing radiation includes X-rays, gamma rays, high energy particles known as cosmic rays, alpha and beta rays, and neutrons. In general, X-rays, gamma rays, neutrons, and cosmic rays are the most penetrating, thus the most dangerous to life. Cosmic rays occur mostly at great altitudes, but there is concern over the risk each astronaut takes with respect to developing cancer through exposure to cosmic rays high above the earth. Alpha and beta rays, which do not have an energy of penetration equal to that of X-ray and gamma ray, can, however, be just as dangerous should they find their way inside the body via drinking water, foodstuffs, and medicines. Thus weaker radioactive substances, producing particles deep inside the body, because of their location, will have sufficient penetrating power to cause mutation, severe damage to internal tissue, and in some cases cancer. The relationship between the wavelength of an electromagnetic wave and its penetrating power and ability to cause a mutation or cancer is indicated in Table 13–2.

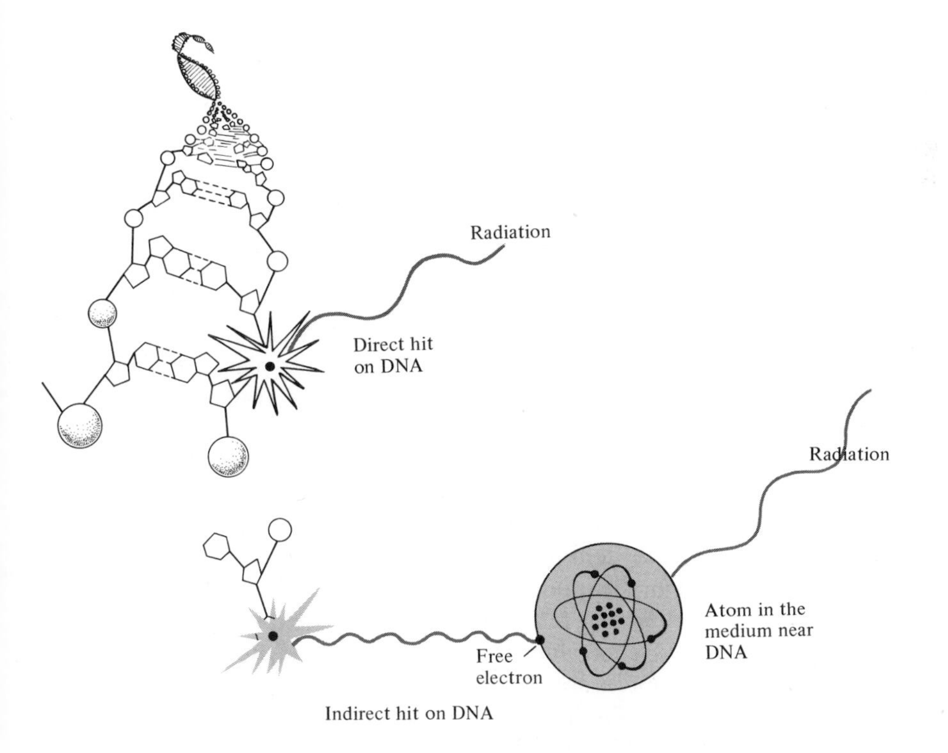

Figure 13–2. Ionizing Radiation. The effect of ionizing radiation is to dislodge an electron from its orbit around the nucleus of an atom. The electron goes off at a high rate of speed and strikes the first object in its path. Radiation can cause genetic damage by either a direct or an indirect hit of the DNA.

Table 13–2 The Spectrum of Electromagnetic Waves (The shorter the wavelength, the higher the energy)

Wavelength, nanometers		*Type of Radiation*	*Use and/or Effect*
10^{11}			
10^{10}			
10^{9}	(= 1 meter)		
10^{8}		Radio waves	Communication
10^{7}	(= 1 centimeter)		
10^{6}	(= 1 millimeter)		
10^{5}			
10^{4}		Infrared	(Heat) energy, cooking
10^{3}	(= 1 micron)		
		Visible light	Vision
10^{2}			
		Ultraviolet	Sunburn (somatic mutation), skin cancer
10^{1}			
10^{0}	(= 1 millimicron)		
		X-rays	Mutation, radiation sickness, cancer, cancer therapy
10^{-1}	(= 1 Ångstrom)		
10^{-2}			
10^{-3}			
		Gamma rays	Mutation, radiation sickness, cancer therapy
10^{-4}			

measurement of the amount of radiation

Various units are used to measure the amount or dose of radiation and the measurement of these amounts is called dosimetry. The first international unit for the measurement of radiation is the roentgen, abbreviated and represented by r, after Wilhelm Roentgen, the man who discovered X-rays. A roentgen represents the number of ions produced in air by a given amount of radiation.

To measure the amount of ionization that takes place in biological materials, we use a more convenient unit called a *rad* (radiation absorbed dose). A dose of 1 rad delivers 100 *ergs* of energy to 1 gram of material (in our case, human tissue). An erg of energy is a very small amount of energy. Perhaps an example will make clear just how little energy there is in one erg. If we drop a standard size golf ball (weight 55 grams or about 1.5 ounces) from a height of about 1 inch (2.54 centimeters) onto the skin of our arm, the *energy* from the fall of the golf ball would not be forceful enough to cause a bruise. Yet the energy delivered by the fall of that golf ball is equal to 100,000 ergs! The same 100,000 ergs is equivalent to 0.0024 calorie of heat, which is only enough to raise the temperature of 1 gram of tissue 0.0025°C, a rise in temperature much less than the increase in your body temperature that occurs after a couple of deep breaths (Frigerio, 1968).

Given that these values are not impressive with respect to the amount of potential energy available, what is the reason for the destructive effect of radiation? The destructive effect of radiation, as stated above, is the potential to ionize an atom of material and therefore change that material. If the change is in a molecule of DNA, the mutation, if not lethal, may be transmitted to future generations.

Many biological effects of ionizing radiation depend on the volume of tissue exposed. In humans whole body irradiation with a dose of 300 to 500 rads is usually fatal, but in the treatment of a malignant tumor, as much as 10,000 rads may be given to a very *small* volume of tissue without serious effects to the life of the patient.

Because humans can be exposed to a mixture of radiation particles and rays, we use a measurement of radiation called a *rem* (roentgen equivalent man). The rem is a biological, rather than physical, unit of radiation damage. A rem of radiation is that absorbed dose which produces the same effect as 1 rad of X-rays in the same tissue. By expressing the dose of radiation in terms of rems, we can compare the quantities and effects per gram of tissue, of different types of radiation to which humans are exposed.

sources of ionizing radiation

Ionizing radiation can be classified according to source, natural or artificial. The natural sources are commonly referred to as "background" radiation. It is believed that background radiation has been present since the beginning of life on earth. There is now the possibility of increasing the background radiation as a result of storing nuclear wastes and from "tailings," the residual material remaining from the mining and processing of radioactive materials. Background radiation of low intensity as a part of our natural environment produces an estimated 20,000 deaths per year in the United States by genetic damage, introduction of cancer, and other shortening of life spans. Sources of background radiation include uranium and thorium from the soil, cosmic rays from space, high energy particles from the sun, and radioactive potassium, phosphorus, and calcium atoms in the body. All of these sources contribute to the total amount of radiation absorbed by an individual during a lifetime.

The intensity of this background radiation varies with geographic location. For example, inhabitants of Harrisburg, Pennsylvania, receive only half as much natural background radiation as do people living in Denver, Colorado.

Artificial sources of ionizing radiation include all forms of nonradioactive elements that have been made radioactive (radioisotopes) for use in universities, hospitals, and industry. Also included are X-rays and gamma rays used in medicine and industry and fission products resulting from explosion of nuclear bombs (fallout) and from nuclear reactors.

Alan Emery (1971) states that radioactive fallout from explosion of nuclear weapons is less of a hazard than the radioactive elements used in medicine and research. This comparison, of course, depends on the number of nuclear weapons discharged. It does not mean that fallout is not dangerous; it simply means that each danger should be seen in perspective and that *all* sources of radioactivity should be considered with respect to their potential for damage.

genetic damage from ionizing radiation

It has been shown that high doses of radiation cause lethal and semilethal mutations, cancer, and death. With this knowledge, the government has placed some control on the amount of radiation to which humans can be exposed from medical equipment, work conditions, or release of radioactive substances into the environment. However, there are a great many sources of low level radiation still with us. For example, radioisotopes are used in medicine and dentistry as tracers to be tracked as they travel throughout the body, and in therapy for the destruction of diseased tissue. In industry there is the low level emission of radiation from the use of nuclear reactors in the production of electricity, from TV sets, microwave ovens, low voltage X-ray machines, and self-luminescent paints used on clocks and watches.

From the literature on the danger of radiation to humans, it can be concluded that the greatest radiation threat to Americans is in their routine medical and dental X-rays. For example, it has been estimated that medical X-rays are responsible for about 3300 deaths per year in the United States and may lead to as many as 46,000 deaths in future generations because of genetic damage. In contrast, statistics indicate that nuclear industries are responsible for only 18 deaths per year and may lead to 140 deaths in subsequent generations. And all evidence indicates that the more dosage a person receives, the more the damage done to his body and genes.

Prêtre (1973) states that half of all diagnostic X-ray examinations are routine chest controls and that during chest X-ray "many important organs are hit seriously while others, and particularly the gonads, are not in the beam and get only *some* scattered radiation." The point to be emphasized in the use of ionizing radiation is that *there can be no threshold dose such that the probability of serious damage is zero.* No dosage of X-rays or any other kind of radiation is completely safe. The medical use of X-ray in detection and treatment is based on a risk/benefit ratio. Although some undesirable tissue damage may result from the use of X-ray in destroying, say, a cancer tumor, without radiation the patient may have little or no chance for recovery. Thus use of radiation in medicine is one thing, but if there is no "safe" dose with respect to the amount of radiation that is required to cause a mutation, why do we continue to build nuclear reactors? Why not stop building nuclear reactors and rely on fossil fuels for generating electricity?

First we must look at a philosophy that is central to any further discussion. That is: Man could free himself of the anxiety in meeting his material needs essentially forever if he had an inexhaustible source of energy. Thus humankind faces a critical decision. If nuclear energy, which produces environmental radiation, is not used, we must rely on highly polluting and also genetically harmful fossil fuels. However, to increase our radiation exposure is (presumably) to increase our genetic load more rapidly than by burning fossil fuels. Yet, an inexhaustible source of energy is absolutely essential to the survival of humankind, where will it come from, solar energy or nuclear breeder reactors? Using only the United States uranium-238 supply, the yet untested fast breeder nuclear reactors could provide our electricity for the next 64,000 years. To be sure, this is a brief interval when we remember that humanity has already been on earth for over 2 million years. However, 64,000 years is still far better than the calculated 300 years for the United States coal reserves. Perhaps we had best get on with using *clean* inexhaustible solar energy?

If technology does not soon make the use of solar energy feasible, the demand for energy may become so great that we will have no alternative but to use nuclear reactors. We may have to trade; accept some deleterious effect in exchange for other socially desirable results. If we must have nuclear reactors, the sensible question at this time is, will we have to have an increase in low level radiation in our environment and for how long? The answer at this time is yes, with current technology there will be an increase in background radiation; for how long is uncertain, as is the ultimate damage to the human gene pool.

ultraviolet radiation

Not all forms of radiation that cause mutations produce ionization. For example, in 1934 Altenberg demonstrated the mutagenic activity of ultraviolet (UV) radiation. UV rays are of a longer wavelength than ionizing radiations and have low penetrating power because of their low energy. One can screen out UV radiation by simply holding up window glass—you won't get a suntan if there is a sheet of common glass separating you from the direct rays of the sun. The glass blocks the passage of the low energy ultraviolet radiation that causes suntan.

Because UV radiation is unable to penetrate tissue to a significant depth, it is not considered as dangerous to humans as are the ionizing radiations. However, UV radiation is of importance as a mutagenic agent in single-celled organisms (the beginning of our food chain) that are exposed to UV. The major problem of human exposure to UV at the moment includes severe sunburn and the possibility of UV-induced skin cancer. It appears that the primary mechanism of UV-induced mutations in human skin tissue is the dimerization (joining together of pairs) of pyrimidine bases in the DNA. The dimers distort, bend, or twist the double-stranded DNA molecule. This results in errors in the DNA during replication or during the cell's attempts to correct the distortion. The normal skin cell can produce an enzyme that excises the joined pyrimidine bases. In short, the cell has a repair system that excises the damaged DNA and replaces the excised region with new bases, a kind of molecular plastic surgery.

Humans who lack the UV repair process because of a genetic mutation are extraordinarily sensitive to sunlight and develop skin cancer (Figure 13-3). Sunburn occurs for those who disregard the effects of UV radiation. Skin cancer occurs in those who are genetically sensitive and expose themselves to UV radiation. Both sunburn and skin cancer are serious disorders that can be caused by UV radiation. However, individuals can protect themselves by avoiding exposure to the direct rays of the sun.

There is, however, a more severe problem facing humankind, the exposure of a more intense UV radiation because of the thinning of the ozone layer of the stratosphere due to our use of aerosol deodorants and hair sprays. Ozone, O_3, is produced by the action of ultraviolet (UV) rays on oxygen O_2. The layer of ozone in the stratosphere that surrounds the earth absorbs the most lethal

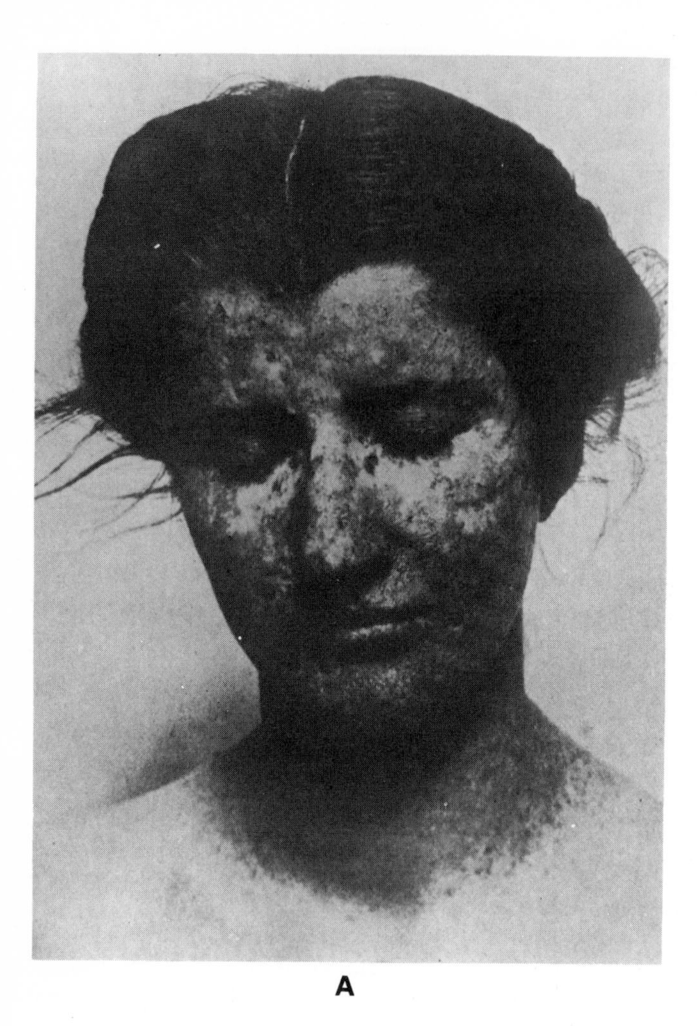

A

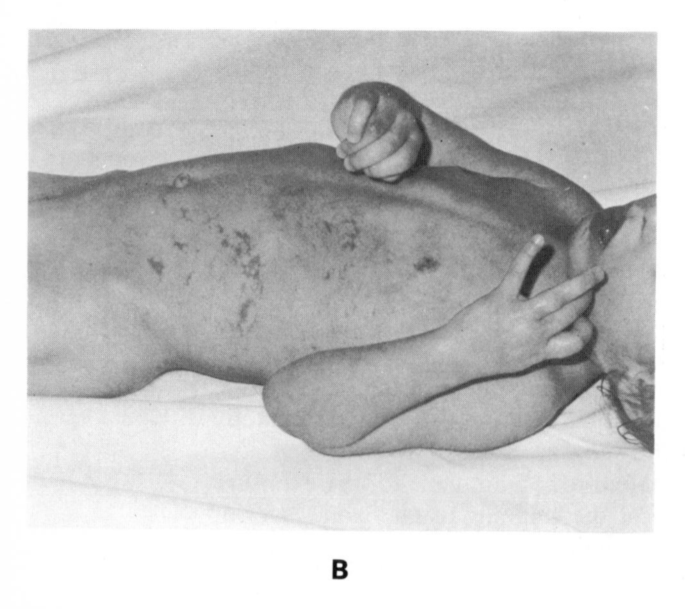

B

Figure 13–3. Xeroderma Pigmentosum, Extreme Sensitivity of Skin to Ultraviolet-Radiation-Caused Cancer. **A.** A female showing the more severe form of xeroderma pigmentosum, an autosomal recessive condition. **B.** The milder form of xeroderma pigmentosum an autosomal dominant. If the condition is expressed early, there are stunted growth, occasional deafness, and epilepsy. Biochemical tests on skin fibroblast cells indicate that a DNA repair enzyme is missing in the affected person. The absence of the repair enzyme results in sensitivity to the ultraviolet portion of sunlight, which can cause the joining of pyrimidine bases (dimers). The patients show a strong tendency to develop skin cancer. (**A** by permission of Munksgaard International Publishers Ltd., Copenhagen; **B** by permission of Charles H. Carter, M.D.)

wavelengths of ultraviolet radiation from the sun. Removal of the ozone layer would allow a narrow band of UV (between 295 nm to 320 nm) to reach the earth. The carcinogenicity or cancer-causing effect of UV radiation has been established in mice and rats. In humans, there is good evidence that UV radiation causes skin cancers such as basal and squamous cell cancer. Complete removal of the ozone layer would have drastic effects on most if not all living systems exposed. An interim report of the National Research Council, studying the effects of UV with respect to depletion of the ozone layer, states that a 10% reduction in ozone would result in about a 30% increase in the incidence of skin cancer (*Medical News*, 1976). There would be a tremendous impact on climate, and this in turn would affect all forms of plant and animal life even if such life were resistant to intense UV radiation. But we know that *much* of our plant life, as well as those microorganisms in the top few millimeters of the oceans and earth, would be directly affected by intense UV radiation.

It has been reported in numerous scientific journals and news media that the primary concern of scientists studying the earth's atmosphere is the *fluorocarbons* used in aerosol cans as the propellant for such products as hair sprays and underarm deoderants. The fluorocarbons in the aerosol cans are chemically inert; that is, they do not react with the chemical (deodorant or hair spray) in the pressurized container (*Medical World News*, 1975). But the fluorocarbon gases are carried slowly upward into the stratosphere where the UV rays decompose the inert fluorocarbons into *very* reactive chlorine atoms, which then interact with ozone (Figure 13–4). The use of fluorocarbons in the United States is enormous, about 1 billion pounds being manufactured per year for use in some 1.5 billion aerosol containers. By 1976 some 16 billion pounds of fluorocarbons had been produced. There are indications, however, that public concern is having some effect; perhaps we can reduce the use of fluorocarbons in aerosols.

Yet the fluorocarbons are not the only threat to the ozone layer. The U.S. Department of Transportation's three-year Impact Assessment Program reported that jet engine exhaust gases of the proposed supersonic transports would have a serious effect on the ozone layer. Because of such environmental considerations the United States government stopped work on building the United States version of a supersonic jet craft. However in 1976 the government approved, on a trial basis, the landing of the British and French supersonic Concorde at east coast airports.

Thus, although we do not feel that UV radiation is a severe threat to human survival now, with the depletion of the ozone layer, it may become as dangerous to life as ionizing radiation.

chemically induced mutations

The problems we now face with regard to chemical mutagenesis in many ways resemble those initially presented by radiation mutagenesis. However, as a result of over forty years of intensive investigation, the problems of radiation mutagenesis are well known and many of its dangers are understood by the public.

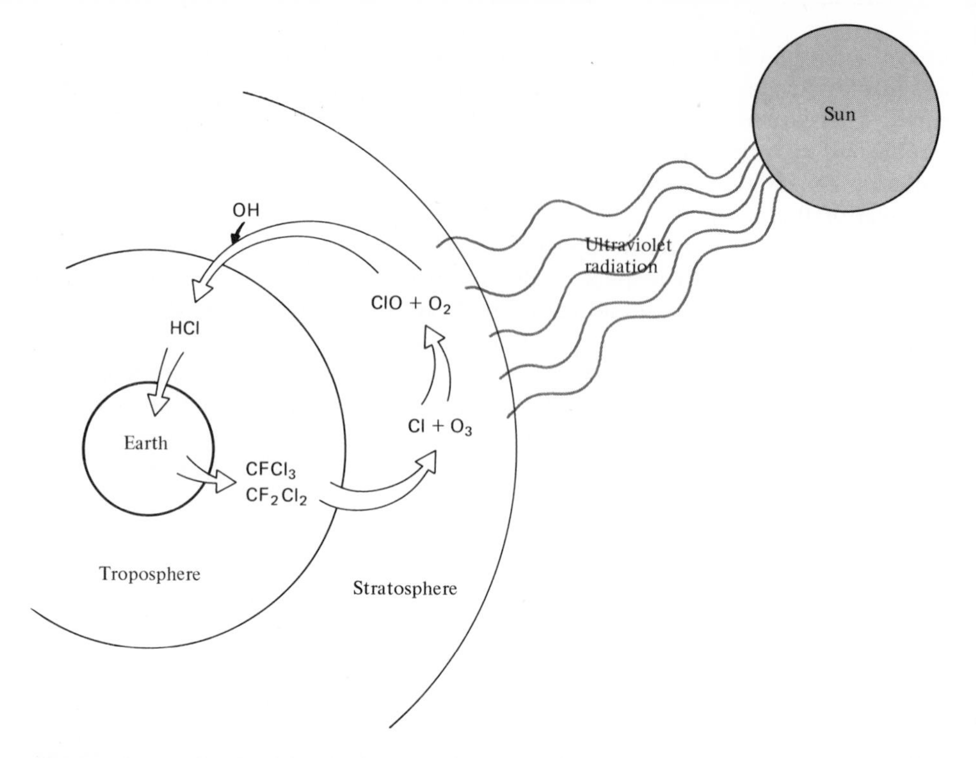

Figure 13–4. The Reaction of Fluorocarbons with Ozone. The fluorocarbons, for example, $CFCl_3$ and $CFCl_2$, are inert near the earth's surface. In the stratosphere, UV radiation causes the release of chlorine atoms. The free chlorine atoms interact with the ozone, reducing the ozone layer.

By comparison, chemical mutagenesis is a new field of study with unique problems, particularly because each new chemical compound, its use and action, poses many questions. Does the compound metabolize to a more potent mutagenic form? Or is its action weakened in the process? What is the effect of the chemical compound when it reaches the intestinal bacteria? And most serious of all, does the compound actually reach the germ plasm in a mutagenic form? Research in many laboratories leads to the conclusion that the mutagenic mechanism, especially the steps leading from ingestion or injection of the chemical compound to the resulting genetic effect, is considerably more complicated than might be expected. There are about 300 publications a month on the mutagenic action of chemicals.

how many chemical compounds are mutagenic?

The first successful use of a chemical to produce an intentional mutation was reported in 1942 (*Life Sciences Research Reports*, 1973). By the end of World War II 39 such compounds were known. Today the exact number of mutagenic chemicals in our environment is unknown, but the number of chemicals man has identified or synthesized is awesome. There are presently over 3 million registered chemicals, and many new ones are added each year. Of these about 500,000 compounds are commercially important, so restricting their use, should they be mutagenic, would be difficult.

Human populations are now exposed to a wide variety of chemical compounds never before encountered because they are newly synthesized chemicals for possible use in industry and research. Many of these compounds are clearly mutagenic to lower organisms, and there are good biological reasons to conclude that at least some are also mutagenic to humans. Because the vast majority of detectable mutations are deleterious, increasing the human mutation rate artificially by adding mutagens to our environment would be expected to be harmful in proportion to the increase.

As a result of increasing numbers of chemical mutagens entering our environment, Alexander Hollaender formed the Environmental Mutagen Society in 1969, and one result was that H. V. Malling began the Environmental Mutagen Information Center (EMIC). EMIC is located in the Biology Division, Oak Ridge National Laboratory, Oak Ridge, Tennessee. The center collects and indexes worldwide information resulting from chemical mutagenic research programs and offers the information to scientists and public information groups throughout the world.

In 1974 EMIC had records on testing of over 5000 compounds for mutagenicity. Of these, almost 300 had been shown to be mutagenic in a variety of biological systems. Mutagenicity tests have been carried out in organisms ranging in complexity from viruses to mammals, and although no chemical has been shown to be mutagenic in humans, the ubiquity of the genetic material, DNA, makes these chemicals a potential threat to humans.

A prime source of chemical mutagens which are of a direct concern to humans are the food additives and pesticides.

Among the food additives that are mutagenic in animal and microbial systems are sodium nitrite, which is used as a preservative in meat, fish, and cheese; sodium bisulfite, used as a bacterial inhibitor in wines and fruit juices and as a fruit preservative; and the artificial sweetener cyclamate. The cyclamates were banned, however, not on the basis of mutagenicity, but on the basis of positive results in carcinogenicity tests.

Animals treated with possible carcinogenic or mutagenic compounds, such as pesticides, antibiotics, tranquilizers, and fatteners, may accumulate these chemicals in their tissues, which are subsequently consumed by humans. Prescribed and over-the-counter chemicals, such as stimulants, tranquilizers, or hallucinogens, are directly consumed by humans and may be mutagenic. The number of compounds that have been shown to cause mutations in one animal test system or another is expanding rapidly. They include pesticides such as captan and maleic hydrazide; the nitrites used as chemical preservatives; the acridines, antibacterial agents; phenanthridines, treatment for trypanosomes; lysergic acid diethylamide (LSD); cyclamates; aflatoxin; aromatic hydrocarbons; and antibiotics such as actinomycin D, mitomycin C, and streptozotocin (*Life Sciences Research Reports*, 1973).

Mutagenic chemicals are used in the processing of all types of canned meats to enhance the taste and color. For example, the nitrites convert the blood or hemoglobin of the raw meat you buy into *nitrosohemoglobin*, which makes the meat look red like freshly killed flesh. It so happens that *nitroso* compounds are also highly mutagenic. It's very disquieting that the Federal Food and Drug Administration (FDA) has no standard test system for measuring the mutagenicity of the various compounds used in your foods as preservatives, etc. However, it does control the chemicals with respect to carcinogenicity.

relationships between carcinogenic and mutagenic activities of chemicals

Many geneticists and cancer researchers now believe that many or most carcinogens are potential mutagens, and vice versa. This view comes from observations that carcinogenesis and mutagenesis are grossly alike in that each *process* leads to inheritable changes in the phenotype. Carcinogens and most chemical mutagens are electron "poor"; their atoms can share electrons or bond with an electron-"rich" substance. Thus the electron-poor (electrophilic) carcinogenic and mutagenic chemicals react with electron-rich sites in DNA and protein. Although electrophilic mutagenic compounds react with DNA, it is not clear with *what* or *how* this same compound reacts to cause cancer. This is to say that although most carcinogens are potential mutagens and a number of mutagens are potential carcinogens, there is *no* simple correlation between carcinogenicity and mutagenicity (Miller, 1973). Miller and Miller (1971) pointed out that many carcinogenic chemicals need activation before they become mutagenic. Some 18 chemical carcinogens became mutagens after becoming "activated" in human liver homogenates.

As new information on the active forms of chemical carcinogens is analyzed, a better correlation between the carcinogenic and mutagenic chemicals can be expected.

comparing radiation and chemical mutagenesis

There are a number of parallels between mutations caused by radiation and those caused by chemicals. Because deoxyribonucleic acid (DNA) carries the genetic information, it follows that a mutation, regardless of cause—radiation or chemical—must involve some alteration of DNA, a change in the shape or structure of DNA or of one of its constituents. One common mutagenic effect is a submicroscopic alteration of the DNA usually involving one or a very small number of the nucleotides. A second is a gross chromosomal alteration that can be seen with the light microscope and can include such aberrations as chromosomal breaks, deletions, and translocations. Although both radiation and chemicals cause chromosome gross damage, it appears that some chemicals react with the DNA to cause mutations in a manner different from the way radiation may cause a similar mutation. Radiation causes point mutations and chromosome breakage by ionization. But the mechanism of chemical action that causes a mutation can be classified, according to the *Life Sciences Research Reports* (1973), into four categories:

(1) *Substitution by base analogs.* Some compound, such as 5-bromodeoxyuridine, is substituted by mistake for one of the four normal bases found in DNA. As a result, improper information is obtained from the DNA.

(2) *Chemical alteration of the DNA.* Some compounds are able to chemically change the structure of the nucleotides such that inaccurate information is provided by the DNA. These changes can be the result of reactions such as alkylation, deamination, or modification.

(3) *Physical binding or intercalation with the DNA.* Some compounds, such as the acridines and aromatic hydrocarbons, are known to interact with DNA in such a way as to induce structural alterations, thus leading to misinformation.

(4) *Depolymerization of DNA.* A number of chemicals are able to induce breakage in the DNA molecule. Some of the compounds that can cause alteration (see 2 above) can also cause breaks in the DNA. In addition, peroxides and free radicals can cause breakage of the DNA molecule.

In radiation-induced mutation one can easily control the dose and rate of exposure and extrapolate from one organism to another. This is not true for chemicals.

Tracing the distribution of, say, a drug and its metabolic products in complex systems is very difficult. The problem is compounded when substances that are not in themselves mutagenic can be metabolically converted into a mutagen or can combine with another factor to become mutagenic. For example, caffeine (found in your coffee and coke) is not a strong mutagen in bacteria, but in the presence of ultraviolet radiation it is highly mutagenic because it inhibits the genetic repair mechanisms. Also, dimethylnitrosoamine, which is nonmutagenic in bacteria, is metabolized into mutagenic products in mammals.

A major problem in studying radiation or chemical mutagenesis is: How can one really determine the consequence when using a source of radiation or new chemical? We know, for example, that the older criteria of chromosome breakage in somatic cells is no longer adequate. Many lymph cell chromosomes carry breaks without apparent harm. Flu shots, colds, and virus infections also result in chromosome breaks in lymphocytes. If such breaks are found to occur in the gametes, the chemical agent is usually rejected because altered genes and chromosomes in the gametes have detrimental results. In general, they cause abnormal development in the zygote or during the embryonic or fetal period of gestation.

On the basis of over 75 years of scientific investigation, it appears that an increase in the human mutation rate (by any means, radiation or chemical) would prove detrimental to humankind—both present and future generations. On the basis of such knowledge, the greatest caution must be exercised in avoiding human exposure to mutagens. Furthermore, the investigation of mechanisms, causal factors, and effects of mutations, in order that we avoid the suffering of inherited diseases, must be continued.

references

Emery, Alan E. H. 1971. *Elements of Medical Genetics*, 2nd ed. Longman, New York

Frigerio, Norman A. 1968. *Your Body and Radiation.* U.S. Atomic Energy Commission, Division of Technical Information, Oak Ridge, Tenn.

Life Sciences Research Reports. 1973. Chemical Mutagenesis and the Safety of Man. *5* (Jan.): 1–6. Stanford Research Institute, Stanford, Cal.
Medical News. 1976. Investigators Study "Long" and "Short" of Ultraviolet Radiation Effects. *Journal of the American Medical Association*, *235*: 1831.
Medical World News. 1975. Hair Sprays and the Ozone Layer. *16* (July): 59–62.
Miller, J. A. 1973. Relationships Between the Carcinogenic and Mutagenic Activities of Chemicals. *Mutation Research*, *21*(4): 195–96.
Miller, E. C., and J. A. Miller. 1971. The Mutagenicity of Chemical Carcinogens: Correlations, Problems, and Interpretations. In A. Hollaender (ed.), *Chemical Mutagens: Principles and Methods for Their Detection*, vol. 1. Plenum Press, New York.
Neel, J. V. 1962. Mutations in the Human Population. In W. J. Burdette (ed.), *Methodology in Human Genetics*. Holden-Day, San Francisco.
Prêtre, S. B. 1973. We Are Being Fooled with These "Genetically Significant" Doses from Diagnostic X-ray. *Health Physics*, *25*(2): 201.

general references

Fishbein, L., W. G. Flamm, and H. L. Falk, *Chemical Mutagens: Environmental Effects on Biological Systems*, Academic Press, New York, 1970.
Hollaender, Alexander (ed.), *Chemical Mutagens: Principles and Methods for Their Detection*, vols. 1–3, Plenum Press, New York, 1971–1973.
Sutton, H. Eldon, *Mutagenic Effects of Environmental Contaminants*, Academic Press, New York, 1972.
Vogel, F., and G. Roehrborn (eds.), *Chemical Mutagenesis in Mammals and Man*, Springer-Verlag, New York, 1970.

part 6

aspects of human genealogy

14

consanguinity: mating of persons with a common ancestor

A fact in itself is nothing. It is valuable only for the idea attached to it, or for the proof which it furnishes.

Claude Bernard

Today genetic research indicates that humans on an average carry some five to eight recessive lethal mutations. In the case of recessive genetic diseases, a defective gene from each parent is required for the expression of the defect, and children produced by related parents have a greater chance of receiving the same defective gene from both parents. In fact, there is a high probability that everyone is a carrier for many genes (has a number of recessive genes masked by normal genes) that would cause harmful effects if present in homozygous combination. Because most harmful genes originated as rare mutations, and are limited to a small percentage of the population, the chance is low that two unrelated persons who mate will be carrying the same harmful recessive genes. The chance is much greater that two closely related persons will be carrying the same harmful recessive genes because they received them from a common ancestor. In short, close inbreeding increases the percentage of homozygosity.

The human race has a long history of restriction concerning consanguinity, the mating of "blood relatives," father-daughter, mother-son, brother-sister, first cousins, second cousins, etc. Recent studies offer some scientific support for this age-old taboo. Seemanova recently reported a study that involved 161 children born to women who had become pregnant by their fathers, brothers, or sons. The same 161 women also produced 95 children fathered by men who were not related to them. The 95 children were used as a control group with which to compare the children of the consanguineous unions. Fifteen of the 161 were stillborn or died within the first year compared to five children of the control group. Sixty-four of the 161 children were physically impaired or mentally retarded compared to four in the control group. The defects included dwarfism, brain abnormalities, heart deformities, profound deafness, and kidney and intestinal disorders. These data appear to support the wisdom of strict religious and government control of consanguineous marriages (*Newsweek*, 1972).

Studies in France and Japan have suggested that the overall infant mortality rate among children whose parents were first cousins was about double that found for the children of parents who were unrelated. Serious malformations occurred in 17 of 1000 children born to parents who were first cousins or closer in blood relationships. Similar malformations occurred in 12 of 1000 children born to parents who were unrelated. Other evidence that consanguinity may lead to an increase in birth defects comes from the increase in the frequency of neural tube malformations (abnormalities involving the formation of the brain and spinal cord) among children born from consanguineous marriages. For example, in Alexandria, of the defective children born to parents who were first cousins 50% had neural tube malformations (Carter, 1967).

prohibition of consanguinity

It was recognized very early in human history that persons or families showing a mental or physical defect were likely to conceive additional defective children. Perhaps it was also recognized that there was a greater chance a malformed child would result if the child's parents were close relatives. Such observations, if they occurred, could account for the enactment of the early social laws prohibiting consanguinity. In Jewish religious law 43 kinds of marriages of relatives are forbidden; 17 are mentioned in the Pentateuch (Leviticus 18:6–18) and 26 were added in Talmud. Some of these laws are biologically sound, whereas others, primarily those affecting marriages between in-laws, have no bearing on bringing together rare genes. However, all laws and taboos prohibiting consanguinity were based on simple, although incomplete, observations that later became part of religious sanctions. The Roman Catholic Church, for example, requires a special dispensation for marriages as close as first cousins, although Orthodox Judaism, Buddhism, and the Japanese religion Shintoism do not inhibit first cousin marriages.

In the United States social restrictions make consanguineous marriages uncommon except in a few tightly knit religious groups such as the Amish in Pennsylvania and Ohio. Bruce Wallace (1972) of Cornell University states that "upon encountering someone afflicted with a rare genetic defect, it is a safe bet in most societies to predict that the individual's parents are indeed cousins." This assumption is made because most homozygous recessive conditions are rare. There is a very low probability that a husband and wife, who select each other from the general population, would both carry the same recessive genes. However, third generation relatives (first cousins), because they have a *common ancestor*, have an eighth of their genes in common; thus the chance that two recessive genes can come together in their children is increased (Figure 14–1). Children of a first cousin marriage (generation IV in Figure

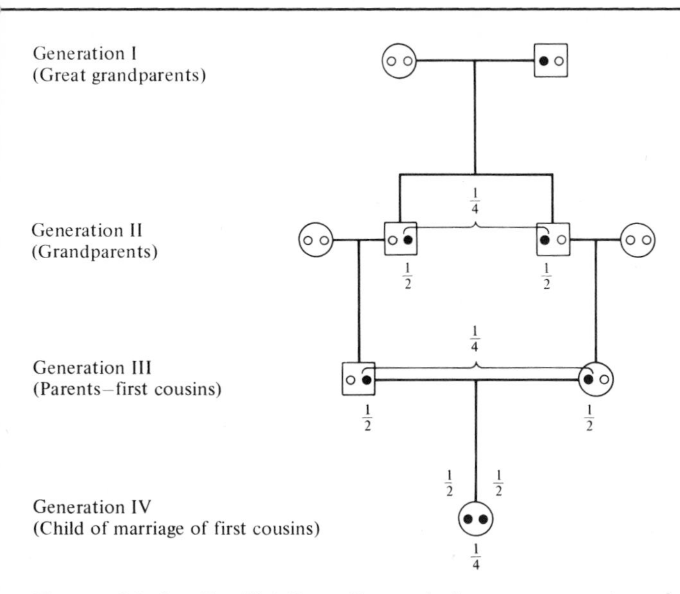

Figure 14–1. Familial Gene Transmission. • represents a defective (or recessive) gene and ○ the normal allele carried in each generation. Note that the defective gene was present in the greatgrandfather. In each generation the defective gene has a 1 in 2 chance of being transmitted from the carrier parent to the next generation. In generation II there is one chance in four that both children would receive a defective gene ($\frac{1}{2}\times\frac{1}{2}$) from generation I. Likewise, there is a 1 in 2 chance that generation II will pass a defective gene to either of the cousins and a 1 in 4 chance that both cousins in generation III have a defective gene. Should the cousins in generation III mate and have a child, the chance is 1 in 2 that the defective gene will be transmitted from one parent and there is a 1 in 4 chance that the child in generation IV will receive two defective genes, one from each parent. The probability of the child being born in generation IV as diagrammed is $\frac{1}{4}\times\frac{1}{4}\times\frac{1}{4}=\frac{1}{64}$.

14–1), because of the chance distribution of all genes in the formation of the gametes of the first cousin parents, will on an average have one sixteenth of their genes in common (I,$\frac{1}{2}$ × II,$\frac{1}{2}$ × III,$\frac{1}{2}$ × IV,$\frac{1}{2}$ = $\frac{1}{16}$).

increasing the chance of bringing family genes together

The increased chance one takes in producing a defective child by mating with a first cousin can be presented by an example. One in every 100 white persons carries the recessive gene for phenylketonuria (PKU is described in Chapter 10). Thus, every white American who marries another white American has a 1 in 100 chance of marrying a person who carries the PKU gene. Thus, 1 in every 10,000 ($\frac{1}{100}\times\frac{1}{100}$) marriages will occur between parents who are heterozygous for the PKU gene. However, if a person decides to marry a first cousin, the chance of these people being heterozygous for the PKU gene is 1 in 6400 ($\frac{1}{8}\times\frac{1}{8}\times\frac{1}{100}$). Recall that first cousins have an eighth of their genes in common. The chance that both recessive genes will come together in a child once in 6400 births is much higher than once in 10,000 births. Thus, when

Table 14–1 Increase in Recessively Inherited Disorders due to Consanguinity		
Inherited Disorder	*Description*	*Cases from Consanguineous Matings, %*
Albinism (European)	Absence of melanin pigment in hair, skin, and eyes; sensitivity to sunlight	20
Fanconi's syndrome	Retarded growth and mental development, tendency to leukemia, deafness, heart defects	20
Friedreich's ataxia	Degeneration of spinal cord, foot deformity, death before 30	12
Hereditary microcephaly	Small head, severe mental retardation, short stature	?
Leprechaunism (Donohue's syndrome)	Elfin face, wide-spaced eyes, abnormal metabolism in liver, thymus, kidney, pancreas, prostate, breasts, testes or ovaries	?
Laurence–Moon–Biedl syndrome	Mental deficiency, progressive blindness, defective tissue development, genital hypoplasia	27
Phenylketonuria	Elevated levels of phenylalanine in blood, mental deficiency (see Chs. 10, 15)	5
Progeria	Small face, baldness, small jaw, beak nose, death in childhood	?
Pyknodysostosis	Underdeveloped facial bones, blue sclera, large head	30
Sjögren–Larsson syndrome	Spastic arm and leg movement, profound mental retardation, ichthyosis, decrease in sweating, seizures	Frequent
Tay–Sachs disease	Lipid disorder, death before 5 (see Chs. 10, 15)	
Ashkenazic Hebrews		2
Non-Hebrews		40
Werdnig–Hoffman disease	Infantile muscular dystrophy	5

Inherited Disorder	Description	Cases from Consanguineous Matings, %
Wilson's disease	Improper copper metabolism, greenish ring around cornea, cirrhosis of liver, progressive tremors, progressive dementia-paranoia, early adolescent death	50
Xeroderma pigmentosum	Growth retardation, inability to repair UV damage in DNA, skin cancer	20

14*/ *consanguinity: mating of persons with a common ancestor

parents produce a child with a rare disorder like PKU, we know that there is about an even chance that consanguinity was involved.

The medical literature lists a number of recessive disorders where consanguinity has resulted in a significant number of defective children. For example, 20% of all cases of xeroderma pigmentosum, 30% of the cases of pyknodysostosis, 20% of the cases of Fanconi's syndrome, 27% of Laurence–Moon–Biedl syndrome (Figure 14–2), 20% of European albinos, and 40% of white Gentiles

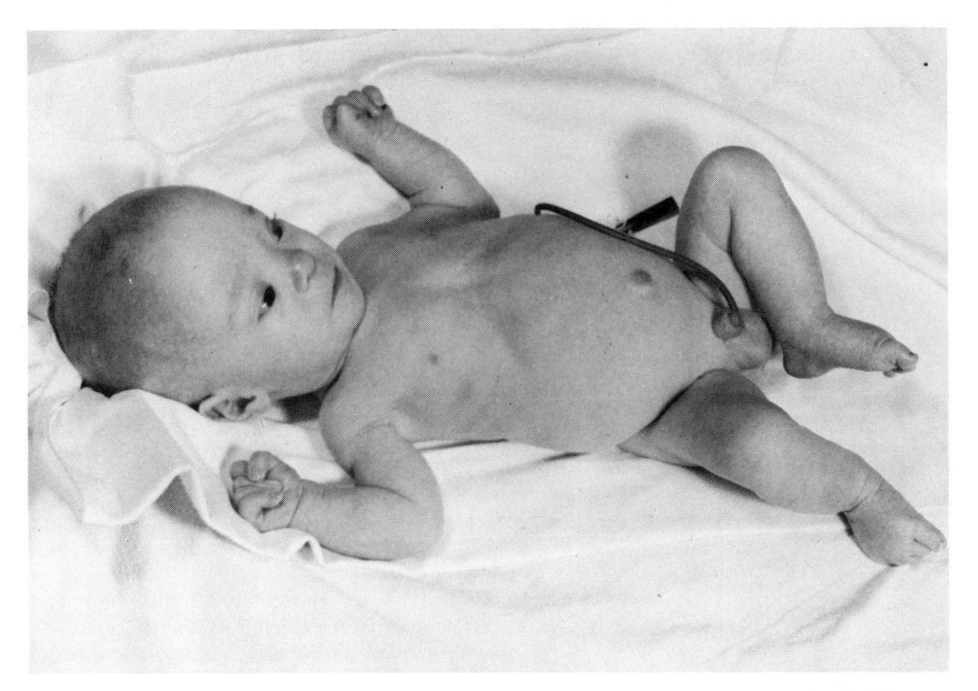

Figure 14–2. Laurence–Moon–Biedl Syndrome. Patients usually have retinal degeneration. Most cases show digital abnormalities such as extra fingers or toes. They are short, generally fat, and have subnormal intelligence. (By permission of Charles H. Carter M.D.)

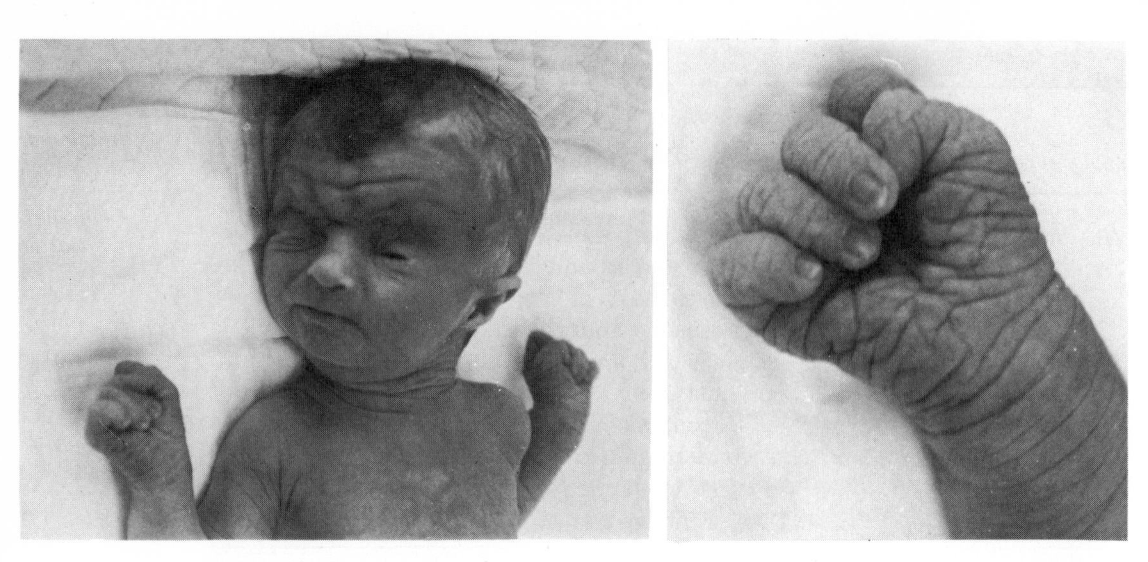

Figure 14–3. Progeria. Usual features are congenital absence of hair and increasingly brittle nails, weight loss, stiffening joints, stunted growth, and severely wrinkled skin. By age five atherosclerosis begins. The average life expectancy is 14 years. (By permission of Charles H. Carter, M.D.)

with Tay–Sachs disease resulted from consanguineous matings (Table 14–1). Other inherited diseases that occur with greater frequency from consanguineous matings than from nonconsanguineous matings are progeria (Figure 14–3), hereditary microcephaly (Figure 14–4), and osteogenesis imperfecta (Figure 14–5). The fact that the parents of Toulouse-Lautrec were first cousins is one of the leading arguments for the belief that the French painter suffered from

Figure 14–4. Hereditary Microcephaly. The affected person usually has a small head with a sloping forehead, large ears and nose, and less than normal intelligence. The condition is inherited as a recessive trait. (By permission of Charles H. Carter, M.D.)

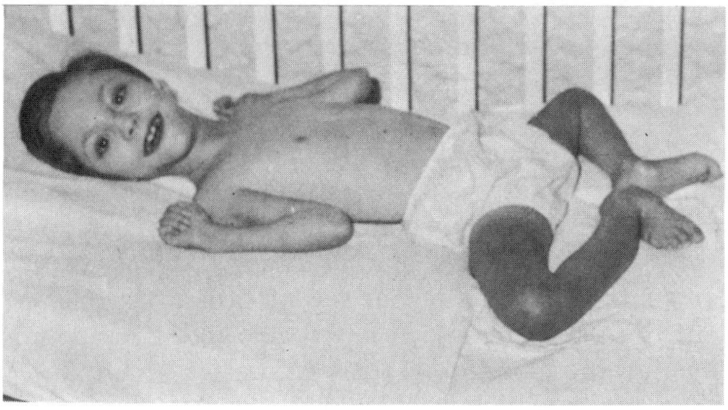

Figure 14–5. Osteogenesis Imperfecta. Thought to be a collagen defect that affects the length and width of the leg and arm bones. The person has short stature due to a marked bowing of the leg bones. The bones are brittle and often malformed. Osteogenesis imperfecta can be inherited as a recessive or dominant gene. (By permission of Charles H. Carter, M.D.)

pyknodysostosis, a rare autosomal recessive, rather than, as had been widely believed, achondroplasia.

A study reported by Hammond and Jackson (1958) is an outstanding example of consanguineous parents increasing the chance of recessive genes coming together in their offspring. There were 14 cases of a recessive progressive muscular dystrophy born to seven couples, six of which were "consanguineous." The incidence for this particular dystrophy in the general population is 38 per 1 million live births.

In general, a rare and unexpected disorder in a child or consanguineous parents strongly suggests autosomal recessive inheritance. First, a dominant trait is more easily detected because one of the parents or grandparents is usually affected; second, sex-linkage can be ruled out because of its special pattern of inheritance; third, recessive defects have a relatively low frequency of occurrence because one must carry both defective genes for the trait to be expressed. Even within a family lineage, many generations can go by without expression of a recessive trait. This would not be the case if in that family lineage the defect were dominant.

If a defective gene is located on the X chromosome (see Chapter 11 for a discussion of sex-linkage), consanguinity will have no effect on its incidence in males. Males get the gene only from the mother, so it does not matter how many common ancestors the mother or father have. The idea that hemophilia was prevalent in the royal houses of Europe as a result of "inbreeding" cannot be verified, and the X-chromosome pattern of inheritance does not support it.

general health of children born to consanguineous marriages

It has been established that an inherited defect has a greater chance of appearing among offspring produced by the mating of close relatives, providing a defective gene is being carried in the family lineage. But, does this mean that all consanguineous marriages or matings will produce noticeably abnormal children? The answer is a definite no. We cannot simply state that inbreeding is good or bad. We only know that such marriages are less desirable if defective genes are known to be carried in the ancestral lineage and the partners insist on having children. In other words, because there is an increased probability of having a defective offspring over the normally accepted chance that accompanies having children, we tend to say such marriages are bad. However, some of these marriages may not produce children! Further, many instances can be cited where the offspring of consanguineous marriages were brilliant or in some manner exceptionally talented. In these cases, it is probable that consanguineous marriages brought together the more "desirable" genes.

We should ask, however, if there is a general tendency for children of consanguineous marriages to have less than equal health when compared to children born to nonconsanguineous parents. Schull and Neel (1965) reported their results on a study of Japanese children born to cousin and noncousin marriages. The offspring of cousin marriages showed an increase of about 15% in early death and a higher frequency of major and minor defects. The age at which they first walked or talked was delayed, physically they were a bit smaller, and their school grades were somewhat lower. These defects are so small that measurements on hundreds of children were necessary to demonstrate them conclusively. Nevertheless, they amount in most cases to a decrement of 1 to 3%.

Collectively, then, there is an increased probability that consanguineous matings will result in an inherited disorder in the offspring as well as affect the general health of the nondefective child. It would appear that those persons who have held firm on the "evils" of consanguineous marriages throughout history have been vindicated.

references

Carter, C. O. 1967. Congenital Malformations. *WHO Chronicle, 21*:287–92.

Hammond, D. T., and C. E. Jackson. 1958. Consanguinity in a Midwestern United States Isolate. *American Journal of Human Genetics, 10*:61–63.

Newsweek, 1972. Children of Incest. October 9, p. 58.

Schull, William J., and James V. Neel. 1965. *The Effects of Inbreeding on Japanese Children.* Harper & Row, New York.

Wallace, Bruce. 1972. *Disease, Sex, Communication, Behavior. Essays in Social Biology*, vol. 3. Prentice-Hall, Englewood Cliffs, N.J.

It must be emphasized that this discussion is to illustrate and explain diseases and conditions prevalent in various racial and ethnic groups. *It is not intended in any way to reinforce certain prejudices that are already a part of our culture.*

During the past few decades many Americans and our judicial system have tried to shed racial and ethnic prejudice. Socially, the intent is to create equality of political and governmental treatment of individuals. However, it must be recognized that each of us is born with a unique set of genes that makes each one, first, a member of the human species and, second *biologically unique.* But, if each of us is a unique individual, different from all others, how can we be divided into racial or ethnic groups? Further, how can the knowledge of our racial or ethnic heritage be of value? The former question can be answered by defining *race* and *ethnic group.* The latter question can be answered by discussing the various inherited race and ethnic group diseases.

The identification of a specific race and ethnic inheritance can expedite the proper diagnosis of a variety of inherited diseases and lead to the proper therapy. Further, great waste in manpower, money, and time can be avoided in the mass genetic screening programs currently developing in the United States (see Chapter 18), by realizing that certain racial and ethnic groups have extremely low frequencies for certain inherited disorders and high frequencies for others. For example, the Negroid race has a very low incidence of cystic fibrosis—on the order of 1 case in 17,000 births—compared to 1 in 1000 or 1500 births in the Caucasoid race. Conversely, the Negroid race has a very high frequency for sickle cell anemia, 1 in every 400 live births, while the incidence in the Caucasoid race, as a whole, is so low that the combined data do not yet allow for a meaningful calculation.

Many of our inherited diseases, including Marfan's syndrome, albinism,

After all there is but one race—humanity.

George Moore

inherited race and ethnic group disorders

growth hormone deficiency, and phocomelia, are without apparent racial or ethnic associations. These diseases can be found with about equal frequency in all races and ethnic groups. Then there are certain diseases like Tay–Sachs, sickle cell anemia, and tuberculosis that appear to have a higher frequency of occurrence within a particular racial stock or ethnic group.

what is a race?

The term *race* itself is frequently misused and misunderstood. Persons who speak languages with common linguistic roots like Latin or Aryan, who share common religious beliefs like Jews or Catholics, who are of the same national origins like Chinese or Russians are often mistakenly thought of as races. However, the cultural characteristics that people share have little, if anything, to do with "race." Scientifically, race pertains solely to biological qualities, not social or political ones.

A race is a statistical aggregate of persons who share a composite of genetically transmissible physical traits (Rose, 1964). Physical anthropologists are in general agreement that any listing of racial features should include such external characteristics as skin pigmentation, head form, facial features, stature, and the color, distribution, and texture of body hair (Haring, 1957). Estimates of racial types range from three—Caucasoid, Mongoloid, and Negroid—to more than thirty. In this chapter, the simplest and most widely accepted division of three major racial types is used—Caucasoid (white), Negroid (Black), and Mongoloid (Yellow).

Thus, the major races of humankind all share "the" human gene pool, but because individuals generally mate within their group, maintain the mutations that have occurred within their group. For example, all humans have genes for the production of melanin, which, depending on the amount of melanin produced, provides different shades of skin pigmentation from white to black. Although it is not known whether the skin pigmentation genes *originally* allowed for the production of dense pigmentation (black) or light pigmentation (white) or some intermediate shade, it can be assumed that in a group with densely pigmented skin the genes transmitted from parent to offspring maintained a population of dark-skinned people. Persons of a separate group whose genes for skin pigmentation did not allow for dense pigmentation were lighter skinned, and they transmitted genes that produced less pigmentation. Similarly, a variety of genes common to all humans, after undergoing changes and transmission in a particular group, cumulatively gave rise to the different races.

Those people possessing light skins, long narrow noses, thin lips, and a profusion of body hair ranging in color from black to blond and in texture from straight to curly are referred to as Caucasoids. Of European origin, they have migrated to the New World, South Africa, Australia, and New Zealand.

Those persons with dark brown complexions, broad noses, prognathic jaws, somewhat everted lips, and black hair which may be straight, wavy, curly, or kinky, are called Negroids. Their original habitats are the African continent and the islands of Oceania. Negroids have been transplanted to the United States, Europe, and Caribbean islands, and South America.

Mongoloids have light brown or reddish brown skin, straight black hair, paunchy cheeks, and noses that often appear to lack a pronounced bridge. Some also have upper eyelids that extend downward toward the nose, giving the eyes an almond-shaped appearance. Mongoloids are the aboriginal inhabitants of Asia and the Americas. Included in this group are those known as Orientals, the Eskimos, and the American Indians (Rose, 1964).

Although we have basically defined the three races that will be discussed in this chapter, there are few people living in the United States who possess *all* the traits prescribed for assignment to a particuar race category. In racially heterogeneous societies, such as our own, miscegenation (or race mixing) has largely eliminated *pure* races, and it becomes increasingly difficult to "place" individuals with precision. Nevertheless, individuals *are* placed into various racial groupings.

what is an ethnic group?

An ethnic group is a group whose members share a similar *social* ancestry. For example, in the Caucasoid race the Polish, Italians, Armenians, Hungarians, Hebrews, and WASPs (White Anglo-Saxon Protestant Americans) each have a unique social heritage and form separate ethnic groups.

Over long periods of time, because of geographical selection or social restrictions on marriage between ethnic groups (for example, religious restrictions against Hebrews marrying non-Hebrews or Catholics marrying non-Catholics, a rare gene mutation that arose in a member of an ethnic group tended to remain within that ethnic group. Thus, an inherited defect over a period of time would tend to have a higher frequency in a closely knit ethnic group than in the general population. A classic example of just such an event appears to have occurred in the Ashkenazic Hebrews. The Ashkenazim have a relatively high incidence of Tay–Sachs disease, whereas the Sephardic Hebrews have the same low incidence as other white ethnic groups. Many examples of inherited diseases and their incidence within different races and ethnic groups are presented in the tables, where the ethnic groups and related ethnic-associated diseases are listed and briefly described. Several socially important ethnically related diseases are discussed in greater detail.

the caucasoids, the white race

The Caucasoid or white race is characterized by Goldsby (1971) as having pale white to brown skin. The hair may be yellow (blond), red, brown, or black. The lips are usually thin, while the nose is high and narrow to broad and snub. Males often have heavy beards and abundant body hair. Frequently balding of the head occurs early, and grayness may begin in the thirties or forties. Body build varies greatly from those very frail and thin, to the muscular, and the obese. The disorders most commonly associated with the Caucasoid race and ethnic groups are given in Table 15–1.

Table 15–1 Genetic Disorders Most Often Associated with the Caucasoid or White Race

Disorder	*Clinical Features*	*Mode of Inheritance**
Hebrews in General		
Niemann–Pick disease‡	Defect in lipid metabolism; enlarged liver and spleen	AD, AR
Usher's syndrome	Defect in ear causes deafness; later in life retinal degeneration occurs	AR
Ashkenazic Hebrews (Central, Eastern, Western Europe; North, South America; Australia)		
Buerger's disease	Blockage of arteries and veins of limbs, especially legs; usually affects males between 15 and 50	PG
Congenital pyloric stenosis	Poor development; death follows if narrowing of pyloric opening is not corrected	PG
Diabetes mellitus	Failure of (beta) cells to secrete enough insulin (juvenile), inhibited insulin action (adult); elevated sugar levels in body	Unclear†
Dystonia musculorum deformans (torsion dystonia)	Bizarre posture; dementia follows	F
Factor XI (PTA) deficiency	Deficiency of blood factor plasma antecedent necessary for proper clotting	F
Familial dysautonomia (Riley–Day syndrome)	Drooling; excessive sweating; subnormal intelligence; tastebud-less tongue; deterioration of autonomic nervous system; insensitivity to corneal pain, burns, and pain caused by sharp objects	
Gaucher's disease	Enlargement of liver, lymph nodes, and spleen; severe psychomotor impairment; swelling of body; liver damage	AR
Hypercholesterolemia (type II lipoproteinemia)	Raised levels of cholesterol in plasma	AD
Hyperuricemia (gout)	Abnormal amount of uric acid in blood	AD
Kaposi's sarcoma	Cancer	PG
Leukemia	Rapid and abnormal growth of leukocytes in blood-forming organs; immature leukocytes in circulation	PG

Disorder	*Clinical Features*	*Mode of Inheritance**
Pemphigus vulgaris	Blisters develop, suddenly heal without scarring	PG
Pentosuria	Deficiency of xylulose dehydrogenase; xylulose (a pentrose sugar) found in urine	AR
Polycythemia vera	Increased number of red blood cells; weakness; vertigo; enlarged spleen; redness and pain of extremities	PG
Spongy degeneration of brain	Brain degeneration	F
Stub thumbs		F
Tay–Sachs disease (amaurotic idiocy)	Cherry-red spot in eye macula; after 6–9 months rapid degeneration of vision and motor skills due to accumulation of lipids; death at about 2–4 years	AR
Ulcerative colitis and regional enteritis	Ulcer formation in colon	PG
Wilson's disease	Improper copper metabolism, greenish ring around cornea, cirrhosis of liver, progressive tremors, progressive dementia-paranoia, early adolescent death	AR
Sephardic Hebrews (Mediterranean Sea area, parts of Turkey, USA)		
Cystic disease of lung	Cysts in lungs	PG
G-6-PD deficiency, Mediterranean type	Quantitative hemolysis of red blood cells following administration of barbiturates and some antimalaria drugs or ingestion of fava beans	X-LR
Oriental Hebrews (Asia Minor, Iraq, Iran, Yemen, Aden)		
Dubin–Johnson syndrome	Chronic benign intermittent jaundice; leads to liver malfunction and later renal failure	AD
Bloom's syndrome‡	Red blotches over face; increased incidence of leukemia; sensitivity to light; narrow head; diminished immunoglobulin	AR
Rendu–Osler–Weber syndrome (hereditary hemorrhagic telangiectasia)	Facial features similar to Bloom's syndrome, but diseased blood vessels more prone to rupture; cirrhosis of liver common	AD

(continued)

Table 15–1 continued

Disorder	Clinical Features	Mode of Inheritance*
Mediterranean peoples—Greeks, Italians, Armenians		
G-6-PD deficiency, Mediterranean type	(See above)	
Thalassemia major (Cooley's anemia)	Markedly insufficient beta chain hemoglobin synthesis; premature death in many cases	AR
Northern Europeans		
Lactose intolerance	Inability to digest lactose	AD
Pernicious anemia	Progressive decrease in red blood cells; muscular weakness; gastrointestinal and neural disturbances	PG
Irish		
Major central nervous malformations:		
Anencephaly	Absence of part or all of brain	PG
Encephalocele	Protrusion of brain through cranial fissure	
Phenylketonuria (PKU)	Excess phenylalanine in blood; severe mental deterioration; eczema in 33% of patients	AR
Amish and Icelanders		
Ellis–van Creveld syndrome	Disproportionate dwarfism; appendages short, body normal; polydactyly; about half have congenital heart defects	AR
Icelanders		
Glaucoma	Fluid pressure in eyeball leads to blindness	AD, AR, F
Eskimos		
Deafness		PG
Kushokwin	Bone deformities; protein deficiency	AR
Otitis media	Inflamed condition of ear	PG
Salivary gland tumors		PG
Whites in General		
Alkaptonuria	Incomplete metabolism of amino acids tyrosine and phenylalanine, presence in urine shown by dark color when alkalinated; pigmentation of cartilage; dark eye spot	AR

Disorder	Clinical Features	Mode of Inheritance*
Anencephaly	Absence of part or all of brain	PG
Breast cancer		PG
Congenital erythropoietic porphyria	Skin lesions in infancy; skin rash; pink stain in teeth; pain; headache; weakness; not sensitive to barbiturates	AR
Coproporphyria	General weakness; sensitivity to barbiturates; excessive coproporphyrin in feces	AD
Cystic fibrosis	Respiratory difficulties, thick mucus; reduced life expectancy (see Ch. 10)	AR
Erythropoietic protoporphyria	Skin sensitive to sunlight; excessive protoporphyrin in red blood cells, plasma, and feces	AD
Lactose intolerance	(See above)	
Meningomyelocele	Hernia of spinal cord and membranes	PG
Oculocutaneous albinism	Subnormal vision; skin and eyes sensitive to sun; tendency to develop skin cancer	AR
Porphyria, acute intermittent	Recurrent attacks of abdominal pain, vomiting; personality changes; excessive aminolevulinic acid in urine; sensitivity to barbiturates	AD
Porphyria variegata	Abdominal pain; sensitivity to barbiturates, vomiting; muscle weakness; personality changes	AD
Rh-negative blood type	Hemolytic disease in Rh-sensitive newborn	AR
Schizophrenia	Disintegration of personality	PG
Spina bifida	Failure of spinal column to close	PG
Tay–Sachs disease, "O" variant	Form of Tay–Sachs disease in non-Hebrews	AR

*AD, autosomal dominant; AR, autosomal recessive; F, familial, but inheritance pattern unclear; PG, polygenic; X-LR mod, sex-linked recessive trait modified by autosomal genes.
†Has been reported as AD, AR, F, PG, and incomplete penetrance (see CH. 10).
‡Bloom's syndrome and Niemann–Pick disease have highest incidence among the Askenazim; incomplete penetrance means the gene is present but is not expressed.

hebrew ethnic groups

It appears that the Hebrews show a higher frequency for a larger number of known inherited diseases than do other Caucasoid ethnic groups (Table 15–1). For example, Table 15–1 lists 26 diseases that are inherited with a higher frequency in people of Hebrew background than non-Hebrew stock. Even within the Hebrew ethnic group, there are two subgroups, the Sephardic and Ashkenazic, each with its own separate frequencies for the same inherited diseases. Of the 26 diseases listed for the Hebrew ethnic group in Table 15–1, 21 have their highest frequency among the Ashkenazim.

The hereditary disorders affecting the different Hebrew tribes are among the most outstanding examples of ethnic-group inherited diseases. With religious regulations that prohibited Hebrews from marrying outside their religion, the Hebrew gene pool became concentrated (Sheba, 1970) even before the dispersal from Israel. Later, as in the case of the Hebrews living in Central Europe, government regulations prohibited Hebrew–Christian marriages. Collectively, various imposed restrictions on mating resulted in limiting the gene pool of the isolated groups of Hebrews for 50 to 100 generations. During this prolonged period of ethnic isolation, inheritable diseases may have occurred by very rare mutation and perhaps, with reduced selection, been perpetuated in the gene pool of the separate isolated Hebrew groups. If, for example, a new very rare defect arose among the Ashkenazim, it would have been isolated because they were isolated and restricted by social tradition and law. The frequency of a given newly formed inherited trait, in a small group, could increase in the particular gene pool of that ethnic group, purely by chance.

Following the Babylonian capture of Jerusalem around the sixth century B.C. the Hebrews, among whom the X-linked disorder glucose-6-phosphate dehydrogenase (G-6-PD) deficiency was known to be present, were exiled to areas like Iraq and Yemen (Marcus and Cohen, 1971). Because of their religious restrictions, the Iraq Hebrews did not intermarry with the non-Hebrew native populations. Later, some of the Hebrews of Iraq migrated to Persia and Kurdistan where they continued their ethnic isolationism. As a result of isolation and ethnic intermarriage, the frequency for G-6-PD deficiency increased among the Kurdistanian Hebrews. However, those who migrated to Yemen intermarried with the local natives and as a result, the incidence of G-6-PD deficiency is now very low compared to those who did not outcross or marry local natives.

Hebrews living in Asia Minor, Iraq, Iran, and Yemen are referred to as the Oriental Hebrews. They have been isolated also from the Sephardic and Ashkenazic groups (Figure 15–1). The Sephardim are Hebrews who migrated to countries surrounding the Mediterranean Sea and the European section of Turkey; it is said that Hebrews traveled to Spain with the Phoenicians before 500 B.C. The separation of the Ashkenazim began about 70 A.D. after the Romans captured Jerusalem. It is believed that the Romans enslaved Hebrew males to row their ships. Exiled with no women of their religion they converted women to Judaism. These Roman slaves are believed to be the forebears of the Ashkenazic Hebrews. They followed the Romans into western and central Europe.

With respect to the original frequency of the inherited G-6-PD deficiency, if the slaves carrying the X-linked defect married non-Hebrew women, then the frequency of that gene would be expected to be lower in the Ashkenazim than in the Oriental and Sephardic groups—and it is (Table 15–1).

The three sub-ethnic sub-groups were genetically isolated from each other

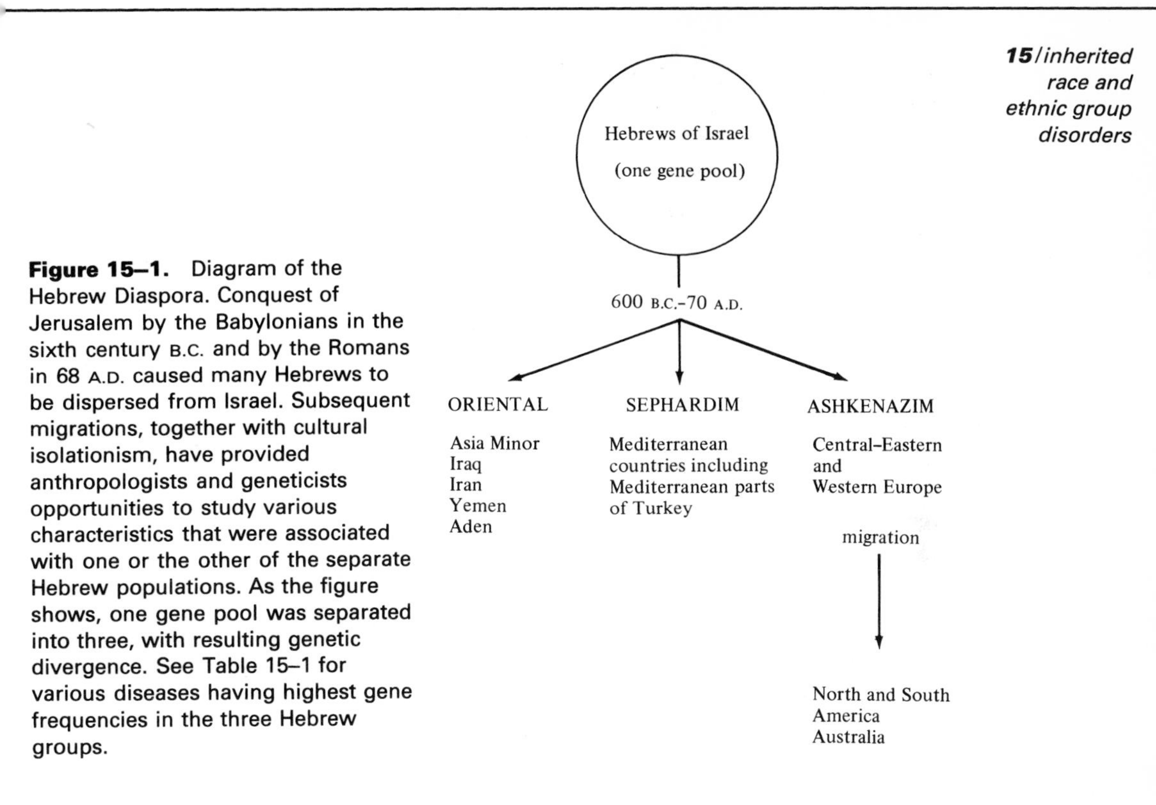

Figure 15–1. Diagram of the Hebrew Diaspora. Conquest of Jerusalem by the Babylonians in the sixth century B.C. and by the Romans in 68 A.D. caused many Hebrews to be dispersed from Israel. Subsequent migrations, together with cultural isolationism, have provided anthropologists and geneticists opportunities to study various characteristics that were associated with one or the other of the separate Hebrew populations. As the figure shows, one gene pool was separated into three, with resulting genetic divergence. See Table 15–1 for various diseases having highest gene frequencies in the three Hebrew groups.

and from non-Hebrews to varying degrees for a period of 2000 years. This separation may have allowed for the processes of natural selection and random genetic drift to change the frequency for the genes that occur in each group.

the ashkenazim

Wilson's disease. Wilson's disease, a hepatolenticular degeneration, is an autosomal recessive trait that occurs most commonly in Ashkenazic Hebrews. However, it also appears in Italians from Sicily and southern Italy and in the Japanese people. It is estimated that 1 in every 500 individuals of these groups is a heterozygous carrier. The disease results in a high storage level of copper in the liver and is fatal unless treated. Daily administration of D-penicillamine started early and continued indefinitely helps in some cases. Wilson's disease usually begins with tremors and awkwardness. It shows up from the late teens to age thirty. Mental deterioration is common, and personality changes with acute schizophrenic episodes are not unusual (Figure 15–2).

Riley–Day Syndrome or Familial Dysautonomia. This dysautonomia is an autosomal recessive disorder that occurs once in 10,000 to 20,000 live births "exclusively" (as far as is known) in the Ashkenazic Hebrews of Eastern European lineage (*Medical World News,* 1975a). A child born with familial dysautonomia suffers from a deterioration of the autonomic nervous system and loses a significant amount of his sensory system. For example, corneal injury is common because of insensitivity to corneal pain. Thus, it is common

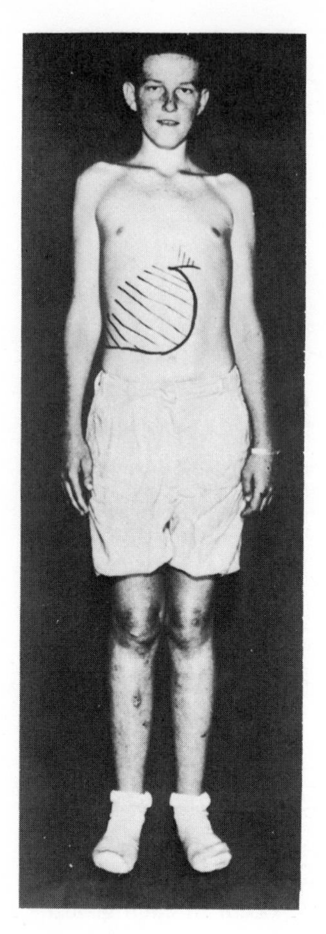

Figure 15–2. Wilson's Disease. The patient usually has an expressionless face and undergoes personality changes He retains copper, and it accumulates in the liver (body area marked in photo) and eventually enters the bloodstream; high levels of copper are stored in the brain and eye tissue. A Kayser–Fleisher ring forms a copper color as it is deposited in a circle around the pupil. Wilson's disease is due to a recessive gene, of highest frequency in Ashkenazic Hebrews, Italians from southern Italy, and Japanese. (By permission of Charles H. Carter, M.D.)

for particles to enter the eyeball and cause permanent eye injury. Defective persons also lack sensitivity to burns and pain caused by sharp objects. Many suffer from chronic pneumonia and lung disease, and life threatening episodes of vomiting. Because the tongue lacks fungiform papillae, they have little or no sense of taste (Figure 15–3).

It was believed that dysautonomic patients were sterile and that less than half survived to age 4, with the remaining patients surviving to age 13. Today, through better supportive care, over 50% survive into their teens and at least one defective female has given birth to a normal son (*Medical World News,* 1975a).

Familial dysautonomia has been compared with five other inherited diseases that have their highest incidence in the Ashkenazic Hebrews: Tay–Sachs, Niemann–Pick, Bloom's syndrome, torsion dystonia (dystonia musculorum deformans), and Gaucher's disease. The comparative data can be seen in Table 15–2. A photograph of a young Niemann–Pick patient is seen in Figure 15–4.

Bloom's Syndrome. Bloom's syndrome occurs among the Hebrews in general with a higher incidence than in other white ethnic groups. Of the Hebrews, the Ashkenazic Ukrainian Hebrews display the highest incidence of

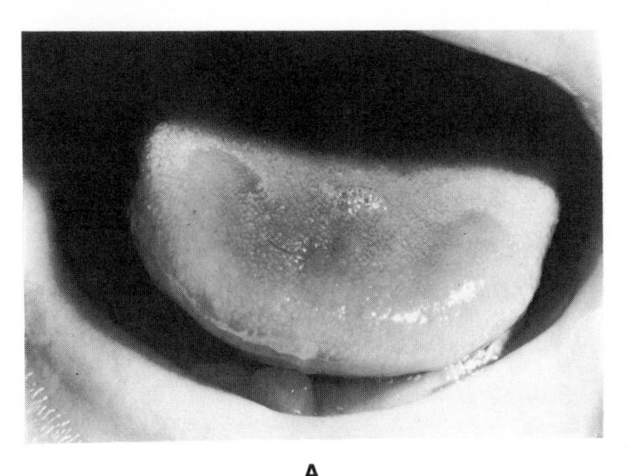

A

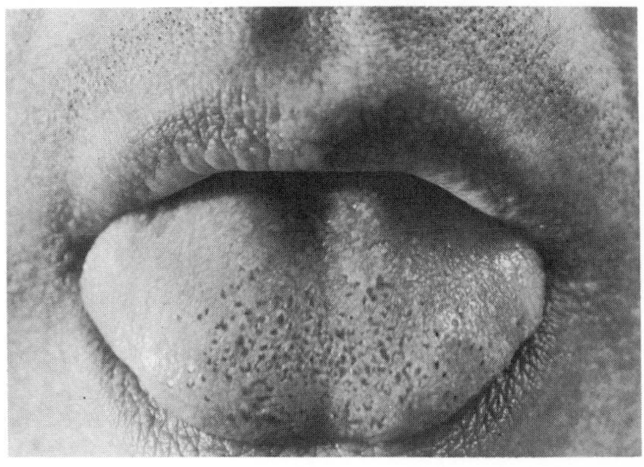

B

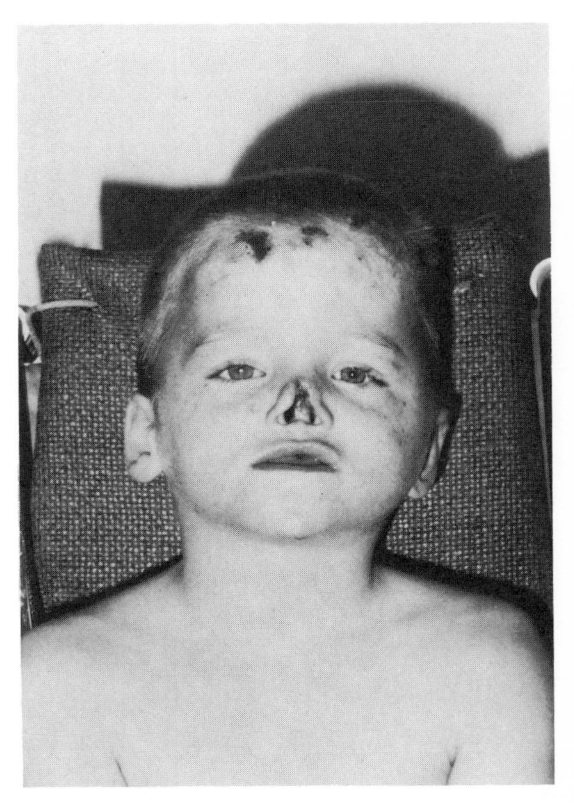

C

Figure 15–3. Tastebud-less Tongue in Familial Dysautonomia. **A.** Smooth, tastebud-less tongue (fungiform papillae absent) of FD patient. **B.** Normal tongue. **C.** A child with familial dysautonomia showing cleft nose (tastebud-less tongue not visible). (**A** courtesy of Dr. Felecia B. Axelrod; **C** by permission of Charles H. Carter, M.D.)

the autosomal recessive condition. In Bloom's syndrome there is a prenatal onset of shortness of stature. Facial and digestive blood vessels have a tendency to hemorrhage. The skull has a long anterior to posterior diameter with defective development of the cheekbones. Malignancy or cancer is the major cause of death in Bloom's syndrome; the fact that leukemia has been cited in several deaths suggests some relationship to Fanconi's pancytopenia syndrome where there is also an increased frequency of leukemia. There is chromosomal breakage in cells of victims of birth disorders.

Table 15–2 The Six "Jewish"* Genetic Diseases†

	Riley–Day Syndrome (Familial Dysautonomia)	*Tay–Sachs Disease*	*Niemann–Pick Disease*	*Bloom's Syndrome*	*Torsion Dystonia*	*Gaucher's Disease (Adult)*
Inheritance	Exclusively Ashkenazic Jews	85.90% Ashkenazic Jews	Predominantly Ashkenazic Jews	Predominantly Ashkenazic Jews	Primarily Ashkenazic Jews	Predominantly Ashkenazic Jews
Carriers in risk population (gene frequency)	1 in 100	1 in 30	1 in 200	1 in 200 (?)‡	1 in 200	1 in 200
Occurrence of disease	300–400 in U.S.	1 in 1000 live births	1 in 100,000 live births	52 cases world-wide‡	1 in 20,000 live births	1 in 20,000 live births
Etiology						
Enzymatic defect	?	Hexosaminidase A	Sphingo-myelinase	?	?	beta-Glucosidase
Metabolic defect	Deficiency in unmyelinated nerve fibers	Accumulation in neurons of ganglioside GM_2	Storage of lipids and cholesterol	?	?	Glycolipid storage

Salient pathology	Autonomic nervous system dysfunction; CV liability; scoliosis; absent pain and taste	Blindness; paralysis; seizures; dementia; psychomotor deterioration	Abnormal psychomotor development; hepato-splenomegaly	Stunted growth; cranial elongation; facial lesions; solar hypersensitivity; linked to leukemia	Bizarre spasmodic distortions of trunk and limbs	Severe episodic rheumatoid-like pain and swelling, splenomegaly
Detection Of carriers	No	Yes	Yes	No	No	Yes(?)
Prenatal	No	Yes, in 4th month of pregnancy	Yes, in 14th week	Presumably by amniocentesis	No	Yes, in 4th to 5th month
Clinical	At birth	At 4th to 6th month	Before 6th month	At birth	At 4th to 16th year	From 2nd year
Clinical	50% mortality by age 4; 75% by 13	Fatal before 5th year of life	Fatal by 3rd year of life	Disabling, not lethal	Disabling, not lethal	Disabling, not lethal

*As characterized by the National Foundation for Jewish Genetic Diseases.
†Courtesy of *Medical World News*, 1975.
‡Unpublished data provided to MWN by Dr. David Bloom.

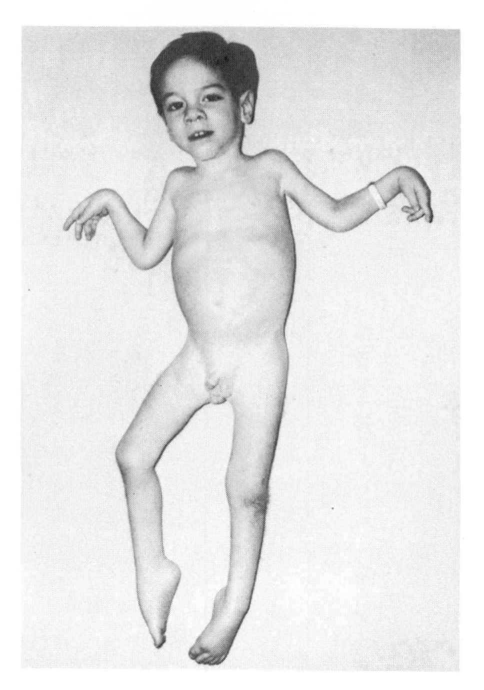

Figure 15–4. Niemann–Pick Patient. This child lacks the enzyme sphingomyelinase, and as a result has an enlarged liver due to improper lipid and cholesterol storage. The defective patient lacks psychomotor development and usually dies by the third year. (By permission of Charles H. Carter, M.D.)

Tay–Sachs Disease. Tay–Sachs disease (infantile amaurotic familial idiocy) was presented in Chapter 10 as one of the ten inherited lipid storage diseases. About 60% of the afflicted infants have a cherry-red spot in the macula of the eye. But there are at least a half dozen other storage disorders in which the cherry-red spot also occurs. In 1962 it was determined that a lipid, ganglioside GM_2, accumulates in the nerve cells of the brain and invariably leads to a progressive degeneration of brain function. When one lacks the ability to synthesize the enzyme hexosaminidase A (hex A), ganglioside accumulates to as much as 300 times the amount found in normal brain cells. Usually the affected infant loses its skills, one by one—the ability to sit up, to reach for things, to recognize parents, and even to smile (Figure 15–5). From the onset of symptoms, at about six months of age, degeneration is general and rapid, usually ending with death before the child's fourth birthday. An even more rapidly progressive form of Tay–Sachs disease has been encountered and is referred to as Sandoff's disease or a Tay–Sachs variant. Tay–Sachs and Sandoff's diseases are essentially the same, except that Sandoff's progresses at a more rapid pace.

Tay–Sachs disease is inherited as an autosomal recessive. If both parents are identified as heterozygous carriers they have a 1 in 4 or 25% chance of producing a Tay–Sachs child with each pregnancy. If a noncarrier mates with a carrier, all their children will be normal; but, statistically, half the children will be carriers.

If we sample populations, we find that the defective Tay–Sachs gene is most frequent in Hebrews and particularly in the Ashkenazic Hebrews. If you are of the Ashkenazic group, you have 1 chance in 25 of being a carrier compared to 1 chance in 400 if you are a Sephardic Hebrew or non-Hebrew. The total

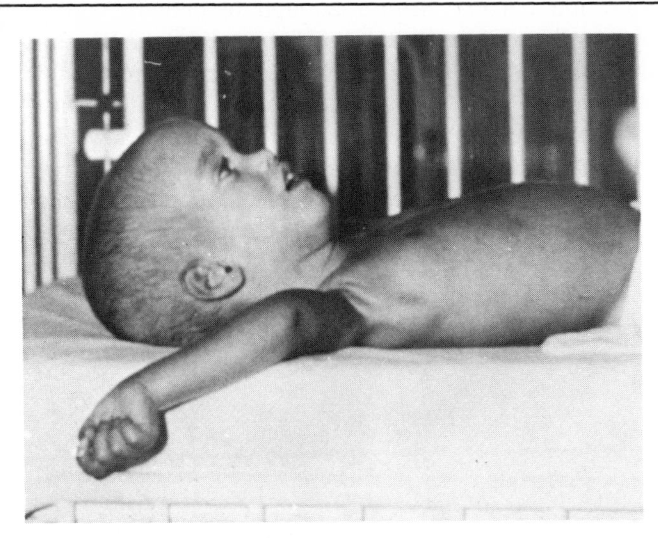

Figure 15–5. Tay–Sachs Child. This child will undergo progressive degeneration of brain function characterized by paralysis, epilepsy, and blindness. Symptoms begin in early infancy and progress to death. The infants are very sensitive to noise, and convulsions are common. In general, the disease can usually be diagnosed by a cherry-red spot with a white halo in the macula of the eye. Tay–Sachs disease is caused by a recessive gene that is most common in the Ashkenazic Hebrews. (By permission of Charles H. Carter, M.D.)

frequency of the disease in the United States is estimated at 1 in 6000 Hebrew births and 1 in 600,000 non-Hebrew births. In the United States of America there are an estimated 6 million Hebrews and approximately 85% of all Tay–Sachs children are born to Hebrew parents.

In New York, 1 in 900 married couples are heterozygous for Tay–Sachs and it is expected that 20 Tay–Sachs children will be born per year in that state. In the entire United States it is estimated that 50 Tay–Sachs children are born annually, 40 of them to Ashkenazic parents (Kaback and O'Brien, 1973).

Caring for the Tay–Sachs child is expensive; the average cost for a single child is estimated at $25,000 per year (O'Brien, 1972). If, as estimated, 50 Tay–Sachs children were born this year, the cost of their care would come to $1,250,000 per year for the care of these children. And since each child is expected to live 3.4 years, a total of $4,250,000 will be spent for care of the 50 Tay-Sachs children born in this one year. As in all human afflictions, what cannot be assessed is the toll in terms of emotional and physical strain on the parents, other children, relatives, and friends.

There is no treatment for Tay–Sachs disease. All that can be done at this time is to improve public knowledge through genetic counseling and screening programs, and hope that people will repond to that knowledge. Procedures and details on genetic counseling and screening for carriers are presented in Chapters 17 and 18.

the oriental hebrews

The Hebrews of Asia Minor, Iraq, Iran, Yemen, and Aden have a high frequency of the Dubin–Johnson syndrome, which is manifested by kidney problems that in late adulthood become severe enough to demand kidney dialysis. The Oriental Hebrew also demonstrates a low frequency of Bloom's syndrome and the Rendu–Osler–Weber syndrome (hereditary hemorrhagic telangiectasia). The latter is similar to Bloom's syndrome in facial telangiectases (thinness in walls of blood vessels); however, in Rendu–Osler–Weber disease the blood vessels in any organ may be affected and the mortality rate due to epistaxis (hemorrhage from the nose) is 4%. Rendu–Osler–Weber syndrome is an autosomal dominant condition and occurs to some extent in other Hebrew populations. Cirrhosis of the liver is also a common problem with this syndrome. Iron and estrogen therapy, skin transplants, and surgical intervention for gastrointestinal or pulmonary hemorrhage have helped those with this condition.

the amish

The Hebrews are not the only ethnic group to have a history of and a higher incidence of certain genetic diseases than those found in the general population. The Amish migrated to eastern Pennsylvania, Ohio, and Indiana between 1720 and 1770 from the Canton of Berne, Switzerland. Today 50% of the Amish live in three countries: Lancaster County, Pennsylvania; Holmes County, Ohio; and Lagrange County, Indiana. The group is now extinct in Europe. Like the Hebrews, the Amish are ideal for ethnic-group study. Their numbers are small, their ancestry is known, and they isolate themselves even more than do the Hebrews. The Amish of Pennsylvania have a higher than usual incidence of Ellis–van Creveld syndrome (chondroectodermal dysplasia, or dwarfism), an autosomal recessive trait. Fifty percent of the stricken individuals die in infancy, usually from heart or respiratory problems. Those surviving are short, usually between 3½ and 5 feet, with disproportionately short extremities. They are polydactyl (have additional fingers and occasionally toes), and their nails are hypoplastic (showing incomplete or arrested development). Investigators have found 43 cases of dwarfism in 26 families. In all families except one, the parents were both normal. In one marriage, a normal 58-year-old man married his normal cousin. They most certainly were heterozygotes, however, since their two children were dwarfs. In tracing the Amish pedigrees back, it turns out that all the 52 parents of the 26 families were the descendants of Samuel King and wife. It is assumed, then, that one of them carried the defective gene.

the alaskan eskimos

A group of Alaskan Eskimos who live in the delta of the Kushokwin River are the only known people who transmit the recessive gene for the Kushokwin syndrome. At birth this deforming disease can be identified by contractures (contraction of a muscle) of the knees, which can be treated. Other musculoskeletal deformities can be detected by X-ray. Hypoplasia of the first or second lumbar vertebrae causes a humped back in infants and progressive elongation of the pedicles (projections from the vertebra that form the beginning of the vertebral arch) of the fifth lumbar vertebrae produces hip deformities. There also are softening and destruction of the humerus (upper arm bone).

mediterranean peoples

thalassemia major (cooley's anemia)

Thalassemia is inherited as a Mendelian recessive gene. The homozygous condition, *th th*, is referred to as thalassemia major, Mediterranean anemia, or Cooley's anemia. In thalassemia major, the anemia is usually very severe. Untreated, patients tend to remain in a precarious position, not having sufficient hemoglobin to transport oxygen. Even with the help of transfusions and antibiotics, many patients die by their late teens. Other complications result in congestive heart failure. The heterozygous condition, *Th th,* is referred to as thalassemia minor and is usually less severe. However, there are a significant number of heterozygotes who demonstrate symptoms equal to that of the homozygous recessive or thalassemia major.

People with thalassemia major make very little beta chain hemoglobin and suffer severe anemia. The effect of anemia includes bone changes due to compensatory bone marrow activity and enlargement of the spleen and liver due to premature destruction of abnormal red blood cells. Thalassemia major usually becomes evident at about age 2. From that point, tranfusions every three to four weeks are necessary to combat the anemia. Thalassemia is usually fatal within the first 20 years of life. Statistically, as discussed in Chapter 10, if two parents are heterozygous carriers and have four children, one of the four would be "expected" to have thalassemia major. The heterozygotes also have a reduced rate of beta-chain hemoglobin synthesis, but in general they have sufficient hemoglobin to meet most or all body requirements. These people may have some anemia and morphologic changes in their red cells, but they do not need the continuous transfusions.

Socioeconomic Implications of Thalassemia. On August 29, 1972, the United States Congress passed a law called the National Cooley's Control Act (Thalassemia Major). This act provided for grants to public and nonprofit organizations for research in diagnosis, treatment, and prevention of thalassemia. Thalassemia screening, treatment, and counseling programs were also included in the federal legislation. Through facilities of the Public Health Service, effective and inexpensive thalassemia screening tests, counseling, and treatment were to be given to any person requesting them.

It is interesting that this law was passed back in 1972 when other diseases, higher in incidence and more severe in nature, were still not covered by federal programs. For example, the Huntington's disease legislation initiated in the late 1960's is still pending in 1976. Diseases like sickle cell anemia were not properly funded until 1972 and incidence is 1 in 400. The question you may wish to raise in class is how and why was thalassemia singled out for political action. However, even though the law was passed in 1972, many provisions have not been carried out, according to the New York Children's Blood Foundations, the only complete thalassemia care center in the United States.

Clinical Costs and Who Pays the Bill. Total care for a thalassemia patient would involve both physical and psychological services. However, as of 1972, there was only one total care service clinic operational for thalassemia in the United States. Most thalassemic individuals are treated as outpatients similar to hemophiliacs (discussed in Chapter 11), and their required transfusions are very expensive. The patients receive transfusions on various schedules, usually one or two units of blood every two to three weeks. The annual cost is $1500 to $4500 depending upon the number of complications and the transfusions, clinic visits, extra tests, and X-rays that are necessary during the year.

Since federal funds do not appear to be available to pay for treatment of thalassemic patients, various organizations and individual families pay the bill. In most of the world, treatment is available without direct cost only to the tiny minority of patients who happen to be living in a community that provides a free health service. Most thalassemia patients live in areas distant from transfusion services and hospital staff skilled in the transfusion of small children. In other places, where and when the service is available, the cost is prohibitive for all except the wealthy or well-insured.

Not only is there the financial burden for the parents, but since thalassemia major is a fatal disease, there is a great emotional suffering for the families. The disease is chronic and debilitating and drastically inhibits the child's normal educational and social activities. Even with our best treatment, most patients of thalassemia major die in adolescence.

examples of generally inherited disorders of caucasoids in general

meningomyelocele

A meningomyelocele is a hernia of the meninges, which are the membranes surrounding the spinal cord. A saclike growth holds a portion of the spinal cord or nerve roots outside the body at some point along the midline of the back. Some meningomyelocele patients also have hydrocephaly (water on the brain) and in addition may have congenital hips and club foot deformities.

Depending on the particular case, surgery may be very successful. Usually, however, there are a great number of associated problems, and the child may be mentally retarded or physically disabled for life.

The mode of inheritance for this disorder is unclear. It still appears that neural tube disorders, such as meningomyelocele (Figure (15–6), are polygenic. In general, if one of the neural tube disorders occurs in a family, there is a 5% risk of recurrence. If there are two defective children, the risk of recurrence increases to 10% for the next child (Lorbeer, 1971). The more severe the

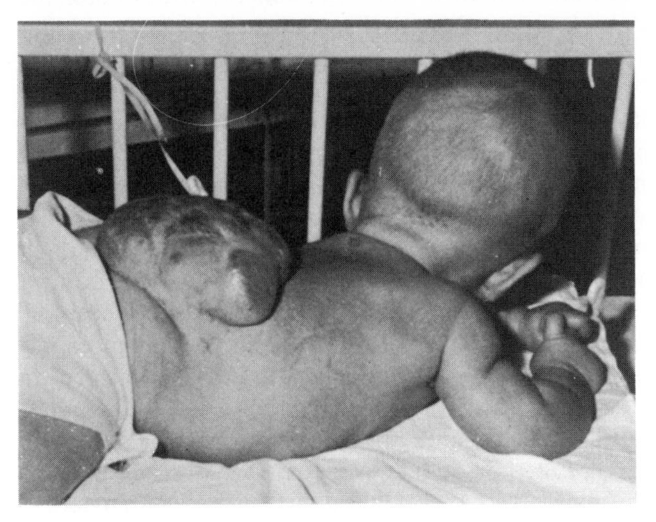

Figure 15–6. Neural Tube Disorder. Meningomyelocele with spina bifida, a congenital defect in the walls of the spinal cord. The lumbar region of the spine is most often affected. The spinal cord membranes and nerves along with spinal fluid create a large tumor on the back. Surgery may be very difficult and is not too successful. (By permission of Charles H. Carter, M.D.)

defect, the higher the recurrence risk. For example, with respect to cleft lip and palate, another trait thought to be polygenic, the recurrence risk in a family is about 3% if the affected child has a unilateral cleft lip but 6% if the child has the more severe bilateral cleft lip with cleft palate. The disorders of meningomyelocele, anencephaly, cleft lip and palate also occur in various chromosomal abnormalities such as trisomy 13 and trisomy 18 (Chapter 6). The incidence of neural tube disorders is about the same for all Caucasoid ethnic groups.

porphyria

Recently, a young lady decided to lose weight by reducing her intake of carbohydrates. This a very common event in the United States. However, in this particular case, the lady did not know she needed the carbohydrate to help offset the expression of inherited porphyria, a disease caused by metabolic disorders of porphyrin. Porphyrin is part of the *heme* portion of hemoglobin. In persons affected with porphyria, excessive amounts of porphyrin are found in their urine and stools. Soon after starting to diet, our young lady began to suffer abdominal pains and changes in her behavior. "Because she was not diagnosed properly and there seemed to be no organic or physical reason for her pain, her doctors thought she must be a mental case and committed her to the psychiatric ward of a local hospital" (*Newsletter,*1974). She was given barbiturates and tranquilizers, the very drugs she should not have taken in view of her condition. Fortunately, the correct diagnosis was eventually made. Soon after she was taken off the barbiturates and tranquilizers, her pain and mental symptoms disappeared.

Such a case history is not uncommon for persons afflicted with one of the five forms of inherited porphyria. It is a difficult disorder to diagnose, and it has a low frequency of occurrence in most geographical regions. There are at least five types of inherited porphyria: porphyria variegata, intermittent acute porphyria, coproporphyria, erythropoietic protoporphyria, and congenital erythropoietic porphyria. The first four are dominant gene expressions, and the fifth is inherited as a Mendelian recessive. The actual symptoms of the disease are highly variable. Persons may go through their lives without realizing that their occasional abdominal pain, easily bruised skin, or bouts of emotional disturbance are associated with an inherited defect. Conversely, others have severe clinical expressions of porphyria, including severe headache, vomiting, abdominal pain, and behavior or personality changes. In the dominant conditions, the ingestion of barbiturate or sulfonamide drugs and the use of certain anesthetics bring on the expression of abdonimal pain, vomiting, a rapid pulse, muscle weakness, and emotional disturbance. In fact some cases of porphyria would not have been detected if the person had not reacted to drugs prescribed to treat the symptons of headache and abdominal pain. Once the diagnosis of porphyria is made, as in the case of the lady trying to lose weight, the withdrawal of drugs reduces the emotional and physical symptoms. The error in porphyrin metabolism that makes one sensitive to barbiturates and other drugs is unclear (see Table 16–1). In the recessively inherited congenital erythropoietic porphyria, the patient becomes extremely sensitized to porphyrin. Sunlight brings on a severe skin disease known as hydroa vacciniforme. The eyelids become hardened and wrinkled, the cornea may become ulcerous, the teeth are stained deep pink from excess porphyrin, and the urine is a deep red color. It appears that more males than females demonstrate the congenital recessive form of porphyria (Waardenburg, Franceschetti, and Klein, 1961).

The defective gene for the disorder porphyria is rare in most regions of the world. Its highest incidence occurs in South Africa among descendents of the Dutch and French immigrant free burghers (lower middle class) who settled in Africa late in the seventeenth century. A study of the pedigrees of South Africans afflicted with porphyria variegata indicates that about 1 million of the 3 million white people in South Africa, or 1 in 3, have one of 40 different family names, derived from the 40 original free burghers. Geoffrey Dean (1969) traced the ancestry of 32 separate families that demonstrated porphyria variegata back to one of these burghers, Gerrit Jansz. Gerrit married Ariaantje Jacobs, an orphan of the Rotterdam orphanage who was sent over to Africa to become a wife of a free burgher. Evidently, either Gerrit Jansz or his wife had the dominant gene for porphyria variegata or a fresh mutation occurred in one of their offspring. Today, in Port Elizabeth, South Africa, the incidence of porphyria variegata is 1 in 250 white people; for all whites living in South Africa it is 1 in 400. Through racial mixing, the gene is also fairly common among the black population of South Africa but not to the same extent as in the white population.

In interviewing a porphyric female patient named van Rooyen, Dean found that her father had sensitive hand skin that separated easily when injured. The father referred to his skin as "van Rooyen skin." His grandfather, father, and three brothers all demonstrated similar skin sensitivity. The great great grandfather was Gerrit Reiner van Rooyen, born in 1814. Dean's trace of van

Rooyen's living descendants yielded 574 people; 16 had died of porphyria and 74 of those still living were porphyric. Thus, family surnames and porphyria variegata are common in the white and black ethnic groups of South Africa.

some negroid or black diseases

In the Negroid or black race, the skin, eye, and hair colors range from brown to black. The hair on the head is often tightly curled and the body hair is sparse. Common facial characteristics are full lips, broad nose, close set eyes, small ears, and a relatively rounded head. Height varies from the five-foot pygmy to the seven-foot Watusi. The body build is generally long-limbed with relatively large hands and feet (Goldsby, 1971). Some genetic disorders associated with the black race are given in Table 15–3.

sickle cell genes / high incidence among blacks

Recently two separate cases involving the sickle cell gene received national recognition. In one, a National Merit Scholar semifinalist holding athletic letters in basketball and football was refused his appointment to the U.S. Air Force Academy in Colorado. The other involved a defensive backfield football player for the University of Colorado who collapsed during a practice session and died 16 hours later of an acute sickle cell crisis. Both scholars were apparently normal. They carried the sickle cell trait in which only one of a pair of genes for producing hemoglobin was defective. One still lives and suffers the social stigma of having been identified as a carrier. The other died, probably because of the increased stress of physical exercise in a geographical location where oxygen concentration is reduced. How can the same genetic defect express itself in such dramatically different ways? To try to answer this question, we review the inherited disease sickle cell anemia.

What Is Sickle Cell Anemia? As discussed in Chapter 5, sickle cell anemia is a disease that results when the body produces abnormal hemoglobin that does not efficiently combine with oxygen and transport oxygen from the lungs to all parts of the body. Because the hemoglobin molecules are abnormal, they combine with one another to form a crystal-like structure. As a result, the red blood cells tend to become rigid and sickle-shaped and are less able to pass through the very small blood vessels called capillaries and arterioles. The sickle shape of the red blood cells gives the disease its name (Figure 15–7). The blocked or reduced blood flow results in a reduced supply of oxygen reaching the tissues. If the blockage of the capillaries and arterioles is widespread and prolonged, a "crisis" occurs that may last from several days to a week and may result in death.

Persons with sickle cell anemia typically show the usual symptoms of severe anemia. Many are moderately jaundiced, the whites of their eyes turning greenish yellow. Chronic punched-out-looking ulcers often appear about the ankles, along with a pallor in the palms of the hands, lips, nailbeds, and tissue

Table 15–3 Genetic Disorders Most Often Associated with the Negroid or Black Race

Disorder	*Clinical Features*	*Mode of Inheritance**
Africans		
Cervical cancer		PG
Esophageal cancer		PG
Duffy blood group Fy[ab]		AD
G-6-PD deficiency	Quantitative hemolysis following the administration of barbiturates and some antimalarial drugs or the ingestion of fava beans	X-LR mod
Hemoglobino-pathies	Hemoglobin variants especially sickle cell anemia and Hb C,	A co R
Hypertension	High blood pressure	PG
Polydactyly	Extra fingers or toes	PG
Sarcoidosis	Formation of tubercle-like lesions in organs, usually skin, lymph nodes, lungs, and bone marrow	PG
Systemic lupus erythematosus	Destruction of capillaries and small blood vessels and other changes in the vascular system	PG
Tuberculosis	Lesions that may heal by fibrosis and calcification; usually affects the respiratory system but may affect other parts of the body, e.g., gastrointestinal	PG
Uterine fibroid	A fibrous, encapsulated, connective tissue tumor in the uterus	PG
Blacks in General		
Galactosemia	Inability to convert galactose to glucose; tendency to bleed; susceptibility to infection	AR
Lactose intolerance	Inability to digest lactose	AD
G-6-PD deficiency	(See above)	
Oculocutaneous albinism	Subnormal vision; skin sensitive to sun; tendency to develop skin cancer	AR
Polydactyly	Extra fingers or toes	AD
Sickle cell anemia	Presence of hemoglobins resulting in crescent-shaped red blood cells; anemia; causes early death	A co R
Sydactyly	Fusion of two or more fingers or toes	AR

*AD, autosomal dominant; AR, autosomal recessive; A co R, autosomal codominant recessive (both protein products present in the heterozygous case; homozygous recessive case is semilethal—death after age 1); PG, polygenic; X-LR, sex-linked recessive; X-LR mod, sex-linked recessive trait influenced by autosomal genes.

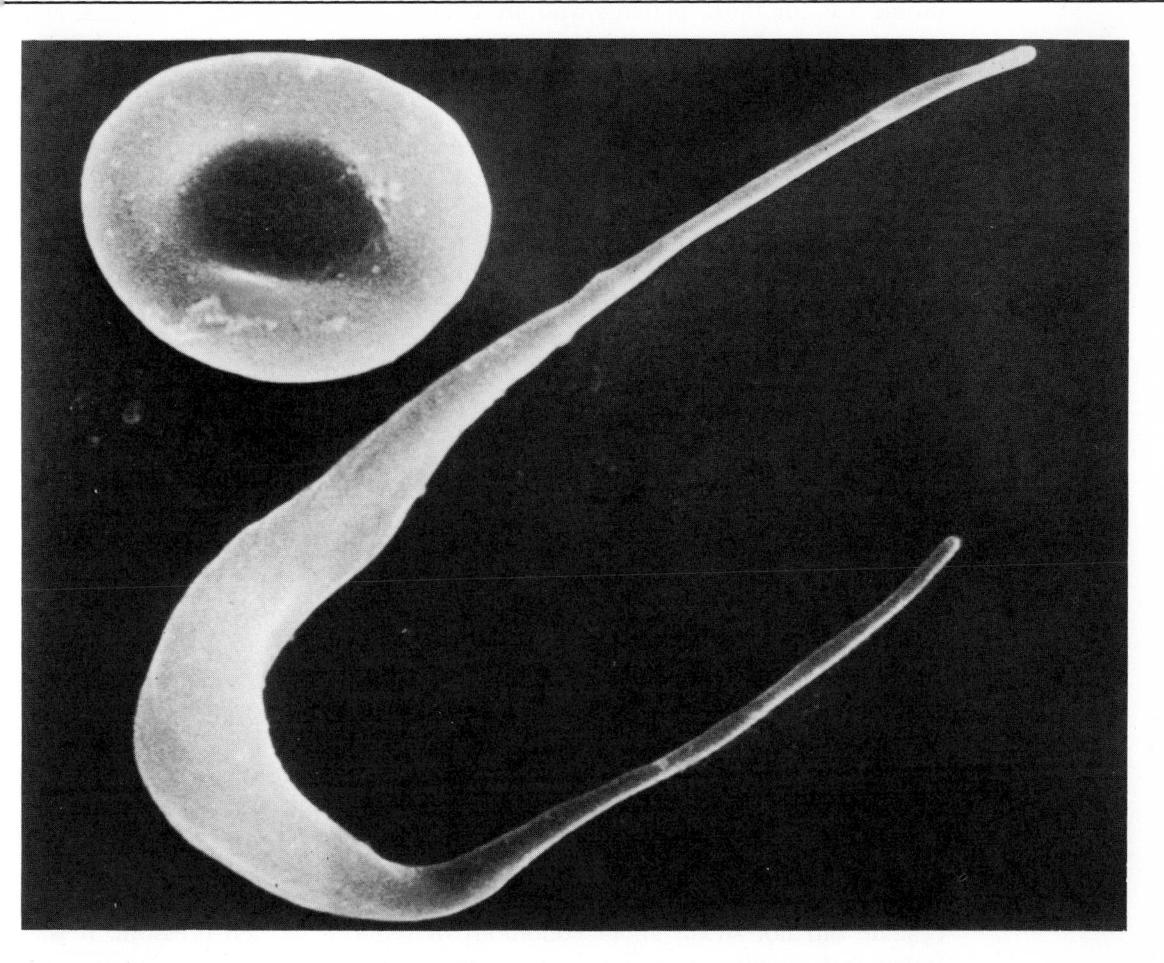

Figure 15–7. Normal and Sickled Red Blood Cells (×6200). Both cells are from a patient with sickle cell anemia. (From B. Cameron and R. Zucker, *Annals of the Sciences, 244:* 61.)

linings of the mouth. Severe pain in the abdomen and in the knees, elbows, and other joints is experienced from time to time in almost all patients with the disease.

In more severe cases, other symptoms include weakness, headache, dizziness, ringing in the ears, and spots before the eyes. Patients are sometimes drowsy and irritable. Some patients become accustomed to their chronic anemia and in spite of some weakness can carry on daily activities except during painful episodes. In a carrier of the disease trait—that is, a heterozygote with one defective gene—the symptoms and anemia are absent except under circumstances of unusual stress, such as physical exercise at high altitudes where a moderate lack of oxygen may cause abdominal pain, nausea, and vomiting, and even death (as in the case of the football player).

Sickle cell anemia is difficult to identify by clinical observation alone because the symptoms are similar to those found in other diseases, such as some abdominal and nerve disorders and rheumatic fever. Blood tests must be made to determine whether the red blood cells will sickle. The greatest problem is for those patients who are anemic, have two defective genes, and periodically undergo crisis.

There are three types of crisis, each of which may occur at any time.

1. Thrombotic crisis, the most common, occurs if the red blood cells assume the characteristic sickle shape and clog the small capillaries, cutting off the supply of blood to tissues. This condition causes swelling and pain but is not fatal unless it occurs in a critical area, such as the central nervous system. Thrombotic crises in the central nervous system are the second most common cause of death in sickle cell children.
2. Sequestration crisis affects one of every five children under the age of five with sickle cell anemia. The blood becomes trapped in the spleen, which becomes enlarged, while the volume of blood in the rest of the body decreases to about half its normal level. Many times this proves fatal.
3. Aplastic crisis is relatively rare. The bone marrow is unable to produce vital new red blood cells. Since the red blood cells of sickle cell anemia patients have a shorter life expectancy than normal, the hemoglobin may be reduced to a fatal level within several days.

Infections are commonly associated with sickle cell crises. Pneumonia, as well as other secondary infections, claims many sickle cell victims. Prostaglandin, a hormone-like lipid found in all animal tissue, is produced in greater quantities during periods of inflammations resulting from infection and may be a crisis-triggering agent. It has been found that prostaglandin causes sickling in suspensions of red blood cells from sickle cell victims (Dyer, 1970).

A number of chemical compounds have been tested to stop the sickling of red blood cells. One, urea, proved to be a failure. Recently, cells treated with cyanate did not sickle under conditions of reduced oxygen pressure. It was found that cyanate reacted with the terminal amino group of the hemoglobin molecule. This reaction, according to Cerami and Peterson (1975), is irreversible. That is, cyanate stays on the hemoglobin molecule for the life of the molecule, which might be harmful to the patient. A new antisickling agent, dimethyl adipimate (DMA), is also being tested. DMA appears to prevent sickling of red blood cells. For example, red cells from sickle cell anemia victims who have have been treated with the drug do not show the potassium loss and viscosity increase that usually accompany sickling. Oxygen affinity is also increased, and so far no ill effects have been observed when DMA-treated human erythrocytes were injected into rats (*Medical World News*, 1975b). Also under investigation is the use of nitrilosides; Houston (1975) reports that the nitrilosides, which are commonly found in fruit seed, bitter almonds, cassava, lima beans, millet, and sorghum, are very effective as an antisickling agent.

The first crisis usually occurs after the first year of birth. An individual who lives through years of individual crises commonly has strokes, heart attacks, permanent heart and kidney damage; in some cases, limbs may have to be amputated.

If the anemic patient is a female, her chance of giving rise to a normal child is reduced because of her inability to produce sufficient hemoglobin to transport oxygen for both. For example, in one study of 34 anemic mothers who became pregnant a total of 50 times, 16 pregnancies or 32% resulted in spontaneous, induced, or required abortion. Another five children were stillborn (dead at birth), and four died in the first year of life. Thus, 50% (25 out of 50) of the pregnancies resulted in spontaneous or medically required abortions, stillbirths, or early deaths.

Sickle Cell Hemoglobin and Its Pattern of Inheritance. Persons with normal hemoglobin, as discussed in Chapter 5, produce hemoglobin A (Hb A). Since we have two genes for each trait, normal homozygous individuals would be Hb^AHb^A; the heterozygous normal carrier, Hb^AHb^S; and the homozygous recessive anemic patient, Hb^SHb^S.

The genotypes, as presented, imply dominant and recessive inheritance. However, for sickle cell anemia, the heterozygote Hb^AHb^S produces both normal and abnormal hemoglobin. Usually, the person produces about 60 to 70% normal hemoglobin and 40 to 30% abnormal hemoglobin. The normal hemoglobin is in sufficient amount so that the carrier or heterozygous person, except for rare cases, does not know that he or she carries the sickle cell trait. In such cases, when both states of the gene produce a product, the inheritance pattern is referred to as codominant. The homozygous recessive case is semilethal because the sickle cell anemic usually dies sometime before his 25th birthday. So, for this inherited disease, when one receives both a normal and a defective gene, the heterozygous case, the gene is expressed in hemoglobin synthesis as a codominant, while phenotypically the recessive homozygote is eventually lethal. One can, however, predict, as in a recessive trait, that there is a 1 in 4 chance that each child of a heterozygous × heterozygous mating will have sickle cell anemia (Hb^SHb^S). The presence of some abnormal hemoglobin in the red blood cells of a heterozygote who appears phenotypically normal is what precipitates tragedies like that of the black football player at Colorado referred to earlier. Under conditions of unusual stress, such as reduced oxygen tension, the red blood cells of heterozygous people may sickle. The greater the percentage of abnormal hemoglobin, the greater the effect of environmental stress that causes the cells to sickle. The range of abnormal hemoglobin varies between individuals. Those most easily affected by lowered oxygen tension show levels of abnormal hemoglobin closer to that of the homozygous recessive. Heterozygotes least affected by oxygen stress may have only 10% abnormal hemoglobin and never suffer adverse effects due to elevation.

Incidence of Sickle Cell Hemoglobin. It is estimated that among the African Negroes transported to the United States the frequency of the sickle cell trait was 15.5%. The reason for such a high incidence (approximately 16 in 100 black slaves) appears to be that the heterozygote had an advantage in an area where malaria was prevalent. Thus, the relatively high rate of sickle cell children born to black parents today is indirectly a consequence of the protective advantage offered against another disease, malaria. Studies in Africa and the Mediterranean basin, where malaria has long been a problem, show that people with some sickle cell hemoglobin are less susceptible to lethal malaria than normal individuals. The plasmodium parasite (*Plasmodium falciparum*) does not readily infect red blood cells with lowered amounts of normal hemoglobin; and when it does, its reproduction is inhibited. Perhaps a combination of abnormal hemoglobin and the infective plasmodium causes many of the red blood cells to sickle, and such sickled cells may be phagocytized (that is, the phagocytes of the body digest the mildly sickled cells and destroy the malarial parasite infecting them), thus preventing the heterozygous host from getting severe malarial infection. This, however, is only speculative; the real mode of protection is unknown.

Whatever the reason, in those areas where malaria has been prevalent for generations, many have sickle cell anemia because they are descended from trait-carrying ancestors who survived malaria. In the United States and other areas where malaria is under control, sickle cell hemoglobin is no longer useful, and is a serious health problem in the black population.

In the United States today, some 300 years after the first black people arrived, the incidence of carriers is down to about 10%. In this environment where there is no malaria to serve as a positive force of selection for the gene, the full force of selection against the homozygous recessive and the limited selection against heterozygotes are manifest.

In the United States, with a 10% frequency of carrier blacks, there is 1 heterozygote in every 10 black persons. For example, a recent survey of the black National Football League players revealed that approximately 1 in 10 was a carrier. The chance of two heterozygous persons mating is $1/10 \times 1/10$ or 1/100. And each child of a mating between heterozygous carriers has a 1 in 4 chance of being a sickle cell child. So, since $1/100 \times 1/4 = 1/400$, we can expect that 1 in every 400 live black births in the United States will have sickle cell anemia. According to the 1971 Census, there are 23 million blacks in this country, which means that if all black anemics survived, there would be about 57,000 anemic individuals. In addition, 2.3 million are carriers.

In a recent study in New York, 1 in 200 whites was a carrier of the trait. By similar calculations ($1/200 \times 1/200 = 1/40{,}000$), there is a 1 chance in 40,000 that two white heterozygotes will mate. If they do, then, with the same 1 in 4 chance of giving rise to an anemic child, we expect that 1 in every 160,000 ($1/4 \times 1/40{,}000$) live white births will have sickle cell anemia. Based on the same Census, we would expect some 1000 white anemics in the United States. C. J. McGrew (1975), however, reported the outcome of screening nonblack Naval recruits for the sickle cell trait. A total of 65,751 male and female recruits of all races were screened for hemoglobins A and S. The incidence of heterozygous blacks was 8% with no significant difference between black males and females. Of the 57,665 nonblack recruits (45,500 men and 12,165 women) 25 males and 2 women showed the sickle cell trait, an incidence of 0.046% in the nonblack population, which is equivalent to 1 heterozygous white per 10,000. McGrew points out that in previous reports there was an implication that the presence of the sickle cell trait in the nonblack population indicated "obvious Negro admixture." From the vast difference in figures presented in the two independent studies, it should be clear that the data are not sufficient to state with assurance the real frequency of the sickle cell gene in the white population.

Identification of white anemics is a problem because the disease is difficult to diagnose and because most texts and medical literature refer to sickle cell anemia as the "black man's disease," creating the idea that whites do not have it. Because the cases are so few, the diagnosis of a white sickle cell anemic is frequently missed. Of course, many of the patients who die young never reach the doctor in the first place.

We may expect the incidence of the sickle cell trait and cases of anemia to rise steadily in the apparently white American population as a result of interracial matings. Also, there is a moderate incidence of the sickle cell trait in the Greeks and Sicilians living in and around the Mediterranean basin. Many of those migrating to the United States most certainly carry the gene. In brief, the sickle cell trait *is* present in our white population.

Social Ramifications of Sickle Cell Hemoglobin. With the advent of better medicine, the anemic is living to ages of employment. Statistics on employment of anemics reveal that the anemic spends 35 working days in a hospital yearly. The time lost without hospitalization was not reported, but it probably was considerable. Further, the crisis can occur at any time, making employment of such afflicted persons difficult. This contributes to the high unemployment rate among those anemic patients who are otherwise able to hold jobs.

The sickle cell patient has severe handicaps. Perhaps the most severe are that the lifespan is considerably shortened and the treatment required to stay alive is expensive.

The social problems of the heterozygous carrier are of a different nature. Because many show symptoms of slight sickling and accompanying problems, they are refused work and must pay higher insurance premiums than noncarrier blacks. Several social problems of this nature are presented in Chapter 22. In short, although screening programs to identify the black carrier are honorable and truly concerned with the counseling of the black carrier, when the identification is made, a social stigma may arise.

A number of important social questions related to sickle cell anemia need discussion. Consider, for example, the following:

1. The heterozygote has a lowered ability to carry oxygen since there is a certain percentage of abnormal hemoglobin in his cells. We acknowledge that many of our black ghettos are located near heavily traveled highways where the level of carbon monoxide is much higher than in the suburban areas. How does this apply to sickle cell hemoglobin? Carbon monoxide (CO) produces a poisonous effect by crowding out oxygen molecules that normally attach themselves to the hemoglobin molecules in red blood cells; it has been shown to have an affinity for hemoglobin more than 200 times that of oxygen. Thus carbon monoxide can cause illness and sometimes death—even in normal people—as in the case of the suicide who runs a hose from the engine exhaust to the inside of his car. But what of the sickle cell anemic with very little hemoglobin to carry oxygen to begin with? What of the carrier? We already know that a reduced oxygen tension will cause the carrier's cells to sickle. Perhaps, as the sickle cell trait becomes more widespread and pollution increases, many more people will suffer sickle cell symptoms due to man-made environmental pressure.

2. Consider the psychological and developmental impairment or maladjustment. Sickle cell anemia is a classic example. If you were an individual with sickle cell anemia and knew that your chances of living past the age of 20 years

were nil, do you think this would affect you tremendously psychologically, not to mention the pain and crises that you would have to endure physically?

3. The disease also has great social ramifications because sickle cell anemia is largely limited to the black race. What happens to the blacks who cannot afford the care of their children who have sickle cell anemia? Are they to be supported by the taxpayer?

4. There are also political ramifications. Those with the sickle cell trait or sickle cell anemia could be manipulated to vote for a certain political candidate who promises to support legislation helping carriers and anemics. In the fiscal year 1975–76 Congress authorized $115 million to investigate the sickle cell problem.

5. Although first described clinically in 1910, this genetic defect captured the attention of only a small group of physicians and medical researchers prior to 1970. The medical profession saw sickle cell anemia in blacks as an "interesting pathology" to demonstrate to their medical students; it was not considered interesting enough, however, to justify a significant research effort in treatment and prevention (Doman, 1974). The neglect of sickle cell research has been recognized publicly by a number of prominent people; for example, Dr. Paul Wolf of Stanford Medical School (*Time*, October 1971):

> If it had been a white disease, there would have been money for research and if money had been there, the interest would have been there.

and Senator John V. Tunney of California (*Medical World News*, December 1971):

> It is fair to say and research figures prove the fact that if sickle cell anemia afflicted white people instead of blacks, we would have made a commitment long ago to end this disease.

6. Recently great strides have been taken in testing for sickle cell trait and anemia. Should everyone be screened? Who should be told if a carrier is identified?

hemoglobin C disease, a second hemoglobin disorder in the black population

Sickle cell hemoglobin was the first case in which the effect of a mutation on the structure of a protein was determined (see Chapter 5). The second most common abnormal hemoglobin is hemoglobin C. As discussed in Chapter 5, this variant, like Hb S, is caused by an amino acid substitution in the beta chain at position number six. In Hb C, lysine replaces glutamic acid (see Figure 5–15).

The incidence of the homozygote ($Hb^C Hb^C$) is 1 in 6000 black Americans, and, for reasons unknown, three times as many black females as black males have Hb C disease. The symptoms of Hb C disease are similar to those of sickle cell anemia but not nearly as severe. The heterozygote usually has 70 to 80% normal hemoglobin and seldom knows he is a carrier. Unlike sickle cell carriers, the Hb C carrier is, from all data to date, symptom free.

some specific mongoloid or yellow diseases

The Mongoloid or yellow race is generally characterized as having skin color from nearly white to brown and hair shafts that are black, round, and of large diameter. The cheek bones are generally prominent. The eyes are usually black or brown with an almond shape. The average height is just over five feet, the torso is relatively long, the arms and legs short, and the hands and feet small (Goldsby, 1971). The genetic disorders most commonly associated with the Mongoloid race are found in Table 15–4.

Table 15–4 Genetic Disorders Most Often Associated with the Mongoloid or Yellow Race

Disorder	*Clinical Features*	*Mode of Inheritance**
American Indian		
Congenital dislocation of hip	Without treatment one leg remains shorter than the other	PG
Rheumatoid arthritis	Inflammation of joints; cartilage increases in size	PG
Tuberculosis	Susceptibility to lesions, which may heal by fibrosis and calcification; usually affects the respiratory system but may affect other parts of the body, e.g., gastrointestinal	PG
Chinese		
Alpha thalassemia	Hemoglobin synthesis interfered with	AR, AD mod
G-6-PD deficiency, Chinese type	Quantitative hemolysis of red blood cells following administration of barbiturates and some antimalarial drugs or ingestion of fava beans	X-LR mod
Nasopharyngeal cancer	Cancer of pharynx and nose	PG
Cuna Indians		
Oculocutaneous albinism	Subnormal vision; skin sensitive to sun; tendency to develop skin cancer	AR
Filipinos (in U.S. only)†		
Hyperuricemia (gout)	Abnormal amount of uric acid in blood	AD
Japanese		
Acatalasia	Inability to break down hydrogen peroxide due to lack of catalase enzyme	AR
Cerebrovascular hemorrhages	Brain hemorrhages	PG

(continued)

Table 15–4 continued

Disorder	Clinical Features	Mode of Inheritance*
Cleft lip-palate	A fissure of lip and roof of mouth	PG
Dyschromatopsia universalis hereditaria	Color blindness (imperfect color vision) See Ch. 11	X-LR
Gastric cancer	Cancer of stomach	PG
Spina bifida	Failure of vertebrae to close; spine usually protrudes	PG
Trophoblastic disease	Disease of outermost layer of developing blastocyst of an embryo	PG
Wilson's disease	Progressive tremors; cirrhosis of liver; copper in urine; progressive dementia-paranoia; early adolescent death	AR
Jemez Indians		
Oculocutaneous albinism	(See above)	
Koreans		
Acatalasia	(See above)	
Dyschromatopsia universalis hereditaria	(See above)	
Malays		
Liver cancer		PG
Pima Indians		
Diabetes mellitus	Insufficient or blocked insulin; elevated sugar levels in body; see Ch. 10	See Table 15–1
Polynesians		
Club foot	Deformity of foot	PG
Coronary heart disease	Disease of blood vessels of heart	PG
Diabetes mellitus	(See above)	
Taiwanese		
High rate of ethanol metabolism		PG

*AD, autosomal dominant; AR, autosomal recessive; PG, polygenic; X-LR, sex-linked recessive; X-LR mod (AD mod), sex-linked recessive (autosomal dominant) trait modified by autosomal genes.

†The expression of gout in Filipinos who migrate to the United States is a remarkable example of environmental influence on a given genotype. The same person in the Philippines would be less likely to suffer from gout even though he has the gene. Presumably the higher concentration of uric-acid-forming compounds in the richer diet available in the United States allows for the gene expression.

possible uses of data on ethnic-associated conditions

Among the disorders listed in Tables 15–1, 15–3, and 15–4, Icelanders have the highest incidence of glaucoma; the American Indians, congenital hip; and the Japanese, cleft lip and palate. Although the tables list many disorders found in their highest frequency within a given ethnic group, it is also interesting that there are diseases of unusually low incidence. Among the Kurdish and Yemenite Hebrews returning to Israel there is a very low frequency of diabetes, much lower than expected based on total population figures. There is also a low frequency of diabetes among the Navajo and Alaskan Indians and among Eskimos. The black population has a low frequency of Rh-negative women, PKU, and cleft lip (the form without cleft palate). Knowledge of the genetic disorders most common to a race or ethnic group can be useful to the anthropologist studying the migration of people and their racial mixtures and to a geneticist in his studies of inherited diseases and patterns of gene flow in human populations.

The proper ethnic identification of a patient and his geographic origin may be useful in the prompt and accurate diagnosis of a disorder. For example, if a patient suffers from severe abdominal pain, fever, headache, and nausea, and is black, he may be in the early stage of sickle cell crisis; if white, and born in the United States, he may have appendicitis; if born in the Mediterranean basin, he may be reacting to a drug or to eating the fava bean; or if white or black, and from South Africa, he may be having an attack of porphyria. Further, for those diseases with a particularly high incidence, such as sickle cell anemia in the blacks, Tay–Sachs in the Ashkenazic Hebrew, or cystic fibrosis in whites, medical resources can be distributed into those areas that are populated with people demonstrating the particular defect.

A rather different concept is possible development of a biological ethnogenetic weapon—the use or exploitation of human genetic differences among races and ethnic groups. For example, it is now well known that the ability to digest the milk sugar lactose, which is found in high concentration in cow's milk, is inherited as an autosomal dominant gene. People with the gene for the synthesis of the enzyme lactase, which hydrolyzes or breaks down lactose into usable sugars such as glucose, tolerate the ingestion of lactose very well. However, those genetically deficient in this enzyme suffer on drinking cow's milk, from various degrees of stomach cramps, pain, nausea, vomiting, and even death. Studies indicate that most humans can tolerate lactose in cow's milk or in milk products up to about age four, after which there appears to be a gene shutdown.

In the United States, about 5 to 10% of the white population and some 70% of the black population are intolerant to various degrees. Other studies indicate that there is intolerance to lactose among Eskimos and South American Indians. By 1970 enough data had been accumulated to indicate that most racial and ethnic groups the world over are intolerant to lactose in varying degrees. Adult tolerance to lactose has been observed in 90% of northern Europeans, and in 80% of the members of two nomadic pastoral tribes in Africa.

Using these data developed on ethnic sensitivity to milk lactose, Larson (1970), in writing on the use of ethnic weapons by the military, noted that "Innate differences in vulnerability to chemical agents between different populations has led to the possible development of ethnic weapons." He suggested the possibility of developing tolerance to toxic substances and the recruitment of troops from naturally resistant populations when using toxic substances in a military manner. However, he acknowledged that "Enzyme inhibitors could turn these troops into a state of paralysis." The point to all this is that there is a theoretical possibility of developing such weapons based on findings that many proteins are present in different genetically controlled forms in human populations. Larson sounded a warning that there might be military applications and apparently intended the article to warn against the development of ethnic weapons, but it has often been interpreted as a somewhat approving progress report on how development is proceeding. Since the appearance of the article, the U.S. Department of Defense has routinely denied any investment of effort or interest in the subjects, but, DOD's credibility being somewhat tattered, the denials have not quieted the speculation.

It is difficult to sit in judgment of the proper use of scientific data. Many times justification is based on winning or losing and in the case of a large-scale war almost anything is considered morally acceptable when it comes to survival. Let us hope that as a world community we never see the day anyone has to resort to the use of ethnic weaponry.

references

Cerami, Anthony, and C. M. Peterson, 1975. Cyanate and Sickle-Cell Disease. *Scientific American, 232*:44–50.

Dean, Geoffrey, 1969. The Porphyrias. *British Medical Bulletin, 25*:48–51.

Doman, Elvira, 1974. Sickle Cell Anemia and Ethnic Chauvinism. *The Sciences, 14*(1):29.

Dyer, K. F., 1970. How Man's Genes Have Evolved. *Science Journal, 6*:29–30.

Goldsby, Richard A., 1971. *Race and Races.* Macmillan, New York.

Haring, Douglas G., 1957. Racial Differences and Human Resemblances. In M. L. Barron (ed.), *American Minorities.* Knopf, New York.

Houston, Robert G., 1975. Sickle Cell Anemia and Vitamin B_{17}, a Preventive Model. *American Laboratory, 7*:51–63.

Kaback, M. M., and J. S. O'Brien, 1973. Tay–Sachs: Prototype for Prevention of Genetic Diseases. *Hospital Practice, 8*:107–16.

Kretchmer, Norman, 1972. Lactose and Lactase. *Scientific American, 227*:70–78.

Larson, C. A., 1970. Ethnic Weapons. *Military Review, 50*(Nov):3–17.

Lorbeer, J., 1971. Results of Treatment of Myelomeningocele, an Analysis of 524 Unselected Cases, with Special Reference to Possible Selection for Treatment. *Developmental Medicine and Child Neurology, 13*:279–303.

McGrew, C. J., 1975. Sickle Cell Trait in the Nonblack Population. *Journal of the American Medical Association, 232*:1329–30.

Medical World News, 1975a. A First: FD Mother Bears a Child. *16*(May):91–93.

Medical World News, 1975b. New Antisickling Agent. *16*(March):21.

Newsletter, 1974. January: 1. National Genetics Foundation, 250 W. 57 St., New York, N.Y. 10019.

O'Brien, John S., 1972. Tay–Sachs Disease: From Enzyme to Prevention. *Symposium on Contributions of Neurochemistry to Neurology and Psychiatry.* American Society for Neurochemistry, Seattle, Wash., March 23.

Rose, Peter I., 1964. *They and We.* Random House, New York.

Sheba, Chaim. 1970. Gene-Frequencies in Jews. *The Lancet, 1*: 1230–31.

Waardenburg, P. J., A. Franceschetti, and D. Klein, 1961. *Genetics and Ophthalmology.* Royal Van Gorcum, Assen, Netherlands.

susceptibility to disease

To be susceptible to a disease means that one has little resistance to a particular disease-causing environmental agent. It has often been stated that we inherit a genetically determined susceptibility to diseases. Genetically determined susceptibility to a disease in turn is referred to as our "genetic predisposition" toward contracting a certain disease. We can define genetic predisposition as physiological variation at the level of molecules due to small but specific differences in the individual's genetic constitution or, more precisely, differences in the individual's DNA. Thus it appears that a part of the individual's biological "uniqueness," which results from the process of sexual reproduction (see Chapter 4), includes his inheritance of gene combinations that "predispose" him to the effects of various environmental agents. The environmental agents may be of a biological, physical, or chemical nature.

The concept of a predisposition to a disease caused by an environmental agent comes from the general observation that in any given population exposed to an environmental agent, not all will be affected. Thus, in a given population, all of whom were exposed to the bacillus that causes diphtheria, those who inherited an immunity would not contract the disease, while those who inherited the susceptibility would develop diphtheria. Under these conditions the variable factor is heredity, which allows for differences in resistance to the bacteria. In another population where all the individuals happened to inherit susceptibility, only those exposed to the bacillus would develop the disease,

Can such things be . . . without a special wonder[?]

Shakespeare

genetic variation, susceptibility, sensitivity, and association

while those not exposed would not develop it. Here the environment would be the variable because even though one is susceptible, he has to come in contact with the organism. In separate settings one investigator might conclude that diphtheria was the result of hereditary influences, while another investigator might equally conclude that the environment was responsible.

The point is, both hereditary and environmental influences are at work. We can make a person artificially immune to diphtheria by means of toxin-antitoxin even though he inherited the susceptibility. But this does not alter the fact that he inherited the susceptibility nor does it change the fact that he will transmit the susceptibility. Our ability to alter the expression of the hereditary character does not change the fact of a hereditary background. Neither our knowledge of the causative agent for a condition nor our ability to cure the disease it causes invalidates the roles of heredity and environment as factors that influence one's susceptibility. Of those who are infected with diphtheria-causing bacteria only the most susceptible will die; others will survive. And those who were resistant to the bacteria at a particular concentration may, if the concentration is increased, become susceptible. Collective observations clearly indicate that alterations in the environment may not only initiate the expression of but also affect the severity of a disease.

Identical and fraternal twin studies and pedigree analyses of afflicted families have been used to detect a genetic predisposition to a particular disease. An excellent example is this case of identical twins. One, a bartender, became heavy and had a poor diet; he developed diabetes. His twin kept his weight down, did not drink alcoholic beverages, and was happily married; he did not develop diabetes. Because they were identical twins we can be reasonably sure that both had the same inherited potential to develop diabetes, but a modification in the environment resulted in clearly different expressions of the genes. Nevertheless, both twins, regardless of their physiological condition, will transmit some portion of their "sensitive" genotype to the next generation. Prior to the development of diabetes, both twins appeared normal, but both carried the genetic susceptibility to the development of a disease. In other words, changing the environment does *not* change one's inherited "condition," but environmental modification may initiate and modify the expression of a disorder.

We do not know to which of environmental agents an individual may be susceptible. On one end of the spectrum, almost all humans are susceptible to the spirochete that causes syphilis, if they come into contact with the causative agent. Because the constellation of genes in the human gene pool makes most humans susceptible to syphilis, it is not possible to determine the manner of inheritance for those genes that make us susceptible. At the other end of the spectrum are the inherited diseases like Huntington's disease on which alterations in the environment, at present, have no effect. Here the responsible gene is clearly transmitted and expressed in the offspring of a family.

Between the two extremes respresented by syphilis, which *appears* to be solely environmental, and Huntington's disease, which *appears* to be solely inherited, are the majority of human diseases, which are caused by variation in our heredity and the environment. In these cases, the *mode* of inheritance is questionable, but the fact that our inheritance is involved in our susceptibility to a disease is not in question. For example, we know that many diseases are caused by a combination of faulty genes *and* an adverse environment. These disorders include duodenal ulcer, ischemic heart disease, congenital dislocation of the hip, spina bifida, clubfoot, pernicious anemia, tuberculosis, cholera, typhoid fever, diarrhea, influenza, dysentery, leprosy, and pneumonia (Gowen, 1961; Carter, 1969; Marx, 1974). Thus genetic susceptibility in this chapter will deal with conditions where the mode of inheritance remains in question. We will discuss situations where there appears to be either a familial or a polygenic tendency for susceptibility or resistance (genetic disposition) to a particular disease. The inheritance of genetic diseases transmitted by single Mendelian genes in the dominant, recessive, or sex-linked condition is presented in Chapters 8, 9, 10, and 11.

With the advent of the "germ theory," the theory that each disease or physiological maladjustment of the body was caused by a specific microorganism, came the idea that science need only find ways to destroy the disease-causing organisms to eliminate most, if not all, human diseases. However, it was soon recognized that to blame germs for all human disease was an oversimplification of the cause of disease:

1. Studies of affected people showed many cases in which no germ or causative organism was present.
2. Studies of populations during past and present epidemics revealed that the severity (based on a number of deaths) of any given disease was always modified with time by a change in the frequency of susceptible persons remaining in the population.

The question is: Why are certain people more susceptible to a given disease than others?

The black population, for example, is generally resistant to yellow fever, but susceptible to tuberculosis. Measles is comparatively mild among the white race, yet it has had devastating effects on the North American Indians and natives of Hawaii. Smallpox is mild in Mexicans, but deadly to the North American Indians. Japanese are relatively resistant to scarlet fever. The Chinese are resistant to tetanus. In New York, the Russians, Poles, and Hebrews are comparatively resistant to tuberculosis, while the Irish are very susceptible. About one of every two people in the United States is naturally resistant to diphtheria (Whiting, 1929). Malaria, as well as other tropical fevers, causes a high mortality rate among Caucasoids, yet natives of malaria-infested countries are resistant. In the United States and Europe each year a number of people succumb to influenza; others are only mildly affected; and, of course, the majority do not get sick at all.

In many cases, the assessment of genetic susceptibility and resistance to a particular disease is very difficult because a gene may not be active throughout the life of an individual. Particular genes functioning in the fetus become inactive at birth; those that become operational may perform for only hours or minutes depending on the specific tissue and the particular cell within that tissue. We have obtained some of our knowledge about genetic susceptibility to particular diseases through the use of human twin studies and model animal systems, which have established that the individual's heredity plays a large role in his general state of health. The goal in clarifying the relationship between people's inherited constitutions and their susceptibility to disease is to detect susceptible people early and to be able to apply preventive measures.

susceptibility to contagious diseases

poliomyelitis and paralysis—a host viral susceptibility

Physicians have commented on the enigma of paralytic poliomyelitis affecting the robust and lively individuals more often than the frail since the nineteenth century. Further, the distribution of paralytic polio is not random. The number of multiple cases within families and communities is of greater occurrence than would be expected from chance alone. Family records show individuals of a family contracting polio many years after their siblings had the disease. Moreover, although the polio virus is common in the environment, only a small fraction of any community is susceptible, and of those who are detectably infected, only about 1% become paralyzed (99% are resistant to paralysis).

A number of characteristics, such as pigmented body spots, long uneven eyelashes, large incisor teeth, and an internal eye fold (Porter, 1968), are correlated with an individual's susceptibility to paralytic polio. Herndon and Jennings (1951) analyzed gathered data on afflicted twins who had been living together during the illness between the years 1940 and 1948. Data concerning the twins' parents, brothers and sisters, were also included. Monozygotic twins (identical) were found to be more susceptible to the disease than dizygotic (nonidentical) twins. Also, it appears that all persons in a given family where paralytic polio is encountered have an increased tendency for infection when compared to the population in general.

tuberculosis

Tuberculosis or "consumption," as it was previously called, is a dreaded disease that causes progressive wasting. Although the bacterium *Mycobacterium tuberculosis* is known to be the agent, the disease is contagious, shared in families, and appears to be passed on from generation to generation in particular families. Continuation of a known bacterial disease within family lineage created confusion because if the disease was caused by a known organism, why did not everyone exposed become tubercular? And why did only certain families continue to produce tuberculotics in each generation? In 1883, Hirsch suggested that people were born with a congenital predisposition toward the disease and that the only requirement for expression in the predisposed person was exposure to the causative agent. The most important data supporting the premise that tubercular disease is expressed by those with hereditary susceptibility were gathered by Kallman and Reisner (1943). These investigators looked for the amount of concordance among mono- and dizygotic twins (what percent of the time were both, or neither, diseased?). They collected data on twins for a five-year period in the state of New York.

The concordance from the identical twin studies strongly indicated an inherited disposition for susceptibility to tuberculosis. Not only did identical twins tend to manifest the disease (if one twin contracted TB, there was a high probability the other would), but the clinical course of the disease was similar. Later studies by Meyer and Jensen (1951) showed that individuals in a population displayed various degrees of allergenic reaction to the tuberculin stain test, and the reaction varied significantly between children of different families in the same community. Yet twin study data indicated heredity has a definite role in one's susceptibility to tuberculosis.

leprosy

Leprosy, like tuberculosis, is caused by a bacterium (Hansen's bacillus), but its transmission requires a longer period of contact. Leprosy is rare in the United States and Europe. The treatment of the disease has produced questionable results at best, and lepromatus leprosy is still considered terminal once a person has become infected. The question is: Since preventive and postleprosy treatment is rather ineffective, why doesn't the disease increase in frequency throughout the world? Additionally, why do only certain people express the symptoms of leprosy? There are "hot spots" of leprosy in the world, and investigations of these "hot spots" have shown an apparent association between an individual's heredity and his susceptibility to leprosy. For example, according to Neel and Schull (1954), there are two major centers of concentration for leprosy in North America: New Brunswick, Canada, and the state of Louisiana. Lepers of the two centers have a common origin.

In 1775, Norman immigrants to the Acadian region of Canada were forcibly relocated. Some were settled in New Brunswick and others were sent to Louisiana. Although leprosy was unknown in the original Canadian colony, the disease became prevalent in certain families whose ancestors were among the original colonists. Thus, the Acadian immigrants appear to have passed on the susceptibility for leprosy to certain families. The case for an inherited susceptibility among lepers is not as strong as for the inheritance of a predisposition to tuberculosis, yet the familial tendency for the development of leprosy cannot be denied.

susceptibility to malignant diseases

A father burst into tears when his 13-year-old son asked: "Dad, will they bury me in braces?" This son was the third of three male children to be stricken with cancer in perhaps a unique family history involving cancer in the United States. The first son died at the age of four after suffering 18 months with lymphatic leukemia. The second son, who was nine, died of schwannoma, a very rare cancer of the nerve linings. The third son, 13 years of age, had bone cancer and wore a leg brace. Were these children heritably more susceptible to cancer *per se*, each expressing it in a different form because of slight individual genetic variation, or were these cases solely a matter of coincidence? These questions remain unanswered. In a recent study at Ohio State University Clinical Research Center, investigators are trying to pin down the cause of an "inherited susceptibility" to a certain type of thyroid cancer (in which the level of the hormone thyrocalcitonin is increased) that has appeared in 29 members of one family. The research personnel have determined that if either parent has that particular thyroid cancer, 50% of the children are "likely" to develop it.

In the author's review of the history of the genetics of cancer in man, the most striking observation was the skepticism and vacillation concerning the importance of heredity in the development of cancer. The complexities involved in determining whether a particular cancer is heritable have been acknowledged since 1950. Yet we do know that a number of conditions that ultimately lead to cancer and several cancers are "directly" inheritable (Table 16–1).

The inherited forms of cancer listed in Table 16-1 usually affect a particular organ, for example, retinoblastoma (cancer of the eye), Paget's disease of the bone, and Gardner's syndrome (polyposis of the colon). Those forms of cancer associated with other inherited traits occur in organs remote from the site of the diagnosed inherited defect; in tylosis of the hands and feet, for example, cancer occurs in the esophagus. In such cases, we only surmise a degree of predisposition or genetic susceptibility. Neoplasms (tumor growths) that seem to show a familial tendency or predisposition are categorized under the general heading of immunologic deficiency diseases (Anderson, 1969). The better known autosomal recessive neoplasms in this category are Bloom's syndrome, Fanconi's aplastic anemia, ataxia-telangiectasia, and Chédiak-Higashi syndrome. The sex-linked disorders Burton's agammaglobulinemia and Wiskott–Aldrich syndrome are associated with an increased incidence of lymphomas or leukemias (Good and Finstad, 1969).

Table 16–1 Inheritance and Cancer

Heritable Forms of Cancer	*Inherited Traits That Involve Increased Tendency to Cancer*
Gardner's syndrome	Ataxia-telangiectasia
Nevoid basal cell carcinoma	Bloom's syndrome
Paget's disease	Burton's agammaglobulinemia
Pheochromocytoma medullary carcinoma (thyroid)	Chédiak–Higashi syndrome
	Down's syndrome
Pheochromocytoma mucosal neuromas	Fanconi's aplastic anemia
Retinoblastoma	Klinefelter's syndrome
Sipple syndrome	Multiple exostosis
	Neurofibromatosis
	Philadelphia syndrome
	Turner's syndrome
	Tylosis
	Werner's syndrome
	Wiskott–Aldrich syndrome
	Xeroderma pigmentosum

In addition to the increased susceptibility to cancer associated with the immunologic deficiency diseases, studies also indicate an increased susceptibility to certain cancers in patients with chromosome number anomalies (see Chapter 6 for discussions of these anomalies): Down's syndrome, a G group trisomy, Klinefelter's syndrome, Turner's syndrome, and the Philadelphia syndrome (Lynch, 1967).

It is important at this point to recognize the dangers involved in making sweeping statements about heredity in the susceptibility to cancerous tumor formation. In any discussion of hereditary in tumor development, one must be quite careful to specify the type of tumor. In some cases, heredity may be extremely important, whereas in others the genetic factor is difficult to assess.

Medical literature contains many case reports of families in which there was a high incidence of some particular tumor. The significance of these reports is difficult to evaluate. Tumors of all types are common. Approximately one out of every seven deaths in the United States each year results from a tumor formation.

Because the disease called cancer is actually a number of different diseases, demonstration of viral involvement in the etiology of one type of cancer, or of several types, does not mean the problem of cancer etiology for all types is solved. According to some views, the majority of cancers may be chemically induced. Alternatively, a virus may be a necessary, but not a sufficient, factor for the development of cancer.

Thus, as different phenotypes interact with the environment, each displays its own particular sensitivity. This sensitivity to various environments and subsequent behavior with respect to the type of cancer produced is what we have referred to as one's inherited susceptibility. As we gain ability to alter the environment, both internally and externally, we may free people of disease. We hope one day to free the patient of cancer by modifying the environment. Steps in this direction have already been initiated by banning the use of chemical mutagens and placing restriction on the amount of radiation we release to the atmosphere.

In summary, we must remind ourselves of one elementary fact. By natural selection species develop biological characteristics that render them very well suited to the conditions to which they were exposed as they evolved. Species perform optimally, both physiologically and behaviorally, under such conditions. If the conditions of life deviate from the normal range experienced by the species during its evolutionary history, maladjustment will almost inevitably occur. I believe this to be the case with cancer. Deviations brought about by rapid environmental change, the production of industrial pollution of all types, chemicals, particulate matter, and radiation have challenged the human species, which refuses to yield to natural selection; so we fight to keep the naturally susceptible alive.

There has been insufficient time for natural selection to produce a genetically new breed of man better adapted, physiologically and mentally, to the conditions of life that now prevail in modern cities. The pace of cultural change has outstripped the rate of possible evolutionary adaptation. If the changes persist, the species will eventually either become extinct or become "genetically adapted" to the new conditions. Thus, we inherited a susceptibility to cancer if only because biological selection has not yet had sufficient time to multiply those most resistant and eliminate those less resistant.

pharmacogenetics / inherited sensitivity to drugs

In examining the effects of drugs on those who have an inherited sensitivity, it is not so much the mode of inheritance we are reviewing as pharmacogenetics, an individual's inherited adverse reactions when exposed to certain drugs. As can be seen in Table 16–2, a number of inherited autosomal dominant, autosomal recessive, and X-linked disorders render the individual sensitive to

Table 16–2 Pharmacogenetic Diseases and Conditions

Disease or Condition	*Mode of Inheritance**	*Drugs to Which Response Is Abnormal*
Slow inactivation of isoniazid	AR	Isoniazid, sulfamethazine, sulfamaprine phenelzine, dapson, hydralazine
Suxamethonium sensitivity or atypical pseudocholinesterase	AR	Suxamethonium or succinylcholine
Acatalasia	AR	Hydrogen peroxide
Acetophenetidin-induced methemoglobinemia	AR	Acetophenetidin
Diphenylhydantoin toxicity due to deficient parahydroxylation	AD, X-L	Diphenylhydantoin
Bishydroxycoumarin sensitivity		Bishydroxycoumarin
Glucose-6-phosphate dehydrogenase deficiency, favism, or drug-induced hemolytic anemia	X-LR mod	Analgesics: acetanilide, acetylsalicylic acid, acetophenetidin, antipyrine, aminopyrine. Sulfonamides and sulfones: sulfanilamide, sulfapyridine, N^2-acetylsulfanilamide, sulfacetamide, sulfisoxazole (Gantrisin), thiazolsulfone, salicylazosulfapyridine (Azulfadine), sulfoxone, sulfamethoxypyridazine (Kynex). Antimalarials: primaquine, pamaquine, pentaquine, quinacrine (Atrabrine).

Table 16–2 continued

Disease or Condition	*Mode of Inheritance**	*Drugs to Which Response Is Abnormal*
Glucose-6-phosphate dehydrogenase deficiency (cont.)		Nonsulfonamide antibacterial agents: furazolidone, nitrofurantoin (Furadantin), chloramphenicol, *p*-aminosalicylic acid.
		Miscellaneous drugs: naphthalene, vitamin K, probenecid, trinitrotoluene, methylene blue, dimercaprol (BAL), phenylhydrazine, quinine, quinidine.
Drug-sensitive hemoglobins		
Hemoglobin Zurich	AD	Sulfonamides
Hemoglobin H	A co R	Same as listed for G-6-PD
Warfarin resistance	AD	Warfarin
Glaucoma due to abnormal response of intraocular pressure to steroids	UC (MA?)	Dexamethasone
Porphyria variegata	AD	Barbiturates
Porphyria, acute intermittent	AD	Barbiturates, sulfonamides
Coproporphyria	AD	Barbiturates, chloroquine diphosphate
Erythropoietic protoporphyria	AD	Barbiturates, griseofulvin
Congential erythropoietic porphyria	AR	
Aplastic anemia		Chloramphenicol
Agranulocytosis		Aminopyrine

*AD, autosomal dominant; AR, autosomal recessive; AcoR, autosomal codominant recessive (both proteins present in heterozygous case); UC (MA?), unclear, perhaps multiple alleles for a single locus; X-LT, sex-linked recessive; X-LR mod, sex-linked recessive trait modified by autosomal genes.

one or more drugs. Should a person with a drug sensitivity take that drug, adverse reactions occur that may lead to death. For example, persons inheriting the dominant gene for porphyria should not be given barbiturates, since they cannot tolerate the drug. Persons who lack the enzyme glucose-6-phosphate dehydrogenase (G-6-PD) are sensitive to the Italian broad bean, certain analgesics, sulfonamides, antimalarials, nonsulfur antibacterial drugs, and various dyes. In many cases, G-6-PD persons undergo severe red blood cell lysis on exposure to these drugs.

Pharmacogenetics is based on the underlying principle of individual variations in drug response. In the first historically documented treatment more than 4000 years ago, astute physicians recognized individual differences in drug response. It was noted that a drug administered in the same dose and by the same route produced markedly different therapeutic and toxic effects in various individuals.

In the early part of this century, E. A. Garrod (1909), a British physician, worked on several rare metabolic diseases. Garrod recognized the variation in biochemistry among individuals and concluded that people would differ in both their reaction to drugs and their natural resistance against infections. However, it was not until the middle of the twentieth century when the use of drugs and antibiotics as medications had become widespread that we realized the accuracy of Garrod's prophesy with respect to our individual response to drugs (personal hypersensitivities).

drugs and fetal development

Currently, 250,000 American babies a year are born with mental or physical defects. Annually 500,000 defective fetuses are lost in spontaneous abortion and stillbirths. Birth defects account for approximately 560,000 deaths. Many children surviving with birth defects are afflicted with blindness, hearing impairments, heart and circulatory defects, mental retardation, bone and muscle disease, and digestive, endocrine, or urinary impairments. It must be mentioned that certain drugs have been definitely linked to human birth defects, and others are under suspicion. Even aspirin, readily available for mass consumption, has been associated with congenital malformation in experimental animals and with jaundice in infants.

Fetus pharmacology, the study of drugs as they affect the developing fetus, is an area of national concern. Surely no one wants a tragedy similar to the one resulting from the use of thalidomide. Over 6000 children have been born with abnormalities of the hands, arms, feet, and legs (Figure 16–1). However, many more mothers took thalidomide in the first trimester than had deformed children, leading one to ask why certain children (the defective) were sensitive to the drug while others were not.

There is some agreement that a drug must pass from the placenta to the fetus. Important factors enabling a drug to pass through the placenta are pH (acid or base conditions), molecular weight, and solubility. Perhaps most important are the individually inherited differences in the placenta itself. Once a drug has passed through the placenta and into the environment of the fetus, the susceptibility of the fetus for that drug most likely determines its effect on fetal development.

Another consideration is the possibility that a drug directly affects the fetal genes. Some drugs may cause mutations, which can later be transmitted if the mutation occurs in the gametes. Should the mutation occur in cells other than the sex cells, it can lead to a serious, although not genetically transmittable, disease.

glucose-6-phosphate dehydrogenase deficiency

Prior to the middle 1950's, pharmacogeneticists had recognized only a few hereditary drug-metabolizing enzyme disorders. The disorder most widely

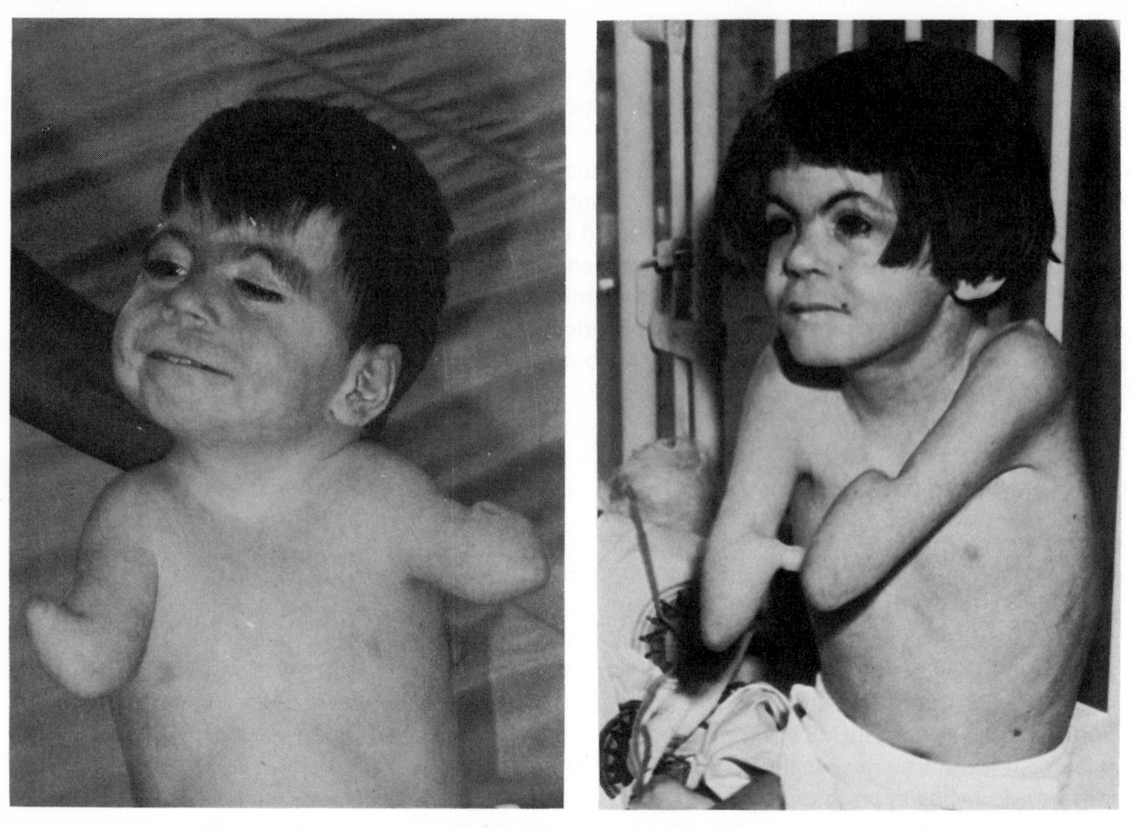

Figure 16–1. Phocomelia. **A.** Child who inherited the recessive gene for seal-like limbs. **B.** Child displaying "phocomelia" as a result of the action of the drug thalidomide taken early during the mother's pregnancy. (By permission of Charles H. Carter, M.D.)

recognized at this time was the severe reaction brought about in certain persons coming in contact with or ingesting drugs such as naphthalene, the common ingredient in moth balls; sulfanilamide, an antibacterial agent; and primaquine, a drug used in combating malaria. The sensitivity to these drugs is an inherited condition characterized by the loss of a gene that controls the enzyme glucose-6-phosphate dehydrogenase (G-6-PD) in our red blood cells. Without this enzyme the red blood cells tend to break down when they come into contact with the above reagents. A condition known as drug-induced hemolytic anemia (lack of red blood cells) results. Studies have shown that there are at least 80 variants of this particular enzyme. The different alterations of this enzyme can account for a gradient of expression going from mild to the most severe reaction to the antimalarial drugs and subsequent anemia.

Persons inheriting the X-linked-recessive, autosomal-modified defective gene for glucose-6-phosphate dehydrogenase make very little glucose-6-phosphate dehydrogenase. The enzyme G-6-PD is the first of some 15 enzymes necessary for the metabolism of sugar into necessary cell material and energy. Tests show that normal red blood cells deprived of glucose behave as though they did not have the G-6-PD enzyme. Because the G-6-PD enzyme is missing from the person's red blood cells, these cells become sensitive to the action of a variety of drugs (Table 16–2) and, for reasons yet unclear, to the pollen or raw tissue of the Italian broad bean, *Vicia faba*. In many persons with G-6-PD deficiency, exposure to the fava bean, antimalarial drugs, sulfonamides, naphthalene, large doses of aspirin, etc., causes the immediate destruction of red blood cells and leads to severe anemia, dizziness, nausea, fever, and vomiting. Persons with the G-6-PD defect who do not expose themselves to such environmental agents do not demonstrate any unusual reactions. However, careful measurements of the life-span of red blood cells indicate that the longevity is reduced from the normal 120 days to 80 or 90 days. This is apparently not a significant drawback because our system is normally capable of generating sufficient red blood cells to overcome the shortened life-span.

During World War II the British developed the drug pamaquine to prevent malaria. However, when the Greek, Hebrew, and black soldiers (specific ethnic groups) began using it, cases of hemolysis (breakdown of red blood cells) were reported. During this same period pamaquine was given to the natives working on the Panama Canal, and 10% of them became ill because of red blood cell hemolysis. Attention was again called to the use of antimalarial drugs when the troops returning from Korea took the drugs to prevent recurrent attacks of malaria. Primaquine, a chemical relative to pamaquine, it was hoped, would prove to be a malarial cure without the side effects of pamaquine. However, red blood cell hemolysis was equally severe. Research revealed that the red blood cells of sensitive individuals contained slightly smaller amounts of a sulfur-containing compound, glutathione (GSH), and were missing the enzyme glucose-6-phosphate dehydrogenase. Fava beans and primaquine cause the formation of hydrogen peroxide in the red blood cells. Normally the GSH and G-6-PD break down the peroxide, but with reduced GSH and no G-6-PD, the red blood cells lyze (break down). Degradation products of the released hemoglobin produce a characteristic dark urine. Individuals with G-6-PD deficiency can now be identified by mixing a blood sample with a hemolyzing drug. In the United States approximately 10% of the black population and many Hebrews and Persians of Mediterranean descent are sensitive to the various antimalarial drugs.

Early History of G-6-PD Deficiency. Dr. Chaim Sheba (1970) believes that the early Hebrews and Phoenicians, who originated in the same geographic region, both showed the inherited X-linked G-6-PD deficiency.

The Phoenicians carried the defect to the Mediterranean islands of Sicily, Sardinia, Malta, and Minorca, where they established communities to support their seamen when they arrived in port. Since these colonists lived in relative isolation and did not intermarry with their neighbors, they preserved a high incidence of G-6-PD deficiency. The Hebrews, as indicated previously, carried the defective gene to Yemen, Iraq, Persia, Kurdistan, the Mediterranean, parts of Turkey, central Europe, North and South America, and Australia. Today, G-6-PD deficiency is spread worldwide, but does not appear in the North American Indian, the Alaskan Eskimos, or the Japanese.

The Riddle of Pythagoras. Pythagoras, a distinguished Greek philosopher, migrated from Greece to Southern Italy where he founded the mystical mathematical numbers cult. Members of the Pythagorean cult were forbidden to walk near or through fields of the Italian broad bean, *Vicia faba*.

In fact, Pythagoras himself, rather than cross a field of the growing broad bean, stopped and gave up his life to the pursuing Greek soldiers. According to Marcus and Cohen (1971), one couple that lagged behind were captured by the Greek soldiers and interrogated by the emperor who wanted to know why the Pythagorean followers feared the fava bean. Rather than reveal their secret the husband chose to die and the wife bit out her own tongue. Behavior such as this by Pythagoras and members of the Pythagorean cult long remained a mystery.

Perhaps Pythagoras recognized that many persons became very ill from inhaling the pollen or eating the raw fava bean. To prevent such illness he simply forbade their use. In keeping the secret to himself and incorporating it into his cult as a sign of the gods, he maintained a certain mystical power. Today we understand the problem as the inheritance of a sensitivity to the bean (favism) via a reduced level of G-6-PD in the red blood cells.

slow inactivation of isoniazid

Isoniazid is one of the few drugs effective in the treatment of tuberculosis. It is metabolized or inactivated in the body, but the rate of inactivation varies among individuals using the drug. Analysis of human excretory products containing isoniazid shows that persons requiring the drug could be categorized as rapid metabolizers and slow metabolizers. One would predict, if the drug itself were the effective agent against *Mycobacterium tuberculosis*, that the slow inactivators would respond to treatment more quickly than patients rapidly inactivating the drug. This was the case as reported by Harris (1961).

Slow inactivator patients with pulmonary tuberculosis placed on standardized isoniazid treatment generally responded earlier in treatment than did the rapid inactivators. The inherited ability to inactivate the drug seems to have no effect on the patient's susceptibility to contracting the disease. Apparently the genes are involved with the patient's response to the drug and not in the complex issue of a person's susceptibility to tuberculosis. In this case there is a clear distinction between genes involved in an individual's susceptibility to tuberculosis and genes associated with the metabolism of a drug used in the treatment of that disease.

sensitivity to suxamethonium (succinylcholine)

In 1949 the drug known as suxamethonium, suxethonium, scoline, anectine, or succinylcholine was first used as a muscle relaxant. The drug is still in use today in general anesthesia and in treatment of certain disorders, such as tetanus. The administration of the usual dose produces muscle paralysis for approximately two minutes. However, in some patients it took over two hours for the drug's effect to subside. Investigation of drug metabolism demonstrated it was normally hydrolyzed by an enzyme called plasma pseudocholenesterase and the sensitive patient recovered from the drug more slowly because of a lower level of the enzyme available for drug hydrolysis (breakdown). After extensive analysis of atypical patients, it became clear that these patients inherited a genetic defect that caused the production of a structurally different pseudocholinesterase, which in turn lengthened the time required for the hydrolysis of suxamethonium. The administration of the drug suxamethonium or succinylcholine to genetically sensitive persons may lead to death (Bourned, Collier, and Somers, 1952; Evans et al., 1952).

resistance to warfarin, a coumarin anticoagulant drug

Warfarin is the substance used in the best rodent poisons. Some rats inherit resistance to warfarin as an autosomal dominant. Warfarin is used in humans as an anticoagulant (interferes with normal blood clotting). However, certain people are resistant to it. In other words, persons needing an anticoagulant drug of the coumarin type required much higher concentrations to reduce the synthesis of clotting factors in the blood (O'Reilly, Pool, and Aggeler, 1968). Resistance to warfarin in humans appears to be inherited as an autosomal dominant.

Other coumarins that behave in a manner similar to warfarin are dicumarol and bishydroxycoumarin. Sensitivity to the anticoagulants can be increased by altering the environment, as in vitamin K deficiency and numerous diseases of the liver.

chloramphenicol and aplastic anemia

The drug chloramphenicol is an effective antibacterial agent. However, after several years of administration in the United States starting in 1952, an association was recognized between chloramphenicol and various side effects. The most serious side effect, which led to intensive investigations, was the charge that chloramphenicol caused aplastic anemia. In a recent article, Nagao and Mauer (1969) offer substantial evidence that sensitivity to the drug chloramphenicol is an inherited trait. Their work was done with identical twins and both twins treated with the drug exhibited the same response. The site of action is in the bone marrow cells where improper metabolism causes aplastic anemia. Although most United States citizens are sensitive to side effects after the administration of chloramphenicol, the Hebrew ethnic group is resistant to the side effects of chloramphenicol.

Another drug causing side effects in the United States and currently substituting for chloramphenicol is aminopyrine (Pyramidon), a respectable antifever drug. However, it is allegedly associated with agranulocytosis, a marked decrease in blood granulocytes.

inherited traits and association with diseases

phenylthiocarbamide taste

The ability to taste phenylthiourea, or phenylthiocarbamide (PTC), is inherited as an autosomal dominant trait. Thus, persons who taste can be either homozygous dominant (*TT*) or heterozygous (*Tt*) tasters. Even in the animal world we can find tasters and nontasters; about 90% of chimpanzees are tasters.

Recognition of the differential ability to taste phenylthiourea came about in a rather curious manner. Apparently Fox, who chemically synthesized the compound, was unable to sense the bitter taste of the chemical. Another person, however, mentioned the compound's bitterness to Fox (1932). Further studies disclosed that what is tasted is the chemical group N—C—S. Females are more sensitive than males—that is, they are frequently able to detect the bitter taste at lower concentrations of phenylthiourea—and the ability to taste decreases with age.

Associations have been made between the ability to taste and certain diseases. For example, nontasters are more likely to have goiter than tasters (Harris and Trotter, 1949), and in male nontasters there is an increased frequency of multiple thyroid adenomas (cancer) (Kitchin et al., 1959). Nontasters are more susceptible to athyreotic cretinism, a type of dwarfism, while tasters are more susceptible to diffuse goiters (Fraser, 1961). Becker and Morton (1964) reported that nontasters develop open-angle glaucoma more frequently and angle-closure glaucoma less frequently than tasters.

The matter of tasting phenylthiourea is one of the least serious drug sensitivities. Whether you are a taster or not can be easily determined; ask your instructor for an inexpensive strip of PTC paper commercially available from nearly all biological supply houses. First, chew a piece of neutral paper (a piece of tablet paper will do). Then, taking out the neutral paper, chew a piece of PTC paper. If you "taste," a bitter (or just a sour) taste should develop as you roll the paper over and under your tongue.

human ABO blood types

The word *association* as used with respect to blood groups and disease means that a given disease occurs in people of a given blood type more often than would be expected to occur by chance. The discovery of the association of blood groups and diseases dates back to 1920 (Alexander, 1920–1921; Buchanan and Highley, 1920–1921). However, the more recent work stimulating research into blood group and disease associations was the work of Aird, which showed persons having gastric cancer were most often of blood type A (Aird et al., 1953 and 1954). Since Aird's initial article, numerous studies have been made on the association of the ABO blood groups and various diseases (Table 16–3).

There are, however, problems with such attempts to associate a disease with a particular blood group. To begin with, it must be realized that, although a certain form of cancer may be associated with blood type A because it is found more frequently among persons of blood type A, the same form of cancer can be found in those of blood types AB, B, and O. There are just relatively fewer persons in these categories. Moreover, an association based on statistical evidence does not in itself represent conclusive evidence for a cause-and-effect relationship.

In fact, there are a number of scientists who argue against a causal relationship for most, if not all, of the associations listed in Table 16–3. The critics of an assumed causal relationship find fault with the methodology of the studies and therefore the conclusions of the studies. If there is a correlation (an association greater than due to chance), then there may be a causal relationship. It is for this reason that blood types and associated conditions are reviewed in this chapter.

We know blood type antigens are inherited, but we also know that the majority of the conditions listed in Table 16–3 are not. What is suggested is that persons of specific blood types are more susceptible to certain conditions than persons of other blood types. You may read the listed references which refute such causal association (Weiner, 1962; Otten, 1967; Cohen, 1970; Vogel, 1970; Weiner, 1970) at your convenience. Those investigators presenting evidence for blood type associations are presented beside the appropriate condition in Table 16–3.

Patients with blood type A have been reported to be more susceptible to cancer of the stomach, the cervix, the salivary glands, pancreas, and ovaries. Associations have been reported for blood type A persons and conditions such as diabetes mellitus in males, arrhythmias (lack of rhythm) of the heart, the ischemic (constriction of heart blood vessels) heart, pernicious anemia, bronchopneumonia, smallpox, adenoviral infection, myocardial infarctions, gallstones, cirrhosis of the liver, arteriosclerosis obliterans, female venous thromboembolisms, high serum cholesterol, and the tendency for higher than normal diet of fatty foods.

Diseases associated most predominantly with blood type O are cancer of the skin, duodenal, gastric, and stomal ulcers, manic depressive illness, erythroblastosis fetalis where the mother is of blood type O, and susceptibility to the A_2 influenza virus.

Persons belonging to blood group B may find comfort in the reports to date in that they are the least susceptible to those conditions which have been investigated with respect to the blood group association. However, it may also be of interest to know only 10% of the white and 21% of the black population in the United States are are blood type B (Stine, 1973).

Those papers indicating significant associations between a condition and a particular blood type also note that there was no association between a given blood group and certain other conditions. For example, there was no association of any of the ABO blood groups with cancers of the colon, rectum, or breasts (Matsunaga, 1962) or with increased susceptibility to infection with certain microorganisms, such as *Microtilaria loa.* Neither was there an association for diseases such as leprosy or pulmonary tuberculosis, nor was there any association established with respect to personality disorders such as schizophrenia and other common psychiatric illnesses. There was no correlation for the fast-moving band of serum intestinal alkaline phosphatase, serum uric acid levels, or body weight. The point is that we do not always find an association between a specified phenotype (trait) and the ABO blood types. This means (for example, in the United States with approximately 225 million people) that certain diseases appear with equal frequency regardless of their blood type, whereas certain other diseases appear to occur more frequently in persons of a particular blood group. There is, as referred to earlier, a great deal of

Table 16–3 Human Blood Types (ABO) and Associated Human Conditions				
	Blood Type			
Condition	*A*	*B*	*O*	*Investigator*
Cancers of				
Stomach	×			Newman et al., 1961; Doll et al, 1960 and 1961; Janus, Bailer, and Eisenber, 1967
Cervic	×			Matsunaga, 1962; Janus, Bailer, and Eisenber, 1967; Bergsjo and Kolstad, 1962
Pancreas	×			Aird, Lee, and Roberts, 1960
Salivary glands	×			Cameron, 1958; Osborne and DeGeorge, 1962
Ovary	×			Osborne and DeGeorge, 1963
Corpus of the uterus	×		×	Osborne and DeGeorge, 1963
Lungs (undifferentiated tumors)	×		×	Ashley, 1969
Lungs (gland ulcer differentiation)	×		×	Ashley, 1969
Lungs (squamous differentiation)	×		×	Ashley, 1969
Skin (xeroderma pigmentosum)			×	El-Hefnawi, Smith and Penrose, 1965
Ulcers of				
Duodenum (50% higher in O nonsecretors)			×	Merikas, Christakopoulos and Petropoulos, 1966
Gastric or peptic (highest in types A, B, and O nonsecretors and type O secretors)			×	Merikas, Christakopoulos and Petropoulos, 1966
Stomal or anastomotic			×	Merikas, Christakopoulos and Petropoulos, 1966
Diabetes mellitus				
Males only	×			Clarke, 1961
Females only			×	Clarke, 1961
Manic depressive			×	Kopec, 1956
Digitoxin-induced arrhythmias	×	×		*Laboratory Management*, 1973
Erythroblastosis fetalis (mother's type O, child type A or B)			× (mother's)	Matsunaga, 1962

Table 16–3 continued

Condition	Blood types A	B	O	Investigator
Ischemic heart disease (decreased blood supply due to constriction and blocking of blood vessels)	×	×		Levine, 1943
Pernicious anemia	×			Matsunaga, 1962
Bronchopneumonia	×			Billington, 1956
Rheumatic fever (carditis)	×	×		Clarke et al., 1960
Smallpox	×	×		Pettenkofer et al., 1962
Bubonic plague	×		×	Pettenkofer et al., 1962
Adenovirus	×			McDonald and Zuckerman, 1962
A_2 influenza virus			×	McDonald and Zuckerman, 1962
Myocardial infarction	×			Allan and Dawson, 1959
Cirrhosis of the liver	×			Billington, 1956
Arteriosclerosis obliterans	×			Weiss, 1972
Gall stones	×			Kjolbye and Nielsen, 1959
Female venous thromboembolic disease (from oral contraceptives)	×	×		*Laboratory Management*, 1973
Elevated serum cholesterol levels	×			Oliver et al., 1969
Electrophoretically slow moving serum alkaline phosphatase		×	×	Langman et al., 1969
High carbohydrate diet		×		Clarke, 1961
High fat intake	×			Clarke, 1961

controversy in this area as to the validity of the concept of blood group association conditions. Perhaps the real problem stems from the fact that we cannot, at the moment, explain *why* one blood group should be more susceptible or resistant to a given disease. Many times, by our failure to understand the cause–effect relationship, while awaiting the birth of new ideas for analysis and measurement, we dismiss the basic fact that there are correlations.

In concluding the discussion of inherited susceptibility or resistance to diseases, we can state with certainty that both the environment and our DNA are responsible for whatever susceptibilities we may have. Because we can only measure and recognize certain expressions and not others, we make relative choices on whether a disease is primarily a result of heredity or environment. A hereditary disease, given the proper environment, is not a disease, and would not be recognized. So it is that as the environment changes, we recognize new "hereditary" diseases. Conversely, one day we may be able to change the gene to fit the then current environment and the disease will be gone. It is susceptibility, not the disease, that is transmitted.

Susceptibility or resistance therefore is a potential that is inherited and studies in the future must take into account physiological conditions that encompass everything from nutrition to emotional stress and how these affect one's *susceptibility* to disease.

references

Aird, I., H. H. Bentall, and J. A. F. Roberts. 1953. Relation Between Cancer of Stomach and the ABO Blood Groups. *British Medical Journal, 1*:799–801.

Aird, I., et al. 1954. Blood Groups in Relation to Peptic Ulceration and Carcinoma of Colon, Rectum, Breast and Bronchus. *British Medical Journal, 2*:315-21.

Aird, I., J. A. H. Lee, and J. A. F. Roberts. 1960. ABO Blood Groups and Cancer of Esophagus, Cancer of Pancreas and Pituitary Adenoma. *British Medical Journal, 1*:1163–66.

Alexander, W. 1920–1921. Distributions of Blood Groups in Patients Suffering from Malignant Disease. *British Journal of Experimental Pathology, 1*(2):66–69.

Allan, T. M., and A. A. Dawson. 1959. ABO Blood Groups and Ischaemic Heart Disease in Man. *British Heart Journal, 30*:377–82.

Anderson, D. C. 1969. Genetic Varieties of Neoplasia. In Anderson Tumor Institute, *Genetic Concepts and Neoplasia.* Williams and Wilkins, Baltimore.

Ashley, David J. B. 1969. Blood Groups and Lung Cancer. *Journal of Medical Genetics, 2*:183-86.

Becker, B., and W. R. Morton, 1964. Phenylthiourea Taste Testing and Glaucoma. *Archives of Ophthalmology, 72*:323–27.

Bergsjo, P., and P. Kolstad. 1962. The ABO Blood Groups and Cancer of the Female Genital Tract. *Acta Obstetricia et Gynecologica Scandinavica, 41*:397–404.

Billington, B. P. 1956. Note on Distribution of ABO Blood Groups in Bronchiectasis and Portal Cirrhosis. *Australasian Annals of Medicine, 5*:20–22.

Bourned, J. G. H., O. J. Collier, and G. F. Somers. 1952. Succinylcholine (Succinoylcholine)—Muscle Relaxant of Short Action. *The Lancet, 1*:1225-29.

Buchanan, J. A., and E. T. Highley, 1920–1921. The Relationship of Blood-Group to Disease. *British Journal of Experimental Pathology, 1*(2):247-255.

Cameron, J. M. 1958. Blood Groups in Tumors of Salivary Tissue. *The Lancet, 1*:239.

Carter, C. O. 1969. Genetics in the Aetiology of Disease. *The Lancet, 1*:1014–16.

Clarke, C. A., et al. 1960. ABO Blood Groups in Rheumatic Carditis. *British Medical Journal, 1*:21–23.

Clarke, C. A. 1961. Blood Groups and Disease. *Progress in Medical Genetics, 2*:81–119.

Cohen, Bernice H. 1970. ABO and Rh Incompatibility. I. Fetal and Neonatal Mortality with ABO and Ph Incompatibility: Some New Interpretations. *American Journal of Human Genetics, 22*:412-39.

Doll, R., B. F. Swynnerton, and A. C. Newell. 1960. Observations on Blood Group Distribution in Peptic Ulcer and Gastric Cancer. *Gut, 1*:31–35.

Doll, R., H. Drane, and A. C. Newell. 1961. Secretion of Blood Group Substances in Duodenal, Gastric, and Stomach Ulcer, Gastric Carcinoma and Diabetes mellitus. *Gut, 2*:352–59.

El-Hefnawi, H., S. M. Smith, and L. S. Penrose. 1965. Xeroderma pigmentosum—Its Inheritance and Relationship to the ABO Blood Group System. *Annals of Human Genetics, 28*:273–90.

Evans, F. T., et al. 1952. Sensitivity to Succinylcholine in Relation to Serum-Cholinesterase. *The Lancet, 1*:1229–35.

Fox, A. L. 1932. The Relationship Between Chemical Constitution and Taste. *Proceedings of the National Academy of Sciences USA, 18*:115–20.

Fraser, G. R. 1961. Cretinism and Taste Sensitivity to Phenylthiocarbamide. *The Lancet, 1*:964–65.

Garrod, A. E. 1909. *Inborn Errors of Metabolism.* Oxford University Press (reprinted 1963).

Good, R. A., and J. Finstad. 1969. Inherited Immunologic Deficiency States and Lymphoreticular Neoplasia. In Anderson Tumor Institute, *Genetic Concepts and Neoplasia*, Williams and Wilkins, Baltimore.

Gowen, John. W. 1961. Genetic Factors in Diseases. In R. I. C. Harris (ed.), *Symposium on Problems in Laboratory Animal Disease.* Academic Press, New York.

Harris, H. W. 1961. High Dose Isoniazid Compared with Standard Dose Isoniazid with PAS in the Treatment of Previously Untreated Cavity Pulmonary Tuberculosis. In *Transactions Chemotherapy of Tuberculosis Conference, 20*:39–68.

Harris, H. D., and W. H. Trotter. 1949. Taste Sensitivity to P.T.C. in Goitre and Diabetes. *The Lancet, 2*:1038–39.

Herndon, C. N., and R. G. Jennings. 1951. A Twin-Family Study of Susceptibility to Poliomyelitis. *American Journal of Human Genetics, 3*:17–46.

Janus, Z. L., J. C. Bailer III, and Henry Eisenber. 1967. Blood Groups and "Uterine Cancer." *American Journal of Epidemiology, 86*:569–78.

Kallman, F. J., and D. Reisner. 1943. Twin Studies on Genetic Variations in Resistance to Tuberculosis. *Journal of Heredity, 34*:269–76, 293–301.

Kitchen, F. D., et al. 1959. P.T.C. Taste Response and Thyroid Disease. *British Medical Journal, 1*:1069–74.

Kopec, Ada C. 1956. Blood Groups in Great Britain. *The Advancement of Science, 13*: 200–203.

Laboratory Management. 1973. Adverse Drug Reactions, *11*(April):29.

Langman, M. J. S., et al. 1969. ABO and Lewis Blood-Groups and Serum-Cholesterol. *The Lancet, 2*:607–609.

Levine, P. 1943. Serological Factors as Possible Causes in Spontaneous Abortions. *Journal of Heredity, 43*:71–80.

Lynch, H. T. 1967. *Hereditary Factors in Carcinoma.* Springer-Verlag, New York.

Marcus, Judith R., and Gerald Cohen. 1971. The Riddle of the Dangerous Bean. In James D. Ray, Jr., and Gideon E. Nelson (eds.), *What a Piece of Work Is Man.* Little Brown, Boston.

Marx, Jean L. 1974. Viral Carcinogenesis: Role of DNA Virus. *Science, 183*:1067–68.

Matsunaga, E. 1962. Selective Mechanisms Operating on ABO and MN Blood Groups with Special Reference to Prezygotic Selection. *Eugenics Quarterly, 9*:36–43.

McDonald, J. C., and A. Z. Zuckerman. 1962. ABO Blood Groups and Acute Respiratory Virus Disease. *British Medical Journal, 2*:89–90.

Merikas, G., P. Christakopoulos, and E. Petropoulos. 1966. Distribution of ABO Blood Groups in Patients with Ulcer Disease. *American Journal of Digestive Diseases, 11*:790–95.

Meyer, S. N., and C. M. Jensen. 1951. Significance of Familial Factors in the Development of Tuberculin Allergy. *American Journal of Human Genetics, 3*:325–31.

Nagao, Takeshi, and A. M. Mauer. 1969. Concordance for Drug-Induced Aplastic Anemia in Identical Twins. *New England Journal of Medicine, 281*:7–11.
Neel, James V., and William J. Schull. 1954. *Human Heredity.* University of Chicago Press, Chicago.
Newman, E., et al. 1961. Secretion of ABH Antigens in Peptic Ulceration and Gastric Carcinoma. *British Medical Journal, 1*:92–94.
O'Reilly, R. A., J. G. Pool, and P. M. Aggeler. 1968. Hereditary Resistance to Coumarin Anticoagulant Drugs in Man and Rat. *Annals of the New York Academy of Sciences, 191*:913–31.
Oliver, M. F., et al. 1969. Serum-cholesterol and ABO and Rhesus Blood Groups. *The Lancet, 1*:7621–22.
Osborne, R. H., and F. V. DeGeorge. 1962. ABO Blood Groups in Parotid and Submaxillary Gland Tumors. *American Journal of Human Genetics, 14*:199–209.
Osborne, R. H., and F. V. DeGeorge. 1963. ABO Blood Groups in Neoplastic Disease of the Ovary. *American Journal of Human Genetics, 15*:380–88.
Otten, C. M. 1967. On Pestilence, Diet, Natural Selection and the Distribution of Microbial and Human Blood Group Antigens and Antibodies. *Current Anthropology, 8*:210–26.
Pettenkofer, H. J., et al. 1962. Alleged Causes of the Present-Day World Distribution of the Human ABO Blood Groups. *Nature, 193*:444–46.
Porter, Ian H. 1968. Genetic Susceptibility. *Heredity and Disease.* McGraw-Hill, New York.
Stine, G. J. 1973. Human Immunogenetics. Ex. 9 in *Laboratory Exercises in Genetics.* Macmillan, New York.
Sheba, Chaim. 1970. Gene-Frequencies in Jews. *The Lancet, 1*:1230–33.
Vogel, F. 1970. Controversy in Human Genetics: ABO Blood Groups and Disease. *American Journal of Human Genetics, 22*:464–76.
Weiner, A. S. 1962. Blood Groups and Disease, a Critical Review. *The Lancet, 1*:813–15.
Weiner, A. S. 1970. Blood Groups and Disease. *Americal Journal of Human Genetics, 22*:476–83.
Weiss, Noel S. 1972. ABO Blood Type and Arteriosclerosis obliterans. *American Journal of Human Genetics, 24*:65–70.
Whiting, P. W. 1929. *Heredity and Human Problems.* Radio publication 50, University of Pittsburgh, pp. 32–41.

part 7

application of counseling, screening, and medicine to genetic problems

genetic counseling

> *There is no road or ready way to virtue.*
>
> *Sir Thomas Brown*

The term *genetic counseling* was coined by Sheldon Reed to replace the term *genetic hygiene*, which had acquired unpleasant eugenic implications. Originally genetic counseling consisted of a single interview in which the pedigree was taken and a prediction of recurrence risk given, and the literature dealt mainly with methods of arriving at estimates of recurrence risks. But merely to recognize a hereditary disease and diagnose the type of inheritance is no longer enough.

The modern counselor must ethically deal with emotions of his own and those of the consultee (person seeking advice). Now, because of the use of techniques such as amniocentesis and fetoscopy during pregnancy, emphasis has been placed on the moral, ethical, and philosophical concepts involved in genetic counseling. Numerous articles are appearing on the techniques, philosophy, psychodynamics, and ethics of counseling in relation to the impact of inherited diseases in the family. Suddenly the consultee and the counselor are faced with choices that were once left to fate. Should the genetically defective be aborted? Do parents have a *right* to produce defective children? Is it wrong to select the sex of your offspring? The methodology of counseling is in a highly experimental stage; it will no doubt be some time (if ever) before there is any general agreement on "the optimal procedure" for counseling (Fraser, 1974).

what is genetic counseling?

Genetic counseling applies the knowledge gained in the field of human genetics. The first goal of the genetic counselor is to determine whether the disorder in question is inheritable. For example, environmental agents such as the German measles virus, the chemical thalidomide, and radiation cause infant malformations or birth defects during the first trimester of pregnancy. Yet, if no changes have occurred in the malformed persons' gametes, the deformity or birth defect will not be genetically transmitted to their offsping.

The counselor is concerned about all types of birth defects. He has to be, because if the defect is not hereditary, his advice to the consultee will be quite different than if the disorder is hereditary. Through the application of the patterns of inheritance the counselor can help parents prevent the occurrence of a number of inherited disorders that may occur in their families. To date there are no cures for inherited diseases. A cure can come only through the use of genetic surgery whereby a faulty gene is replaced with a new one. There are, however, a number of inherited disorders that can be treated with various degrees of success. Some therapies for genetic defects are discussed in Chapter 19.

An enormous volume of facts has been collected in the field of human genetics in the past 25 years. Collectively these facts offer the opportunity for new insight into a number of age-old genetic diseases. With the increase in knowledge of human genetics came an increase in demand for genetic counseling. With the increased use of genetic counseling we have come to realize that we do not yet have the optimal method of presenting genetic data to the layman nor do we have a sufficient number of counseling centers to cope with the numbers of people that could benefit from the service. To date, for many different reasons, less than 5% of the people who could benefit from the counseling service have talked with a genetic counselor. The National Foundation/March of Dimes directory lists 680 genetic counseling centers for a world that will soon have 5 billion people (Bergsma and Lynch, 1971). The United States has at least 270, or approximately one third of the relatively few that exist, but with a population of approximately 215 million, this translates into one center per 800,000 people. In the United States, between 5 and 10% of the population show an inherited disease. Some 25% of all beds in hospitals and institutions are occupied by people with inherited disorders. Still, these data on inherited defects do not include simple susceptibilities to other agents.

Recent findings suggest that 20% of all heart attacks occurring in men before the age of 60 are caused by one of three genes that regulate the body's fat metabolism. One gene controls the level of triglycerides; the second controls the level of cholesterol in the blood, and the third is responsible for combined hyperlipidemia. It is estimated that 1 in every 160 Americans has one or another of these three genetic conditions that predispose an individual to heart attack, making them "among the most common disease-producing genes in our population." The frequency is one of the highest known for an inherited defect (DHEW, 1974).

application of counseling, screening, and medicine to genetic problems

A study at Montreal Children's Hospital (Scriver, 1974) revealed that approximately 11% of all children admitted had an inherited single gene defect, 0.5% had a chromosomal aberration, and 19% had polygenic disorders. Collectively, approximately 30% of the children, or about one in three, admitted to that hospital had a genetic disease. The same hospital records showed that one in every eight adults had been admitted for a genetic disease.

In a survey in Newcastle-upon-Tyne, it was found that no less than 42% of childhood deaths in hospitals could be attributed to genetically determined disorders (Emery, Watts, and Clark, 1973) In the United States, based on an annual birthrate of approximately 3 million, conservative calculations indicate that 250,000 children have one or more of the 2000 known inherited disorders. Many of these disorders are listed in Chapters 8, 9, 10, and 11. Those chapters discuss the various modes of inheriting genetic diseases as well as many of the emotional and financial aspects of certain specific disorders.

who does genetic counseling?

A recent issue of the *Genetic Counseling Newsletter* (1973a) indicates that if we include all known genetic counselors listed in state health department records with those persons counseling at the 270 centers, there are approximately 700 active genetic counselors in the United States or one genetic counselor for 300,000 people. M. D. Herzbert (1972) states that only 2% of those requiring genetic counseling are receiving it. In spite of this short supply of professional counselors, there are only two colleges in the United States (Sarah Lawrence and Rutgers)—and none in Canada—that have a "genetics associate" program for counselors in their academic curriculum. Fraser (1974) points out that there is no prescribed genetic counseling training program in North America. As indicated, because of the lack of college and university programs, most often the counselor has had no formal academic or medical genetic training in the techniques of counseling.

Depending on the counseling center, the interviewing and counseling are often done by one person; in other centers the counseling is performed as a multiple interview, multidisciplinary process. While the multidisciplinary counseling center offers a variety of skilled professionals, each performing one phase of the overall interview and diagnosis, the single counselor approach offers a better opportunity for establishing confidence between the counselor and the consultee. However, the overall concern must be complete accuracy in diagnosis, best possible therapy, and follow-through with support services. In view of these criteria, it would appear that the best possible situation is a counseling team, preferably associated with a hospital. It would be difficult, if not impossible, for an individual counselor in private practice to provide immediate accurate service without the aid of a cytogenetics laboratory and the proper biochemical diagnostic facilities. Screening for the various disorders takes the skills of many professionals.

On the premise that the genetic counseling center approach is the best possible method of delivering genetic medicine to the patient and the family, the National Genetics Foundation has linked 46 of the leading medical institutions throughout the United States and Canada into a dynamic, collaborative network of Genetic Counseling and Treatment Centers that provide what is undoubtedly the greatest competence in genetic disorder therapy available in the world today. A central referral system directs patient or physician to the appropriate center.

what should be the genetic counselor's role?

Most geneticists agree that a genetics counselor should advise the consultee on the probability of a defective birth, explain the various patterns of inheritance, and work up a pedigree chart for the consultee. But after the counselor collects the facts, *how* does he relate the knowledge he has gained to the consultee? In other words, what is "the" method of genetic counseling? Quickly we realize that there is no "best" method for genetic counseling because counseling is an "understanding" of a set of facts according to the counselor's frame of reference, background in the science of genetics, and previous training and experience in effectively communicating with consultees.

Thus, the question "What is the best method of genetic counseling?"—because it deals with personal judgments and the interpretation of facts—has a relative answer. Many times a counselor, in making a decision, is faced with a number of valid alternatives, but he must select the one *he feels most appropriate* for the situation. Certainly, most would agree that a counselor should have up-to-date knowledge. But in explaining the medical facts, can he be certain that he has the latest information? How sure can he be that he knows the best possible treatment available? Even when the counselor is up-to-date in his scientific knowledge, he must judge the best possible way to communicate the information to his consultee. He can never be sure just how or if a consultee has understood what has been said. It is most difficult to evaluate the meaning the consultee derived from the counseling session. Further, the circumstances that usually bring about the counseling session result in the consultee being apprehensive. Learning that he may be involved in nature's game of hereditary roulette can be very unnerving and lead to confusion and misunderstanding as to what has been said by the counselor. It should also be realized that the public, in general, is unaware of specific genetic disorders. In other words, when the consultee reaches a counselor's office, he receives a great deal of new information on top of the emotional burden he may already have and he may add an emotional response that serves as a barrier to further counseling.

In a recent survey of parents who had children with the disease phenylketonuria, 61% did not know that the disease was inherited, 58% did not understand the importance of early diagnosis, 56% said they never received professional counseling and 56% were unaware that a special diet was available that might help their child (Frankel, 1973).

In order to communicate effectively, the counselor must consider the educational background of the consultee. In many cases the counselor deals with consultees of low IQ. Even those consultees with normal or higher than normal intelligence rarely have a working knowledge of human genetics, chromosome aberrations, or hereditary disease. The counselor *decides* how he will tell the consultee of his findings. The ethical decision by the counselor on *how* and *what to tell* the consultee may limit the ways in which the counselor can communicate with the consultee. Such decisions, although made with the best interests of the consultee in mind, limit the knowledge of the consultee as he prepares to make decisions. Leonard, Chase, and Childs (1972) studied the response to and effectiveness of genetic counseling. They found that 44% of the 61 families investigated generally misunderstood the material presented either because the counselors too often "selected" what they would tell the consultees or because of the educational background of the consultee. The effectiveness of counseling was measured by the way consultees had used the information they had received and the extent to which they had comprehended it. The studies concluded that the principal obstacles to the effective use of genetic counseling were religion, denial based on emotional conflicts, and a lack of knowledge of genetics and biology.

An equally difficult assignment for the counselor is presenting his knowledge in an unbiased manner. It is highly improbable that any counselor can impart information in a particular case without some personal bias. The reason most likely arises from the fact that in order to counsel, the counselor must first have access to the consultee's personal and family history. The history includes parental age, ethnic background, reproductive history in the family (abortions, stillborns, or dead siblings), and the age, sex, and health of all living children.

Bearing in mind that a consultee may have one or all of the above obstacles to understanding a counselor, counselors in many instances choose a *nondirective* rather than a *directive* approach to genetic counseling. The counselor who is directive takes part in the decision-making process of the consultee; he "fails" if the consultee does not take his advice. The nondirective counselor "fails" only if he does not succeed in allowing the decisions to be made by the consultee.

the nondirective approach and truth telling—an ethical decision

The nondirective approach involves the presentation of facts in an unbiased manner, with a sincere, honest, and sensitive attitude toward the consultee as an individual, and places the entire responsibility for decision making in the hands of the consultee. The counselor tries to be continually aware of the consultee as a person and to mantain a balance between scientific objectivity and sympathetic sensitivity (Fraser, 1963).

The counselor's neutrality in the decision-making process should be a primary goal. This point cannot be overemphasized, especially when one considers that although our knowledge in human genetics is increasing rapidly, it is far from complete. Counselors can be, and have been, fooled with respect to certain inherited conditions because of improper measurements and observations and/or because a number of genetic diseases have very similar symptoms.

Kenneth Vaux (1974) refers to a case of a lady with strong religious convictions who had a child with Tay–Sachs disease (discussed in Chapter 10). When this lady accidentally became pregnant again, she sought amniocentesis and other tests to determine the "genetic health" of the fetus. Several tests gave what appeared to be conclusive evidence that the child had the disease. After much contemplation and consultation, the family decided on abortion. *The abortus was completely normal.* The shocked family had been cautioned regarding the inconclusiveness of the tests used. Still they had not been prepared for this result. From traditional faith they affirmed the sanctity of life and decided on abortion only with great hesitation. Family and friends suggested both immediate abortion and allowing the pregnancy to go to term, hoping for a normal child against the evidence. Already having one Tay–Sachs child weighed strongly against having another.

The case vividly points up the limitations of genetic knowledge and technology. It highlights the deep conflict of values accompanying decisions in this area. It points to the awesome weight of the decisions made, and finally highlights the importance of presenting the facts in an unbiased manner and of letting the parents make the decision. However, as stated before, the counselor probably cannot completely disassociate himself from his own values and present the information in such a way that the recipient is completely free to make his own judgment. For example, in interpreting the probability even for a simple single gene disease, the counselor, depending on his level of personal emotional involvement in the particular case, may bias or slant the data. The counselor may not change the "truth," but his tone of voice, manner of speech, and other facial and body gestures can influence the information transfer, counselor to consultee. In a case where a counselor feels that a pregnancy might be best for a family he could say to Mr. and Mrs. X, there is *only* one chance in four that your child would be affected. Your chances for a healthy birth are *very* high, three chances out of four or 75%. For family Y with the same inherited defect, but a different family social history, the counselor could say, your chance of a normal child is *only* 75%; there is a 25% chance that any pregnancy will be defective, a relatively *high* risk. In both cases, the counselor presented truthful data, but the wording was slightly changed to place the emphasis on the birth of a normal child in the first case and on the birth of a defective child in the second.

In still another situation of "truth telling," the counselor may be faced with a particularly difficult problem of how much to tell a consultee and when. How much "truth" does a counselor owe a patient? Should this question even be asked?

Robert M. Veatch (1972) reviewed the case of a 41-year-old pregnant schoolteacher. She realized the increased chance of the child being a mongoloid (Down's syndrome) owing to her age (see Chapter 6), yet she did want the child if normal. She was in the 18th week of pregnancy had two previous normal children, and decided to seek counseling. The genetic counselor recommended amniocentesis for a chromosome analysis for Down's syndrome.

The results of the karyotype analysis of fetal chromosomes showed the child would be normal in this regard. However, the cytologist found that the child carried an extra Y chromosome, that the child was XYY instead of XY. As discussed in Chapter 6, some research suggests that XYY males may be inclined to violent acts, including sexual offenses, although other recent studies do not support this.

Considering the inconclusive nature of such research, the possible danger to society and the consultee's family, the impact of the information on the way the consultee might treat the child, the counselor was faced with the dilemma of whether to tell the consultee the child carried an extra Y chromosome. After all, the consultee only asked to know whether the child would have Down's syndrome. And that question was answered by the counselor. However, is it not a counselor's moral duty to do "no wrong to his patient." But which is wrong, telling or not telling of the XYY condition? Since, in fact, the evidence for XYY-associated abnormality is weak at best should the counselor create anxiety in what would have otherwise been a very happy consultee? But what if she later learns of her child's extra Y chromosome? She may, in fact, be less able to handle the situation then, and her emotional upset would be compounded by her loss of trust in her counselor. Certainly, there are many arguments for and against telling the consultee about the extra Y chromosome in the developing fetus. Veatch, in offering his opinion, states that the counselor "ought to share his discovery with the consultee," Do you agree?

In responding to the article written by Veatch concerning the XYY case, a Boston Lawyer revealed a similar situation (Annas, 1972). It appears that an Assistant District Attorney in Massachusetts threatened proceedings against a doctor to revoke his license to practice medicine when he discovered that the doctor had withheld information that his six-month-old son was XYY. He only withdrew the suit when the doctor agreed to advise him on how to raise his son and to inform all other XYY parents of the genetic composition of their children within a reasonable time after he learned of it. Should the Assistant District Attorney have coerced the doctor into informing all other patients?

Many genetic counselors, like medical doctors, attempt to operate on the principle of nondirective counseling. But it is difficult not to involve the value system of the counselor in the counseling situation. A recent survey of practicing genetic counselors in the United States concluded that counselors hold widely varying opinions on how directive they should be, yet most tend to see themselves as nondirective. The data suggest that the majority of counselors do in fact influence decisions. Only 54% of all the counselors queried stated that they leave all decision to the patient while 7% tell consultees what they themselves would do or advise them. Thirty-nine percent show personal bias with respect to the consultee's decision (*Genetic Counseling Newsletter,* 1973b). In brief, it appears that the consultees' decisions to marry, to have an abortion, or to adopt children will be guided by their educational background and personal needs and also by the degree of directiveness of their genetic counselor.

Whether the counseling is directive or nondirective, certain criteria are valid:

1. Making the consultee feel at ease.
2. Ascertaining that the consultee understands what the counselor has said (Herzbert, 1972, says that most adults give a positive response when asked if they understand, even though they have little or no comprehension).
3. Keeping the counseling session in confidence.
4. Stating the mechanics of inheritance to the consultee in the simplest of terms.
5. Using layman's vocabulary where possible.
6. Setting time for a return visit.

The major difference between directive and nondirective counseling is in whether or not the counselor actively participates or helps the consultee to make a particular decision. Directive counseling has a positive influence on the consultee's decision. Whether one receives directive or nondirective counseling depends on the *moral* judgment of the counselor.

Counselors *can aid* the consultee in weighing the risk in personal terms so that a rational plan of action can be formulated. But the counselor should *never coerce* the consultee. The consultee with a nondirective counselor will always be the ultimate decision-maker.

A question that has been raised many times is: Can society afford the price of allowing the individual the ultimate choice when it comes to giving rise to severely defective offspring? In trying to answer the question, we immediately become involved in personal as opposed to group ethics and moral behavior. A number of topics that can be discussed by counselors with their consultees in order to prepare the consultee to make a decision are: the religious views of the consultee, the economic burdens of the particular defect, the effects on family life, the social stigmas, the economic burdens to society, the alternative forms of parenthood, the notification of the extended family, the consultee's view of the disease and its influence on limiting the family, methods of contraception, the consultee's attitude toward sterilization, and the possible effects on the consultee's sex life. Consideration of these criteria with the consultee by a counselor implies that he is making moral judgments for and perhaps about his consultee.

consultee adjustment

The impact of a diagnosis on the consultee can cause a variety of reactions. The common reactions include anger, guilt, and blame. In the event a decision is to be made by the consultee a psychological adjustment may also have to be made. Three stages of consultee adjustment in coping with their problems have been described in the literature.

The first stage is characterized by shock and disorganization and is relatively short. During this stage the counselor should be particularly sensitive to the consultee's feelings.

The second stage involves rationalization, denial, and blame casting. During this period the parents are likely to reject the counselor and what he is imparting. For example, Shore (1973) presents the following picture of parent denial:

> A father sits happily holding his severely retarded five-year-old son who is unable to walk, stand or sit unaided. The child cannot speak and he has repetitive myoclonic seizures. The child's mother devotes her full time to exquisitely detailed care of the boy, neglecting her seven-year-old daughter who is rebellious and in danger of dropping out of school. The father, when asked about the son, says cheerfully, "Fred is such a good little fellow, we'd like to have another one just like him . . . ."

The emotional burden varies according to whether the patient has a defective child, is carrier of a defective gene, or is a possible defective. Many women blame themselves for the birth of a defective child because of some trouble during their pregnancy. Often couples or individuals feel that the defect was brought on by some indiscretion in their past. Consultees may need support in facing relatives who want to place the blame on one side of the family (Sly, 1973). For example, a dominant defect usually appears in only one side of the family. The father's side of the family may blame the mother's and vice versa. Emotional problems may involve the entire family structure.

Franz J. Kallman (1965) points out that in no situation calling for genetic counseling should it be taken for granted that the person will be able to attain a realistic attitude without the counselor's help. Even the person of high intelligence may be emotionally shaken when presented with frightening factual information.

The third stage is "equalization." If this phase is reached, then the parents are able to act more realistically and function more effectively.

One last point should be made with respect to the performance of the genetic counselor. The responsibility assumed by the counselor is, in a psychological-medical sense, awesome. In medicine an error may have dire consequences for the individual receiving the treatment, but an error in genetic counseling may last for generations.

reasons for seeking genetic counseling

The recognition of the Mendelian principles of inheritance in pea plants and their eventual application to human heredity gave birth to the field of genetic counseling. Today, with rapid progress in the fields of biochemistry, molecular biology, cytology, and medicine, we approach the "coming of age" of the genetic counselor. The shortage of counselors, the wide range of possible problems, and the increased awareness of the benefits from counseling may soon lead to the problem of who shall be counseled. The idealistic answer is: those for whom it is most meaningful, or to whom the counseling is of most value. That, of course, is an ideal. In reality, it may be that those who can afford it will be most apt to be counseled.

Without discussing why the service of genetic counseling is in short supply or the ethical issues of who shall decide *who* receives counseling where it is in short supply, we can discuss the reasons people most often seek genetic counseling.

One or both prospective parents may have a disorder and either know of its hereditary nature or suspect that their problem may be transmitted to their children. Parents in this group, when the illness is actually inherited, usually demonstrate a dominant disorder.

Dominantly inherited diseases are characterized by the fact that only one of the two parents is defective (Figure 17–1). Autosomal dominant inheritance, as explained in Chapters 8 and 9, generally affects individuals through two or more sequential generations. Individuals who are heterozygous for an autosomal dominant trait have a 50% chance of producing a defective offspring at each pregnancy. The variability (penetrance and expressivity, explained in Chapter 8, Table 8–3) of the phenotypic expression may make it difficult to determine if a person carries the responsible gene.

If the disorder has variable penetrance, the counselor must calculate the difference between the expected 50% occurrence of the trait by chance and the

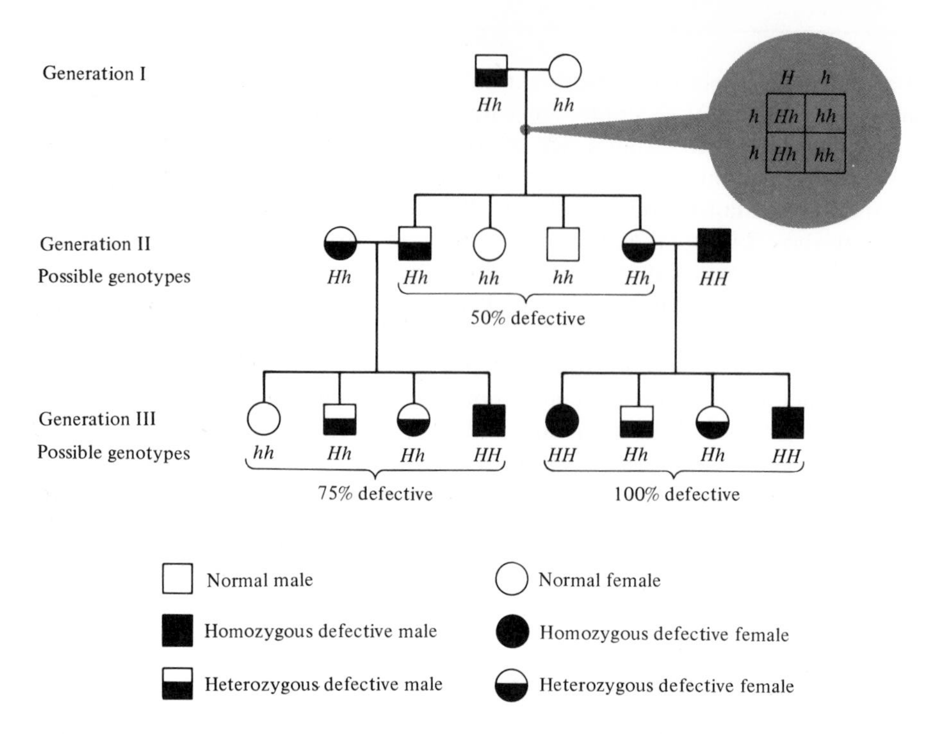

Figure 17–1. Human Pedigree Analysis for Huntington's Disease. Huntington's disease is a Mendelian domiant trait; that is, in the heterozygous condition the defective gene dominates over the normal gene. In generation I the father is heterozygous and defective while the mother is homozygous normal. In generation II there is a one in two chance (50%) that any child will be defective, carry one defective (*H*) gene. In generation III the chance of being normal or defective for any given birth is given. Note that in generation III the chance that any child on the left side of the pedigree will be defective is 75%, while on the right side the chance that any child will be defective is 100%, that is, each child will carry one dominant gene (*H*). For other examples of dominant defects see Table 9–2.

actual observed frequency of occurrence of the trait within the particular family being studied. For example, the dominant defective gene for congenital ptosis (drooping of the upper eyelid due to paralysis) has a penetrance in the general population of approximately 60%. An affected consultee should therefore be given a risk factor of 30% (0.6×0.5) for the probability of having a child that will *express* the defect. The consultee should realize, however, that the chance of the gene being transmitted to any child is still 50% and if a child receives that gene he will have the same 30% chance of producing a child showing the disease. However, for most of our severe dominant gene defects, penetrance is nearly 100% and the person who does not express the disease will not transmit the disease to his children.

application of counseling, screening, and medicine to genetic problems

A partial list of the over 1000 dominantly inherited traits is presented in Table 9–2. It can be seen that there are dominantly inherited diseases known to affect the skeleton, skin, muscle, organs, and nervous system. The dominant diseases that create the more severe emotional and financial problems are those whose expression occurs or persists after the first year of birth (see Table 9–1).

To demonstrate how severely one can be emotionally involved, Clark Fraser (1963) cited a 27-year-old woman who sought genetic counseling because her mother and five of her aunts and uncles had developed Huntington's disease. The young woman, after being informed of the full implication of the disease and her chance of expressing its symptoms, was visibly shaken. She feared the development of the disease, and the transmission of it; consequently she avoided marriage. After some time, she began to develop the symptoms she observed in her mother and aunts. The genetic counselor referred her to a neuropsychiatrist who found, as often occurs in cases of such emotional impact, that the symptoms were psychosomatic. She watched her mother and affected aunts gradually become suspicious and stubborn and develop increasingly short tempers. Her mother was terribly hard to live with; she could hardly walk, and her speech had become unintelligible. As counseling continued, the counselor could only help the young woman to deal with her bitterness and hate. Earlier in life she had wanted to marry and have children, but she feared for her children and for her husband if she should develop the disease. Even though she was free of the disease, she developed psychosomatic symptoms, and the continuing fear of Huntington's disease deprived her of a normal, happy life.

If an individual with Huntington's disease decides to marry, or finds out about the implications of the disease only after getting married, the genetic counselor can discuss the alternatives to producing children. Possibly the couple will decide they do not need children for a happy life, or they may adopt children, or with modern technology donor insemination may be an acceptable alternative.

In cases where there is a sporadic occurrence of a dominantly inherited disease, the first appearance in a family may indicate a new mutation in one of the two normal genes or in the gamete produced by one of the parents. The parents should be counseled according to their emotional acceptance of that birth. Although the chance of a recurrence from the same parents is negligible, the parents must deal with the burden of the affected child who will have a 50% chance of passing the error on just as for any other case of a dominantly inherited disorder.

In general, those persons most often seeking genetic counseling already have one child with a birth defect. The parents are usually referred to a counseling center by the family physician, obstetrician, or pediatrician. The afflicted child may be the result of a new dominant mutation as described above. However, more likely, the unexpected genetically afflicted child results from the chance uniting of gametes carrying recessive genes, one from each parent (Figure 17–2; also see Chapters 8 and 10). The question in the minds of parents is whether subsequent children are in any danger of inheriting the disease. For a recessive trait there is a one in four chance that each child resulting from the mating of heterozygous parents will exhibit the disorder. The heterozygous parent is also referred to as a carrier because the defective gene is present, though not expressed. For rare recessive traits, one parent is usually homozygous normal, so that many generations may pass without the expression of the defective gene. The fact that inherited traits of a recessive nature can skip many generations and thus appear sporadic in a family pedigree makes it difficult in some cases for one to distinguish a recessive trait from the ocurrence of a new mutation. However, a detailed pedigree analysis, provided one can be

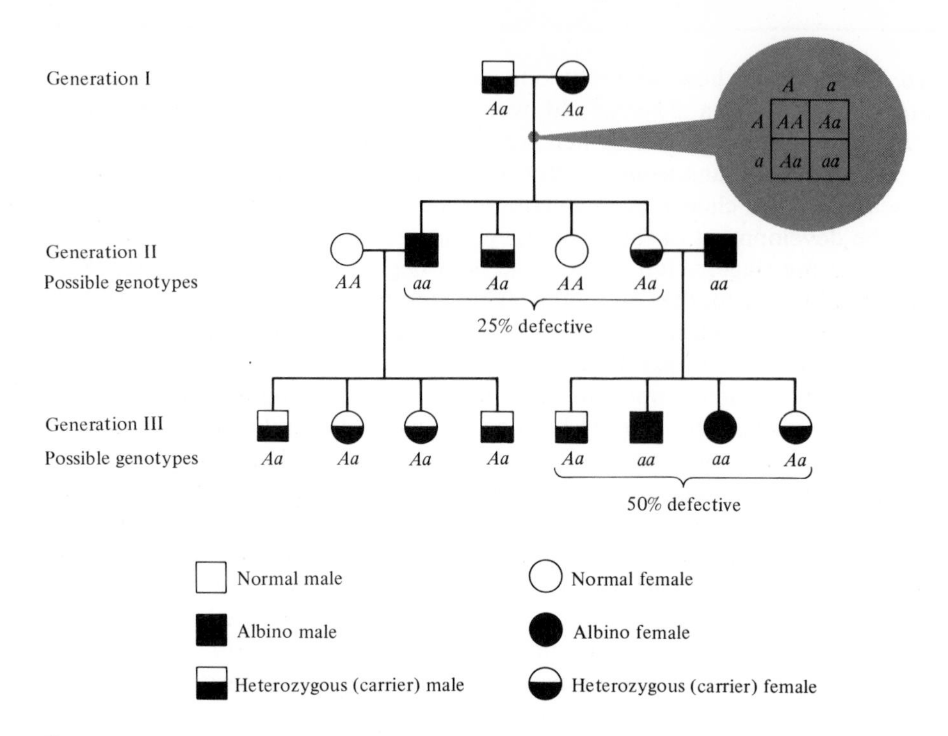

Figure 17–2. Human Pedigree Analysis for Albinism. Albinism is a Mendelian recessive trait that is expressed only in the homozygote. In generation I both parents are heterozygous normal, each carrying only one recessive gene. In generation II the chance of having a defective child is 1 in 4 (25%). In generation III, based on the parents with genotypes as shown on the left, all new births will be normal and heterozygous; on the right 50% will be normal and 50% defective. For other examples of recessive defects see Table 10–1.

worked up (if sufficient accurate and detailed information is available), may allow the counselor to make the distinction.

One aspect of a recessively inherited disease that differs significantly from the dominant mode of inheritance is that one half of the children produced by two heterozygous, but phenotypically normal, parents are expected to be heterozygous carriers like the parents. That is, the chance of any pregnancy producing a heterozygous carrier child is one in two.

For example, a mother whose first baby was born with Tay–Sachs disease would have a one in four chance of having another Tay–Sachs child (if she mated with the same male parent), a one in two chance of producing a heterozygous carrier, and a one in four chance of producing a homozygous normal child. If she became pregnant and prenatal tests showed the child to be normal, that child would have a 67% chance (two in three) of being a carrier (since we already know it is not a Tay–Sachs child). However, at conception the chance of being a carrier was 50%. The relative increase in probability of producing a carrier will occur in every recessive disorder that can be diagnosed during development if every homozygous abnormal is sacrificed. There are about 800 known recessive disorders, a number of which are described in Chapter 10 and Table 10–1.

application of counseling, screening, and medicine to genetic problems

Explaining the inheritance of sex-linked disorders to a couple seeking counseling is different from explaining the inheritance of autosomal traits. Associated with X-linked dominant inheritance is the crisscross pattern of inheritance, father to daughter to grandson. Because males give their X chromosome only to female offspring, affected males can be assured of normal male offspring, although all their daughters will be affected. Females have the probability that one half of their children, regardless of sex, will be affected (Figure 17–3).

If the X-linked trait is inherited as a recessive, the heterozygous female is a carrier while the hemizygous male expresses the defect. For example, if the phenotypically normal mother of a child affected with the X-linked recessive disorder hemophilia also has an affected brother or uncle, she is presumed to be a carrier; that is, she has a recessive gene on *one* of her X chromosomes, and each male offspring she may bear will have a 50% chance of being affected. The heterozygous female, carrying a normal allele, does not manifest this disease clinically. Occasionally, however, careful examination may reveal an increase in blood-clotting time (see Chapter 11).

Affected males cannot have affected male offspring unless the mother is heterozygous. In this case the defective gene came from the mother, not the father. Thus, an affected male can be assured (provided the mother is homozygous normal) of having clinically normal children, although all of his daughters will be carriers and thus could present him with affected grandsons. Parents of affected males have a probability of 50% that each subsequent male offspring will be affected (the mother is heterozygous). Normal males have no risk of passing the disease on to subsequent generations. A sister of an affected brother has a 50% probability of being a carrier, just as each brother has a 50% chance of receiving the defective gene from the mother. The combined probability that the sister will have an affected male child is 25% (0.50×0.50). Of course, if clinical or laboratory examination reveals that the sister is a carrier or if she has proved her carrier status by having an affected child, the risk immediately reverts to 50%, similar to that which existed in her mother. There are at least some 150 known X-linked disorders, a number of which are described in Chapter 11 and Table 11–1.

People with polygenic inherited traits—for example, cleft lip and palate, spina bifida, congenital hip, and birth defects caused by chromosome abnormalities—arrive at the counselor's office for various reasons, but both polygenic defects and chromosome abnormalities are slightly different from what has been discussed. Polygenic or multifactoral traits (see Chapter 8 and Table 8–4) have a lower frequency of initial occurrence and recurrence in a family, and most counseling for such traits is based on empirical data. Disorders that do not follow simple patterns of inheritance, and in which no chromosomal abnormality can be found, cannot accurately be predicted to occur or recur within a pedigree and represent difficult problems for the genetic counselor. Often a polygenic cause is assumed, but the degree to which both the environment and defective genes interact to produce the phenotype is unknown. For polygenic disorders, it is especially important to pursue every aspect of the family history in order to rule out the possibility that the defect is a subtle expression of a more generalized, simple inherited disorder. If the

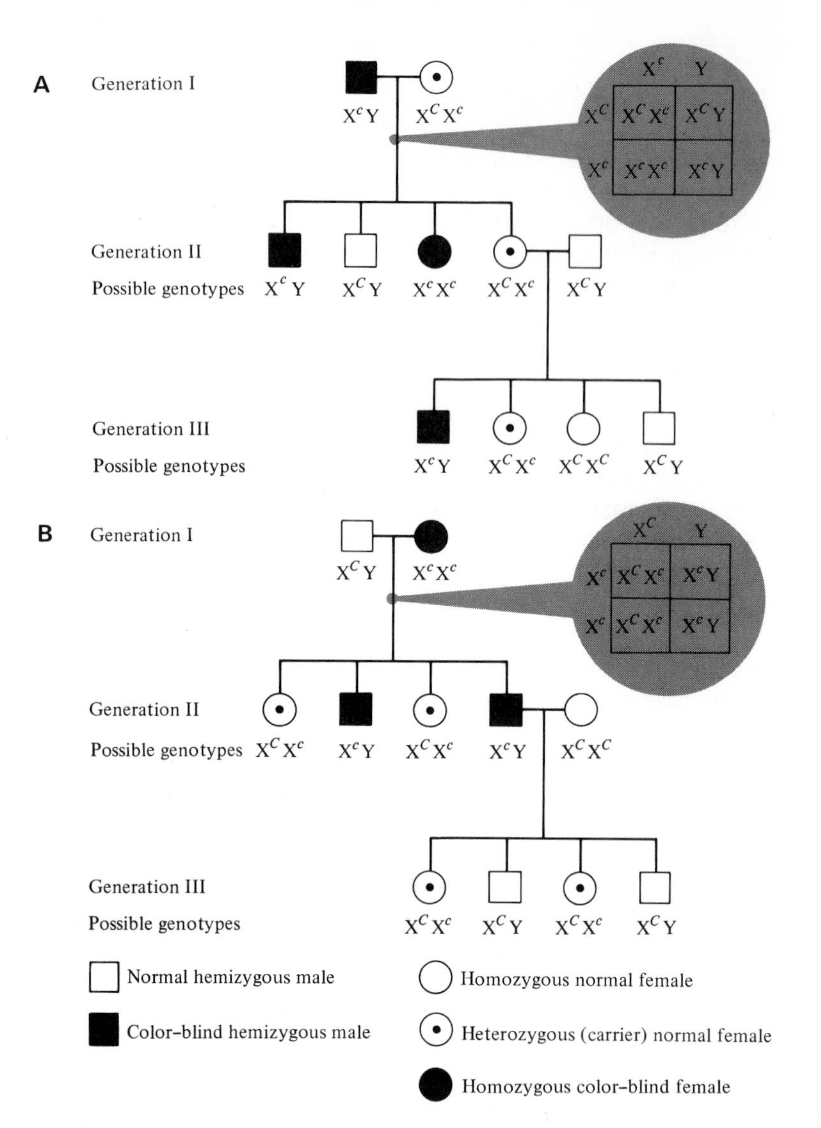

Figure 17–3. Human Pedigree for Color Blindness, a Sex-Linked Recessive Trait. **A.** Color-blind father and heterozygous normal mother (generation I) could give rise to normal and defective sons and daughters (generation II). Marriage of carrier female to normal male in generation II can give rise in generation III to defective males, normal males, heterozygous and homozygous normal females. In generations II and III there is a 1 in 2 chance (50%) that each male will be defective. **B.** Normal father and color-blind mother (generation I) give rise to all color-blind sons and heterozygous carrier daughters (generation II). The marriage of a normal generation-II female with a color-blind male gives rise to heterozygous normal daughters and normal males in generation III. For examples of other sex-linked defects see Chapter 11 and Table 11–1.

pedigree and history are extensive enough, it may be possible to estimate the recurrence risk from the distribution of affected individuals (Cross, 1972).

Chromosome abnormalities in general are not inherited, but parents of a defective child should seek counseling in order to determine if a chromosome abnormality occurred. If the defect is due to a chromosomal abnormality, as presented in Chapter 6, the patient can then be counseled as to the type of aberration and its significance for the next generation in a fashion similar to that of gene-related defects. The recurrence risks for parents having a child with chromosomal abnormalities will vary with the sex of the carrier parent and the type of abnormality. For example, female carriers of a 21/21 translocation have a 100% risk of producing a mongol child, whereas a male carrier of a 21/21 translocation has a risk of approximately 2%. In order to calculate risk figures for each type of chromosomal imbalance, it is necessary first to determine the chromosome combinations that are possible in the zygote and then to compute the proportion of such possibilities that can cause the disease.

Sporadic cases of chromosomal abnormalities (that is, when parental karyotypes are normal) require a different approach. The cause is usually unknown, making accurate counseling difficult. In the syndromes most often studied, such as trisomy E, trisomy D, and trisomy 21, advanced maternal age seems to increase the risk (see Chapter 6).

Often, prospective parents are aware of one or more relatives with an inherited disorder and want to know the chance that they or their children will demonstrate the disease. Another group to be considered are parents that are closely related, for example, first or second cousins, have heard it is "wrong" to marry each other because they increase the chance of giving rise to a defective child. In general, the risk factor for consanguineous marriages is overemphasized, especially where each of the prospective parents is normal and no obvious inherited defects or other severe medical problems have occurred in their familiy lines. Nevertheless, there *is* an increase in the overall chance of bringing rare defective recessive genes together in cousin marriages (the genetic implications of consanguinity are presented in Chapter 14). The risk factor, however, is lower than the traditional emphasis placed on the presumed drastic effects of such marriages suggests.

A large number of states are carrying out large-scale genetic testing programs for as many as 15 separate inherited diseases. In many cases the program calls for testing the newborn. If they are defective, then the normal parents are tested to determine whether they are heterozygous carriers. In other programs, only the adult population is screened; for example, the screening program for Tay–Sachs, a recessive defect, was set up to detect carriers. This is also the case for diseases such as sickle cell anemia, cystic fibrosis, porphyria, and a number of other inherited diseases as well as the Rh blood factor condition. In general, when heterozygous carriers are identified, they will want to learn more about the defective gene, the way in which it is inherited, their chance of demonstrating the disease, and whether they will transmit it to their children.

Some inherited diseases have their highest incidence in a particular race and ethnic group (see Chapter 15). In many states, various ethnic groups are involved in their own screening programs to identify people of high risk, the carriers. As such persons are identified, they are directed into counseling sessions. Wide-scale screening and counseling programs have been established, depending on the particular ethnic group, for diseases such as phenylketonuria, hemophilia, cystic fibrosis, sickle cell anemia, thalassemia, and Tay–Sachs disease.

interpretation of the data by the counselor

Mathematically stated risk figures have little meaning for the average individual because he has little if any basis for comparison. The counselor should be prepared to explain the meaning of a 2%, 50% or 100% risk figure. The counselor should place genetic risk figures in language that will enable the consultee or family to understand. For example, if a child is born with Down's syndrome due to a trisomy 21, and the mother is young, say age 25, she should be told that the initial chance of having a child with Down's syndrome by simply becoming pregnant was about 0.16% and that even though she did have such a child, by chance, her *chance* of the event happening on her next pregnancy is still less than 1%; that trisomy 21 is not an inheritable condition; that having this child is "a chance accident of nature"; and that all pregnant females have the same low (0.16%) chance of having a child with Down's syndrome. This kind of counseling may provide some reassurance to a young mother whose first child has trisomy 21. For most sporadic types of congenital malformations, the recurrence risk is small, often less than 1%. Most couples with a malformed child have fears of much higher risks. A great deal of the emotional anxiety can be prevented by the dissemination of the information already available in the field of human genetics.

the psychodynamics of genetic counseling

Very little is known about the psychodynamics (mental action) of genetic counseling. There are a number of obstacles that prevent the consultee from understanding and applying risk figures to his particular situation. Fear, denial, guilt, anger and a sense of frustration must be reduced to manageable proportions before the consultee can rationally plan for the future. Feelings of guilt may arise because the mother smoked, was overweight, or initially did not want the baby, perhaps considering or attempting abortion. A father may worry about the defect being related to a previous venereal disease. One parent may blame the other, and whole families may blame each other, the child, and parents. Some parents may feel guilty because they carry a defective gene and have passed it on to the child. Especially if the defective gene comes from a *known* carrier, as in cases of sickle cell anemia or Tay–Sachs disease, the parents may need help in accepting responsibility for the child's disease.

Fraser (1974) has received recent reports dealing with the impact of specific diseases on the family and on the subsequent reproductive behavior of the parents of defective children. For instance, one report stated that following the birth of a child with Down's syndrome, reproduction virtually ceased among the brothers and sisters of the parents of the affected child, whereas no evidence of such an effect was found in a second report. Fraser also stated that parents differ widely in their reaction to a given risk. The same risk figure may look formidable to one person and negligible to another, and the same defect may look trivial to one person and tragic to another. To provide optimal counseling, we need to know more about the psychodynamic effect of a defective child on the life style and reproductive behavior of families.

the effect of a defective child on the parents

Numerous articles have been written for the lay public describing the moral and unifying strength a defective child brings about in a family. These same articles rarely mention and never adequately describe the destructive aspects of such situations. Case studies are replete with broken homes and high suicide rates resulting from the birth of one or more defective children. Certainly some malformed and retarded children are loved and do make positive contributions to the family; but to suggest that their positive value is such that it outweighs their negative impact is to imply that the birth of a defective child is a desirable event. Is it? To whom? To the child? To the parents? Parents, given an opportunity to choose, will elect to produce a normal child.

According to MacIntyre (1973) one of the most ego-involved behavior patterns in our society is the decision and actions on the part of a married couple that lead to the production of a child.

> This behavior represents not only an expression of each person's love for the other, but also the fulfillment of a deep and basic individual need to provide continuance of oneself into posterity. It is no wonder, then, that such a couple looks forward to the birth of the child with pride and satisfaction and with the expectation that this baby will represent a combination of all that is best in each of its parents and will be one of the most beautiful children ever produced.

Thus, if a child is born with a serious genetic defect, physically deformed or mentally retarded, the parents undergo an emotional shock that cannot be really understood by parents of normal children. Since there is a general lack of understanding of the actual emotional and financial burden of a seriously defective child, our society manages to promote what *should be* the appropriate behavior of the defective child's parents. Society expects parents to love their children, and it may not recognize the problems faced or the lack of love felt by parents of severely malformed children. According to MacIntyre,

> it is not unnatural for the parents of a malformed child to experience feelings of revulsion or even hate for that child because of the terrible blow to their egos that its birth has produced. Furthermore, it is not unnatural for such parents to harbor the secret wish that the child would die.

Parents having mixed emotions concerning their defective children, regardless of cause, will do well to discuss their feelings with a sensitive and sympathetic counselor.

the effect of a defective child on brothers and sisters

Although individual cases may convince us that the child with Down's syndrome taught his brothers and sisters what "love" is, that the experience for his normal brothers and sisters is a rewarding lesson in self-sacrifice and tolerance, it is difficult to believe that in our society acceptance is the general pattern of reaction to a defective sibling. Normal children are cruel, at best, to each other during play at young ages. Toward defective children playmates are even more cruel, even though in many cases cruelty is not meant. It may be nothing more than childhood curiosity, but the mother, brother, and sisters of a defective child many times feel the stigma. In fact, the normal children, when they learn that the disease of their brother or sister is inherited, may feel that they may also be defective in some way. Further, many kinds of family activities may have to be curtailed either because of money spent on the defective child or because the child cannot make trips or be far from a medical center. Taken together the emotional distress of the parents, the financial burden, social stigma, and curtailment of family activities mean that such defective children, if kept at home, affect the psychodynamic development of all members of the family.

We feel compelled to ask what factors promote high risk families to continue having severely defective or high risk children? For example, what prompted a New Jersey family, with a one in two chance of having hemophilic males, to have three sons. In this case, two of the three males lost the parental gamble—they have hemophilia. The same point can be made in any instance when the risk factor is greater than 25% or one chance in four. For some people any risk over that which occurs by chance (between 5 and 10%) is too large. How do we identify the motivating forces within widely differing groups of people with regard to the risks they are willing to accept? What do the differences in parental attitudes mean in terms of individual rights versus the right of others or even the right of the child to be born? (Good points for class discussion.)

references

Annas, G. J. 1972. XYY and the Law. *The Hastings Center Report, 2 (April): 14.*

Bergsma, Daniel, and Henry T. Lynch (eds.). 1971. *International Directory of Genetics Services,* 3rd ed. National Foundation/March of Dimes, New York.

Cross, H. E. 1972. Genetic Counseling. Symposium: Genetics Applied to Ophthalmology. *Transactions of the American Academy of Ophthalmology and Otolaryngology, 76: 1203–13.*

DHEW. 1974 *What Are the Facts About Genetic Disease?* Publication No. NIH74–370. Department of Health, Education and Welfare, Washington, D.C., p. 21

Emery, A. E., M. S. Watts, and Enid Clark. 1973. Social Effects of Genetic Counseling. *British Medical Journal, 1*: 724–26.

Frankel, Mark S. 1973. Genetic Technology: Promises and Problems Policy Studies in Science and Technology. *George Washington University, Monograph No. 15.*

Fraser, F. C. 1963. On Being a Medical Geneticist. *American Journal of Human Genetics, 15*: 1–3.

Fraser, F. C. 1974. Current Issues in Medical Genetics. *American Journal of Human Genetics, 26*: 636–59.

Genetic Counseling Newsletter. 1973a. Counselors: A Self-Portrait. *1*: 29–33.

Genetic Counseling Newsletter. 1973b. Counselors: Indirect Direction. *1*:32–33.

Herzbert, M. D. 1972. Genetic Counseling: A Counselor's View. *The New England Journal of Medicine, 287*:1304–1306.

Kallman, Franz J. 1965. Some Aspects of Genetic Counseling. In James V. Neel, Margery W. Shaw, and William J. Schull (eds.), *Genetics and the Epidemiology of Chronic Disease.* Publication No. 1163 (February). Department of Health, Education and Welfare, Washington, D.C.

Leonard, C. O., G. S. Chase, and Barton Childs. 1972. Genetic Counseling: A Consumer's View. *The New England Journal of Medicine, 9*:433–39.

MacIntyre, M. Neil. 1973. Prenatal Diagnosis, An Essential to Family Planning in Cases of Genetic Risk. In Donald V. McCalister et al. (eds.), *Readings in Family Planning.* Mosby, St. Louis.

Scriver, C. R. 1974. Inborn Errors of Metabolism: A new Frontier of Nutrition. *Nutrition Today, 9*(Sept.–Oct.): 4–15.

Shore, M. F. 1973. Normal Mourning and the Family of the Affected Child. *Genetic Counseling Newsletter, 1*: 2.

Sly, William S. 1973. What Is Genetic Counseling? *Birth Defects—Original Series Contemporary Genetics Counseling, 60*(April). National Foundation/March of Dimes, New York.

Vaux, Kenneth. 1974. *Biomedical Ethics.* Harper & Row, New York.

Veatch, R. M. 1972. The Unexpected Chromosome . . . A Counselor's Dilemma. *The Hastings Center Report, 2*(Feb.): 8–9.

Genetic screening is the testing of fetal, child, or adult populations in order to determine which of those tested may be defective and which of those who appear normal may be heterozygous and carry a recessive gene. More specifically, screening has the following goals:

1. *Detection of those actually affected (resulting from a dominant gene, homozygous recessive, or in a few cases, polygenes).*

2. *Detection of heterozygous carriers of recessive genes.* If it were possible to identify all heterozygous recessive individuals, it would be possible, if matings between two such individuals were avoided, to prevent the occurrence of diseased offspring except for cases of new mutation. When the trait is a severe disease, many people voluntarily restrict their reproduction. Normal siblings of affected individuals, as a result of close contact with the disease, are particularly concerned over the prospect of transmitting disease to their children. Since, statistically, one third of the siblings of affected individuals are homozygous and two thirds are heterozygous normal, it can be very important to them, for psychological reasons, to know their genetic status.

3. *Detection of persons who, although having the genetic constitution for disease, do not develop it, either because of the presence of other genes or because of environmental factors.* In such individuals the genes are said to be nonpenetrant. Nevertheless, such genes may be transmitted to offspring. Penetrance may involve either a homozygous recessive genotype or a heterozygous dominant genotype. Although individuals in whom deleterious genes have failed to be expressed cannot be classed strictly with carriers, there is a need to recognize them. Present techniques for recognizing these people, through a pedigree analysis or in some cases biochemical tests for a given protein, are not totally satisfactory.

I believe in the supreme worth of the individual and in his right to life, liberty, and the pursuit of happiness.

John D. Rockefeller

genetic screening

4. Detection of individuals who may later develop disease. There are many inherited diseases whose effects appear only in later life. Among inherited diseases that show late expression are xeroderma pigmentosum, between ages 10 and 20; adult gout, after age 20; and polyps of the colon, usually after age 30 (see Table 9–2 for other examples of late onset gene expression). Recognition of affected individuals prior to the age of reproduction could greatly reduce the incidence of these and other late expression diseases. Hopefully, if the condition could be recognized early, a new form of therapy could be found to lessen the severity of the disorder.

5. Detection of those defects caused by numerical or morphological chromosome aberrations.

the value of genetic screening

Two principal benefits of screening programs are bringing knowledge to the public and providing therapy. The knowledge provided to screened parents may allow personal choice in planning parenthood and in delivering normal rather then abnormal children in cases where detection is possible. Screening data also are useful for detecting new mutations and discovering where they occur in greatest frequency. In this manner the genetic constitution of a population can be monitored with respect to environmental effects on the human gene pool. In general, then, the screening of large populations (mass screening) should be done to promote public health, to provide therapy for those in need, and to provide people with knowledge that allows them to make their own decisions. These are certainly noble goals; unfortunately, a review of the facts as they appear in the literature suggests that these fine goals are being socially prostituted.

Originally, screening for inherited defects was begun with good intentions and perhaps vague notions of preventing genetic defects such as phenylketonuria (PKU) and sickle cell anemia. Indeed, the enthusiasm for screening has become so great that certain states have passed laws making it mandatory. It turns out, however, that our previous understanding of the prognosis for PKU and the sickle cell trait is based on dubious scientific evidence. In the meantime, many people have been labeled as potential risks and, in some cases, even denied employment and subjected to needless anxiety (see Chapter 22).

The initiation of mass screening dates from 1961 when Guthrie reported a method for detecting high levels of phenylalanine in the blood of humans. This condition, as we later learned, is not always associated with mental retardation. Yet, today, because of early work that showed an association of mental retardation with high phenylalanine levels in the blood, 43 states have mandatory testing of newborns for high levels of phenylalanine and the other 7 provide the tests on a volunatry basis (Guthrie, 1972). Some 90% of the approximately 3.5 million annual live births in the United States are screened for high levels of phenylalanine. However, it can be concluded from the reports of Bessman and Swazey (1971) and Nitoswsky (1973) that laboratory tests do not detect PKU, but instead, high blood phenylalanine levels, which can have causes other then PKU. Furthermore, the tests suitable for the mass screening required by law are subject to misinterpretation and error. The tests are not accurate; they miss a number of cases of PKU and yield false positive reaction in an even greater number. Given a positive test, the physician may put the child on a low phenylalanine diet. But a child who does not have PKU is endangered by the diet because the normal child needs the daily requirement of phenylalanine for building body protein. The amino acid phenylalanine makes up about 5% of all body protein. Thus, a normal child on a diet restricting his intake of phenylalanine will suffer physical deterioration and a number of non-PKU children have died after being placed on the diet (Buist and Jhaveri, 1973). Nevertheless, mass screening of infants for PKU is still practiced in all 50 states and is still an emotion-charged issue (Bessman and Swazey, 1971; Nitowsky, 1973; see Chapter 22).

sociological questions concerning genetic screening

Various screening programs have already been initiated across the United States by individual investigators associated with medical institutions, through state legislation based on public approval, and by ethnic group organizations. To help alleviate the number of different problems (moral, financial, medical, legal) that have come up in the field of genetic screening, The Institute of Society, Ethics and The Life Sciences at Hastings-on-Hudson, N.Y. 10706, has drawn up the following list of criteria for a screening program (Hastings Research Group, 1972):

1. Well-planned attainable program objectives.
2. Involvement of all communities in designing the program objectives.
3. Provision of equal access by all.
4. Adequate and accurate testing procedures.
5. Absence of compulsory measures.
6. Procedures for obtaining informed consent.
7. Safeguards for protecting test subjects.
8. Open access of communities and individuals to program policies.
9. Provision of follow-up counseling services.
10. Education as to the relation of screening to realizable or potential therapies.
11. Protection of family and individual privacy.

Although this list is commendable, it represents an "ideal" program that is ambitious and not generally possible. Each point is subject to moral and ethical judgments that vary with respect to what *ought* to be done, by whom and when, and lacks precision as to just what is sufficient.

In order to assess the Institute's proposals for implementation of genetic screening programs, certain questions will have to be answered. The answers to those questions directly affect the individual, the family unit, and subsequently the whole society. These questions may serve as a basis for student research reports, class and panel discussions, and student projects such as polling various segments of the community on their understanding and opinion of the issues.

1. *Have the potential advantages resulting from collecting the data been weighed against the potential harm to the individual, family, or community?* The prevention of serious defective births is considered to be a very noble and worthy cause by many scientists, clergy, and laymen. The implications of having fewer severely defective children to support can be evaluated in terms of reducing emotional suffering and financial stress. However, in order to achieve this end, what can we do about the impact of identifying the carrier. The identification of carriers of certain inherited traits has the potential for supporting and creating social bias and subsequent hardship on those identified, especially in diseases that occur most frequently in a particular ethnic group (see Chapters 15 and 22).

2. *Screening in the sunshine, or who has the right to know?* Even if society should choose to retain freedom for individuals to decide what is best for themselves, there always remains the thorny question of who should be allowed to see the results of screening programs. For example, when there is continuing controversy over whether the heterozygous carriers are healthy (sickle cell carriers or persons who are genetically sensitive to certain drugs or foods), should a potential employer be allowed access to such information? What about a life insurance company? What of potential marriage partners, credit departments, blood banks, or the public at large, your neighbors? What of high school, college, or professional athletic teams? Can they demand that black athletes be screened for the sickle cell trait? In 1966 the Kansas City Chiefs of the American Football League began screening their black team members for the sickle cell trait. Soon thereafter the policy was adopted as a League rule. However, in 1973 it was decided that screening would be voluntary and would proceed on player request.

In certain states that sponsor drives to hire the handicapped or disabled, employers are legally responsible for accidents caused by epileptics! Although excellent medication is available for epileptics, under such legal restrictions, who will hire them? What if such action is adopted with respect to the hiring of other handicapped? Should their files, recorded in confidence in the doctor's office, be available to industry? (Certain of these issues are discussed in Chapter 20.) Should a genetic counselor be compelled to report hereditarily defective persons to local or central data collection banks? Can or should a genetic counselor be held responsible for an incorrect diagnosis? If so, in what way? At what age should a child be told he carries a defective gene or that he will demonstrate a severe incurable genetic illness by age 30? And what of men with an extra Y chromosome? Should they receive special consideration for their behavior in a court of law?

3. *Should genetic screening be mandatory or voluntary?* Of all possible questions, the question of civil rights as guaranteed by the Constitution of the United States is the most volatile. This text addresses the issue by providing information that can be used to help in deciding whether anyone or everyone should be screened for various genetic defects by law. History indicates that a variety of communicable diseases were brough under control by law. School children, except in special cases, have no choice but to receive immunization prior to attending public schools. Blood tests to detect syphilis must be

administered before a marriage license is issued. In view of these precedents, why not have legislation to help in the elimination of severe hereditary diseases, especially since they are incurable and usually do not respond to therapy? One can argue that because the communicable diseases, although contagious, *are* treatable, let each choose whether or not he wishes to receive preventive medicine. Genetic diseases, on the other hand, are not contagious, they are not curable, and they directly threaten the health of others in those cases in which they have been determined to be inheritable. However, because laws do not provide for personal decision in cases of communicable diseases, why in the case of the more severe inherited diseases are parents allowed to determine the fate of the unborn—to be born defective—especially in such cases as chromosome trisomies 21, 18, and 13, where prognosis has been shown to be reliable. Some argue that we have legislated better protection for our pet animals than for our children. On the other hand, every law restricts personal freedom, and freedom, once lost, is difficult to reclaim. Further, techniques for detection of serious disorders are not 100% accurate so there is the chance of *destroying a normal fetus.*

Clearly the detection of a fetal genetic disorder and the possibility of selective abortion raise important medical, legal, and moral questions. It seems impractical to propose amniocentesis (see Chapter 6) and other forms of fetal evaluation in all pregnancies in the hope of assuring parents of having only normal children, for a variety of abnormalities cannot yet be detected. By seeking the ultimate means for having only "healthy" children, we may lose respect for human life and individuality. In short, the present tolerance for some degree of imperfection might be displaced by the impossible goal of seeking fetal perfection.

Another important consideration is the number of genetic disorders that, if untreated, may result in death or severe disability but are nevertheless compatible with life and normal function by environmental manipulation such as dietary restriction and therapy. For example, how many of you reading this sentence are diabetic? Would you sanction prenatal diagnosis and selective abortion in such instances? How do you choose which inherited disorders ought to be screened if the only solution is aborting the fetus? If the majority submits to the idea that certain disorders *must* be detected for the benefit of society, a law to that effect may be passed. Then, does it not follow that the next step would be to enforce abortion? What good is detection if there is no treatment and the parents still have the ultimate decision? Further, should there be legal proscription against reproduction when a genetic disease is possible? If so, what genetic diseases should be included in such an injunction? What would you do with those parents who violate the law and have a *normal* child?

What would you suggest be done about those who inherit the smoking respiratory syndrome caused by the absence of alpha$_1$-antitrypsin (a dominantly inherited defect, see Table 9–2). Several states now have laws that forbid smoking in many public places, for example, in elevators and supermarkets. Similar laws are pending in 39 states. Is it right for the alpha$_1$-antitrypsin-deficient people to smoke or work in areas of high pollution when their inherited defect puts them at exceptionally high risk of having severe lung and other respiratory problems? The nonsmokers can argue, as suggested by

Powledge (1975), that it is unfair to charge them high premiums on health insurance that is paid out most often in respiratory cases to those in the high risk category. Before we think that this example is out of line, it can be stated that Sweden has already screened 108,000 newborns for alpha$_1$-antitrypsin deficiency and plans to follow their development (Powledge, 1975). What will happen to the information gathered from this screening study? Will it later be used to eliminate the defective from certain job opportunities or used to increase their insurance premiums. Will it be used by the "militant" and nonmilitant nonsmokers to expand their success of obtaining no-smoking sections in airplanes, trains, buses, theaters, and college classrooms and to demand that *all* smoking be prohibited? Will those without the alpha$_1$-antitrypsin deficiency insist that a prenatal detection system be used to identify the defective and force legislation to the effect that these persons never be allowed to smoke or work in jobs or live in areas with a certain pollution index? It should be obvious to the reader that the question of individual rights will remain a major problem in the area of genetic medicine.

Joseph Fletcher (1972), in discussing individual rights, feels that a guideline based on "needs" should be used in developing policy with respect to genetic screening. Fletcher argues that "needs" are the moral stabilizers, not rights. If human rights conflict with human needs, let need prevail. If medical care can use genetic controls preventively to protect people from disease or deformity, or to ameliorate such things, then let so-called "rights" to be born step aside. Fletcher feels that rights are nothing but a formal recognition by society of certain human needs; and as needs change with changing conditions, so rights should change too. The right to conceive and bear children has to stop short of knowingly making crippled children—and a study of genetics can give us that knowledge.

what genetic diseases are being screened?

One of the more exciting discoveries in developmental biology in recent years is the recognition that many of our dreaded inherited diseases have their origins in events that occur during early embryonic development. Using newly developed methods in biochemistry and cytology we can now detect a number of genetic diseases prenatally (during pregnancy, see Figure 18–1) or postnatally (after birth). For example, fetal cells taken from a sample of amniotic fluid can be used for the detection of chromosomal abnormalities (Chapter 6), metabolic defects, the sex of the child, stages of fetal lung and nerve development, and the effects of Rh incompatibility. Such information gives new impetus to research into the field of fetal therapy of inherited disease in hopes that the defect can be corrected before birth. Early diagnosis and treatment of fetal and newborn disorders have been particularly dramatic in the management of erythroblastosis fetalis, or Rh hemolytic disease, which results from an incompatible Rh factor pregnancy. This blood group condition, which once caused the death of as many as 2600 infants a year, is now on its way to being eliminated (see Chapter 12).

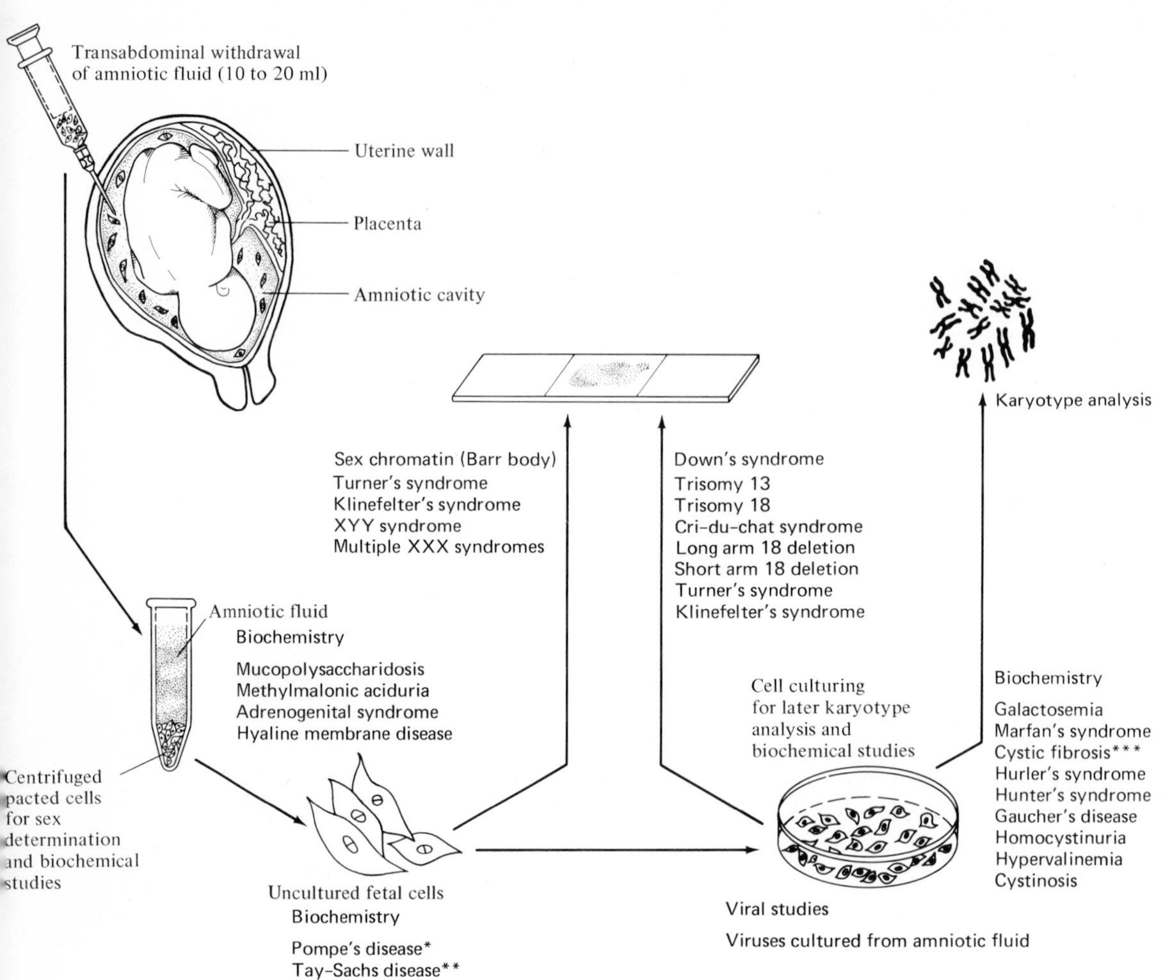

Figure 18–1. Amniocentesis and Prenatal Detection of Genetic Abnormalities. Procedures for prenatal detection of inherited disorders are performed with cells from amniotic fluid. Biochemical studies of cultured cells can reveal metachromasis and enzyme deficiencies. Cytogenetic analysis reveals abnormal chromosome number and configurations.

*Pompe's disease is detectable in amniotic fluid and in uncultured or cultured cells.

**Tay–Sachs disease is detectable in uncultured cells, but more reliable results are obtained from analysis of cultured cells.

***Cystic fibrosis is detectable prenatally. Heterozygote (carrier) is not distinguished from the homozygote (disease).

application of counseling, screening, and medicine to genetic problems

In addition to detecting Rh incompatibility, the technique of amniocentesis has also revealed metabolic errors such as the Lesch–Nyhan syndrome (see Figure 11–7). This disorder involves severe developmental and neurological disturbances and the overproduction of uric acid. Since it occurs in males having a recessive X-linked gene, it is possible, by determining the sex of the fetus, to tell whether the fetus carries a high risk of such a serious disease. Such information aids parents and physicians in deciding what course of action to take.

Table 18–1 Human Metabolic Diseases* Amenable to Prenatal Diagnosis from Enzyme Assay of Cultured Amniotic Cells or Amniotic Fluid†

Acatalasemia	Hyperuricemia, X-linked (gout)
Adrenogenital syndrome	Hypophosphatasia
Anencephaly	I-cell disease
Argininosuccinicaciduria	Isovalericacidemia
Aspartylglucosaminuria (AGU)	Ketoacidosis (infantile)
Ceramide lactosidosis	Ketoaciduria, branched-chain
Chédiak–Higashi syndrome	Krabbe's disease
Citrullinemia	Lacticacidosis (lactose intolerance)
Congenital erythropoietic porphyria	Lesch–Nyhan syndrome
Cystathioninuria	Lysosomal acid phosphatase deficiency
Cystic fibrosis	Mannosidosis
Cystinosis	Maple syrup urine disease
Diabetes mellitus	Marfan's syndrome
Erythroblastosis fetalis (Rh sensitization)	Metachromatic leukodystrophy
Erythropoietic porphyria	Methylmalonicaciduria I
Fabry's disease	Methylmalonicaciduria II
Fucosidosis	Mucopolysaccharidoses
Galactokinase deficiency	Hunter's
Galactosemia	Hurler's
β-Galactosidase deficiency	Maroteaux–Lamy
Gaucher's disease	Morquio's
Generalized gangliosidosis	Sanfilippo's A and B
Glucose-6-phosphate dehydrogenase deficiency	Scheie's
Glycogen storage disease type II (Pompe's disease)	Muscular dystrophy, X-linked
Glycogen storage disease type III	Myotonic muscular dystrophy
Glycogen storage disease type IV	Niemann–pick disease
GM_1 gangliosidosis type 1 (generalized)	Oroticaciduria
GM_1 gangliosidosis type 2 (juvenile)	Progeria
GM_2 gangliosidoses types 1, 2, and 3	Pseudoxanthoma elasticum
Hemolytic anemia	Pyruvate decarboxylase deficiency
Histidinemia	Refsum's disease
Homocystinuria	Sandhoff's disease
Hyperammonemia	Sphingomyelinosis
Hyperglycinemia, ketotic	Spielmeyer–Vogt disease
Hyperglycinemia, nonketotic	Spina bifida
Hyperlysinemia	Tay–Sachs disease
Hypervalinemia	β-Thalassemia
Hyperornithinemia	Werner's syndrome
	Wolman's disease
	Xeroderma pigmentosum

* For a number of the disorders listed, the homozygous and heterozygous states cannot yet be distinguished.

† The table does not include the many chromosomal disorders that also can be diagnosed through amniocentesis.

Similarly amniocentesis allows the detection of the metabolic error of galactosemia. If a child is born with galactosemia, he is unable to metabolize milk galactose. Continued ingestion of milk and milk products by the infant will lead to stunted growth, mental retardation, cirrhosis of the liver, and cataracts. It is now possible to detect over 100 inherited errors in metabolism (Table 18–1) and many chromosome-caused abnormalities (Table 6–3) during fetal development.

cost/benefit analysis in genetic screening

Anyone planning a screening program should keep in mind certain important provisos: that the disorder to be screened must be an important health problem and must be treatable (and/or preventable) and that the screening test must be reliable and cost-effective. The first requirement may be easier to meet than the others; inherited diseases are important health problems, but this is a *relative* statement. Few are successfully treated; most of our current methods of detection are less than 100% accurate; and many of the tests for the detection of prenatal disorders listed in Table 18–1 have been derived at considerable expense.

Most of the disorders are very rare, and about half the states are now taking a second look at the cost of various state-supported screening programs. Based on an individual letter survey of each of the 50 state health departments in mid-1974, all states finance, to some degree, PKU testing. In a sense, the fundamental question being asked now by state legislatures is whether mass screening is financially worthwhile (*Medical World News,* 1971). For example, Rudolf Hormuth, a PKU expert with HEW's Maternal and Child Services Bureau, asks how long we can continue spending $3 to $4 per test to detect less than 200 PKU infants per year out of over 3 million children screened.

If you were a parent whose child's disease was detected and treated as a result of the massive PKU screening program in the United States, you might wish to have the program continued. However, as a taxpayer you might wish to have the same money spent on other national health care programs for "normal" people. Guthrie (1972), who developed the PKU test now in wide use, warns: "Discontinuance of PKU screening would represent not merely medical, but also fiscal irresponsibility of the grossest sort." The detection and treatment of one case of PKU represents an outlay of up to $50,000; but failure to detect that case means a child that must almost certainly be institutionalized for the rest of its life, representing an outlay of at least $250,000 (this assumes an average life-span of 50 years and an annual expenditure for custodial care conservatively estimated at $5000). The $250,000 figure includes no allowance for the lost earnings of the untreated case. We are not talking here about human values, but purely economic ones. Is it better to spend $50,000 now or five times that sum later. There is good reason to expect that the $50,000 figure can be reduced. One approach to cost reduction is through the use of automated equipment. Guthrie and others suggest that the cost of caring for the defective person is many times that of preventing their occurrence via a screening program. And so discussions concerning cost versus benefit go on. However, while such discussions proceed, programs for the screening of other diseases are being introduced across the United States, either through legislative mandate or on the community level.

One of the first examples of community involvement in a particular genetic disease occurred in the Baltimore–Washington area in 1971 with a program for screening the Hebrew population for Tay–Sachs carriers. At the time there were over 240,000 Hebrews in the area, 80,000 of whom were of childbearing age (18 to 43). Between May and September of 1971, 7800 individuals were screened. During that time four couples were identified as at-risk (carriers).

The public reaction to the program has been positive. Prior to the announcement of the program 98% of the Hebrew population in the Baltimore–Washington area had never heard of Tay–Sachs disease. Within a few months 95% of the Hebrew population was aware of it, and a significant number volunteered to be screened.

The cost to initiate the Baltimore–Washington program was approximately $65,000 to provide equipment, laboratory, and office personnel and supplies. All those screened were asked to donate $5 to help cover the cost of the program and only a few were unable to pay.

The counseling services of the program were to help people to make decisions of the highest personal importance. At-risk couples received direct contact with the counselors whose training enabled delivery of test results in a careful and sensitive way. Following this lead a large number of Hebrew ethnic groups across the United States are now involved in community screening programs for Tay–Sachs.

analysis of existing state-supported screening programs

In response to the letters of inquiry to state health departments, it was apparent that all 50 states screen for PKU; 7 states provide screening on a voluntary basis, while 43 states have legislation for mandatory testing of all live births. The majority of states also, either via legislation or through voluntary organizations, screen for sickle cell carriers. Very few states screen for more than these two disorders: 6 states screen for Tay–Sachs, 7 for galactosemia, 3 for cystic fibrosis, 4 for chromosome abnormalites, 6 for diabetes and 2 for hemophilia. (Several states give direct aid to families of hemophiliacs, although they do not support screening programs. Financial help is supposed to be available to all hemophiliacs via federal legislation of 1973–74.) Three states, Massachusetts, Minnesota, and Oregon, appear to have outstanding genetic screening programs.

massachusetts

Massachusetts has one of the top multiphasic genetic screening programs in the country. Having begun with testing for PKU in 1962, the Metabolic disorders

Table 18–2 Blood Specimens and Tests for Metabolic Disorder Screening in Newborn Infants in Massachusetts (1972)

Specimens	*Tests*	*Disorders Sought*
Umbilical cord blood (birth)	Beutler enzyme assay	Classic galactosemia
	Guthrie phenylalanine	Phenylketonuria
	Murphey enzyme assays	Hereditary angioneurotic edema and alpha$_1$-antitrypsin deficiency
Newborn blood (2–4 days of age)	Guthrie phenylalanine	Phenylketonuria
	Guthrie leucine	Maple syrup urine disease
	Guthrie methionine	Homocystinuria
	Guthrie tyrosine	Tyrosinosis
	Paigen assay	Galactosemias

Laboratory of the Institute of Laboratories, Massachusetts Department of Public Health, now conducts the largest newborn testing program for metabolic disorders in the world.

From each baby born in Massachusetts the following specimens are obtained:

1. Umbilical cord blood, at birth
2. Newborn blood at two to four days of age
3. Urine at three to four weeks of age.

Table 18–3 Amino Acid Metabolic or Transport Disorders Detected by the Screening of Urine by Paper Chromotography in Massachusetts (1972)*

Disorder	*Number Detected*	*Incidence*
Group I		
Cystinuria (complete)	16	1 in 13,894
Hartnup's disease	14	1 in 15,879
Histidinemia	14	1 in 15,879
Iminoglycinuria	13	1 in 17,100
Group II		
Argininosuccinic aciduria	1	1 in 222,302
Fanconi's syndrome	1	1 in 222,302
Hyperglycinemia (ketotic)	1	1 in 222,302
Hyperglycinemia (non ketotic)	1	1 in 222,302
Hyperlysinemia	1	1 in 222,302
Hyperornithinemia (with liver disease)	1	1 in 222,302
Persistent mild hyperphenylalaninemia	1	1 in 222, 302

*Specimen taken at three to four weeks of age. Total tested, 222,302.

Table 18–4 Costs of screening (1972)

Specimen	*Total Specimens Tested*	*Total Costs*	*Total Tests*	*Cost per Test*	*Cost per Infant*
Cord blood	75,000	$35,000	300,000	$0.12	$0.48
Newborn blood	90,000	67,000	450,000	0.15	0.72
Newborn urine	75,000	78,000	300,000	0.26	1.04
Totals	240,000	$180,000	1,050,000		$2.24

All specimens are impregnated into filter paper to facilitate handling. Tables 18–2 and 18–3 list the disorders screened by blood and urine analyses and the results of a sample of the tests.

Each year a total of approximately 225,000 specimens are received from 90,000 to 95,000 babies and a total of over 1 million tests are performed. Since 1962 over 215 children with various metabolic disorders have been diagnosed. The cost of the program in the State of Massachusetts is summarized in Table 18–4.

a brief survey of pre- and postnatal screening methods

Methods now in use in the United States for screening for genetic disorders pre- and postnatally can be divided into two major categories, chromosome and biochemical analyses.

chromosome analysis

Chromosome analysis is the direct examination of the chromosomes for the detection of gross numerical or structural abnormalities. The techniques currently utilized for the chromosomal assessment include

1. Microscopic examination of stained buccal smears (scraping from inside of the cheek) or hair follicle roots cells for the presence of X-chromatin material (Barr body).
2. Examination of interphase nuclei in stained buccal smears or hair follicles for the presence of the Y chromosomes by fluorescent technique.
3. Evaluation of the total chromosome complement by karyotype analysis on either leukocytes or fibroblasts obtained from a blood sample, skin biopsy, or cultured fetal cells (see Figure 18–1).

Barr body determination gives an indication of the number of X chromosomes contained in the cells and reveals numerical sex chromosome abnormalities (Chapter 6). In addition, any marked alteration in the size or shape of the Barr body mass may indicate the presence of a structurally abnormal X chromosome.

Thomas and Scott (1973) developed a technique for the examination of interphase cells for the presence of Y chromosomes. Their method is based on the finding that the number of Y chromosomes is equal to the number of fluorescent bodies within a cell that has been treated with a quinacrine stain. Early evidence indicates that the technique may be useful in identification of males with the XYY syndrome and the cytological determination of the sex chromosomes of newborn infants with ambiguous external genitalia. In situations where a gross chromosomal abnormality is suspected, a complete chromosome analysis is done, preferably by one of the more recently developed banding techniques.

Giesma or fluorescent banding is discussed in Chapter 6. Banding techniques identify structural abnormalities that would have been missed by other cytogenetic methods. However, while cytogenetic studies do demonstrate gross structural abnormalities, there is no doubt that minor structural changes still go undetected.

biochemical analysis

Biochemical tests are used to study the relationship between chemical constitution and biological action found in the various body tissues such as sweat, blood, and urine. The majority of metabolic tests are done to measure a specific enzyme, a biochemical compound or metabolite, or biochemical changes that occur at the cellular level.

Assay of a Specific Enzyme or Structural Protein. The most satisfactory laboratory method for establishing the presence of a single gene defect is to analyze the substance that the gene ultimately controls, presumably a protein or enzyme. For example, when a mutation alters the structure or amount of a gene product, it is often possible to identify the defect by measuring the enzyme or protein in question.

Tests suitable for mass screening, when sampling of the gene product requires the use of tissue of the fetus, are not yet available. Many more disorders could be screened in utero if it were possible to obtain tissue, particularly blood, from the embryo without damaging it. A micromethod is now experimentally available for detecting the presence of Hb S (sickle cell hemoglobin) in the fetus at about 14 weeks gestation (World Health Organization, 1972). Postnatally the formed elements of blood or serum can serve as a source of the enzyme or protein and may be useful for purposes of mass screening. This is amply demonstrated by the recently developed mass screening programs for the detection of hemoglobinopathies such as sickle cell and beta-thalassemia.

A number of blood tests are available to measure specific proteins. For example the Beutler tests for galactosemia is used for assessing the activity of erythrocyte galactose-1-phosphate uridyl transferase, the enzyme that is missing in galactosemia. A test that reveals a deficiency of C′ esterase inhibitor in the plasma is used for detecting hereditary angioneurotic edema. The disease rarely causes symptoms in childhood, but adults may be afflicted by recurrent abdominal pain and patches of edema that can be lethal if the larynx is involved (Nitowsky, 1973). Recently, a test has been developed to detect early possibilities of emphysema. It appears that a deficiency of alpha$_1$-antitrypsin in the plasma is associated with the development of emphysema. The majority of disorders listed in Table 18–1 can now be detected through the use of enzyme assays.

Screening for a nonenzyme protein is illustrated by the mass screening now taking place for sickle cell hemoglobin. Sickle cell disease is the most common of many inherited hemoglobin disorders. The most suitable method for screening hemoglobinopathies is the electrophoresis. There are several different types of electrophoresis (cellulose acetate, citrate agar, etc.), but all of them are based on the principle that charged molecules migrate at different speeds when placed in an electrical field. Electrophoresis can also be used to detect other conditions, such as carriers of Lesch–Nyhan syndrome, serum protein disorders, lipoproteins associated with heart disorders, and glucose-6-phosphate dehydrogenase deficiency.

Assay of Metabolites. Perhaps the most widely employed biochemical screening method for detection of a specific genetic defect is an indirect approach in which the serum or urine is screened for metabolites or amino acids. This approach is based on the principle that when the gene product, an enzyme, is involved in a metabolic pathway, alteration in enzyme activity will change the concentration of the metabolic reactants. The metabolite or reactant imbalance may then be detected as an excess or deficiency of the expected reaction product (Nitowsky, 1973).

A number of enzymes and hormones have also been found in the amniotic fetal fluid that, depending on the quantities found, can give clues to certain disorders. One group of investigators established the diagnosis of the adrenogenital syndrome in utero by measuring the level of 17-ketosteroids in amniotic fluid obtained during the thirty-ninth week of pregnancy.

Diagnosis of many metabolic disorders would be easy if abnormal levels of metabolites accumulated in the amniotic fluid However, Prescott, Magenis, and Buist (1972) state that this rarely occurs because metabolites cross the placenta and are degraded by the mother. Tay–Sachs disease is probably the best known disorder that can be detected prenatally by biochemical analysis of culture amniotic cells. The enzymatic defect in Tay–Sachs disease is caused by excess synthesis, diminished degradation, or both, of a specific lipid called ganglioside GM_2. In Tay–Sachs children this lipid accumulates to levels 100 to 300 times normal. Later research revealed that the enzyme hexosaminidase A (hex A) is missing in Tay–Sachs children, and this is now believed to be the cause of excessive accumulation of GM_2. Discovery that this enzyme is missing also made possible prenatal diagnosis of the disorder, since the level of hex A can be established from cultured amniotic cells obtained during the second trimester of pregnancy. Most Tay–Sachs babies appear perfectly healthy at birth and develop normally for the first three to six months. Gradually, however, the central nervous system degenerates because of the progressive intraneuronal accumulation of excess amounts of the sphingolipid ganglioside

GM_2. By eight to twelve months of age, the infant usually shows physical and mental deterioration, and death occurs between the ages of three and five (see Chapters 10 and 15 for further detail on Tay–Sachs disease).

Blood or serum is tested postnatally. In general, an abnormal accumulation of metabolites does not occur until the infant's intake or production of metabolites exceeds his capacity to metabolize or excrete them. This usually takes several days, depending on the severity of the defect and on the volume and composition of the diet.

Finally, the whole process of screening for genetic disorders is far from perfect. Many genetic disorders cannot be diagnosed prenatally. The heterozygous state in parents, in most instances, cannot be determined except through the birth of an afflicted child. There are undetectable genetic disorders such as Huntington's disease, which does not manifest itself until late in life, after the victim has passed the childbearing age. Lack of penetrance and variable expressivity complicate the situation with certain genetic disorders. There is also the problem of distinguishing the homozygote from the heterozygote in prenatal tests for recessive disorders. And there is the problem of educating the medical profession to the involvement of genetics in medicine; according to a recent report in the *Journal of the American Medical Association* (1975) only "slightly more than half of all family practitioners clearly agreed that many metabolic disorders are genetically based, although such a genetic basis has been supported by data accrued since the days of Sir Archibald Garrod." In short, man's knowledge in the field of screening for genetic disorders is still in its infancy.

references

Bessman, Samuel P., and Judith P. Swazey. 1971. Phenylketonuria: A Study of Biomedical Legislation. In E. Mendelson, J. P. Swazey, and Irene Tavies (eds.), *Human Aspects of Biomedical Innovation.* Harvard University Press, Cambridge.

Buist, N. R., and B. M. Jhaveri. 1973. A Guide to Screening Newborn Infants for Inborn Errors of Metabolism. *Journal of Pediatrics, 82*(3): 511–22.

Fletcher, Joseph. 1972. Ethical Aspects of Genetic Controls: Designed Genetic Changes in Man. *The New England Journal of Medicien, 285*: 776–83.

Guthrie, Robert. 1972. Mass Screening for Genetic Defects. *Hospital Practice,* June: 93–100.

Hastings Research Group. 1972. Ethical and Social Issues in Screening for Genetic Disease. *New England Journal of Medicine, 286*: 1129–32.

Journal of the American Medical Association. 1975. Genetic Screening. *234*: 102.

Medical World News. 1971. After Ten Years of PKU Testing, A Reevaluation. November: 43–44.

Nitowsky, H. M. 1973. Prescriptive Screening for Inborn Errors of Metabolism: A Critique. *American Journal of Mental Deficiency, 77*(5): 538–50.

Powledge, Tabitha M. 1975. Genetic Screening vs. Personal Freedom. *Science Digest, 78*: 48–55.

Prescott, G. H., R. E. Magenis, and N. R. Buist. 1972. Amniocentesis for Antenatal Diagnosis of Genetic Disorders. *Postgraduate Medicine, 51*: 212–17.

Thomas, G. H., and C. I. Scott. 1973. Laboratory Diagnosis of Genetic Disorders. *The Pediatric Clinics of North America, 20*(1): 105–40.

World Health Organization. 1972. Genetic Disorders: Prevention, Treatment, and Rehabilitation. *Technical Report Series,* No. 497 : 3–46.

The main purpose of genetic screening is the detection of an inherited disease. The main purpose for counseling is prevention of genetic diseases that impair the physical and mental well-being of the patient. Therapy or medical treatment of genetic diseases is designed to reduce the suffering of the afflicted person and in turn relieve the emotional anxiety of family and friends.

Together, genetic screening, counseling, and therapy can be referred to as *genetic medicine*, which is beginning to emerge as a meaningful clinical discipline. Genetic medicine offers society a means of detection, prevention, and limited treatment for a number of inherited diseases. Approximately 100 serious inherited disorders can be identified during the fetal stage of development through the use of amniocentesis (Table 18–1). Screening of high risk adult populations allows for the identification of carriers of recessive genes. Proper counseling of the carriers allows the prospective parents to make relevant decisions as to whether they wish to take the genetic gamble and try for a normal child. To date, however, there are very few if any inherited defects for which effective therapy is available during the pre- or postnatal period.

available therapy for genetic diseases

The therapies in use or being considered for a number of inherited diseases are listed in Table 19–1. Some types of therapy are discussed below.

Extreme remedies are very appropriate for extreme diseases.

Hippocrates

therapy for genetic diseases

immunology

Tissue typing (to determine if donor tissue is immunologically acceptable to the recipient) is necessary for successful transplants of human organs. The human kidney has been used in transplants where patients demonstrated the inherited form of polycystic renal disease. Kidney transplants have been tried in the treatment of patients with Fabry's disease and cystinosis (discussed below). Bone marrow transplants have also been performed on patients with thalassemia and various forms of immune deficiency diseases (see Chapters 11 and 15). In any case of transplant therapy, genetically compatible tissue is essential and remains the major problem for transplant therapy. A second form of immunologic therapy is the injection of Rh globulin (gammaglobulin) to prevent the destruction of fetal red blood cells that might occur during subsequent Rh incompatible pregnancies (see Chapter 12).

surgery

Surgery is most often used for the removal of an organ or growth. In effect surgery helps to relieve the patient's symptoms. For example, splenectomy (removal of the spleen) is quite effective in correcting the disorder of hereditary spherocytosis, a type of anemia. Removal of part or most of the large intestine does away with polyps of the colon that usually become malignant. Constructive cosmetic surgery has been most helpful in inherited disorders such as Apert's syndrome, gynecomastia (breast enlargement in males), hypospadias (urethra opening on the underside of penis), retinoblastoma (cancer of the eye), and some polygenic central nervous system disorders such as spina bifida and various myeloceles.

enzyme induction

Drugs can be used to induce the production of certain enzymes. For example, in some types of hereditary jaundice, phenobarbital stimulates the production of a missing enzyme and relieves the symptoms of the disease (Hamilton, 1972).

enzyme activation

The administration of large doses of vitamins sometimes restores enzyme function (perhaps through establishing the lost affinity or association between an enzyme and its vitamin cofactor). Administration of vitamin B_{12} has shown limited success in the treatment of methylmalonicaciduria and vitamin B_6 offers some hope in the treatment of homocystinuria. The most successful vitamin therapies to date use vitamin D for inherited refractory rickets (Scriber, 1974) and vitamin B_{12} for pernicious anemia.

Table 19–1 Inherited Diseases and Therapy

Disease	*Therapy*	*Results**
Acrodermatitis enteropathica	Diiodohydroxyquinoline + zinc (e.g., zinc sulfate)	Excellent
Adrenogenital syndrome	Cortisone; induce salt loss	Good
Alpha$_1$-antitrypsin emphysema	No smoking, clean air	Good
Apert's syndrome	Surgery	Poor to fair
Crigler–Najjar syndrome	Blood transfusions, phenobarbital	Fair, short-term life expectancy
Cystic fibrosis	Diet, antibiotics, mist	Good, short-term life expectancy
Cystinosis	Diet—restricted methionine and cystine + increased ascorbic acid	Questionable
Cystinuria	High fluid intake, alkaline, penicillamine	Fair to good for urolithiasis
Diabetes insipidus	High fluid, low salt	Fair
Fabry's disease	Enzyme replacement	Questionable
Familial goiter	Levothyroxine	Good
Familial hyperlipoproteinemia	Diet—restricted to short chain fatty acid	Fair
Familial xanthomatosis	Cholesterol diet	Poor to fair
Favism (G-6-PD deficiency)	Avoidance of specific drugs and fava beans	Excellent
Fructosemia	Fructose-free diet	Good
Galactosemia	Galactose-free diet	Good
Glycinemia	Protein-restricted diet	Poor to fair
Gout (hyperuricemia)	Low uric acid diet	Fair
	Colchicine + allopurinol	Good for pain
Gynecomastia	Surgery	Cosmetically excellent
Hartnup's disease	Nicotinamide	Fair to good
Hemochromatosis	Iron-restricted diet, venesections	Fair to poor
Hemophilia	Injections of blood factor VIII	Good to excellent
Hereditary jaundice	Phenobarbital	Good
Hirschsprung's disease	Surgery	Good
Histidinemia	Histidine-restricted diet	Questionable
Homocystinuria	Methionine-restricted diet + vitamin B_6	Questionable
Huntington's disease	Drugs for behavior control	Good
Hydrocephaly	Cranial shunt, wrap head	Good
Hypercholesterolemia	Low cholesterol diet + drugs	Fair to questionable
Hyperlipidemia	Low fat diet	Fair to questionable
Hyperuricemia	Low uric acid diet	Poor to fair
Hypospadias	Surgery	Cosmetically good
Ichthyosis vulgaris (alligator skin)	Retinoic acid	Fair to good

Table 19-1 continued

Disease	*Therapy*	*Results**
Intestinal polyposis	Surgery	Good if early
Lactose intolerance	Avoidance of milk and other dairy products	Excellent
Manic depressive	Drugs	Fair to good
Maple syrup urine disease	Diet—restricted leucine, isoleucine, valine, and methionine	Poor to fair
Meningomyelocele	Surgery	Poor
Migraine headache	Drugs	Fair to excellent
Oroticaciduria	Uridine	Good
Osteogenesis imperfecta	Calcitonin	Fair
Osteoporosis	Calcitonin	Good
Paget's disease	Calcitonin	Good
Pernicious anemia	Vitamin B_{12}	Excellent
Phenylketonuria	Phenylalanine-free diet	Uncertain
Polycystic renal kidney disease (late)	Kidney dialysis, kidney transplant	Life saving
Porphyria, acute intermittent	Experimental status, IV 3 mg hermatin/kg body wt	Good (initial reports)
Pyridoxine dependency	High intake of pyridoxine	Fair to good
Ragweed hay fever	Desensitization	Fair to good
Renal tubular acidosis	Alkali	Good
Retinoblastoma	Surgery or radiation therapy + cryotherapy	Life saving
Rh incompatibility	Gammaglobulin injection	Good to excellent
Scurvy	Vitamin C (ascorbic acid)	Excellent
Sickle cell anemia	Cyanate injection	Fair to good in crisis
	Nitrilosides	Under study
Spina bifida	Surgery	Poor
Tyrosinemia	Diet—restricted phenylalanine and tyrosine	Questionable
Urea cycle disorders Argininosuccinicaciduria Ornithinuria Citrullinura	Protein intake restricted to 1 g/kg body wt/day	Poor
Usher's syndrome (progressive deafness and blindness)	Vitamin A	Poor
Von Willebrand's disease	Injections of blood factor VIII	Good to excellent
Wilson's disease	Penicillamine, restricted copper	Fair
Wiskott–Aldrich syndrome	Blood platelet transfusion	Good
Xeroderma pigmentosum	Avoidance of direct sunlight, use of medicated creams	Fair

*It should be understood that the terms excellent, good, fair, poor, and questionable are relative evaluations based on a range of responses of the patients. All patients may not give the same response to any one treatment

enzyme replacement

A normal highly purified enzyme can be administered to a patient deficient for that enzyme. This has been tried with minimum success in the treatment of Fabry's disease (Brady et al., 1973). With Hurler's and Hunter's syndromes (mucopolysaccharide diseases) unsuccessful attempts have been made to infuse patients with plasma from normal persons in the hope that the enzyme of the normal plasma would degrade the accumulated mucopolysaccharides in the patients. Children with argininemia, high levels of arginine, have been infected with Shope papilloma virus on the theory that a viral gene might allow for the production of the enzyme arginase to degrade the accumulated arginine. To date these experiments have also been unsuccessful. If a breakthrough occurs in this area, and it does seem promising, various ways can be envisioned for treatment. Possibilities include stabilizing the purified enzyme on a column and running the patient's blood through it or placing the enzyme in time-release capsules, or if all else fails, daily injections of enzymes similar to the use of insulin by diabetics.

supplemental factors

Various factors required by the body, but missing because of a genetic error, can be replaced. Examples are blood factor VIII, the antihemophilic globulin required for blood clotting; calcitonin for the retention of bone calcium in hereditary disorders such as osteogenesis imperfecta; penicillamine to reduce the copper levels in patients with Wilson's disease; anti-Rh globulin to prevent Rh disease; and insulin for diabetes.

diet management

The most used treatment of inherited disorders appears to be regulation of diet. This is not to say that diet therapy alone is adequate; rather, it reflects the present limitations in the treatment of inherited disease. Diet modifications may call for

1. Low level of dietary metals such as low iron for patients with hemochromatosis (excess of iron in liver, heart, and pancreas) or low copper for patients with Wilson's disease.
2. High level of dietary metals, such as high copper intake in patients with Menke's disease (kinky hair syndrome) who have too little copper in their tissues.
3. Low level of various dietary amino acids. In phenylketonuria, a minimal level of phenylalanine has been established; in homocystinuria, cystine is kept low.
4. Elimination of dietary factors. In galactosemia and lactose sensitivity, the milk sugars galactose and lactose, respectively, are eliminated from the diet; for glucose-6-phosphate dehydrogenase deficiency, fava beans are eliminated from the diet and various drugs, especially barbiturates, from medications (see Table 16-2).

5. High levels of dietary vitamins. For the treatment of scurvy there is the use of vitamin C; for refractory rickets, vitamin D; and for pernicious anemia, vitamin B_{12}.

However, there are a number of diets that give at least questionable results, or for which the effectiveness of the diet varies with the individual as in the treatment of phenylketonuria, of the urea cycle diseases, histidinemia, and homocystinuria.

Millions of dollars are spent annually to discover the causes of genetic blindness, muscular dystrophy, Huntington's disease, cystic fibrosis, and many other diseases. Understanding the molecular causes of such disorders is essential because understanding the cause may lead more directly to effective treatment of the symptoms. One inherited recessive disorder, acrodermatitis enteropathica, is effectively treated with diiodohydroxyquinoline and zinc sulfate. This disease is very striking in that the symptoms are scaling and blistering around the body orifices, the eyes, genitals, and anal areas (Figure 19–1). If left untreated, the affected person loses his hair, eyebrows, and eyelashes and passes bulky, putrid-smelling stools. Patients become withdrawn and die at an early age (Roberts, 1970). The administration of diiodohydroxyquinoline brings about a dramatic improvement; hair grows, weight increases, and skin lesions heal. The action of the drug is not understood, but Roberts (1970) says that it is believed by some that there is an enzyme defect in the intestinal lining of such people. Because of the lack of enzyme activity in the intestinal lining, a "toxic" substance is not properly detoxified. The drug, binding to this substance, neutralizes the toxic effect and normal body activities resume.

For those defects—galactosemia, phenylketonuria, maple syrup urine disease, and tyrosinemia—that may be treated with some success, dietary management should begin as soon after birth as possible. Because early diet therapy in these disorders may be of significant help to the individual patient, it is important that the physician, who may only see one or two cases in his career, be able to obtain both the information on patient management and special diets for his patient. To this end, food and drug manufacturers in the United States and Europe produce specially formulated diets for the treatment of particular inherited diseases, and a number of genetic centers provide assistance in diagnosis and management of patients, or admission to a center if required (Table 19–2). One such specially formulated diet food is Lofenalac, a compound that is low in the amino acid phenylalanine for use in treating PKU.

Canada has a National Food Bank, estblished in 1973 "to develop a central facility to improve the importation of food items, to offer a wide range of items to most treatment centers, to share new product information and to consolidate a treatment network for purposes of sharing a pool of necessarily limited information" (*Nutrition Today*, 1974). Patients receive specially prepared diets from the Food Bank centers. By reviewing the registries of the centers, personnel at these centers have identified "hot spots"—locations where the need for a particular diet is highest. This information helps in the allocation of proper medical personnel as well as maintaining a sufficient supply of the special diets.

We have a long way to go in the delivery of effective genetic medicine for the majority of inherited diseases, and especially so in the area of therapy

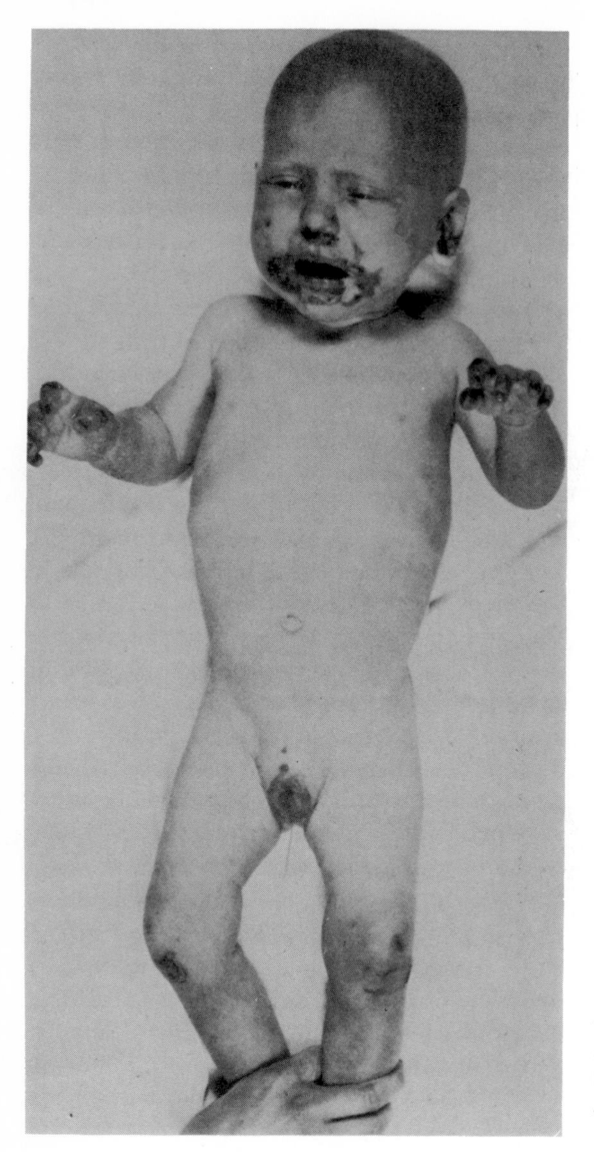

Figure 19–1. Acrodermatitis enteropathica. Early onset of symptoms: blistering and scaling of eyes and of genital and anal areas and lesions occurring on any or all digits (fingers and toes). Treatment with diiodohydroxyquinoline is excellent; all symptoms disappear in a short time and normal health results. (By permission from J. A. F. Roberts, *An Introduction to Medical Genetics,* Oxford University Press, London, 1970.)

application of counseling, screening, and medicine to genetic problems

(Table 19–1). However, as we prepare for the day when we can treat all inherited diseases as effectively as we now treat diabetes, there is a very important point that must be considered. Genetic medicine can only treat the phenotype, it does not alter the defective genotype. For example, in diabetes, insulin helps the diabetic to lead a near normal life, but the insulin has no corrective effect on the defective genes involved. This point leads us to an interesting question for class discussion: what happens if science continues to improve medicines to keep the genetically defective alive to the point where they reproduce? Will the human gene pool become so contaminated that a social collapse will be unavoidable? (Social collapse is defined here as the time when there are insufficient numbers of healthy people to provide for the

Table 19–2 Genetic Centers with Amino Acid Formulas for Diet Therapy*

Children's Hospital Los Angeles, California 90027	Massachusetts Metabolic Disorders Screen Program Forest Hills Boston, Massachusetts 02430
Department of Pediatrics University of Colorado Medical Center Denver, Colorado 80220	Pediatric Neurology Unit University of Michigan Ann Arbor, Michigan 48104
Medical Genetics Section Emory University Atlanta, Georgia 30322	Department of Pediatrics New York University School of Medicine Bellevue Hospital New York, New York 10016
Department of Pediatrics Abraham Lincoln School of Medicine University of Illinois Chicago, Illinois 60680	Children's Hospital Medical Center Cincinnati, Ohio 45229
Division of Human Genetics Johns Hopkins University Baltimore, Maryland 21205	Mental Retardation Center University of Wisconsin Madison, Wisconsin 53706

*List prepared by the American Academy of Pediatrics in cooperation with the U.S. Food and Drug Administration. From *Nutrition Today*, 1974.

defective.) How long do you think it might take for a social collapse to occur in the United States? What would be the necessary social conditions that might lead to such an event?

prospects for prenatal therapy

Although effective therapy is available for some disorders, as indicated in Table 19–2, for most of the known inherited human diseases there is no specific therapy. If science could develop the proper prenatal therapy for the inherited diseases, therapeutic abortions, which are objectionable to a large segment of our population, might for the most part be avoided. This effort to improve prenatal therapy is supported by many physicians, but the lack of progress is discouraging. In fact, genetic counselors, physicians, theologians, and the lay public are now questioning the use of pre- and postnatal screening for genetic diseases because of the limited therapy available. Dr. Carlo Valenti of Brooklyn's Downstate Medical Center feels that hope for or expectation of adequate prenatal treatment of most diseases is "unrealistic and untenable." He concludes "that abortion must remain the solution to inherited diseases" (Powledge and Sollitto, 1974).

Dr. Valenti's statement is a sad but accurate, commentary on today's genetic medicine. The question is: How long will it be before effective therapies are available for the majority of the genetic diseases? The answer may well be determined by state and local governments that are supporting a ban on all human fetal research. For example, it is currently against the law in the state of Massachusetts to experiment on any living human fetus before or after an induced abortion unless the experiment is to save the life of the fetus (Culliton, 1974). If such laws were to be enacted in all states, the search for understanding of gene–environment interaction during fetal development would be set back, with a consequent delay in the field of fetal therapy (see Chapter 22 for examples of legislation and fetal research.)

Dietary manipulation is the most common form of therapy (Table 19–1). This suggests a close relationship between the gene and its environment—more specifically, between the gene and nutrition. The various metabolic alterations that occur in the majority of human genetic diseases as revealed by metal, carbohydrate, and amino acid accumulation or deficiency emphasize the need for further research to gain an understanding of the interaction of genes and nutrients during fetal development.

Why and how, for example, are so many of our inherited diseases related to mental retardation? For many years science has been aware of the association between defective genes, fetal growth, and brain development. Yet, even today science can only guess at the mechanism of gene action that brings about the actual disease. Progress is being made, however slowly. We have learned that genetically determined variation in the function of an enzyme can cause changes in the amount of metabolite available for conversion into required products. Certain genes control the absorption of nutrients (metabolites) through membranes. Such a defective gene literally starves otherwise normal cells of their required nutrients. In X-linked hypophosphatemia the body's inability to reabsorb orthophosphate through the renal epithelial membranes results in rickets and dwarfism. Recent therapy, which appears to be effective, is the supplementation of the diet with phosphate salts, at levels sufficient to maintain serum orthophosphate within the normal range. The deficient transport system is circumvented and the mutation is effectively neutralized. The rickets heal and, perhaps of even greater importance to the patient, normal growth is sustained (Scriver, 1974).

An excellent example of gene–nutrient interaction has recently been clarified in studies of mouse development reported by Lucille S. Hurley (1968). This is a good example of investigation of prenatal development carried out on a substitute species, but the application of such knowledge ultimately requires similar investigations on humans. Mice with a mutant gene, called pallid, show ataxic behavior (muscular incoordination due to improper ear development) indistinguishable from genetically normal mice grown under conditions of manganese deprivation. If ataxic mice are placed in water, they drown; they are unable to stay right side up (Figure 19–2). When the investigators raised pregnant genetically defective pallid mice on a normal diet, 68% of the progeny were ataxic. When pregnant pallid females were given a diet containing high levels of manganese, none of their progeny were ataxic. However, the normal appearing females from this group, if given a normal diet, produced ataxic progeny. Thus, the raised level of manganese altered the phenotype of the mutant mice; they did not become ataxic. However, the manganese did not alter the genotype. Supplementation of the maternal diet with manganese

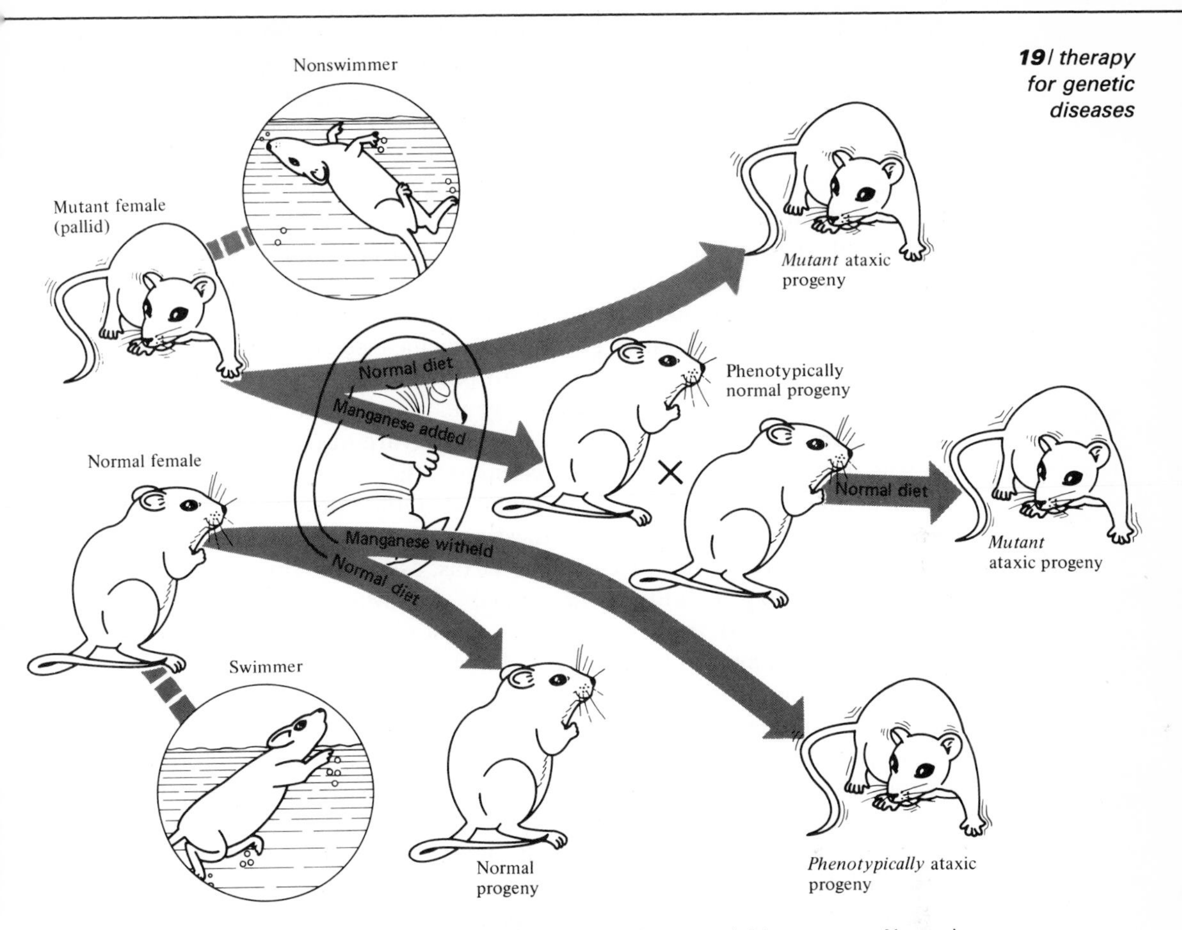

Figure 19–2. Relationship Between the Pallid Mouse Gene and Manganese. Normal (nonpallid) pregnant mice produce normal progeny that can stay upright in water and swim (swimmers). However, if the pregnant female is deprived of manganese, the progeny are ataxic like mice with the genetic mutation, pallid, that results in ataxia; they cannot stay right side up in water, and they drown (nonswimmers). If mutant (pallid) pregnant mice are given increased levels of manganese during the time of fetal development, they produce nonataxic offspring as though they did not carry the pallid gene. On a standard diet matings between siblings of that litter produce ataxic progeny. Because these phenotypically normal appearing offspring give rise to pallid offspring, we can assume the pallid gene defect can by bypassed through manipulating the diet.

during gestation prevented the expression of the inherited genetic defect in the offspring, thus demonstrating an interaction between the pallid gene and the nutrient manganese. To date, this pallid–manganese relationship is the only example in fetal research where withholding of a nutrient from a normal embryo mimics a gene defect while increasing the level of the same nutrient overrides the effect of a mutant gene (Figure 19–2).

This is a demonstration of a possible avenue of therapy during fetal development. There is, as yet, no equivalent case of intrauterine treatment for human inherited diseases. The discovery of analogous situations in humans could establish the use of diet manipulation to prevent congenital malformations (human ocular albinism appears to be similar to the pallid gene defect in mice). Other examples of manipulation of the fetal environmental to correct an inherited disorder involve the use of cortisone and thyroxine in mice. Recent work with mice carrying a defective gene that leads to blindness shows that injecting cortisone into the pregnant female prevents blindness in the progeny. Further, in certain strains of mice administering thyroxine to pregnant females reduces the incidence of spontaneously occurring hare lip (Emery, 1971). Results of studies on fetal mouse development do not mean there *will be* equivalent therapy for human fetal defects, but data from experiments on other animals suggest that effective fetal therapy for humans is possible as we increase our understanding of the action of the genes in their environment.

In the final analysis, genetic medicine—screening, counseling, and therapy—offers prospects that are both promising and frightening. Dr. Phillip Handler suggested that the human gene pool might be cleansed through selective abortion. He feels that "some people are dispensable." This may well be true, but since every individual is unique, they may not be replaceable. But, then, perhaps some need not be replaced. Other prominent scientists admit there is an urgent need to discuss these matters. Margery Shaw, director of the Medical Genetics Center at Houston, Texas, asks: "Do our genes belong to us, or do they belong to future generations?" But Marc Lappe at the Institute of Society, Ethics and the Life Sciences, Hastings-on-the–Hudson, New York, counters: "To maintain that we are adding to the public good by conscientious genetic guidance and screening may be fallacious and ethically dangerous." If there is no agreement among those directly involved in the existing programs of genetic medicine, what can laymen do? They can become informed citizens by discussing such problems and reading from available literature on the subject. Then, and only then, will laymen be prepared to vote or to voice their concern independently on a particular issue. Recently a bill passed the Senate to fund a two-year commission that would evaluate the need for a permanent National Advisory Commission on Health Science and Society to monitor the progress of technology with respect to its effects on society. Also, the World Health Organization and the Council for International Organization of Medical Sciences have, by formal resolution, recommended the establishment of an international committee to consider the moral and social issues precipitated by recent breakthroughs in the life sciences.

Does the application of genetic medicine threaten basic rights? Will freedom of choice be supplanted by coercive measures? In the use of genetic medicine, does man lurk in ambush for man? Or is man about to play a positive and intentional role in his own evolution?

references

Brady, Roscoe O., et al. 1973. Replacement Therapy for Inherited Enzyme Deficiency. *New Zealand Journal of Medicine, 289*:9–14.

Culliton, B. J. 1974. National Research Act: Restores Training, Bans Fetal Research. *Science, 185*:426–27.

Emery, Alan E. H. 1971. *Elements of Medical Genetics*, 2nd ed. Longman, New York.

Hamilton, Michael (ed.). 1972. *The New Genetics and the Future of Man.* William B. Eerdmans Publishing Company, Grand Rapids, Mich.

Hurley, L. E. 1968. The Consequences of Fetal Impoverishment. *Nutrition Today, 3* (Dec.):3–10.

Nutrition Today. 1974. Infant Screening for Inborn Errors of Metabolism. *9*(Nov–Dec.):21–23.

Powledge, T. M., and Sharmon Sollitto. 1974. Prenatal Diagnosis—The Past and the Future. *The Hastings Center Report, 4*(Nov.):11–13.

Roberts, J. A. Fraser. 1970. *An Introduction to Medical Genetics.* Oxford University Press, London.

Scriver, C. R. 1974. Inborn Errors of Metabolism: A New Frontier of Nutrition. *Nutrition Today, 9*(Sept.–Oct.):4–15.

politics, social outcasts, and genetics in law

The title of this brief chapter is meant to imply that certain very important historical events may have been influenced by association of politics, genetics, and individuals. The chapter does not attempt an exhaustive coverage of political involvement in the history of genetics.

There are many examples in the history of science when a particular government or religious order inhibited the advance of science. Recall Roger Bacon's persecution for his criticism of Aristotle's teachings. Then there was the inquisition of the seventeenth century that threatened Copernicus and would not allow him to publish his discovery that the earth is not the center of the universe. And Galileo, who believed Copernicus but feared the same inquisition, recanted in 1616 and dared not speak again of the theory of Copernicus. And in 1925 there was the famous Scopes trial where the state government said it was unlawful to teach evolution in the classroom (discussed in Chapter 3).

In most instances the government or religious order brought charges against the men of science. Rarely did a scientist bring about the demise of fellow scientists through the use of a political power base. There is one outstanding exception: T. D. Lysenko, a plant physiologist, not only became deeply involved in the purge of fellow scientists, but also in his ignorance caused the decline of Soviet agriculture. This particular example of political intervention in science became known as Lysenkoism. In the Soviet Union Lysenkoism led to the abuse of individual rights of geneticists, the abuse of scientific data, the purge of a scientific discipline of genetics, and the revival of the Lamarckian hypothesis that environmental characteristics are inheritable. It may even have affected international politics; the Soviet Union has been forced to turn to the

A politician thinks of the next election; the statesman, of the next generation.

J. F. Clarke

genetics and politics: three case studies

United States for grain that it might be producing at home if Lysenko had not achieved political power.

lysenkoism, politics, and the soviet union

In the early 1900's, geneticists knew nothing of the mechanisms of gene action during development. To many of the early geneticists, populations were to be improved by artificial selection through the practice of culling, separating out the best and destroying the rest; this was and still is the principal way varieties of grain, farm animals, plants, race horses, and so on are improved. During this period, geneticists and physiologists were generally quite skeptical of each other. Physiology appeared strong because of the pioneer work of a Russian, Pavlov, and his studies on conditioned reflexes in dogs, while Mendel's findings in the field of genetics lay undiscovered, perhaps because even in Mendel's time physiologists and experimental embryologists held the limelight.

It is safe to say that Lysenko severely delayed the growth of Soviet agriculture. The important thing to understand is that if this could happen once, perhaps it could happen again, but where? In general, a person's actions are related to his time. It has often been stated that "when the time is right" certain events will occur—discovery, war, famine, etc. A brief review of the times might suggest how T. D. Lysenko achieved the power to purge scientific adversaries and significantly set back Russian genetics.

Prior to the rediscovery of Mendel's work on genetics in 1900, Marx and Engels, the originators of modern Communist theory, supported the scientific theory of Lamarck, that the environment causes physiological and mental change in a person that can be transmitted to their offspring. For example, a father who developed large muscles through hard labor would pass on to his son that body characteristic. Marx and Engels were fond of the Lamarck concept since it provided a device by which the achievements of communism could be transmitted from parent to offspring (Caspari and Marshak, 1965).

In 1918, Paul Kammerer, an Austrian experimental biologist, conducted experiments on the development of the nuptial pads of the midwife toad (*Alytes obstetricans*) that gave support to the Lamarckian argument that hereditary traits could be acquired from the environment. Even though his work was declared a fraud by Bateson, who himself once believed in Lamarckism, and Nobel in 1926, Kammerer was invited by the Soviet Academy to move his laboratory equipment and continue his Lamarckian experiments in the Soviet Union. It would appear that the Russian political structure even then would influence or dictate the manner and method of genetics that would be pursued in the Soviet Union (Koestler, 1972). The great Soviet physiologist Pavlov, whose work on conditioned reflexes in animals had highlighted the importance of environmental influences in behavior, announced in 1923 that a conditioned reflex in mice was inherited. Pavlov's later withdrawal of this claim did not discourage the Lamarckian Marxists in the U.S.S.R.

In 1935, a very successful Soviet plant breeder, I. V. Michurin, held that Lamarckism was correct and that Western belief in the use of Mendelian genetics was incorrect. He pointed out that Western genetics took a great deal of time to produce a suitable plant and cost a great deal in labor and money. At this point in time, the Soviet Government was growing impatient for higher yields in farm production. Further, Michurin was a hero of the Russian Revolution and a personal friend of Joseph Stalin.

Following the Revolution in Russia, the Marxian theory of dialectical materialism was gradually applied to all forms of human activity. According to this theory, the process of material change in a society proceeds in a dialectical manner. That is, a society develops through a sequence of an established order (thesis), the introduction of a challenge to that order (antithesis) in the environment, and the resulting struggle to a new society (synthesis), which in turn becomes a new thesis. It is easy to see that if one believed in Lamarckism, the inheritance of acquired characteristics, then he would be in line with the prevailing political philosophy and with other scientists like Michurin. Thus, the debate between the Soviet Lamarckists and Mendelists had political overtones from the start.

The application of dialectic materialism to scientific debate that took place under Stalin's "reign of terror" made rational discourse and a peaceful resolution of scientific differences in genetics impossible. In the atmosphere of massive repression of the 1930's, the spy hunts and centralized inflaming of passions, and a feverish search after "enemies of the people" in all spheres of activity, any scientific discussion tended to become a struggle for the side represented by the Mendelists (Medvedev, 1969). The stage was set, the time was right, waiting for a Lysenko to come along. Lysenko was someone who would say what the government wanted to hear—that he could increase farm output in a short period of time.

the rise of lysenko

T. D. Lysenko was, in 1927, an obscure plant breeder with a flair for the dramatic. Of peasant origin and lacking in formal education, Lysenko was nevertheless a showman. He made the front page of *Pravda* that August as "the barefoot Professor Lysenko." In January of 1929 Lysenko presented a paper on his new discovery, "vernalization"—the practice of temperature treating of winter wheat for spring planting. He stated that the development of plants can be speeded up or slowed down by environmental conditions such as temperature. If one of the developmental stages is altered by environmental influences, the subsequent stages will also be changed and an organism with different physiological qualities, better adapted to the environmental influences in question, will result. Lysenko claimed that environmentally induced changes are transmitted to the progeny and result in better-adapted plant lines (Caspari and Marshak, 1965). Lysenko made spectacular claims for the improvement of Soviet agriculture, none of which was ever realized.

the great purge

Lysenko never trusted the trained Mendelian geneticists and was resentful of their prestige. He thought of himself as an agriculture breeder to improve plant species. He felt the breeder attempted to improve cultivated plants and animals by selecting the best specimens he could find and interbreeding in the hope of producing better. The geneticist, he felt, was only interested in the study of inherited defects. Lysenko was contemptuous of the defective plants and animals of the geneticist, as well as intimidated by the geneticists' sophistication and education.

In the 1930's, Lysenko formed a "creative partnership" with I. I. Prezent, a self-styled Darwinist and experienced manipulator of the tools of political slander. Prezent had already destroyed the reputation of one notable scientist, Raikov, who spent six years in prison as a result. Lysenko's education in Darwinism and interest in genetics dates from the formation of this partnership. Lysenko had only limited success in applying vernalization as a practical concept, but because of the desperate situation existing in Soviet agriculture and the maximum use of propaganda on his part parleyed this limited success into the directorship of a regional plant-breeding institute. From this base Lysenko and Prezent launched their attack on genetics.

Soviet agriculture was at this time in dire straits. Lysenko and Prezent took advantage of this situation to press the attack on two fronts. First they attacked the science of genetics as bourgeois, reactionary, foreign, and metaphysical. Then, through political manipulation they pressed a series of impossible demands on the All-Union Institute of Plant Breeding (AIPB), headed by N. I. Vavilov, the foremost Soviet Mendelian geneticist of the time.

The first sign that the Soviet Government was involved in the Soviet genetics was the cancellation of the Seventh International Congress of Genetics that was to be held in Moscow in 1937. Then two years later, in 1939, when the Seventh International Meeting was rescheduled to be held in Edinburgh, Scotland, the Russian geneticists failed to appear. Meanwhile, at the urging of Lysenko and Prezent, the Soviet Central Committee assigned the AIPB to do, in essence, ten years' work in four. When Vavilov protested that the assignment was impossible, Lysenko volunteered to do it in two years. By this tactic, Lysenko gained the ear of Stalin, and that feat overshadowed his failure to complete what was indeed an impossible assignment. By promising immediate results, which he knew to be impossible, Lysenko took advantage of the desperate agricultural situation, while Vavilov, working with the slow methods of Mendelian genetics, would not promise what he could not do. Vavilov's honesty sealed his fate. In 1936 he was called upon to defend his views. In 1937, after an address by Stalin about liquidating double-dealers and Trotskyites, Prezent proceeded to identify all opposition to Lysenko. The position of Vavilov continued to deteriorate. Lysenko was able to place men inside the AIPB to forment opposition to Vavilov. Vavilov was being systematically stripped of his power and prepared for the sacrifice to come. He realized what was happening and wrote (Langer, 1967):

> We shall go to the pyre, we shall burn, but we shall not retreat from our convictions. I tell you, in all frankness, that I believed and still believe and insist on what I think is right, and not only believe—because taking things on faith in science is nonsense—but also say what I know on the basis of wide experience. This is a fact and to retreat from it because some occupying high posts desire it, is impossible.

Vavilov was arrested in 1940 as a British spy and died in prison in 1943. His closest friends and collaborators also perished in prison. With this stroke Lysenko established by force what he could not achieve through scientific reason: the conversion of Russia to Michurinist biology.

the new soviet genetics

The tenet of Lysenko's new biology was simple, he contended that the inheritance of characters acquired by plants and animals in the process of their development was possible and necessary. Michurin's teaching shows every biologist the way to regulate the nature of plant and animal organisms. The way of altering it in a direction required for practical purposes was to regulate the conditions of its life by physiological means (Lindegren, 1966). For example, Michurin believed that to increase the sweetness in the fruit of a fruit-bearing tree you need only add a solution of sugar to the ground at the base of the tree. The roots of the tree would take in the sugar and transport it to the fruit. In other words, the fruit would acquire the sweet taste "character" directly from the environment. Seeds from this "sweet" fruit, when planted, would then produce new trees with the "acquired character," sweet fruit. Such inheritance had been suggested by Lamarck (1744–1829) in his theory of the inheritance of acquired characters (Chapter 3). Thus, Trofim Lysenko, a shrewd plant physiologist with little or no knowledge of genetics, said that all phenotypic changes regardless of cause were conditions that could be inherited as previously suggested by Lamarck and Michurin. He felt that physiological changes, regardless of cause, were passed from one generation to the next. He could not appreciate that it was the change in the genes on the chromosomes that controlled change in development. To strengthen his argument concerning the inheritance of acquired characters, Lysenko said that one could change winter wheat to spring wheat by storing germinated grain in the cold. This change in species could not be repeated in the U.S.S.R. or in the United States.

In spite of continued failures to improve Soviet agriculture, Lysenko retained influence over the biological sciences. When questioned, Lysenko resorted to political methods of coverup by which he gained power in the first place. In 1948, threatened by a scientific rebellion, Lysenko went to Stalin to help in organizing a thorough purge of opposing viewpoints from the scientific community. By the time of Stalin's death in 1953, Lysenko supporters held all positions of importance in the biological sciences. With Khrushchev's denunciation of Stalin and the personality cult, some differing viewpoints were allowed to emerge, but Lysenko remained firmly in control. Khrushchev's support of Lysenko assured that any attempt to change the orientation of biological science would fail. Finally the forced resignation of Khrushchev, brought about mainly because of a failing agricultural policy, signalled the end of Lysenkoism by destroying its political power base. Lysenko was relieved of his position in the Soviet Academy in 1965.

In the final analysis, however, it seems clear that it was the basic unsoundness of Lysenkoism per se, plus Lysenko's inability to evaluate critically his own work or that of his associates that proved his undoing. None of the promises made by the supporters of "socialist biology" for the solution of production and technological problems in the Soviet agriculture were fulfilled. Lysenkoism was a unique phenomenon in modern science (Kosin, 1974).

Western progress in the field of molecular genetics since the early 1950's probably hastened the decline of Lysenkoism. Such scientific primitivism could no longer be camouflaged by rhetoric. Biochemists and geneticists in western Europe, the United States, and other countries were rapidly exploring new frontiers in molecular biology.

the penalty

The costs to the U.S.S.R. of Lysenkoism are monumental in terms of wasted agriculture and human resources. The highest long run cost will be the lost potential of an entire generation of miseducated biologists. One of the greatest problems in the reconstruction of Soviet genetics is the lack of qualified teachers. The non-Lysenkoist tradition has been kept alive by men who are aged. Eleven are reported to be in their sixties, and only one, Belazev, is still active in teaching. The result is that while a large number of young people are interested in going into genetics, the Soviet Union does not have an intermediate generation capable of teaching them (Langer, 1967). Further, with the rout of medical genetics that accompanied the Lysenko purge of Mendelian geneticists, there was a loss of gathering statistics on hereditary diseases, professorships and laboratory facilities to study human inheritance and subsequently a lack of genetics courses, genetic counseling, conferences, and exchange of information with the rest of the world. The necessity of restoring medical genetics in Russia has been recognized since the fall of Lysenko, and official recognition for the concepts of Mendelian genetics is doing much to improve Russian medical genetics.

The effect of Lysenkoism is still felt in Russia. In 1972 the Soviet Union purchased about 19 million metric tons (some 700 million bushels) of wheat from the United States and another 10 million metric tons of grain from other countries. In 1975, when the United States produced 5.7 billion bushels of hybrid corn and 2.2 billion bushels of hybrid wheat, the Soviets ordered a minimum of 10 million metric tons of grain from the United States and hoped also to purchase at least that amount from other countries. Since the early 1960's the U.S.S.R. has made yearly purchases of grain from the United States and Canada that by now run into billions of dollars. The selling of our grain to the U.S.S.R. in the early 1970's created a degree of political unrest in the United States. The large sales to the Soviet Union coincided with poor harvests, drought, and sales to other foreign countries to produce a shortage of grain. The price of all grain-related products increased: bread, beef, cereals, etc. In a way, Lysenko also caused problems for the United States by leaving Soviet agriculture dependent on outside purchases.

In 1970 Medvedev, the Russian biologist and member of the Soviet Academy, whose *The Rise and Fall of T. D. Lysenko* was published in 1969, was committed to an insane asylum. He was guilty of expressing his scientific views on topics outside his specialty. Public outcry by fellow scientists across the world prevented his continued incarceration. He was set free, but stripped of his Soviet citizenship and forced into exile. He is now working at the National Institute for Medical Research, London.

Medvedev is known internationally for his work in gerontology, the study of aging. His reputation was established in the West in the 1950's when he put forth the "error theory," the theory that errors in cell reproduction play a role in aging.

He has become more widely known in the West recently for his publication of a book, in Russian, documenting the ten years of Soviet harrassment of Alexander Solzhenitsyn, a Nobel prizewinning novelist who also has been deprived of Soviet citizenship. The two men are close friends.

hemophilia and porphyria—two genetic defects that may have significantly altered the course of history

The diseases of hemophilia and porphyria may have significantly altered the course of global history because the gene for hemophilia may have had a part to play in the current Soviet politics and porphyria may have had its influence on the United States. Both countries are recognized industrial and military powers of the world.

hemophilia

It has been stated that Queen Victoria of England (Figures 11–2 and 11–3) transmitted the gene for hemophilia to her daughter Alice. Alice in turn passed it on to her daughter, who eventually married Czar Nicholas II of Russia (Aronson, 1974). During the reign of the Czar, the peasants rioted in 1905 trying to obtain more land to farm. In industry, people went on strikes for an eight-hour working day, and the Czar created additional ill will by having his soldiers shoot about 1000 of the peacefully marching workers out on strike. This is known in history as "Red Sunday." Stolypin, an imaginative, ruthless chief minister who managed to appease the peasants, was assassinated in 1911, and the Czar reverted to the use of spies to catch troublemakers. During this time of tense political turmoil, the Czar and Empress found that their son had hemophilia. Afraid for his life, they were open to suggestions for his salvation. Help was given by a "half-mad, wholly evil, dirty, ignorant and power-hungry monk from Siberia, Rasputin" (Brinton et al., 1973). It is believed that Rasputin, who had the ability to hypnotize, used his power of hypnosis to prevent the bleeding spells of the Czar's only son. By caring for the hemophilic son Rasputin manipulated the Empress, who in turn manipulated the Czar. Rasputin became the "real" ruler of Russia. The spread of political corruption and political decay that followed under the influence of Rasputin is considered to have been significant in bringing about the Bolshevik Revolution, which was the beginning of the present current political state of the U.S.S.R.

If the Czar's son had not been hemophilic, there may not have been a Rasputin in Russian politics. Would there have been a Lysenko affair?

porphyria

A second historical case involves King George III, the ruler of England in the eighteenth century. Old manuscripts indicate that King George suffered from such classic porphyric symptoms as dark urine, bouts of intense abdominal pain, and periods of insanity for which royal physicians prescribed a strait-jacket. The disease, which has been traced back to Mary, Queen of Scots, was transmitted to members of the houses of Stuart and Hanover and to the Prussian royal lines.

An inherited disease, porphyria is the inability to metabolize properly "used" molecules of hemoglobin. Other symptoms are sensitivity to light, colic, nausea, weakness, visual disturbances, headaches, tremors, convulsions, and periods of incoherence (porphyria is discussed in Chapter 9). King George had several prolonged bouts of mental illness associated with what has now been diagnosed as porphyria. In fact, his fits of mental anguish resulted in the founding of psychiatry as a serious branch of medicine (MacAlpine and Hunter, 1969). It is believed that it was during one of his mental lapses that he imposed the Stamp Tax on tea. This produced such colonial resentment and hatred of the King that the Patriots of Boston held the historic Boston Tea Party, which along with other grievances led to the American Revolution. What if King George had not suffered from porphyria? Would not our history be different? Would we have celebrated the 200th year of our freedom in 1976? While it may be provocative to discuss what history might have been had George III been properly diagnosed and treated, the plain fact is that the disease was not described in the medical literature until 100 years later. Even now, with the advances in our understanding of this disease, the diagnosis of porphyria is still being missed or overlooked.

As discussed in Chapter 23, geneticists of international reputation, from several countries and various disciplines within genetics, are developing a spirit of international cooperation for discussion of experiments with the potential to harm all of humankind. What we need next is an international funding agency to review plans and to support those experiments that the geneticists, other scientists, and laymen feel are of redeeming biological and social value. An international approach to supporting science would minimize the conflict with individual governments and subsequent suppression of knowledge.

references

Aronson, Theo. 1974. *Grandmama of Europe: The Crowned Descendents of Queen Victoria.* Bobbs-Merrill, Indianapolis.

Brinton, Crane, et al. 1973. *Civilization in the West,* 3rd ed. Prentice-Hall, Englewood Cliffs, N.J.

Caspari, E. W., and R. E. Marshak. 1965. The Rise and Fall of Lysenko. *Science, 149*:275–78.

Koestler, Arthur. 1972. *The Case of the Midwife Toad.* Random House, New York.

Kosin, I. L. 1974. Soviet Genetics, History, Commentary. *BioScience, 24*:583–89.

Langer, Elinor. 1967. Soviet Genetics: First Visit Since 1930 Offers a Glimpse. *Science, 157*: 1153.

Lindegren, Carl C. 1966. *The Cold War in Biology*. Planariam Press, Ann Arbor, Mich.

MacAlpine, Ida, and Richard Hunter. 1969. Porphyria and King George III. *Readings from Scientific American.* Freeman, San Francisco.

Medvedev, Zhores. 1969. *The Rise and Fall of T. D. Lysenko.* Columbia University Press, New York.

> They're all here, aren't they? General Tom Thumb and his diminutive bride, the Siamese twins, the albino family—the collection of human oddities that used to make us laugh and shiver as we walked through the carnival midway as kids. For a lot of us, although we didn't know it at the time, this was our introduction to genetics—gawking at some of nature's more dramatic mistakes and feeling very thankful and a little smug at our own apparent normality.
>
> Man's fascination and fear of nature's genetic mistakes stretches back to the dawn of his own awareness—that dim past where everything unusual that happened was an omen from the gods. Tribal customs often demanded that a strange looking child be killed at birth. And in some cases the parents were considered cursed, and banished and killed along with their offspring.

These were the opening lines from a National Educational Television program (Prowitt, 1973).

There are fewer carnivals and circuses today, but malformed humans are still on display. They are on display as an attraction to people in general who have been sheltered and "protected" from nature's deformities. The protection was either by a loving parent who did not wish to expose her child to the gross anomalies that can occur in human development or by civil law. For example, in 1972, Sealo the Sealboy (Stanley Berent, age 74), whose hands grew from his shoulders, and Poo-pah (Norvert Terhune, age 50) the pigmy, a 3-foot 6-inch dwarf, were threatened with arrest under a 1921 law that prohibited freak shows and carried a penalty of a $1000 fine or a year in jail (*Time*,

I've never met a person, I don't care what his condition, in whom I could not see possibilities. I don't care how much a man may consider himself a failure, I believe in him, . . . The capacity for reformation and change lies within.

Preston Bradley

human curiosities and social acceptance

1971). The event occurred in North Bay Village near Miami, Florida, while they were with the 1972 World Fair Freak Attractions, a sideshow then touring the county fair circuit. Sealo, Poo-pah, and World Fair Freak Attractions took their case to the Florida Supreme Court contending that the state was interfering with their right to make an honest living. It was demonstrated that in spite of their normal intelligence, they were not able to obtain work outside the circus. Justice Hal Dekle expressed the opinion of the court that the law was unconstitutionally broad and that the handicapped must be allowed a reasonable attempt to use his talents to earn a living. The court, by 6 to 1 vote, found:

> It may be that certain malformations, perhaps those relating to private areas of the body or some which may be repulsive or vulgar in nature, would so affect the morals and general welfare as to lend themselves to a prohibition.

But this was not the case with the dwarf and the sealboy (Figure 21–1).

The point is that society has been cruel to those who are less fortunate in their inheritance and development. Frequently the mistakes of nature were done away with, as in Sparta where the malformed were taken to a special place and left to die. On the other hand, in ancient Egypt, dwarfs were highly

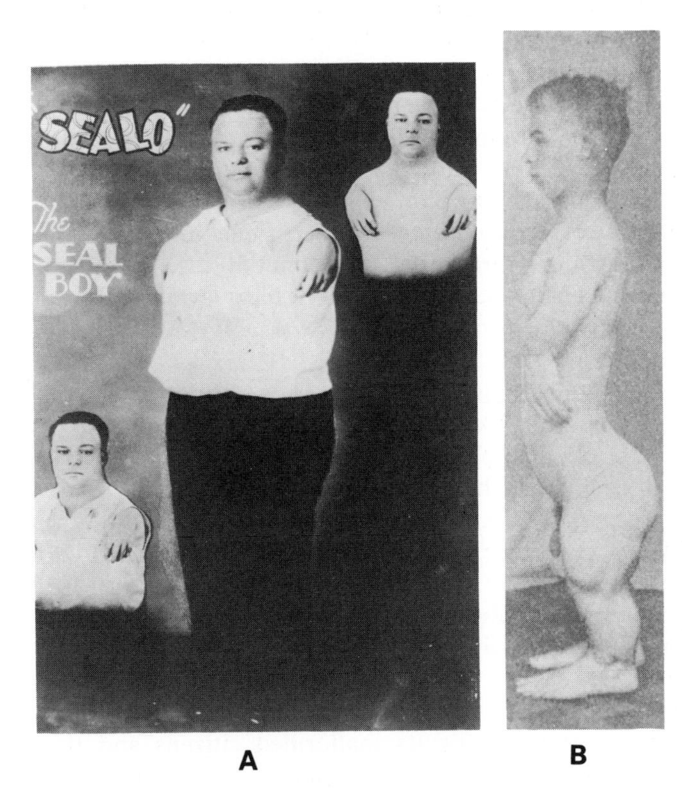

Figure 21–1. **A**. Sealo the Sealboy. Born with short flipper-like arm-hand projections from his shoulders. Possible case of genetic phocomelia (**B**), a rare dominant trait. Environmental mimics are the thalidomide children of the 1960's (see Figure 16–1). (**A** from Bernard Kobel, Choice Circus Photographs; **B** by permission from Tage Kemp, *Genetics and Disease*, Munksgaard International Publishers Ltd., Copenhagen, 1951.)

respected. Bes, an achondroplastic dwarf, was revered as a god. Tutankhamen kept a harem of dwarfs for his pleasure. Later in history, the royalty of Rome enjoyed "freaks" as sexual curiosities. They were prized for their reputation of sexual prowess and were used in mock battles for the emperors. Then, during the Middle Ages (tenth to thirteenth centuries) variously malformed humans were burned at the stake or otherwise done away with as they were thought to be offspring of the devil. During the sixteenth century human malformations again became a curiosity, especially among the royalty.

James IV of Scotland was reported to have kept a two-headed person, Louis XIII of France kept a heteradelphian (a man with extra legs growing out of his abdomen) in court, and the Medicis kept a pit of achondroplastic dwarfs to perform sexually to amuse the court. The presence of dwarfs in the court of Philip II is recorded in the paintings of Velasquez. In England, during the reign of King James II, special ointments were sold that were supposed to retard the growth of children, ensuring them of a bright future in his court. Victor Hugo tells of Chinese dealers who placed a small child in a bottomless vase. They laid it down at night and stood it up during the day. After several years they smashed the vase. Supposedly the child had the shape of the vase (Drimmer, 1973). The child was undoubtedly sold to a collector of human oddities.

However, the fears of evil association persisted among the poor, and something of the medieval attitude toward the malformed exists even today. In many cases it is not only the child who suffers but also the parent. Recently in Reynosa, Mexico, a couple whose first child had been stillborn, was horrified when the second pregnancy resulted in a heteradelphian child. The child had an extra pair of legs. Superstitious peasants in the neighborhood believed the child was cursed and stoned the home, attempting harm to all involved. To add to the family's troubles, a traveling circus offered to buy the deformed child for diplay in its freak show. Afraid the child would be harmed or stolen, the family had to leave their home until corrective surgery was performed in Brownsville, Texas. The heteradelphian birth is a rare event; only about 30 cases have been documented. The most recent birth occurred in Miami, Florida; the child was born with four arms and four legs. Modern surgery was able to restore this child to society.

Had either of these heteradelphian children been born 30 years earlier—and survived—it would most likely be found working in a sideshow of a circus or carnival. Because of society's attitude towards those persons born with serious physical malformations, they have little choice in their destiny, their way of life is predetermined to a great extent by either a genetic or an environmental accident (such as the use of thalidomide by a pregnant woman). Thalidomide is an environmental cause of phocomelia; the child is born with seal- or flipper-like appendages on a normal body trunk.

politics, social outcasts, and genetics in law

In a recent television round table discussion, a group of circus performers who routinely appear in "freak" shows agreed that Soviet Russia is doing the right thing by prohibiting public display of human malformations. The Soviet Government provides special training for its malformed citizens and then employs them in suitable respectable positions (Drimmer, 1973). In the free enterprise system of the United States, however, there are few alternatives to the carnival or circus sideshow, as was pointed out on a National Educational Television program entitled "Freaks and Circus Life." Four "freaks"—a fat man, an albino, a stone lady (whose joints were slowly calcifying), and a giant with Marfan's syndrome—attended and described their hardships and pleasures. About halfway through the program the interviewer remarked on the new medical technologies that permit identification of genetic defects. He further stated that because certain genetic defects can be identified early, parents may have the option of aborting the defective fetus. The interviewer then asked the four guests what their decision on themselves would have been. Their verdict was unanimous: abortion. It would appear from their testimony on these programs that these performers feel they are abused by society for conditions that are beyond their control.

During the nineteenth century the sideshow "freak" became a profitable business. With considerable economic gains to be made from displaying human curiosities, the search was on for new and different "freaks" of nature. In the mid-1800's Julia Pastrana became known as the gorilla or bear woman. She expressed the recessive gene for hypertrichosis, excessive hair growth (Figure 21–2). It has been said that the appearance of Julia and others gave rise to the myth of the werewolf. Except that her face and body were covered by a thick black pelt of hair, Julia was normal. She fell in love with her manager, and they were married in 1860. The manager agreed to the marriage rather than lose her to a rival sideshowman. She became pregnant and her husband-manager

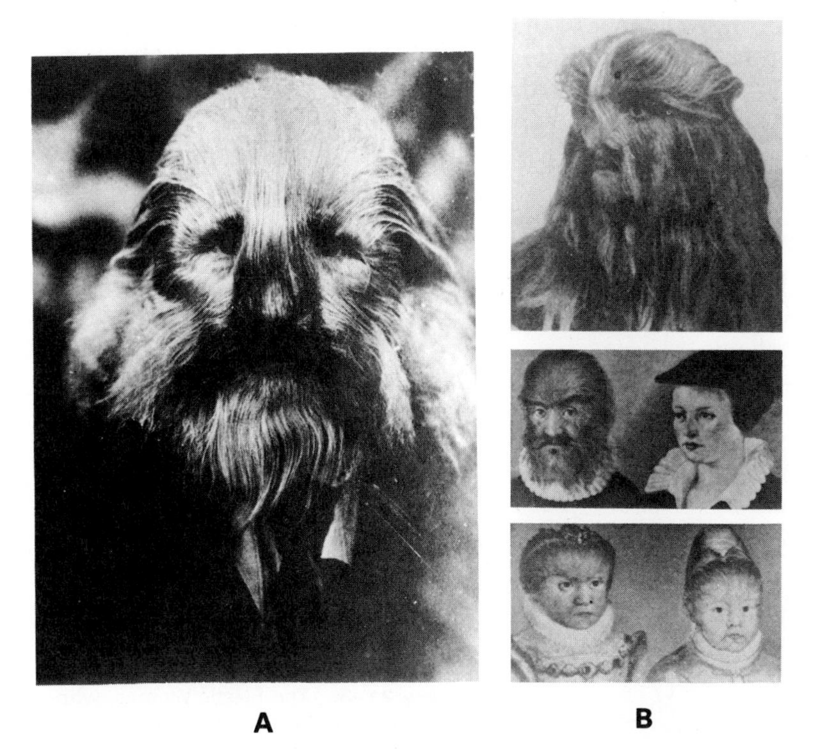

Figure 21–2. **A**. Lionel, the Lion-Faced Boy. In 1907, at age 26, Lionel was earning $500 a week with the Barnum and Bailey circus. He was born in Russia. Like Julia Pastrana, he would be an excellent example of hypertrichosis lanuginosa, a genetic dominant trait. **B**. Persons demonstrating diagnosed dominantly inherited hypertrichosis lanuginosa. (**A** from Bernard Kobel, Choice Circus Photographs; **B** by permission from P. J. Waardenburg, *Genetics and Ophthalmology*, Van Gorcum, Assen, The Netherlands, 1961.)

sold tickets to the birth. The child also expressed the trait but was stillborn, and Julia died five days later. The day after her death, her husband had her and the child embalmed and toured Europe displaying the human curiosities.

During this same period, the American Phineas Barnum rode to riches and fame by exhibiting a wide variety of human curiosities. His favorite, and the most famous of all time, was the 21-inch dwarf General Tom Thumb. The highlight of their association was Barnum's arrangement for the wedding of Tom Thumb to a lovely lady of equal size. The wedding was a spectacular. The Rockerfellers and Vanderbilts sent diamonds as wedding gifts. President Lincoln sent a suite of gold-lacquered miniature furniture, but when General Thumb and his wife later attended a White House reception, there was criticism of the President for inviting "midgets" or "freaks," and Lincoln's oldest son refused to attend the reception because the "circus freaks" were there.

As associates in the circus world learned, many of the small people or dwarfs were very bright. For example, one such dwarf, Pasha Hayati Hassid, was 30 inches tall, weighed 34 pounds, and spoke seven languages.

In 1881 P. T. Barnum joined forces with Anthony Bailey, and together they built the collection of curiosities called the Giant American Museum of Human Phenomena and Prodigies. The collection included the Leopard Girl, a piebald Negress (a dominant gene); a double-jointed albino (albinism, a recessive gene, see Figure 10–4); armless and legless wonders; Queen Mab, a 22.5-inch dwarf; the Telescope Man, who could lengthen and shorten his spinal column at will; the Rubber Man and Elastic Girl, people who could pull their facial skin a good distance from their faces (a skin disease or possibly the inherited dominant Ehlers–Danlos syndrome, Figure 21–3). The circus also

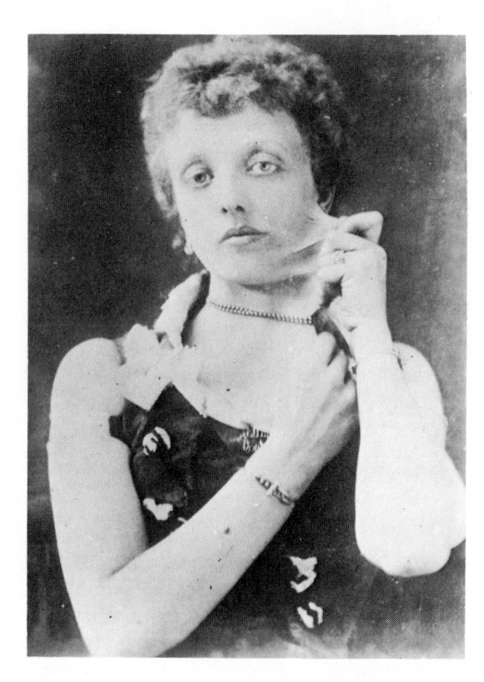

Figure 21–3. The Elastic Woman, who appeared with the King–Franklin circus in 1889. She was able to pull the skin of her face and neck out more than 6 inches. A very similar genetic condition is the genetic dominant Ehler–Danlos syndrome. (From Bernard Kobel, Choice Circus Photographs.)

politics, social outcasts, and genetics in law

displayed people like the Knotty Man, a case that now would most likely be identified as neurofibromatosis, a genetic dominant (Figure 21–4), Mignon the Penguin Girl, who appears to have had the genetic dominant phocomelia, and Toney the Alligator Boy (Figure 21–5).

From a review of a large number of photographs of human circus sideshow curiosities, it appears that the majority expressed dominant traits. A number of circus performers have produced large families. About 90% of the circus human curiosities married normal partners (Drimmer, 1973). On occasion, persons who would ordinarily have sought a circus for a means of self-support have integrated themselves in their communities. Some rose to considerable social heights. Charles Lochart, a possible chondroplastic dwarf (he stood 42 inches tall), became the Texas State Treasurer and served for three terms.

As the reader is well aware, society does change, and it has changed some in its attitudes toward the less fortunate malformed human. This is evidenced by an increased dialogue about malformations, increased financial assistance to those unable to work, and better employment situations for those who can work. But there will always be those who need to look down on somebody; recently, in Detroit, a young man complained that his heavily freckled face kept him from getting a position as a city bus driver. Yet freckling is a rather common dominantly inherited characteristic.

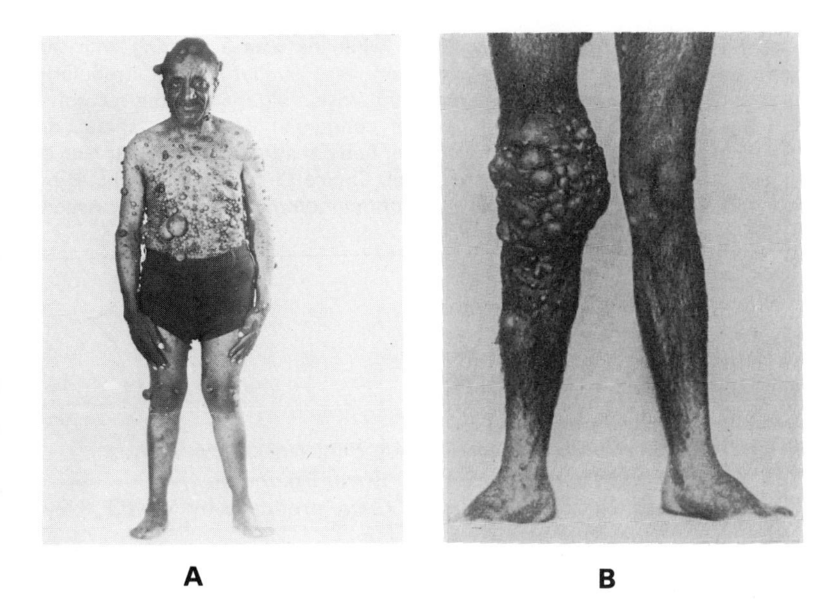

Figure 21–4. A. The Knotty Man. His body was covered with lumpy growths. He was with Kraft Shows in 1947. His appearance with respect to the tumorlike body growths is similar to Von Recklinghausen's disease (neurofibromatosis), a condition inherited as a genetic dominant, illustrated in **B.** (**A** from Bernard Kobel, Choice Circus Photographs; **B** by permission from Tage Kemp, *Genetics and Disease*, Munksgaard International Publishers Ltd., Copenhagen, 1951.)

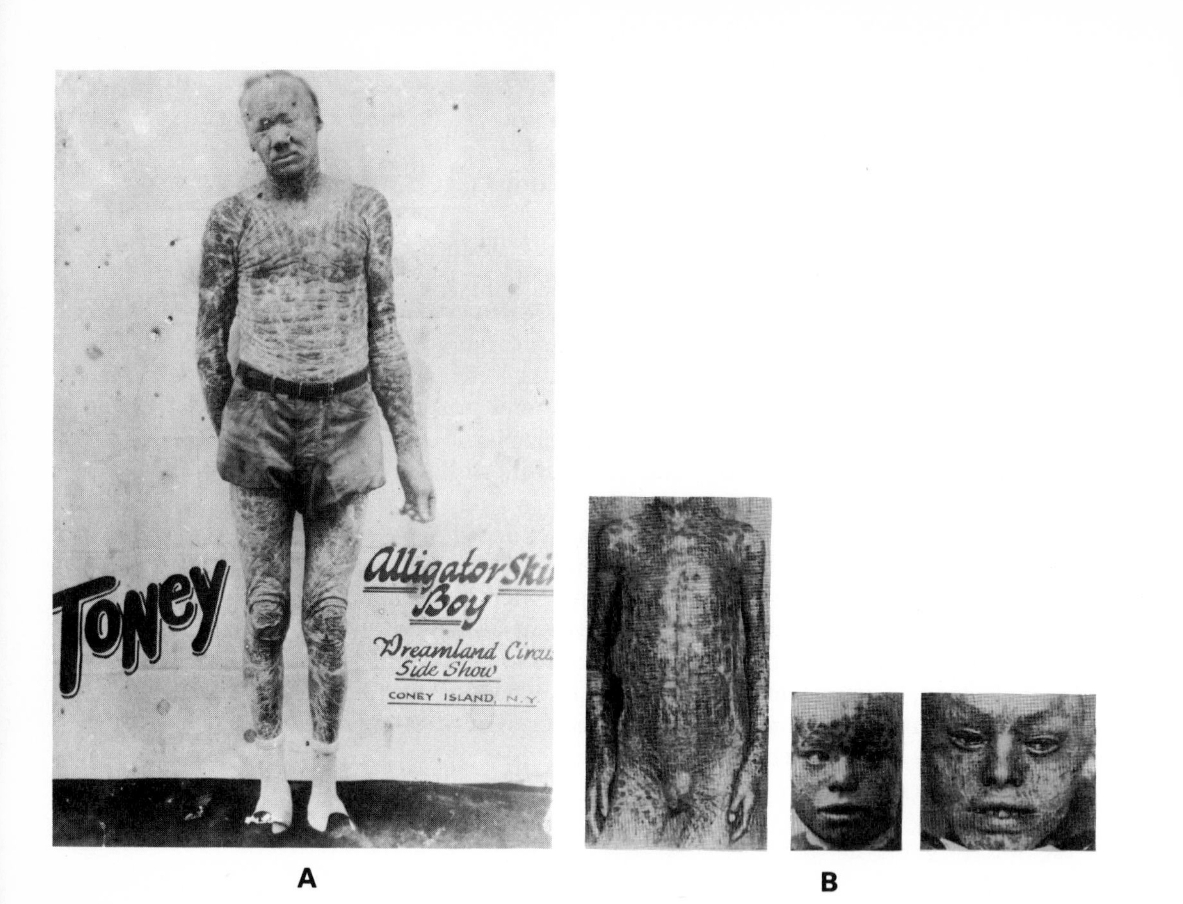

Figure 21–5. **A**. The Alligator Boy. Taken in 1915, while he was appearing with the Dreamland Circus, the photograph shows a condition very similar to the irregularly inherited dominant and X-linked recessive inherited ichthyosis vulgaris and the recessive ichthyosis larvata. **B**. A medically diagnosed case of congential ichthyosis larvata, an inherited recessive disease, in a female patient. *Left*: Her body at age 13; center: her face at age 5; *right*: her face at age 18. (**A** from Bernard Kobel, Choice Circus Photographs; **B** by permission from P. J. Waardenburg, *Genetics and Ophthalmology*, Van Gorcum, Assen, The Netherlands, 1961.)

references

Drimmer, Frederick. 1973. *Very Special People.* Amjon Publishers, New York.

Kemp, Tage. 1951. *Genetics and Disease.* Oliver and Boyd, Edinburgh.

Prowitt, David. 1973. *Genetic Defects: The Broken Code*, produced by WNET, New York, aired on Public Broadcasting Service.

Time, 1971. The Incomplete Twin. June 28, p. 57.

Since the rediscovery of Mendel's work in 1900, information from the science of genetics has been applied in the areas of eugenics, medicine, sexual selection, artificial insemination, cloning, chromosome-based behavioral problems, and law. The courts may soon be faced with legal problems of children born from transplanted ovaries or from foster wombs and, perhaps one day, even children born of eggs fertilized in vitro and developed in an artificial womb or children produced by cloning (the whole person being regenerated from a single body cell).

But how did the science of genetics become involved in our social-legal system? Was there immediate recognition of the relevance of genetics, the study of heredity, to our system of jurisprudence? To appreciate fully the development of genetics as it applied to humans, both in and out of court, let us take a painless excursion back through time to 1865 when a monk, Gregor Mendel, had to leave his vegetable garden because his research material, *Pisum sativum*, the common garden pea, was devastated by the pea beetle. This disaster was closely followed by an equally frustrating problem to a research scientist; he received an administrative appointment as head of his monastary. In reviewing his data collected by simple observation of individual characteristics of his pea plants, Mendel announced to an unattentive world that he could predict the outcome of matings between different plants. Mendel's simple mathematical analysis, which allowed him to predict the number of times a specific trait would appear in the offspring of mating parents, was, in retrospect, one of the more exciting scientific achievements of his day. However, the significance of Mendel's work went unnoticed for the next 30 years, perhaps owing to the social climate created by Charles Darwin and the furor following his pronouncements concerning natural selection and organic evolution. Mendel, like Darwin, had the gift of wonder, the ability to question, the power to generalize, and the capacity for application. He was a capable organizer and excelled at presenting ideas and observations in a precise manner.

From science man gains knowledge, but what will he do with it?

genetics, society, and the law

Mendel's principle of gene segregation offered an explanation for the transmission of traits from generation to generation. The principle of gene segregation became the cornerstone of the genetics of all diploid organisms. The realization that all living organisms had a means by which they transmitted their genes to their progeny meant that heredity had a causal basis and at the same time offered an early indication of the universality of genetics. These two principles offered scientists and legal experts alike a sense of security when dealing with characteristics that are transmitted between parent and offspring. Perhaps unfortunate for legal applicability, many characteristics are dependent on more than one gene expression (polygenic). Others—those most useful in a court of law—are Mendelian traits or traits controlled by genes at single loci (Chapters 9, 10, and 11).

paternity identification

One of the most common and, to date, most important uses of genetics in our judicial system has been in cases of disputed parentage. Such questions as "Is he the father of child X?" or "Is she really the mother of that child?" have long been familiar, and cases such as these have come before the courts. There have also been cases where parents attempt to claim a lost child or claim that there has been an identity mistake at the hospital with respect to their child. In any of these situations we can only attempt in a court of law to determine whether a given individual is *not* the parent (see Chapter 12).

In one very unusual paternity suit a husband in West Germany was granted a divorce on the grounds of adultery. He wanted the court to waive child support for fraternal twins that were born to his wife. The wife maintained, however, that one child was his since she had had sexual relations with her husband and with another man about the time of conception. Blood tests showed that the twins were fathered by two *different* men and that *only one* of the twins could genetically have been his.

In another case, perhaps the first of its kind, paternity was established using chromosome matching. The case occurred in Sweden involving a black couple. The wife applied for an abortion and the husband sued for a divorce on the grounds of adultery with a white man. At the end of the fourth month of pregnancy amniocentesis was performed (see Figure 18–1) and the fetal chromosomes, after being stained with quinacrine dye for fluorescence (see Chapter 6), were examined and compared to the dye-stained chromosomes of the husband. In this case they knew the "other man" and also compared his chromosomes to those of the fetus. The fluorescent pattern of the fetal chromosomes, especially the Y chromosome, more closely resembled that of the husband. In addition, the fetus and husband shared a genetic tissue antigen that was absent from the "other man" and the wife. Five months later the mother gave birth to a black child.

The courts in most states now recognize genetic evidence if the inheritance of the trait in question is a matter of general agreement, that is, if the trait follows a monohybrid pattern of Mendelian inheritance (described in Chapter 8). The inherited traits most often used in court are the blood types A, B, O,

M, N, and Rh and the secretor factors. Other blood types that also have been used are the Kell, Lutheran, and Duffy groups. Race and Sanger have calculated the probability of exoneration if a person is falsely accused in a paternity suit. On the basis of the six blood groups listed and the secretor factor a falsely accused man could expect to be exonerated in 62% of such cases (Neel and Schull, 1954). With typing for all fifteen known blood groups, a falsely accused man would be exonerated in over 90% of such cases.

In at least ten of our states the court has the power to order blood tests in any case where the problem of determining parentage exists. For example, in the 1968 divorce proceedings case of Rose vs. Rose, the Court of Appeals in Marion County, Ohio, granted a divorce to the husband, but awarded support for the minor children. The husband appealed on the ground that he was not the father of one child. It was established that there was no sexual relationship between the husband and wife during the time in question. A blood test further excluded the husband, since the mother had blood type O, the father type O, and the child type B. A medical doctor wrote the court that it is "absolutely impossible" for the child in question, blood type B, to come from the father of blood type O. "By strict Mendelian law, the father would have to be either B or AB type blood" (Table 22–1).

Although the evidence as presented was probably correct, it could be argued that the husband's gamete had undergone a new mutation and that the child was actually his. However, the probability of a new mutation occurring in the father's gamete would be about one chance in a million. In this case, the husband won the appeal and was relieved of child support.

In the 1965 case of Houghton vs. Houghton, the wife took action to divorce her husband on grounds of extreme cruelty. The husband cross-petitioned for divorce on the grounds of adultery and extreme cruelty. The District Judge for Douglas County granted the wife a divorce and custody of the two children and ordered the husband to pay child support. The husband appealed the ruling to the Nebraska Supreme Court. The Supreme Court reversed the lower court

Table 22–1 Possible A, B, O Blood Groups

Blood Groups of Parents	*Expected Blood Groups of Children*	*Unexpected* Blood Groups of Children*
O × O	O	A, B, AB
O × A	O, A	B, AB
A × A	O, A	B, AB
O × B	O, B	A, AB
B × B	O, B	A, AB
A × B	O, A, B, AB	None
O × AB	A, B	O, AB
A × AB	A, B, AB	O
B × AB	A, B, AB	O
AB × AB	A, B, AB	O

*An unexpected could result from a new mutation in one or the other of the parental gametes.

decision on the basis of blood group evidence that was strongly in favor of the cross-petition for adultery. Two different tests were made of each person's blood, one for the Rh system and one for the MN system. Dr. Greene testified at length in regard to these tests and their significance. In his testimony he stated that in the Rh system there are six antigens, these are usually written *C, D, E, c, d,* and *e*. Each person has these antigens in three pairs: a pair of *C*'s, a pair of *D*'s, and a pair of *E*'s. From this, 64 gene combinations and 27 different antigenic phenotypes are possible. One of each pair of antigens (*C, c*), (*D, d*), (*E, e*), in the offspring comes from the father and one from the mother. In the case before the court, the mother has *e* antigen but no *E* antigen; therefore she is *ee*. The husband has *e* antigen but no *E* antigen; he must also be *ee*. The daughter has *e* antigen and *E* antigen; the *E* must have come from another source. Essentially the same thing was found by Dr. Greene for the MN system. The MN blood system is composed of a pair of antigens that can occur in the following combinations: *MM, NN,* or *MN*. The mother had *M* and *N* antigens; she was *MN*. The husband has only the *N* antigen; he was *NN*. The daughter had the *M* antigen but not the *N*; she was *MM*. The daughter could have inherited one *M* from the mother; however, the other *M* could not have been inherited from the father, since he was *NN* and had no *M* antigen. The above findings, except for a chance double new mutation $e \rightarrow E$ and $N \rightarrow M$ in the paternal gamete, which is extremely unlikely, would exclude the husband as the father.

The district judge who dissented in this case felt that the matter of determining paternity or nonpaternity of children born in wedlock is of such importance that the blood tests should not be considered conclusive unless it is so provided by legislative enactment. Does legislation, however, make the test more accurate or the decision more humane or correct? He went on to say that we all recognize the advances made in medical science, and that the courts recognize and accept new procedures. Yet the duty of preserving the home and protecting children born during the marriage relationship from the stigma of illegitimacy is one of the most important responsibilities of society and the courts. Do you agree with the dissenting judge?

genetic screening and legislation

In 1955 sex chromosomes were first identified in fetal cells taken from amniotic fluid (amniocentesis). By 1966 we were able to photograph the human chromosomes and arrange them systematically with respect to their morphological appearance (see Figure 6–8). Now geneticists, physicians, and biochemists working together can detect over 100 genetic defects in the developing embryo by karyotyping or cell chemistry, using amniotic fluid, and by sampling fetal blood. The ability to detect various genetic defects has taken on a new social importance with the recent U.S. Supreme Court decision that a pregnant woman can make a personal decision through the first trimester of pregnancy, and in consultation with a physician in the second trimester, to keep or abort the developing fetus. Thus programs for the early detection of inborn errors of metabolism or chromosome abnormalities are expanding across the

United States. The screening programs are usually conducted by medical centers and recent programs are now screening for as many as a dozen defects per child (Chapter 17). However, counselors must be careful not to encourage mass screening for every defect. For any given defect, it should be asked whether the specific condition would be suitable for mass screening, keeping in mind that early detection may *not* now allow for effective treatment.

phenylketonuria testing programs

A questionable side of mass genetic screening lies in the use of the information to produce legislation that may, in retrospect, be ill-advised. As examples of useful, but perhaps overzealous legislation, we may review preliminary results of two separate screening programs, one for a recessive genetic metabolic disorder, phenylketonuria (PKU), which causes mental retardation, and the second for sickle cell anemia, in which the defective person usually dies before age 30. Within 18 months following the initial Guthrie test for high levels of blood phenylalanine, 30 states followed Massachusetts in passing mandatory PKU screening laws. Through legislation, tests for PKU are now mandatory in 43 states, and tests for sickle cell anemia are mandatory or voluntary in 28 states (Cooper, 1967, Rosenberg, 1970; 1972; Powledge, 1973; Hormuth, 1971). The reasons for making the PKU test mandatory were supposedly (1) the high incidence of the genetic defect in the American population with defects in 1 child in every 10,000 to 20,000 live births (in fact, PKU can be considered as a rare genetic defect); (2) a political move by social-minded politicians; and (3) perhaps most important, some evidence of effective treatment. However, the evidence available on treatment at the time was very limited and very questionable. Cases are known in which *untreated* "phenylketonurics" showed no brain damage. In fact, there are recorded cases of untreated "PKU" children reaching normal or high mentality; for example, of two affected siblings, one later became a student at Oxford (Cooper, 1966).

The problem, as created by rapid legislation in this particular situation, stems from the fact that the screening tests were made compulsory before there was sufficient knowledge gathered from measurement of phenylalanine levels in a sufficient number of normal people. Although legislation occurred as early as 1963, only recently have we learned that there are at least six causes for a child having an elevated level of phenylalanine; only one of these, the deficiency of phenylalanine hydroxylase, is clinically significant in producing the mentally retarded PKU child. The salient point is that a child born with a high level of phenylalanine can be normal, but mistakenly diagnosed and placed on the restricted phenylalanine diet. The normal child, lacking sufficient phenylalanine in the diet, can be injured physically as much as the untreated PKU child. Further, only recently have we learned that for the best diagnosis, the child should not be tested before the second to sixth day after birth. Herein lies another problem; with increasing health costs, fewer hospital beds available, and new ideas on postnatal maternal care, mothers now leave the hospital two or three days after delivery. Given that detection of a PKU child is more difficult once he leaves the hospital, we have still another deficiency in the legislated PKU program. Early legislation, before sufficient facts were in, most assuredly committed a number of healthy normal children to grossly abnormal lives and the child's parents to immeasurable emotional strain and considerable financial loss (Buist and Jhaveri, 1973).

Regardless of the apparent tragedy that occurs with misdiagnosis of high level phenylalanine children in the PKU-legislated screening program, this program was used and will continue to be used as a prototype for the mass screening of other serious genetic defects. With respect to the cost of screening, the uncertainty of preventing mental retardation by dietary control, and the fact that PKU is a relatively rare disease, it is questionable whether legislation should have occurred for this particular genetic defect. Most states have tried to justify their PKU screening programs on the basis of cost/benefit analysis. However, in terms of efficiency, any system that requires the processing of 10,000 to 20,000 children to detect one case of PKU cannot be rated as a highly efficient program. A number of authors have pointed out that perhaps the money spent on this program could have been more effectively spent for better diagnostic and treatment facilities for the 99 out of 100 retarded children who do not have PKU. What are your feelings about this particular criticism? (Class discussion.)

In light of the continuous rise in cost of health care services, screening programs limited to the detection of PKU will be hard to justify. Herein lies still another problem. Rather than give up the elaborate PKU screening system, there is a rush to screen for other genetic defects at the same time one screens for PKU. Should this expansion not be delayed until we have thoroughly reviewed the potential psychological and social stigma associated with mass screening detection of each new genetic defect?

sickle cell testing programs

The second and most recent rush to legislate mass screening (using PKU as a prototype) was for sickle cell anemia (SCA) examinations of the young. In Washington, D.C., SCA is now legally classified as a "communicable disease" with the compulsory testing of school children being administered within the immunization program. In Massachusetts, one is legally classified as a diseased person if he is a carrier of the sickle cell trait (carries one defective gene) regardless of the fact that the individual is healthy. In New York, Massachusetts, Mississippi, Georgia, Illinois, Virginia, and the District of Columbia children must be tested for the presence of a sickle cell gene before they can enter school. In New York, Kentucky, and Virginia, blacks must be tested for the presence of one of the sickle cell genes, along with the usual tests for venereal disease, before they can marry. In Kentucky a fine of "not less than \$100 nor more than \$300" is assessed for failure to comply. This genetic screening before a marriage license is issued may be thought of as compulsory medical awareness inasmuch as the person is not prohibited from getting married. The law is to make prospective parents aware of a potential problem. However, such a law does represent an increasing governmental involvement in establishing eugenic (genetic) criteria for human reproduction.

In 1974 the Chicago Bar Association drew up the Illinois Domestic Relations Act and placed it before the state legislature. The law, if passed, would require all applicants for marriage licenses to be tested not only for the standard venereal diseases, but also for "any other diseases or abnormalities causing birth defects." The proposed law does not list any specific defects, nor does it offer any guidelines for defining them. The lawyers who drew up the proposed law suggested an additional idea for the future: "to require that a stated abnormality peculiar to certain races, for example, sickle cell anemia (affecting primarily the black people) or Tay–Sachs (affecting primarily the Hebrew people), should be corrected as a condition preceding the issuance of a license. Meritorious as this may be, it is at the present time impossible to impose such a requirement" (Ausubel et al., 1974). Is it time to question governmental wisdom in determining the rules for marriage and procreation? Or is it too late? Should we eliminate *all* laws restricting marriage (mating)?

The states of Georgia and Kentucky have new laws specifying that the newborn shall be tested for the sickle cell trait (carrier). The idea is admirable except that it is generally held that the carrier condition cannot be detected until there is sufficient synthesis of adult hemoglobin, and this takes some 15 weeks (Nathan, 1973; *Newsweek*, 1973). However, it has recently been reported that we will soon be able routinely to detect the carrier at birth (*Laboratory Management*, 1974). Yet what can one hope to accomplish by gaining knowledge that a baby has sickle cell anemia or is a carrier of the trait? If the child is found to be a carrier, there is little that can be done. For all practical purposes he will grow up and, except for sudden deprivation of oxygen, will develop normally. If a child is anemic, there is no cure; thus far, treatment only relieves pain. As for laws that require school children to be tested it should be realized that children who are anemic will have a crisis before they reach school age. Thus, such detection is too late; the children will already have had crises. If the legislation is to detect carriers and offer genetic counseling, it is too early. How can a healthy six-year-old carrier understand that he may give a fatal disease to 25% of his children if he marries a normal, healthy-looking (carrier) black girl. And what of the social stigma of being labeled a carrier? Implementation of the law may also create feelings of guilt and anxiety among parents whose children are carriers of the trait and fear in those children who do not understand the difference between having the trait (one defective gene) and having anemia (having both defective genes).

The legislation for compulsory testing may identify the carriers for SCA. But identification increases the social dangers of public stigma, the creation of overprotective parents, psychological problems of the child, charges of discrimination, black genocide rather than genosave, and racism. Publicity to insure the success of SCA legislation has caused a great deal of misunderstanding; a Red Cross publication refers to sickle cell anemia as the black scourge. Newspapers and magazines refer to SCA as the killer disease and repeatedly call it the black man's disease. However, according to Dr. Joseph Robinson, New York State Health Department, 1 in 200 whites carries one defective gene for sickle cell anemia (*Newsweek*, 1973). Airlines had refused to hire black stewardesses who are carriers. United Air Lines announced in 1973 that persons with the sickle cell trait, the carriers, were now eligible for cabin positions. Life insurance companies refuse coverage or use high-risk rate policies for those blacks carrying the trait, and it has been alleged that blacks have been fired once they have been identified as a carrier. There is even a documented case of the death of a black child because his mother told the doctor he had SCA; the crisis stage the doctor thought he was observing turned out to be an attack of acute appendicitis (Powledge, 1973; Reilly, 1973).

Nobel laureate Linus Pauling (1968) said:

> I have suggested that there should be tattooed on the forehead of every young person a symbol showing possession of the sickle-cell gene or whatever other similar gene, such as the gene for phenylketonuria, that he has been found to possess in single dose. If this were done, two young people carrying the same seriously defective gene in single dose would recognize this situation at first sight, and would refrain from falling in love with one another. It is my opinion that legislation along this line, compulsory testing for defective genes before marriage, and some form of public or semi-public display of this possession, should be adopted.

Evidence is accumulating that hastily legislated screening may not have been the best way to deal with sickle cell carriers. There are two bases on which the mandatory sickle cell screening statutes may in fact be vulnerable to constitutional attack. First, it is difficult to identify any legitimate governmental interest in the screening of infants and children. Second, to the extent that these screening programs are directed only toward the black population, either by statutory language or by implementation of the statutes, it can be argued with force that they violate the equal protection clause of the Fourteenth Amendment (Green, 1973). As of February 1973 nine states had passed mandatory legislation requiring testing for sickle cell anemics and carriers. Since then, as noted by Dorothy O. Blackburn of the National Sickle Cell Disease Program, U.S. Department of Health, Education and Welfare, six states (New York, Illinois, Massachusetts, Maryland, Georgia, and Virginia) and the District of Columbia have passed legislation repealing the mandatory sickle cell screening law.

other screening programs

During the summer of 1972 President Nixon signed a bill authorizing the expenditure of $11 million to ascertain the effects and incidence of thalassemia major (Cooley's anemia) in the United States. Another pilot screening program is being carried out in Maryland and Washington on people of Hebrew origin to determine the incidence of heterozygous carriers of Tay–Sachs disease. Thalassemia major (Cooley's anemia) has its highest incidence among Italians of Mediterranean extraction, and Tay–Sachs mainly affects Ashkenazic Hebrews of East European ancestry. These diseases, ethnic in origin and inherited as genetic recessives, and hemophilia (bleeder's disease), a sex-linked recessive trait, are in a screening bill pending in the New Jersey State Assembly. This same bill, if passed, will give the Health Commissioner of New Jersey the discretion of placing additional hereditary defects on the list for mandatory screening. Similar bills are pending in the states of Maryland and Massachusetts (see Chapter 18).

It is obviously uneconomical to screen everyone for everything. Racial and ethnic backgrounds may be relevant. Some genetic diseases are present predominantly in specific ethnic groups. For example, PKU is most common in the Irish, Tay–Sachs in Hebrews of eastern European origin, and thalassemia major in people of Mediterranean origin (see Chapter 15). But we must decide whether, if possible, we want to legislate for the screening of the 2000 genetic diseases now known. Do we want legislation that promotes mass screening for the identification of those people who are genetically ill when in fact the possibility of therapy or corrective gene surgery is nil? People of all races, colors, and creeds must have the *option* of knowing they are carriers of a defective gene and of choosing freely among the medical, social, and legal remedies available to them. Or should they? If one is poor, does he have any options without public support? Will the public support programs where the cost is high? What happens if the options are lost? Society must proceed with caution. Legislated mass screening to identify carriers might be used to promote another piece of social eugenic legislation calling for enforced abortion and sterilization in order to "preserve" the human gene pool. Of course, today this last statement may appear to be "alarmist," but history is marred by misapplication of "good science" and "good intentions."

artificial insemination donors and the question of fatherhood

In 1971 there were approximately 25,000 reported cases of artificial insemination with some 10,000 resulting in live births (*Time*, 1971), and these figures may be conservative (Yussman, 1975). The use of AID is increasing yearly in spite of contradictory court rulings on whether it is an act of adultery and can be used in divorce proceedings and whether the child of artificial insemination is legitimate or illegitimate. During the trial of Orford vs. Orford, the Canadian court offered the opinion that adultery was "the voluntary surrender to another person of the reproductive powers or facilities of the guilty person" (*Newsweek*, 1972; Stone, 1973). In 1954 the Superior Court of Cook County, State of Illinois, ruled that artificial insemination was adulterous in the case of Darnboos vs. Darnboos.

In another case, Sorasen vs. State of California (1967), a divorced defendant was convicted for failure to support a child by the Municipal Court, County of Sonoma. The defendant, the nongenetic (nonbiological) father, appealed and the higher court reversed the decision even though the husband had consented to the act of artificial insemination. The court held that it could not prove in any way that the husband was the father of the child. Therefore, he should not be compelled by law to support a child not of his own making.

In the case of Strnad vs. Strnad (1948) the New York Supreme Court granted "modest visitation" rights to an ex-husband to see the child of his artificially inseminated wife. The court's decision was based on the fact that the child's best interests were involved in the decision and that the husband did "semi-adopt" the child during their marriage. The court did not make a statement concerning the other legal property rights of the child. Here the court stated, however, that the child was not illegitimate.

In a later case before the same court, Gursky vs. Gursky (1963), the husband requested annulment and separation; the court held the nongenetic father liable ($25.00 per week for wife; $20.00 per week for child) for child support since he had given his written consent to the act of artificial insemination. The court ruled that the husband had implied promise of furnishing support. However, the court ruled that the child was illegitimate. This decision casts doubt on the legal status of thousands of children already born of artificial insemination.

In another case, a wife, separated from her husband, was awarded custody of their child conceived by artificial insemination. In trying to deny the nongenetic father visiting privileges, she said that the child had been conceived with the semen of an anonymous donor. The husband protested, stating that the insemination had taken place with his full knowledge and consent. Like any father, he said, he loved the child as his own and considered himself the child's father. The court granted him visiting privileges. However, when the mother moved from New York, where the decision had been handed down, to Oklahoma, she reopened the case. The Oklahoma court ruled that the husband had no parental privileges, since he was not in any sense the child's genetic father. If this man is not the child's real father, who is? The anonymous stranger who donated the semen? Could he now claim the child? Can the child claim legal rights of social inheritance? What of the attending physician who performed the artificial insemination? In those states where it is still illegal and considered an act of adultery, is the physician not guilty of tort (wrong doing)? In such a case the consent of both parents would not protect the physician. Since the Oklahoma case, the states of Oklahoma, California, Georgia, and Florida have legitimized children conceived by artificial insemination.

Artificial insemination can be used to foster eugenic goals or simply to fertilize the female in cases of male sterility. Regardless of the reason, the child never knows his real genetic father. The genetic father and mother have for centuries been considered the biological parents. However, with the large number of artificial insemination conceptions and with the inability of our courts to respond consistently, we could suggest that a new definition of parenthood is in order, a definition dealing with the "rearing" responsibility for the child regardless of method of conception or gestation. There are precedents in the legal rights granted to adopted children and adoptive parents.

legal rights of children

Consider this situation. The parents were genetically counseled that there was a 50% or 100% chance that a current pregnancy could or would result in a malformed child. The parents decided to have the child, and it was malformed. Later in life the malformed offspring learns that his suffering could have been avoided if his parents had accepted the advice of the genetic counselor and aborted the fetus. His case then is that his parents knowingly committed him to a life of the malformed. He sues for damages. How should this case be judged?

In 1973 the Supreme Court of the United States and the State Supreme Court of Massachusetts rules that a fetus must be born alive before being considered a legal person. In 1972 the New York Supreme Court, Appellate Division, in Bryn vs. City of New York, held "the child begins a separate life from the moment of conception, but the child in the first 24 weeks of gestation is not a person within the protection of the Fifth and Fourteenth Amendments of the Federal Constitution." Once born, however, does a child have the legal right to sue his parents for physical, emotional, or other social handicaps such as genetically caused mental deficiency, other inherited defects, illegitimacy, or various congenital diseases? Until recently, it was universally accepted that a child had no legal right to recover damages from his parents, since it was felt that the parents could not reasonably be held accountable for events that led to injury of the fetus. The reasoning was that without the parents' willing cooperation, there would be no child; therefore, there would be no cause for legal litigation.

The parents of a mongoloid child born in 1972 refused to allow corrective surgery on the child's digestive system. There was no opening between the stomach and the intestine. The child starved to death in a Baltimore hospital, but it took 11 days (Gustafson, 1973). There was no legal action in this case. Three years later, parents in Portland, Maine, made a similar decision concerning their deformed infant and the hospital officials asked the court for a decision. The judge held that the deformed child had a right to life and ordered the doctors to operate on the child. The child died 15 days after birth regardless of the surgery. These cases point up the confusion with respect to parental and child rights. Should retarded or severely deformed infants have their lives prolonged at the cost of parental anguish, expense, and use of valuable medical personnal? On the other hand, we must ask, had a *legal* wrong been committed against these children? If so, by whom?

James Watson, a Nobel laureate in biology, has suggested that declaring someone a person at birth may not be the best possible situation. He asks whether a child should be declared a person before it is three days old. The thrust of this question rests on the fact that most birth defects are not discovered until *after* birth. He concludes (Riga, 1973):

> If a child were not declared alive until three days after birth, then all parents could be allowed the choice that only a few are given under the present system.

Still another type of legal situation occurred in the case of Gleitman vs. Cosgrove (Capron, 1973). A New Jersey trial court would not allow for damages to a congenitally defective son (sight, hearing, and speech) regardless of the fact that the physician failed to counsel or advise the mother, who had German measles during the first trimester of pregnancy, that her child had about a 25% chance of being born defective. The family contended that the lack of advice or counseling prevented her from obtaining an abortion. On appeal, the New Jersey Supreme Court stated that the claim of the parents was difficult to reject since an abortion would have reduced the financial and emotional stress involved in raising a defective child. The court ruled that "if the physician had not been negligent as claimed by the parents, the child would not have been born at all" (Parker, 1970; Capron, 1973). However, "it was against public policy to allow the recovery of damages for the denial of the opportunity to take an embryonic life." In rejecting the parents' claim to be compensated for their emotional trauma and the expense of raising a deformed child, the court indicated that if the child had been allowed to choose, it would have taken the option of life along with its handicaps rather than no life at all. Do you agree with this analysis? (Class discussion).

Another question that should be asked is: Does the fetus have the right to a "proper environment" while in the uterus? Granting of such a right would necessitate that females who smoke or take an excessive amount of tranquilizers or alcohol would have a court-appointed guardian to stop such behavior during pregnancy. In such cases who would pay the bill for the guardian? Then, what are the mother's rights? We are legally, though not morally, free of these problems momentarily since the U.S. Supreme Court ruled on January 22, 1973, that a fetus is not a person under the Constitution and therefore has no legal right to life. On the other hand, in the case of Smith vs. Brannan a New York court upheld the right of a child to sue for injuries sustained in utero, saying "justice requires that the principle be recognized that a child has a legal right to begin life with a sound mind and body" (Revellard, 1973), and the U.S. Fifth Circuit Court of New York further established the rights of the unborn child when it found that pregnant indigent mothers and their children are eligible for state welfare benefits. These two cases deal with different issues, but each seems to establish the rights of the unborn child to be born healthy. In keeping with the idea of having a healthy child, the decision of the U.S. Fifth Circuit Court makes the state a party to the well-being of the child by requiring funds to be spent during the prenatal period. The court in these two cases is placing some of the responsibility for the well-being of the child on the state. If the state assumes such obligations, then it follows that legislation may be enacted to protect those for whom the state is responsible.

the XYY genotype—a legal defense?

Normally, if one has, in addition to the 22 pairs of autosomes, an X and Y chromosome, he will be male; if XX, a female. As discussed in Chapter 6, some males have an extra Y chromosome (XYY); this abnormality is sometimes referred to as the "criminal syndrome." By some reports the XYY males are of lower intelligence, taller, and more aggressive than normal XY males, although there has been a great difference of opinion as to whether the XYY male has a greater tendency to commit crime than normal XY males. However, recent articles (Gardner and Neu, 1972; Jarvik, Klodin, and Matsuyama, 1973; Hook, 1974) comparing the incidence of XYY male births in the general population and in selected institutional populations indicate that the extra Y chromosome apparently does affect social behavior. Gardner and Nue (1972), reporting on the contribution of the extra Y chromosome to sociopathic behavior, state unequivocally that "there is at least a tenfold increase in the number of XYY males in institutional populations." They feel that there is a causal relationship between the extra Y chromosome and abnormal social behavior by the carrier and recommend that our judicial system take this factor into account in assessing questions of responsibility. Jarvik, Klodin, and Matsuyama (1973) have concluded that criminals are the only group in which the extra Y chromosome occurs significantly more often than an extra X chromosome and suggest that the Y chromosome predisposes one to aggressive behavior. Hook (1974) states that there is a definite association between having the XYY genotype and being found in certain types of security settings. And, according to Money, Franzke, and Borgaonkar (1975):

> There is today a sufficient body of knowledge available to indicate that the supernumerary Y chromosome creates a population at risk for behavioral disability. Whether that disability materializes or not is another issue. It does not always do so, but when it does, the responsible factor is not known. The present study shows that this factor cannot be correlated simply with social class. Though it transcends social class, the factor that induces behavioral disability in the XYY child's development may, nonetheless, be social. It may have its origins in the dynamics of family life, independent of social class.
>
> There is little doubt that the XYY baby may pose an extraordinary challenge to a parent in that his behavior deviates excessively from the parents' schema of expected baby behavior. Such deviation may so completely subvert a mother's idea and practice of her maternal role that an accelerating, vicious cycle of deviant child–mother and mother–child response is established.
>
> Even with a highly versatile mother experienced in child rearing, a given XYY baby may develop to be behaviorally discrepant. Some intrinsic factor alone might be responsible for the unusual behavior of the XYY child. This issue cannot be settled at the present time.

There have been several instances where the issue of XYY and criminal behavior has been brought before the court. For example, in Paris, France, the court appointed a commission consisting of Dr. Jerome Lejeune (who distinguished himself by his contribution relating mongolism to a chromosomal disorder) and two well-known psychiatrists to establish, if possible, the defendant's responsibility for his crime in view of his XYY condition. The commission reported their results: the defendant's criminal tendencies probably had a substantial biological basis. But the commission was unable to say, with respect to the French definition, that the defendant was "insane." The court rendered a verdict, guilty of murder, and handed down a diminished sentence of 7 years (normal sentence, 11 to 15 years).

The XYY condition was also found in a French animal caretaker, who strangled a prostitute without apparent cause. In Melbourne, Australia, Allen Bartholomew, the psychiatrist at Pentridge Goal, Melbourne, was called as the only witness for the defense of a young man who killed an elderly woman from whom he was seeking aid. In answer to questions during his testimony, Bartholomew related to the jury that an extra Y chromosome is an abnormal condition and therefore "each cell in the defendant's brain was abnormal." The defendant was acquitted on the ground that he was insane at the time he committed the murder.

In the United States the XYY condition has not been successfully used by the defense in order to gain the defendant's freedom. Possibly the first American court decision involving consideration of the XYY chromosome problem and legal insanity was the case of Millard vs. State of Maryland in 1968 (*Atlantic Reporter*, 1970). Defendant Millard was found guilty of robbery with a deadly weapon. He was sentenced to eight years in prison. He appealed the case, but the Court of Appeals affirmed the lower court's decision, based on Maryland's Penal Code. According to Article 59 Section 9a, of that code:

> A defendant is not responsible for criminal conduct and shall be found insane at the time of the commission of the alleged crime if, at the time of such conduct as a result of mental disease or defect, he lacks substantial capacity either to appreciate the criminality of his conduct or to conform his conduct to the requirements of law. As used in the section, the terms "mental disease or defect" do not include an abnormality manifested only by repeated criminal or other antisocial conduct.

The defendant's only medical genetics witness, from the George Washington School of Medicine, testified that "chromosomes [in the cells of the body] are the way that all genetic machinery is passed from one generation to another," that "all things that are passed on from parent to child must go through chromosomes," and that 46 chromosomes constituted the normal complement per cell. Therefore, a person who possessed 47 chromosomes was genetically abnormal. In concluding, the medical geneticist pointed out that the defendant was not fit to stand trial because of his behavior pattern resulting from his genetic defect. (Note: the defendant had tried to commit suicide in jail.)

An extra Y chromosome was also found in the cells of a man who strangled, raped, and mutilitated a woman in New York. His unsuccessful defense was based on the fact that he had no control over his actions because of the presence of the extra Y chromosome. The problem of behavioral associations with the XYY chromosome condition is further clouded by bad publicity such as the *erroneous* labeling of Richard Speck, who killed eight Chinese nurses, as XYY. And more recently, a leading United States geneticist stated at a meeting that "we can't be sure XYY actually makes someone a criminal, but I wouldn't invite an XYY home to dinner" (Culliton, 1974). Such statements are unreasonable and only bias or mislead the uninformed. Lawrence Razavi, a Standord University geneticist, reported at a recent meeting of the American Association for the Advancement of Science that he found that dangerous sex criminals frequently suffer from an addition or loss of a chromosome. He based his studies on 83 men, of whom a large number were serving time for repeated sex crimes.

We will conclude the discussion of the XYY male by asking the following questions:

Should we allow XYY persons to remain at large in our society if such persons, as certain investigations indicate, are antisocial, aggressive, and a potential menace to society? If we presume, as indicated by the medical geneticist from George Washington School of Medicine, that XYY individuals are incapable of distinguishing right from wrong, does it make sense to allow their freedom? But can or should one's freedom be restricted when there is little substantial *proof* that the incidence of XYY people in our penal institutions is any higher than the expected based on the population figures (see discussion of XYY males in Chapter 6). Further, what can you say of people who suffer from *other* organic factors that contribute to their criminal behavior? For example, a single gene change causes the Lesch–Nyhan syndrome where the defective person also shows mental retardation, self-mutilation, and aggressive antisocial behavior. In many cases antisocial behavioral changes have occurred with persons suffering from genetic diseases such as myotonic dystrophy, schizophrenia, porphyria, manic depression, and some cases of alcoholism. There is the possibility that many criminals have a predisposition to crime based on behavior conditioned by genes. If there turns out to be firm evidence that criminal behavior is produced by the extra Y chromosome, then the question becomes: When do we commit these people and to which institutions? Can we let them go free until they commit a crime, and then plead their inability to recognize right from wrong? What of *their* personal rights? How will society react? What new laws will be written?

restrictions on consanguineous marriage

Of all the traditional sexual taboos, the prohibition of consanguineous marriage has gained the most scientific support. Children born of these marriages have long been known to show more mental and physical abnormalities than the general population, and we know that children whose parents are biologically related have a greater chance of having a heritable disease than those born to unrelated parents. In cases of recessive genetic diseases, one can predict with high probability that the child's parents are related (see Chapter 14). The explanation for the increased frequency of inherited genetic diseases among children born from consanguineous marriages is that both parents have received their genetic complements from a limited gene pool.

Laws prohibiting intermarriage in the United States were formulated in the nineteenth century with religion rather than science providing much of the motivation for such legislation. Currently, in all states *except* Georgia, an individual is prohibited from marrying a sibling or his (her) child, parent, grandparent, or grandchild. In Georgia, a male may marry his daughter or grandmother, and an uncle can marry his niece. The uncle–niece marriage is also permitted among Hebrews by state law *only* in Rhode Island. In general, marriage between first cousins is also prohibited (Farrow and Juberg, 1969).

The prohibition against consanguineous marriages, although once based mainly on religious beliefs (the Catholic church still requires special dispensation for consanguineous marriages), is now largely based on the genetic relationship between persons. In civil law the relationship between relatives may be defined by the number of steps separating the relatives in question, counting from either relative up to their common ancestor and then down to the other relative.

It has been estimated that each of us on the average carries a minimum of five genes that could cause death if present in the homozygous state. There may be several thousand additional genes that, in the homozygous state, cause a diversity of genetic defects such as mental retardation, blindness, deafness, nerve degeneration, certain types of cancer, and problems of sexual identity. Thus, current laws prohibiting consanguineous marriages have genetic support; they help to prevent the coming together of defective genes. In passing these laws, we have yielded freedom to our legislatures to decide our social behavior with respect to marriage. The degree of relationship selected is arbitrary (historical). Would you favor a wider prohibition or greater freedom? Should we logically extend the same prohibition to others with a similar probability of defective offspring?

mental ability

Recently, the Supreme Court of Iowa in an 8 to 0 ruling, legally terminated the parent-child relationship between husband (IQ 74) and wife (IQ 47) and their four-year-old twin daughters on the ground that the parents did not have the ability to care properly for the children or the capacity of a person of normal intelligence to love and show affection. We cannot determine from the evidence presented in this case that the parents were mentally defective as a result of direct inheritance. However, there is evidence that about 50% of those with severe mental retardation (IQ of less than 50) result from either an inherited metabolic disorder or a chromosome abnormality (Porter, 1968). A similar ruling was made by this court to place five of the six children of a second family up for adoption. The tragedy of the legal ruling, and why these particular cases are mentioned, is the fact that each set of parents had a new child born during the court proceedings and these children were therefore immune to the court rulings affecting the previous children. Such matters as these cause misgivings about the ability of our judicial system to respond to complex social issues. Although we could explore the pros and cons with respect to the physical, emotional, and financial strain of a new trial, the underlying justification in the first case, for example, cannot be changed by the birth of a new child. The wife will continue to have an IQ near 47. Further, with both parents of such low IQ, there is a greater familial tendency for the children to have less than normal IQ.

As another example, in 1972 a twelve-year-old male killed an 82-year-old female. The boy had an IQ of near 70. The Juvenile Court Judge was confused because the boy had a low IQ, and Missouri had no facilities to deal with this kind of problem. Situations such as these offer insight as to why compulsory sterilization was legislated in the early 1900's. According to eugenically conscious people, the sterilization laws were to prevent the continued procreation of the mentally defective, insane or incompetent, the epileptic and those with criminal tendencies.

problems resulting from sex selection

What if we could select the sex of the child? Recent research has shown that this may be possible through several techniques. For example, centrifugation can separate the heavier X-bearing sperm from the Y-bearing sperm. Also, during electrophoresis of sperm, the X-bearing sperm collect near the anode while the Y-bearing sperm collect near the cathode. These achievements, if technically applied, mean that the husband can have his sperm separated and then the wife could be artificially inseminated with the sperm of choice. In addition, L. B. Shettles has claimed that the sex of a child can be selected if the woman precedes each intercourse with a mild acid douche for a girl or a mild alkaline douche for a boy (Rovik, 1968). It must be made clear, however, that the scientific community has not accepted any of these methods as proved because test data are limited.

Studies by Etzioni (1968) and Westoff and Rindfuss (1974) suggest that society as a whole would prefer boys. They state that sex selection if available, would lead to a 7% surplus of males in addition to the 2.5% surplus that already exists in the population. What would such an increase in male population mean? Would females have a better selection of males from which to choose their mate? Would our society give even more emphasis to male-oriented problems? What of the excess of males who could not find mates? Etzioni relates that males are less religious than their parents, and read less, do less for the moral education of their children, and are less culture conscious than females.

Although all has not yet been written either for or against the use of genetic sex selection, one may wonder whether we should begin to formulate laws to control the use of sex selection.

law and the future

We are confronted by certain scientists and politicians who favor mandatory screening for genetic defects while retaining for the parents the "right" to choose whether or not to have the child. But is there a public health justification for mandatory genetic counseling for the prevention of hereditary diseases? Genetic diseases are not contagious and are usually untreatable. Public health requirements are based on communicability, and genetic diseases *are* communicated—by genes from parent to offspring. Yet public health measures for communicable genetic diseases have never been given the amount of governmental support that other communicable diseases have received. Is it because fewer individuals are affected by inherited disease and the disease is not contagious? Why don't the ethical principles upon which quarantine and inoculation were formulated, the factor of "risk" of contracting a communicable disease, apply to inherited diseases? Is *ethics* governed by the number of people "exposed" to risk? To how high a risk? To how serious a risk? Is the long-accepted ethic of "personal rights" meaningless? What if one cannot provide a pregnant parent facing a genetic dilemma with understandable information at the level of that patient? How do you really know that they realize what is being said? The counseling is of questionable value and may infringe on the individual's rights since a dull or uneducated parent will most often be led by the influence, gestures, tone of voice, and facial expressions of the counselor. On the other hand, suppose the counselor is advising a highly educated patient who is carrying a child that will be born with meningomyelocele (part of the spinal column and spinal nerves extended in a sac or pouch formation on its back). Is it reasonable to assume that even this parent can make a meaningful decision? How can she know of the frustrations of mental retardation, the unproductiveness of endless surgery, the feeble use of crutches, and the disposal of urine bags? Is she really equipped to make any decision? Should we have a law that relieves the burden for the undecided, since *no decision* by the parent *is* a *decision by default* (the defective will be born)? Further, we must face the reality that some couples do not wish to know what the counselor has to tell them even if the law requires their child to be screened for such conditions as phenylketonuria and sickle cell anemia. What then?

Do we need a law beyond the provision for screening programs—a law that will give the parent the right to bear only the child that will be born "normal"? Keep in mind that to arrive at genetic defective-free births, the defective must be sacrificed! But if we expect our current zero population growth to be accepted by parents, do we not owe each couple the right to have two normal children? Should we begin to take into account the fact that in many cases we can determine that the developing fetus is crippled and deformed and enforce therapeutic abortion? Can we successfully legislate a balance between the "rights" and "needs" of people? Can we be sure of not eliminating anyone who could rise above a given handicap? Or does it not matter in an overpopulated world? Should all detected abnormal fetuses be aborted?

We must use discretion in selecting which defects, not only because of the many unanswered moral and ethical dilemmas but also because, through research, certain defects may eventually yield to diagnosis and completely satisfactory medical therapy. Today, screening programs in certain states can evaluate the fetus for some thirteen of our more severe defects through the use of amniocentesis. In the Massachusetts program one child in 3639 had one of the thirteen genetic metabolic or transport disorders tested for. However, the overwhelming majority of inherited defects go undetected in utero because morphologic and biochemical restrictions now placed on the scientists generally confine them to the use of the amniotic fibroblasts. Severe hemoglobinopathies, some of which have a rather high incidence, cannot be detected through amniocentesis; this defect is found only in the red blood cells. But as we improve our diagnostic techniques and detect many of our severe hereditary defects, we will faced with another choice: Which defects can we as individuals afford to treat? Or which will the state, by law, decide to treat? Or perhaps some combination of both? At that time a large and disproportionate part of our total health care program may be changed into a massive national genosave program with the individual abdicating his reproductive function to the test tube, which can more easily be manipulated by law.

Through 1972 approximately 1500 amniocenteses were performed in the United States. In 1974, 3000 women had second-trimester amniocentesis (Culliton, 1975), and in 97.2% of these cases amniocentesis was followed by the birth of a normal infant. In 1973 the U.S. Supreme Court decided that a woman could have an abortion for any reason during the first trimester or an abortion through 24 weeks (second trimester) after consultation with her doctor. This "law" has made it relatively easy, in some places, for the abortion of genetic defectives diagnosed during the first and second trimester. Previously it was not uncommon for a judge to rule in favor of a hospital that denied an abortion even when there was reason to believe the fetus was severely defective. Recall the women who had taken thalidomide and had to leave the

United States for an abortion, or the 1968 case of Steward vs. Long Island College Hospital. In this case Mrs. Steward, who had rubella in her first trimester, was denied a therapeutic abortion by the Chief of Obstetrics. The child was born seriously crippled. Justice Beckinella set aside a lower court verdict that awarded $100,000 to the defective daughter. Justice Beckinella said the court had reached the conclusion that there is no remedy for having been born under a handicap, whether physical or psychological, when the alternative to being born in a handicapped condition is not being born at all. In other words, a plaintiff has no remedy against a defendant whose offense is that he failed to consign the plaintiff to oblivion. Such a cause of action is alien to our system of jurisprudence. Yet at that time the courts allowed abortion of the fetus for cases of forcible rape or in preservation of the mother's life. Is it more acceptable to destroy a fetus due to rape (the child was not responsible and in most cases will be healthy) than due to a severe disease?

Therapeutic abortion, regardless of legislated limitations or leniency, will not be the complete answer to our social problems in dealing with hereditary defects. This can only partly be achieved even by our best possible solution—the prevention of birth defects before conception. Contraception coupled with genetic engineering of the future may one day relieve us of our genetic burden.

It has been suggested that eugenics could free man from genetic defects. Unfortunately this will never be realized if for no other reason than the occurrence of new mutations. If wisely applied, eugenics can increase the frequency of genes most appropriate to a given environment at a particular time in our evolutionary history. Simply stated, the practice of eugenics can increase the survival value of different combinations of genetic material within a given environment. To help in the propagation of the more "desirable" phenotypes of our population the National Foundation/March of Dimes lists over 270 clinics that deal with heredity problems. The problem confronting society is establishing the appropriate legislation that would increase the quality of life for all; the increase for some is at the expense of others. As noted earlier, a mongoloid child (Down's syndrome) was born with an intestinal obstruction that would be lethal if left untreated. Acting in the best interest of their two normal children, the parents of this child decided to deny the lifesaving operation and the baby was placed to one side with the sign "nothing by mouth." It took 11 days for the child to starve to death. Yet, normal feeding would have hastened death—should the child have been fed? Should the parents have been permitted to decide on the quality of life for that child regardless of its future? How then does one treat the religious ethics of denying life-sustaining blood transfusions to the normal or the use of insulin to a diabetic? The question in either case is: Should legislation permit the ending of a human life so that other life can be sustained with reasonable quality? Is there a point at which the newborn should not be considered a human owing to extensive abnormalities? The quality of the human gene pool is determined not only by the number of living genetic defectives but also by the number of carriers (heterozygotes) in the population. To eliminate the heterozygote or keep the level of genetically defective heterozygosity to the minimum level, we look to a future of genetic engineering (discussed in Chapter 23) and corresponding legislation.

Today newspaper headlines are made by theologians, politicians, research scientists, medical doctors, and "ethical theorists." Many of those are looking for sensationalism by writing about test tube babies, cloning, the artificial womb, and the surrogate mother; finally (and almost without reverence) each asks: Who wants to play God? It might be comforting if these people were correct in their predictions of the significant genetic cures in 10 to 25 years. At the moment we have no way to delete defective genes from a genome, or replace them, even if we *could* manufacture the required genes. Perhaps this is a blessing in disguise; society is ill-prepared at this moment to deal with the fantastic power available to man when and if he achieves the ability to perform genetic surgery. Ruling out the potential for completely destroying humanity, nothing since the beginning of humanity has been equal to the enormous potential of genetic engineering, or posed more complex legal, ethical and social problems for mankind.

Our system is not up to the challenge today—will it be when genetic engineering is a common reality? For example, in 1965 Senator Walter Mondale introduced legislation to establish a committee to study the effects of genetic engineering. The Senate unanimously passed this bill in 1971. An initial report of the committee came out in November 1972. When one considers that it took six years to establish the committee, perhaps even thirty-five years to prepare for the legal, social, and ethical implications of commonplace engineering and cloning as means of practicing eugenics is not enough (genetic engineering and cloning are discussed in Chapter 23).

Laws reflect social and cultural change: it is the duty of everyone to help in the continual renovation of social laws. Although laws will never be "the" perfect instrument for a changing society because they *follow* social change, by constant adjustment a more ethically and humane system of jurisprudence may emerge.

We recognize that scientific knowledge is expanding and that some of this knowledge will be used to promote social progress for "national concern." However, programs for genetic counseling, genetic screening, and the practice of eugenics and euthenics create troublesome possibilities because they challenge the dignity of man and his relationships to his society and democratic form of government. Without question, our government must be concerned with the improvement of our people individually and collectively. Through certain legislation, our government has sought to enable us to cope with our environment by providing better medicine through governmentally funded research programs. But now we stand on the threshold of an era in which it will become possible to alter the intrinsic characteristics of man himself.

This possibility, together with other biomedical possibilities, raises staggering problems. The answers are by no means clear. But our free and dynamic society will accept the fruits of expanding knowledge and hopefully seek to use these fruits constructively for the betterment of all. We must assume, moreover, unless we are prepared to abandon our faith in the democratic process, that in a democratic society our government will be responsive to public sentiment. Hopefully it can thus respond to the majority without suppressing the rights of the minorities. We must accept the principle that the public may, if it chooses, permit, guide, or even force its government to open or close Pandora's box.

references

Ausubel, Frederick, et al. 1974. The Politics of Genetic Engineering: Who Decides Who's Defective? *Psychology Today, 8*:30–43.

Atlantic Reporter, 1970. 261 A. 2nd. 2nd series, No. 2, pp. 227–32.

Buist, N. R., and B. M. Jhaveri. 1973. A Guide to Screening Newborn Infants for Inborn Errors of Metabolism. *Journal of Pediatrics, 82*(3):511–2.2.

Capron, A. M. 1973. Informed Decision-Making in Genetic Counseling: A Dissent to the Wrongful Life Debate. *Indiana Law Journal, 48*(4):581–604.

Cooper, I. Q. 1966. The Role of Government and Legislation in Management of Problems in Medicine. In J. A. Anderson and K. F. Swaiman (eds.), *Phenylketonuria and Allied Metabolic Disease.* Proceedings of Conference, Washington, D.C., pp. 168–72.

Culliton, B. J. 1974. Patients' Rights: Harvard Is Site of Battle over X and Y Chromosomes. *Science, 186*:715–17.

Culliton, B. J. 1975. Amniocentesis: HEW Backs Tests for Prenatal Diagnosis of Disease. *Science, 190*:537–40.

Etzioni, Amitai. 1968. Sex Control: Science and Society. *Science, 161*:1107–12.

Farrow, M. G., and R. G. Juberg. 1969. Genetics and Laws Prohibiting Marriage in the United States. *Journal of the American Medical Association, 209*:534–38.

Gardner, L. I., and K. L. Neu. 1972. Evidence Linking an Extra Y Chromosome to Sociopathic Behavior. *Archives of General Psychiatry, 26*:220–22.

Green, H. P. 1973. Genetic Technology: Law and Policy for the Brave New World. *Indiana Law Review, 48*(4):559–80.

Gustafson, J. M. 1973. Mongolism, Parental Desires, and the Right to Life. *Perspectives in Biology and Medicine, 16*(4):529–57.

Guthrie, Robert. 1972. Mass Screening for Genetic Defects. *Hospital Practice, 7*:93–100.

Hook, Ernest B. 1974. Racial Differentials in the Prevalence Rates of Males with Sex Chromosome Abnormalities (XXY, XYY) in Security Settings in the United States. *Journal of Human Genetics, 26*:504–11.

Hormuth, R. P. 1971. Screening-Organization. In H. Bickel, F. T. Hudson, and L. I. Woolf (eds.), *PKU and Some Other Inborn Errors of Amino Acid Metabolism.* George Thieme Verlag, Stuttgart.

Jarvik, L. F., Victor Klodin, and S. S. Matsuyama. 1973. Human Aggression and the Extra Y Chromosome. *American Psychologist*, 28:674-82.

Laboratory Management. 1974. Screening Newborns for Sickle Cell Diseases. *12* (April):9.

Money, John, Alice Franzke, and Digamber Borgaonkar, 1975. XYY Syndrome, Stigmatization, Social Class, and Aggression: Study of 15 Cases. *Southern Medical Journal, 68*:1536–42.

Nathan, David, 1973. Detection of Aberrant Hemoglobin of Fifteen-Week-Old Fetuses. *Bio Medical News, 4* (Feb.):6.

Neel, James V., and William J. Schull. 1954. *Human Heredity*, University of Chicago Press, Chicago.

Newsweek. 1972. Proving Paternity. May 8, p. 64.

Newsweek. 1973. The Row over Sickle Cell. February 12, p. 63.

Parker, Carey W. 1970. Some Ethical and Legal Aspects of Genetic Counseling. *Birth Defects: Original Article Series, 6*:52–57.

Pauling, Linus. 1968. Reflections on the New Biology. *UCLA Law Review, 15*(3):269. (3):269.

Porter, Ian H. 1968. *Heredity and Disease.* McGraw-Hill, New York.

Powledge, Tabitha. 1973. The New Ghetto Hustle. *Saturday Review of Science*, February, p. 38.

Reilly, Philip. 1973. Sickle Cell Anemia Legislation. *Journal of Legal Medicine, 1*:39–48.

Revellard, Muriel. 1973. French Law and Artificial Insemination. In *Law and Ethics of AID and Embryo Transfer* (Ciba Foundation Symposium 17). American Elsevier, New York.

Riga, P. J. 1973. Genetic Experimentation: The New Ethic. *Hospital Progress, 54*:59–63.

Rosenberg, L. E. 1970. Identification of the Defective Patient: Biomedical Techniques. In *Fogerty International Center Proceedings*, No. 6.

Rovik, David M. 1968. *Your Baby's Sex: Now You Can Choose.* Ballantine, New York.

Stone, O. M. 1973. English Law and Artificial Insemination. In *Law and Ethics of AID and Embryo Transfer* (Ciba Foundation Symposium 17). American Elsevier, New York.

Time. 1971. Man into Superman, the Promise and Peril of the New Genetics. April 19, pp. 33–52.

Westoff, C. F., and R. R. Rindfuss. 1974. Sex Preselection in the United States: Some Implications. *Science, 184*:633–36.

Yussman, Marvin. 1975. Private communication. Department of Obstetrics and Gynecology, University of Louisville.

part 9

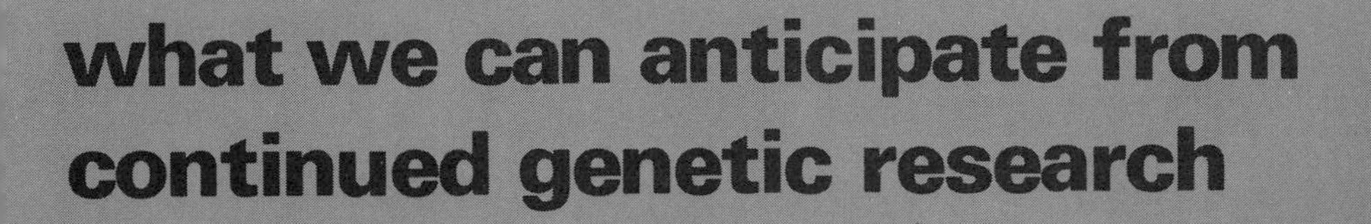

Each generation of humans looks toward the future hoping that they can "socially" provide for their children. They have, however, biologically provided the information for the future "health" of their child. In so many cases, the health of the child is poor because of a defective chromosome or gene. Observations of defective children and affected adults throughout history have undoubtedly fostered the idea of improving man.

From the time that humans began to use empirical observations for the breeding of their best dogs or strongest horses or highest-yielding plants, they must have recognized that the same practice among themselves would produce stronger or brighter or less sickly humans. The Bible contains many references to health problems resulting from consanguineous matings. Certainly observations would have been made that mentally retarded parents gave rise to retarded children and that parents with certain forms of "sickness" gave rise to children who also grew up with that sickness, while parents with better health gave rise to children who were also of better health.

Humans wanted to free themselves of disease in the way cattle and plant breeders produced disease-resistant varieties of plants and animals. However, the methods used in the improvement of plants and animals were not acceptable for use in the improvement of humans. People usually were unwilling to discard malformed children or sick adults. But still attempts were made. As discussed in Chapter 3, prominent men such as Haeckel, Pearson, and Davenport proposed that drunkenness, laziness, vagrancy, mental defects, criminal tendency, and sickness were inherited. It was said that people with

Science is a very human form of knowledge. We are always at the brink of the known, we always feel forward for what is to be hoped. Every judgment in science stands on the edge of error—science is a tribute to what we can know.

Jacob Bronowski

the future: promise or peril?

these tendencies should not be allowed to reproduce because they would only give rise to more of their own kind. These men believed in the genetic improvement of humans (eugenics) by allowing only those who appeared "fit" to reproduce. To insure that the socially "unfit" did not reproduce, many were institutionalized or sterilized (a negative form of eugenics). Such acts of human against human evoked a gathering sentiment of opposition, which included most American geneticists, that dampened the eugenics movement of the early 1900's.

Although the eugenics movement was slowed and never regained the popular support of the early 1900's, it never completely died out. Neither scientists nor laymen gave up the idea of reproducing "beneficial" traits. Suddenly in the twentieth century, as a result of the discoveries in molecular biology of the cell, the mysterious code of life was deciphered. We now know what our hereditary material is made of, how it replicates, how it maintains the evolutionary message of the human species, and how the DNA directs the cell to make specific types of structural and functional proteins (see Chapter 5).

By the application of sound genetic principles we as patients receive donor blood without fear and have babies even though the husband and wife are Rh incompatible. We can modify some of the phenotypic effects of genetic defects for those with defective vision, hearing, physical disabilities, and certain metabolic disorders by surgery, the administration of chemical agents, dietary restraints, or man-made technological hardware. Genetic surgery on humans has been attempted at least twice. In Dr. James Clever's work, patients were injected with virus that hopefully would restore the synthesis of a nuclease necessary for the excision of pyrimidine dimers in hope of curing xeroderma pigmentosum (cancer of the skin). In H. G. Terheggen and Stanfield Roger's work, two German children were injected with Shope papilloma virus in order to increase their level of arginase since the high blood levels of arginine is known to lead to mental defects. Both Clever's and Terheggen and Roger's attempts were unsuccessful.

Scientists can now bring about fertilization (the union of human sperm and egg) in a Petri dish and grow the developing embryo through about 60 days. And an in vitro fertilized egg has already been implanted in a carefully selected female for continued study during embryo development; there is a claim that three such fertilizations have gone to completion with normal infants resulting. Human fetal cells and amniotic fluid can now be routinely withdrawn via amniocentesis during pregnancy, and by culturing the cells, checked for various hereditary defects (see Figure 18–1 and Table 18–1). Complete blood transfusions are now possible in the fetus before birth. Human cells can be routinely grown in a test tube; cell division is normal.

Since 1960 scientists have synthesized a gene and a protein, reconstituted a virus, fused human cells with mouse cells to produce animal-human cell hybrids, performed genetic surgery in bacterial cells, grown a complete carrot plant from a single cell, grown a colony of identical frogs by transplanting nuclei of older intestinal cells into the cytoplasm of enucleated eggs, produced allophenic mice, and successfully transferred frozen-thawed mouse embryos to foster mothers that had normal pregnancies and produced nondefective mice.

what we can anticipate from continued genetic research

The news media, emphasizing the exotic possibilities of rapid advances in genetics, publish sensational articles on such scientific possibilities as test-tube babies, cloning to produce identical masses of humanity, repairing damaged genes, substituting a chosen gene (almost as one would order it from a wholesale catalogue) for one with which he is not particularly pleased, or ordering the sperm of choice to be sure the child will "fit in." It is even suggested that one day, perhaps in the not too distant future, children can undergo gestation in an artificial womb and be raised in a Skinner compartment. And, today, none of these ideas seems merely science fiction fantasy.

Few geneticists today doubt that genetic engineering will be done. To begin with, whole chromosomes may be deleted or added via cell fusion techniques. Or single genes or groups of genes might be taken into specific malfunctioning cells by special viruses that will be created in the laboratory specifically for this kind of genetic engineering. The ability to manipulate single cells genetically or to take individual healthy cells and clone new individuals to the exact size, shape, and temperment desired will give man a God-like power never before equalled in its potential for bringing about cultural and/or biological change. We have yet to see if we can cope with the tremendously complex legal, moral, and biological issues that man unleashed when he split the atom. Yet, splitting the atom, in retrospect, will (if we don't annihilate ourselves first) appear to have been a warm-up exercise for the energies required to respond to and live with the impact of genetics in a society whose social-cultural evolution has not kept pace with science. We must not deceive ourselves by *assuming* that foresight and morality will prevail.

Yes, it appears that humans have acquired "dominion over all living things" and through the use of technology created biomedical environments that successfully neutralize the major forces of natural selection. In fact, it is the biomedical revolution that holds greatest *promise* and *peril* for humankind. Biomedical advances offer us the potential of hope for a life free of pain and at the same time threaten us with such social disaster as the loss of tradition, families, individuality, variety, and the most precious—personal freedom of choice.

Technology brings power, and power brings increased responsibility. What is our responsibility to each other and to all other forms of life? Because we are now, for the most part, controlling our evolution, what principles shall we set down as guidelines for continued human survival? Shall we "hang on" to our current system of ethics? Can we afford to cherish today's morals in tomorrow's world? Must our morals change with new knowledge? What is an ethic, and what is moral? Perhaps we need to define these terms so that we can judge for ourselves whether today's science will create tomorrow's social calamity.

The terms moral and ethical are often used interchangeably. "Morality is what people believe to be right and good and the reasons they give for it while ethics is critical reflection about morality and the rational analysis of it" (Fletcher, 1974). A moral question raised by the rapid advance of molecular genetics is: Should we use genetic engineering, *exchange new normal genes for old defective genes*, in the sperm and egg prior to conception or in the zygote after conception? The ethical problem created by this moral question is: How does society make the decision to genetically redesign a cell? *Why* and *how* should it be done? Thus, as science and technology advance, moral questions and ethical dilemmas appear to be accumulating in every facet of our daily lives. And the frightening observation is that science and technology move ahead "blindly" as though there was a guiding hand that will eliminate error and prevent harm. Prior to the enormous scientific discoveries in the biological sciences of the 1950's, religion played an important role in what *ought* to happen. In fact, religion has always played a major role in the *moral* approval or disapproval of new discoveries.

Since the late 1950's there has been discussion of a "new morality" in response to rapid advances in the technology. How are we to use such powers as the following?

1. *The control of death.* The average life expectancy is being extended through the use of medicines, organ transplants, respirators, pacemakers for the heart, and kidney dialysis at a time when we are being told that simple drugs and medicines have been a major cause of the population explosion.
2. *The control of human potential.* By manipulating human genes, we can create new human potential and make the world free of diseases.
3. *The control of the environment.* Although the limits of human expression may be genetically determined, the manipulation of the environment in which the genes express themselves can determine whether a gene is favorably expressed.
4. *The control of human behavior.* In addition to the control of human behavior that could be achieved by control of the environment and gene manipulation, biochemical manipulation of the brain holds promise of controlling the development of distinctly human traits such as speech, imagination and choice (Kass, 1971).

But is there really a "new morality," or is it simply that a new set of questions demands the same careful consideration given to any moral questions concerning the dignity of humankind? Can we really say that we care more or less about the quality of a given life than our parents or grandparents did or that we think any more or less of our children or of our neighbor?

The fundamental difference between the decision-making process then and now is one of *time*; before the rapid advance of technology, there appeared to be time to consider the effects of major decisions that dealt with life, death and the quality in between. Now there are so many variables of major significance, each of which can individually affect humanity's very survival, that there does not appear to be sufficient time to consider the impact of one before the next event occurs. Unique to the last half of the twentieth century are the population explosion, use of insecticides, the poisoning of the oceans, the fouling of the air, a worldwide food shortage, a shortage of energy supplies, destruction of the ozone layer surrounding the earth, the threat of nuclear war, and the possibilities of genetic control by biological selection, by genetic engineering, and by cloning of the types of people who will inherit the earth.

what we can anticipate from continued genetic research

Of these nine "threats" to human harmony, only the last, the problem of genetic control will be discussed here; the others have received ample exposure in the various media. But only the last, the possible genetic control over the quality of human life, takes us beyond what we *believe* possible to be achieved by human resources.

the peril

Because we have taken on and subdued some of the forces of natural selection, 95% of all children born alive live to age 30 and thus have ample opportunity to reproduce. But reproduce what? Humans for sure, but what of their mental and physical condition? A high death rate keeps down undesirable traits that express themselves in early death. But because of medical advances, bearers of once-lethal genes are no longer selected against and these genes may now cause behavior change and destroy intelligence (Osborn, 1960). Because humans, especially in the developed nations, will not soon give up their standards of living, we must seek methods to correct genetic defects rather than settle for methods of maintenance for the afflicted. It is this very idea—correction of defective genes—that presents both *peril* and promise to humankind. To correct or lower the number of defective genes in the human gene pool implies that there must be some form of reproductive control! To suggest what such control can mean to society, an imaginative scientist, Aldous Huxley, wrote a futuristic book, *Brave New World* (1932), in which he described the processes of reproducing humans without the normal reproductive process. Huxley's Dr. Bokanovsky perfected a method of growing as many identical people as he chose, and he could alter his process such that each individual was destined to be a "happy" soul, that is, satisfied with his or her place in society. Dr. Bokanovsky produced his carbon copies for the "good" of the individual and the "good" of the state. The "Bokanovsky process" produced classes of "humans" (perhaps we should say just "people" without implying "human") called Alphas, Betas, Gammas, Deltas, and Epsilons. Each was distinct from the other in responsibility for the overall good of society. The masses were manipulated by a few who had the power to control the reproduction process. The director of Hatcheries and Conditioning told his attentive students, as he took them on a tour of his *people hatchery*, that under old conditions *one egg* and *one sperm* gave *one embryo* but a bokanovskified egg bud proliferates from 8 to 96 buds and each bud will become a "perfect" embryo and *all* 96 embryos will become "perfect" adults (except for possible mistakes by the technicians in the laboratory tending the bokanovskified eggs). The director of Hatcheries and Conditioning continued his lecture by adding that the Bokanovsky process is a *major* instrument of *social stability* because for the first time in history, it was possible to produce *standard* men and women in quantity.

In the 1930's *Brave New World* must have seemed rather like its contemporary, the Buck Rogers comic strip, in which people flew to other planets in spaceships and used ray guns in battle, etc. The adventures of Dr. Bokanovsky, like the adventures of Buck Rogers, were speculative and exciting to the imagination, yet terrifying in their implications for the future. Today, some half century later, humankind actually faces an ethical dilemma based on the possibility that scientists *will be able to make carbon copies of humans by a*

Figure 23–1. Sexual Reproduction in Humans Versus the Cloning of People. **A.** From a single fertilization of a human egg there is generally produced a single fetus, as seen in the photograph, which shows a fetus aged six to seven weeks. **B.** In cloning, a single fertilized egg, as suggested in Aldous Huxley's *Brave New World*, may bud and give rise to many identical fetuses. By altering the medium it may be possible to direct the final product into the person required by society. (**A** by permission of Robert Wolfe, Senior Medical Photographer, Department of Biomedical Graphics, University of Minnesota Hospitals; **B** courtesy of Instructional Communication Department, University of North Florida.)

process called "cloning"—the process of growing a colony or clone of identical plants or animals from one or a few cells of a single parent (Figure 23–1).

what we can anticipate from continued genetic research

cloning—of humans?

Biological developments have made *Brave New World* credible. Steward (1970) of Cornell has succeeded in growing a complete carrot plant from a

B

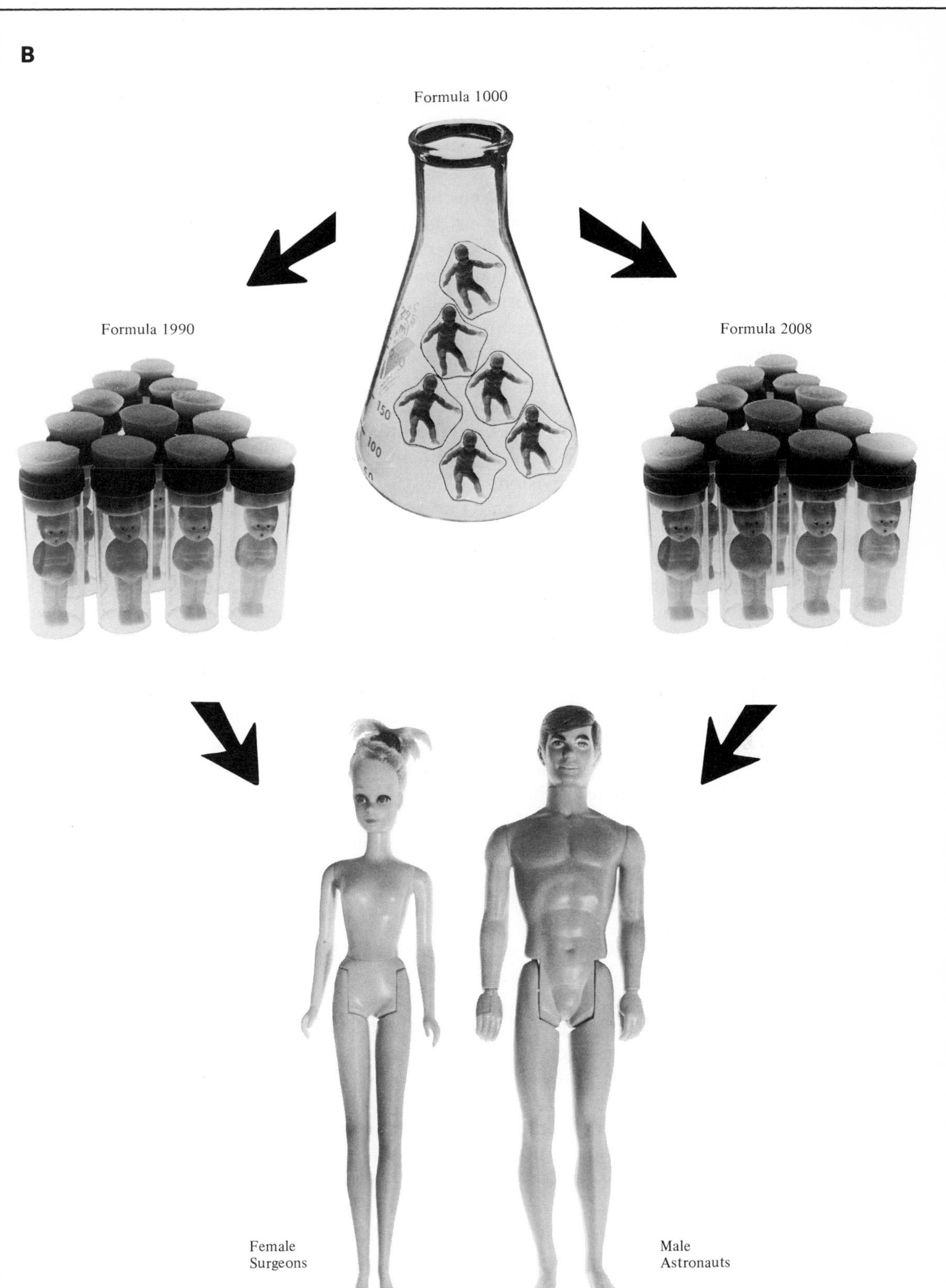
Formula 1000
Formula 1990
Formula 2008
Female
Surgeons
Male
Astronauts

single isolated carrot cell taken from a mature plant. He placed the cell in coconut milk; it dedifferentiated, underwent mitosis, and eventually formed a complete plant (Figure 23–2). Other scientists have succeeded in growing complete tobacco, potato, and sugarcane plants from single cells. Once the right nutrient culture medium is found, that is, one that allows or stimulates the cells to differentiate, a new plant can be formed from a single nonfertilized plant body cell.

Work with animal systems has not yet had equal success, but John Gurdon (1969) of Oxford University has been able to develop a colony of identical South African clawed toads (*Xenopus laevis*). He removed the nucleus from individual cells of the intestinal lining of a tadpole. These cells, like the cells of our own intestines, are specialized in function and make digestive enzymes for the toad's digestive processes. These cells have completed differentiation and under normal circumstances will remain intestinal cells. By removing the nucleus, Gurdon extracted the cell's genetic material. He placed that nucleus into a toad egg cell from which he had previously removed the nucleus. Once in the cytoplasm of the egg cell, the nucleus of the intestinal cell behaved as though it were the original egg cell nucleus. It began to make new DNA in preparation for cell division. Cell division began and continued through the formation of the tadpole and the adult toad (Figure 23–3). How soon the cloning of a human cell—before the year 2000?

If cloning of humans *is allowed* in our society, we must be prepared to deal with very difficult moral issues. Does one ever die if his cells are cloned into an individual identical to himself? And what of clones made from an animal-human hybrid cell? And what happens to genetic variability and the value of human diversity? With the power of cell cloning, should government provide equal opportunity for each of us to be cloned as we approach death? Should preserved bodies—those 50 being kept in cold storage until future research can thaw them out and correct their physical problem—be cloned? What about Egyptian mummies, should we bring back a pharaoh or prince and have them stand trial for some of their social injustice or honored for their insight, as the case may be? What of all those who suffer ill health at older ages—if certain of their individual cells are perfectly normal, should they be cloned? What of those with inherited defects, regardless of disease—do they not have the same right to request cloning as the well? Do we clone persons convicted of murder?

what we can anticipate from continued genetic research

To ask such questions is more than idle speculation; Joshua Lederberg, a Nobel laureate in biology and medicine, recently told a Senate hearing that in *ten to thirty years* we could do "essentially anything that we care to do in the area of biological engineering." Lederberg added, however, that we may not get there because society will not *fund* the required investment. This testimony, if correct, implies that perhaps we have a say in what *will* or what *ought* to happen in the field of biological engineering of humans. *But do we really?* One view is yes because science has been supported mainly by public tax dollars. For example, in the United States the federal government funded 25% of all research and development in 1946–47 and 70% in 1959–60 (Goran, 1974). But do we have much say in how those tax dollars are spent? In any event, perhaps the first question is: "*Should*" we become successful at human engineering, just because we *can* achieve the "success" that Lederberg refers

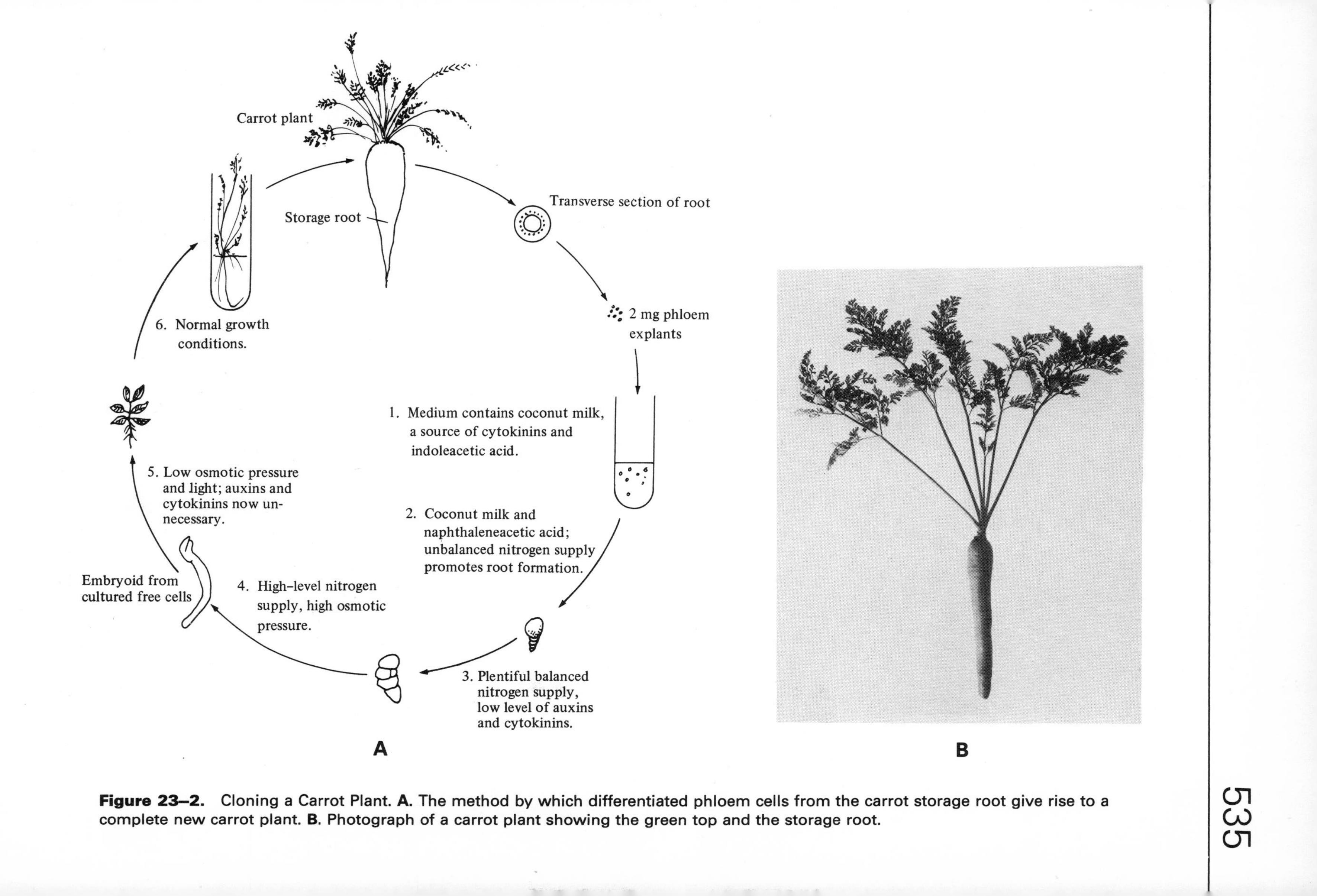

Figure 23–2. Cloning a Carrot Plant. **A.** The method by which differentiated phloem cells from the carrot storage root give rise to a complete new carrot plant. **B.** Photograph of a carrot plant showing the green top and the storage root.

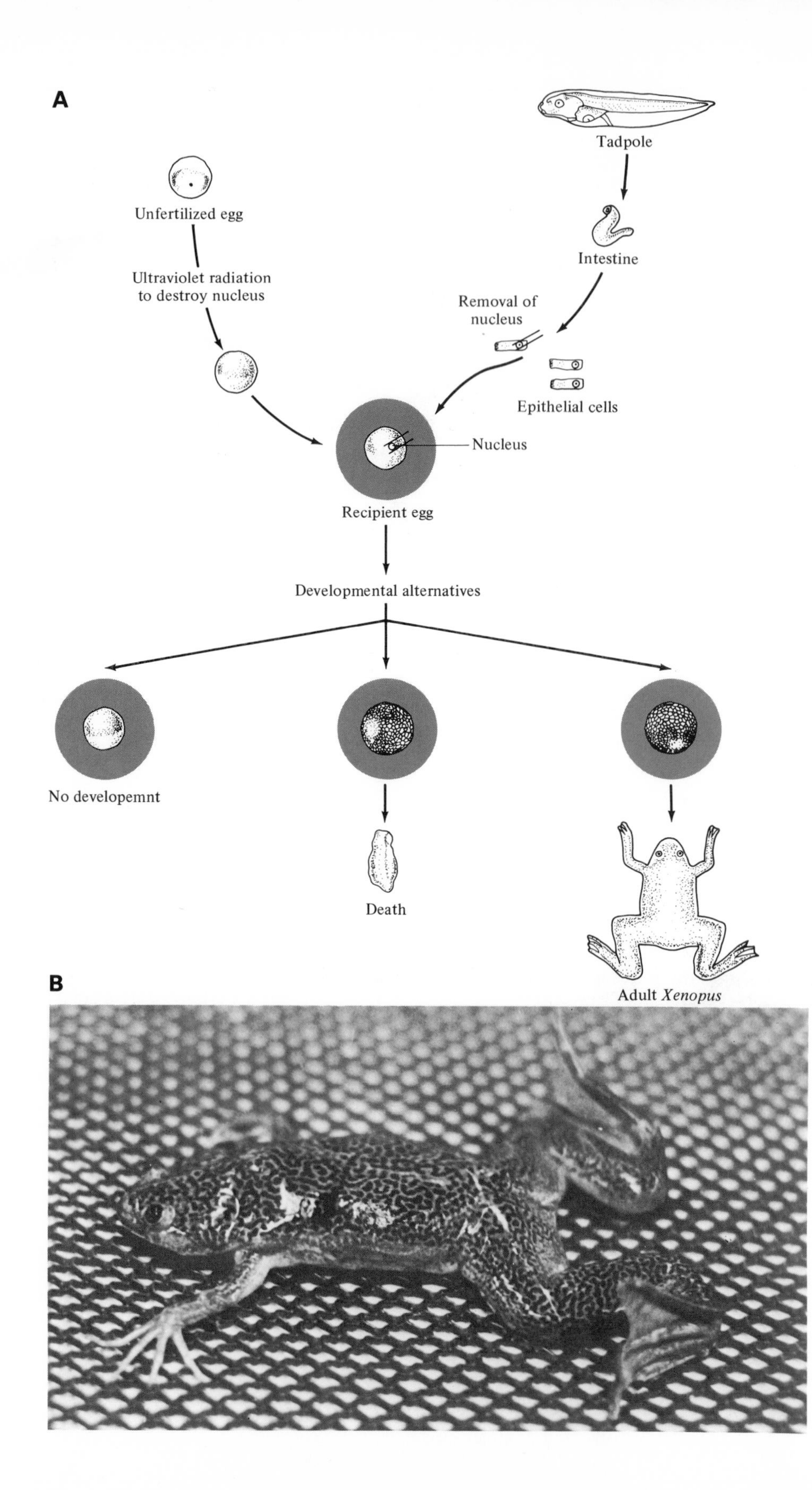
A
Tadpole
Unfertilized egg
Intestine
Ultraviolet radiation
to destroy nucleus
Removal of
nucleus
Epithelial cells
Nucleus
Recipient egg
Developmental alternatives
No developemnt
Death
Adult Xenopus
B

to? Because something *can be* does not necessarily mean it *should be.* Scientific facts are one thing, but values and moral commitment are another.

One of the more perplexing problems of making decisions about whether certain research should be done, especially if it affects the control of humankind, involves knowledge. Although laymen must be involved in the ultimate decision, the laymen should be informed of the issues and be presented with the facts so that they can understand the implications. However, through the process of presenting information, the scientist most likely will bias the layman, one way or the other. Also, scientific data accumulate at such a rapid rate that even scientists have trouble keeping up. Moreover, scientists themselves are uncertain of the impact the results may have on society. Scientists, politicians, laymen, and the clergy have sets of conflicting values, and, as Salvador Luria (1972) says, "Such conflicts are often unrecognized or at least not explicitly formulated, especially when one set of values is so prevalent in society that it is taken as an absolute while conflicting values are considered as 'deviant'." Another thought offered by Luria is that our ethical values are the product of social history and have a profound effect on the direction of change in human affairs. In part, we direct our future by our activities, which are conditioned by our history of ethical values. An example of projecting a value into a society that was ready to listen, if not yet ready to act, occurred in 1965 when the president of the American Chemical Society said at the Society's National meeting that the United States should make "creation of life" in the laboratory a national goal comparable to our space program. In just under a dozen years, research on creating life, engineering of life, and the cloning of life is being supported by government grants, and projected dates for their achievement have been presented by the less conservative scientists.

If we clone, it seems that most people want to know what humankind will be like in the year 2001 or 2050. There are, of course, many extremely serious questions to be asked before we actually clone the first human, and scientists, clergymen, politicians, and laymen are asking them. But it is difficult if not impossible to anticipate the overall religious, political, and economic effects cloning will have. Some possible effects in the economic area might include the following somewhat humorous examples. Perhaps we would see the development of powerful lobbies against the production of people who don't need underarm deodorants because they don't perspire. And what of the agony in powerful cosmetic industry when they learn women will no longer require beauty aids—they're all born naturally beautiful. And pity the razor companies because men will vote out beards or have those that wipe off with a wet cloth. The wigmakers will also be in financial difficulty—no more bald males. Think of the "headache" or "antacid" companies, the alcoholic beverage companies, the psychiatrists, etc.; "perfect humans" won't require those medical and social aids. The answer to what humankind will be like by 2050 would appear to depend on whether population growth continues to increase, whether the

Figure 23–3. (*Opposite*) Cloning of *Xenopus laevis.* **A.** The method by which differentiated specialized intestinal cell nuclei of a tadpole were placed into the cytoplasm of unfertilized eggs that were enucleated by ultraviolet radiation. About 1% of such prepared eggs developed into adult *Xenopus laevis.* **B.** A photograph of *Xenopus laevis* (courtesy of Kevin Inyang, University of North Florida).

energy problem can be solved, and whether widespread cloning of humans occurs. At the moment, the author would like to believe that cloning of human cells simply to reproduce humans or more of the same type human will not be socially acceptable. Experimentally it may be moral to obtain the information on *how* to clone a human, but the arguments as to *why* clone a human are insufficient. However, should scientists find methods by which they could genetically engineer a cell before it was cloned, then the arguments may become overwhelming in favor of producing humans through cloning.

genetic engineering

Genetic engineering has been defined as changing undesirable genes to more desirable ones while the gene resides in the chromosome or exchanging the "defective" gene for one that has been synthesized to be "normal." Scientists hope to change a gene by specific directed mutation of the nucleotides of the DNA in situ using radiation and chemicals, but it is not now possible to direct specific base pair changes in a given location on a particular chromosome, and probably will not be for the next ten to twenty-five years. We do not have a method to protect all base pairs except the one that a "ray" is intended to *change.* And we cannot now either change or replace a gene because we know only the *approximate* location of a very few of the perhaps 100,000 genes found in a set of human chromosomes.

Even if we had the knowledge of precise gene location and could change or replace a gene, there is still a major problem in that we do not really know which cells to engineer. This is not a question of who will receive treatment—we will concede, although we may be a bit optimistic, that when genetic engineering comes about it will be available to all. Rather, there are many complications in identifying a defective gene, a gene that by some measurement needs changing. There are also the technical problems of how to reach and correct *only* those cells that are defective. For example, if you are age 30 and diabetic, there is presumably no need to change the gene or genes in *all* body cells, perhaps only those in the insulin-secreting region of the pancreas.

What do we do about individuals who lack penetrance, who have the defective gene but do not show the disease. If they are at a reproductive age, it is vital that they be changed before they transmit the gene to an offspring who does show the defect. Of course, we can expect that the offspring will be corrected when the defect is recognized, but how much damage will have been done before the defective trait is detected? And what of the fact that most human traits are polygenic? Since we do not know exactly how many genes are involved in a polygenic disease, we do not know how many will need to be changed. Only by experience with genetic engineering will some of this information become available. If society requests the use of cloning, genetic engineering will become very important because all cells carry five to eight recessive lethal genes and many heterozygous recessive defective genes. To insure "genetic health" the cell to be cloned should, for efficiency, be genetically engineered to "perfection" before incubation and growth. Further,

what we can anticipate from continued genetic research

should a gene defect occur during growth, the development stage could perhaps be momentarily stopped and those cells engineered.

Such thoughts on genetic engineering and cloning may appear "far out," but cloning in bacteria has been routinely done over the last forty years. A single bacterial cell is isolated and allowed to reproduce (one cell divides and becomes two, etc.), forming enormous populations of identical genotypes and phenotypes. Genetic engineering of bacterial cells has been in practice since Avery, MacLeod, and McCarty identified Griffith's transformation factor as DNA (experiments discussed in Chapter 5). Since then, the phenomena of bacterial transformation, conjugation, and transduction have been used for genetic engineering of bacterial cells. A brief discussion of transformation, conjugation, and transduction is presented so that the reader can appreciate how basic research on these phenomena in the bacterial system has given rise to speculation about genetic engineering in humans.

Transformation is the incorporation of "naked" DNA of one cell by another and the expression of the trait or traits that are coded in the molecule of "naked" DNA. For example, scientists extract the genes (DNA) of bacterial cells that have the ability (the genes) for producing the enzyme beta-galactosidase (see Chapter 5). The extracted DNA, free of most contaminating materials and referred to as "naked" DNA, is then placed in the liquid medium in which are growing bacterial cells that *cannot* produce beta-galactosidase. A *small number* of the beta-galactosidase-negative cells take in the naked DNA containing the gene for producing beta-galactosidase and incorporate that DNA into their own chromosome. After the incorporation of the DNA, *some* cells begin to produce beta-galactosidase and on cell division *transmit the gene to their offspring.* The beta-galactosidase-negative cell has been genetically engineered and acquired a phenotype it did not previously have.

From this example, it is easy to speculate that we should at least try transformation in humans with genes defective in the production of a particular enzyme, for example, hexosaminidase A, the lack of which causes Tay–Sachs disease (discussed in Chapter 10). If one offered the defective cells DNA containing the correct gene, a similar exchange *might* occur and deficient cells would gain the ability to produce hexosaminidase A. The problem is not in concept, but in application. *When, to what cell or cells,* and *how* does one offer the normal gene for producing hexosaminidase A? Even if we had a test tube containing only genes for the production of hexosaminidase A, there are no guarantees as to how a human cell will behave after the transformation. In one set of experiments, for example, white-eyed *Drosophila* were transformed by using DNA of vermilion-eyed *Drosophila.* An interpretation of the evidence from these experiments suggests that the *Drosophila* maintained both the mutant "white eye" gene and the introduced gene for vermilion eye. However, the "vermilion" gene became associated with the chromosome at a site different from the eye color locus. According to Fox and colleagues, such displaced gene association interferes with the functioning of the normal gene at that site (Early, 1975).

There is some evidence that transformations of human cells may be possible in the near future but only on an experimental basis. Recently, human cells grown in tissue culture flasks, which lacked the ability to produce the enzyme inosinic acid pyrophosphorylase, were genetically transformed to produce the enzyme. DNA extracted from enzyme-producing cells was placed in the tissue culture medium and the deficient cells became transformed apparently in the same manner as that which occurs in bacterial cell transformation.

The problems in attempting to apply the techniques of transformation to human systems are many. Most serious is the lack of specificity. Even *when* we are able to offer a human cell with a defective gene only one gene at a time, we cannot be sure that the cell will take in only the correct quantity of DNA, incorporate it correctly, and express only the desired phenotype. Besides, in terms of engineering of a defective adult, how would we transform defective cells in, say, his kidneys? How could we get the DNA there unchanged and in sufficient quantity? Fox and associates have transformed eleven genes associated with two chromosomes of *Drosophila* by dissolving the chorion membrane of *Drosophila* eggs and immersing the eggs in a solution containing the transforming DNA. Fox's group are also trying similar techniques for the transformation of ova in mice. If successful, then perhaps someone will attempt such work on the human egg (Earley, 1975).

Lucien Ledoux of Belgium reported that he corrected a mutant strain of *Arabidopsis*, a plant of the mustard family, by flooding the roots with a solution of DNA extracted from a bacterium whose DNA was normal for that gene. A German scientist recently reported that when he injected DNA of a red-colored petunia plant into an albino (white) plant *some* red coloring appeared (Galston, 1974). Perhaps injections of DNA will work in humans, but do we have a volunteer recipient? What of donor and investigator—will they agree on how it's to be done? What of failure—who "picks up the pieces?" Such questions always arise in science, and they are overcome. In general, however, transformation does appear to have serious drawbacks, and this author does not believe it will be used routinely in human genetic engineering.

Conjugation is the uniting of bacterial cells during the act of "sexual" reproduction. By studying the transfer of bacterial genes from one cell to another, scientists were able to learn of the events that occur as short and long segments of DNA are transferred from the donor cell to a recipient cell (Figure 23–4). It was learned, for example, that not all DNA taken in is incorporated and not all that is incorporated is expressed. But more important to human design is the fact that specific regions of DNA can be transferred between bacteria, incorporated, and expressed in the restoration of a function lost through gene mutation. The method of conjugation in bacteria is not really a good model for gene transfer in human experimentation, but it offers insight into what may be expected in cell hybridization techniques.

Transduction is the virus-assisted transfer of DNA from one bacterial cell to another. Studies on the life cycles of viruses specific for a particular species of bacteria and the ability of a bacterial virus to carry genes of one bacterial cell to another have excited the imagination of scientists. If one could synthesize "normal" human genes and attach them to human virus DNA, perhaps the virus could exchange the normal gene for an abnormal gene in a human cell similar to the way bacterial viruses exchange genes between different bacterial cells. "If a gene could be synthesized" would appear to be a major obstacle except that H. G. Khorana has synthesized a gene, using "off the shelf" chemicals, that duplicates *one* of the thousands of genes of the bacterium *Escherichia coli.* Since the gene proved to be functional in *E. coli*, we

what we can anticipate from continued genetic research

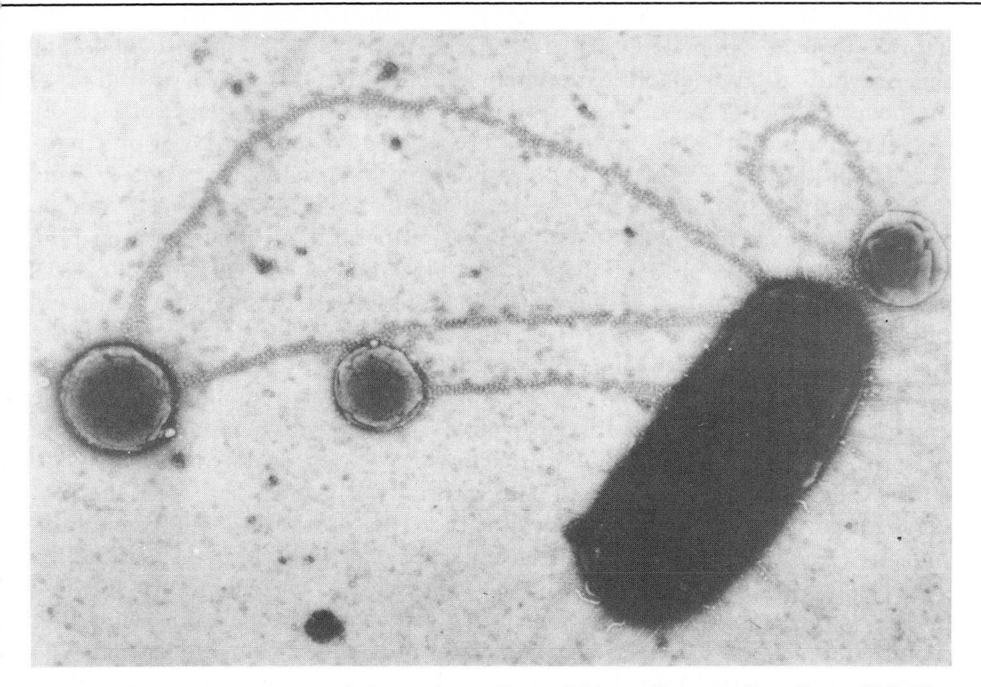

Figure 23–4. Conjugation. A bacterium, *Escherichia coli,* extending "sex-pili." The sex-pili are thought to be the conjugation "tube" through which the DNA of the donor cell is transferred to the recipient cell. Sex-pili are produced only by the donor cell because of the genes carried on the donor sex-factor. The sex-pili are identifiable by the use of donor sex-pili specific virus that can be seen in the photograph, attached to the sex-pili (Courtesy of David P. Allison.)

are technologically ready to synthesize "made to order genes" (*Time*, 1976). It has been shown that the genetic code is universal, that is, the same codon of bases is used by many living organisms, from bacteria to plants and animals, for the incorporation of a given amino acid into protein. Also, DNA of different animals and of different bacteria can be transcribed and translated in human tissue culture cells. But what of using virus as the "errand boy"? In Australia, bacterial viruses were recently used with some success to transfer the gene for the production of the enzyme beta-galactosidase of bacteria into a tomato plant tissue culture. In England the same experiment was tried with recipient cells of a sycamore tree. The results from the sycamore cell culture are controversial because there was no agreement on whether the cells were genetically changed (Galston, 1974).

Attempts have been made to determine whether viruses could carry genetic information that would be expressed in humans. For example, H. G. Terheggen and Stanfield Rogers infected two argininemic human females with rabbit Shope virus. Argininemia is caused by the lack of the enzyme arginase that helps control the level of arginine in the blood. High levels of arginine lead to mental retardation. The reason for infecting the girls with the Shope virus is that it had been previously shown that the Shope virus carried a gene that would produce the enzyme arginase. The results of the experiment were not conclusive. In the work of James Clever, a virus was used to infect a human who had the recessively inherited xeroderma pigmentosum, cancer of the skin.

Persons with this condition lack a DNA excision repair enzyme. Clever was hopeful that the virus would carry in the bacterial gene for the production of this enzyme. Although both experiments apparently failed, such work reflects the belief that virus can be used to replace defective genes in genetic diseases.

Munyon et al. (1971) reported the use of mammalian herpes simplex virus (HSV) for the transfer of the gene responsible for production of the enzyme thymidine kinase to human cells that previously could not produce thymidine kinase. The thymidine-kinase-negative cells, after viral infection, expressed the new phenotype, the production of thymidine kinase, indicating that the virus carried the gene and that the gene was successfully transcribed and translated into the enzyme. Qasba and Aposhian (1971) reported that DNA of mice could be coated with polyoma virus protein, making a *pseudoviron*, and such polyoma pseudovirons could transfer mouse DNA to human and mouse embryo cells. With this type of investigative success it is easy to imagine the use of virus to transport specially made genes to various parts of the body to correct genetically defective cells. The work of Carl Merril reinforces the idea of viral transport of "normal" genes for the replacement of defective genes in humans. In Merril's studies, cells from a patient with recessively inherited galactosemia (lack of the enzyme alpha-D-galactose-1-phosphate uridyl transferase) were cultured in the presence of *lambda virus* (a bacterial virus) that carried the *normal bacterial gene* for the production of the enzyme. The bacterial gene carried into the human cell line by the lambda virus permitted the cells to produce the missing enzyme (Hadsell, 1971). Because scientists can synthesize a gene, because they can construct a virus to carry that gene, and because the demand for "health" is paramount in our society, we will soon be able to alter a cell's genetic potential using a method of transduction. But, again, the results will be limited to single gene defects because of our limited knowledge concerning the specific genes that are responsible for a given trait. Should this limitation be overcome, and should we solve the problem of how many genes are involved in polygenic disorders, we then face the problem of just how much DNA or how many genes one can package into the virus.

What one may think to be a major concern, now that it is apparent that some limited forms of genetic engineering will be available by the year 2001, is the moral implications in using genetic engineering. But genetic engineering of bacteria or plants and animals other than humans does not create moral "crisis." Because morals tend to change with our social needs, it would be presumptuous to set a moral code for future generations. Such questions as whether parents will have a moral right to "burden" society with the birth of a defective child, once genetic engineering is available, will have to wait until such time. Hopefully by then many social "rights" and "needs" will have been reevaluated as they have been throughout history.

what we can anticipate from continued genetic research

Those who fear the misuse of new forms of genetic engineering perhaps can take comfort from the fact that the methods for artificial insemination have been available for at least thirty years and this procedure could have been used to produce quite drastic changes in height, intelligence, or any other quantitative trait with a high heritability if it were widely applied. But it has not been used for any such purpose. Humankind has been blessed with intelligence and therefore the opportunity to choose. So far, the conscious choices made by humankind have, with certain exceptions, shown a sense of responsibility to humans and other life forms. The single major issue that looms forth when we discuss the different forms of genetic manipulation of humans is: How can we avoid the misuse of the application of genetic innovations, cloning and engineering? *Who will be the decision makers?*

an evaluation of danger

To evaluate the danger of genetic advances, two questions should be asked.

1. Are we under *moral* obligation to halt the gathering of knowledge because we *think* it *may* be harmful to humankind?
2. If scientists achieve gene manipulation, will there be *time* to promote its benefits and oppose its abuse?

The answer to the first question probably will be "yes" in certain cases. As for the second question, there will be ample time in the United States because we have a democratic form of government. I do not think scientists can *surprise* us with an announcement of "doomsday" genetic potential comparable to the explosion of the first atomic bomb that was developed under secrecy for military purposes. If it were possible to "build" a genetic bomb, the potential would lie in the area of genetic manipulation of human behavior. But can it "surprise us" like the explosion of a nuclear device? Perhaps it does not have to—to be just as effective. Certainly over time, if man were to allow the genetic engineering or cloning of "behavior types," the nuclear bomb would appear insignificant in retrospect, except as a weapon for death. Such a behavioral device might be "exploded" on the public by constructing a virus with the genes necessary to change human behavior to some fixed state. Such virus could then be put in the public water supplies of all but the "controllers." However, I do not think this is possible because most of man's traits are polygenic. We are fifty years away from any kind of success in exchanging and correcting a single gene defect, even though James Watson, who shared the Nobel prize for his work on the structure of DNA, predicted in 1971 that "test tube" methods of reproduction would be routine in ten years and "cloning" of humans would be achieved in less than 25 years. With the development of procedures for the exchange of single genes will come the determination and social adjustment for other possible applications of genetic manipulation—just as other uses for atomic energy have been found after the bomb was built. Besides, we already have the pharmacological, surgical, nutritional, and psychological methods for controlling human behavior *now*! And those who fear that cloning might be used for building an elite army of fearless, bright, and savage soldiers forget that continued research can make the cloned soldiers vulnerable!

Perhaps the three greatest dangers for the future are

1. Loss of species variability. If we were to allow a process like cloning to occur for most of the population, we would reduce human variability

and thus subject our species to the imminent danger of extinction due to an environmental change.

2. Public panic. Out of fear public pressure might nonselectively or unwisely inhibit genetic research. Humankind *has* a moral obligation to cure disease. Although science may appear to be slow in such matters, we will learn to repair genetic defects and cure cancer—but not if a moratorium is called on scientific research because we *fear* for our *future*. Our parents, by supporting research, provided us with a high standard of living, increased production of improved foodstuffs, and miracle medicines. Having derived these benefits from science for ourselves, should we determine that future generations become stagnant?
3. Distortion of our moral judgment that we are our brothers' keeper. Should humankind suddenly find a new image of itself, one that rejected present morality, there could be drastic changes for the future of man. The charge against the scientist is that he is indifferent to his "brothers." Much of this image comes from the sensational popular press that distorts the dangers of science and ignores its benefits.

Let us briefly review some of the potential benefits of the future advances in genetic research.

the promise

Over 100 years ago, Francis Galton proposed, as many had done before him, that humans control their genetic quality by carefully selecting their mates for physical, mental, and social attributes rather than mating because of the nebulous factors involved in romance (see Chapter 13). The eugenic movement was mechanistic in the sense that the supporters of eugenics believed that good character was solely determined by our heredity. Today we realize that heredity and environment work in harmony to produce the successful phenotype. Humans have circumvented some of the forces of natural selection by discovering medicines, air conditioning, housing, etc. There are at this moment at least three major forces of natural selection (stabilizing, diversifying, and directional) operable in human evolution (see Chapter 3). But even these forces are being seriously challenged by the technological process of new medicines, surgery, the possibility of gene manipulation, and our relentless pursuit to control the weather and guide our own biological destiny.

Because we are *morally* obligated to care for the sick and protect human function, we can look toward the future with its promise of gene therapy and cloning as methods of relieving the inheritable afflictions of man. This is no small detail. For we recognize over 2000 inherited human diseases of which very few are treatable (see Chapter 19). In addition, we realize that any gene in the human gene pool has a certain chance of being mutant at any moment. And if a change occurs, the chance is very high, probably over 90%, that the change will be biologically detrimental. To relieve man of the known afflictions and to *be prepared* to handle those yet to occur through the "miracle" of gene

what we can anticipate from continued genetic research

therapy (genetic engineering) would appear to be of tremendous biological and social value. The truly outstanding advantage of gene therapy lies in the fact that not only is the phenotype corrected, but the genotype is corrected and all future offspring, barring new mutation that can also be corrected, will be genetically healthy.

With the advent of "superior genetic health" in the human species, there would be a tremendous release of personnel, raw materials, money, and institutional resources, now tied up in the care for the genetically afflicted, for use in other aspects of society. Further, there would be a definite lowering of environmentally caused diseases of humans because we would be able to treat the whole set of chromosomes of an individual and thereby reduce their sensitivity to various drugs and other environmental agents.

And what of the possibilities of just cloning? Imagine the problems we would overcome in the field of immunotherapy and transplant rejections. No longer would we have to wait for donor organs to become available or have the problem of maintaining an organ once taken from a donor (Figure 23–5). Organ banks of all types would be established where one could have cells of each of his organs taken and preserved while they are young and healthy. Should an organ malfunction, say, your heart, you simply order a new heart by dialing the organ bank. Then a few heart cells of those being held in deep freeze would be thawed and allowed to develop into a new heart. This heart could be easily "plumbed" into the person's body without fear of the rejection—either the heart by the body or the body by the heart—that causes the loss of most transplant patients now.

If cloning were coupled with genetic engineering of the cell—that is, if cloning of cells only occurred after cells were "correctly engineered"—society

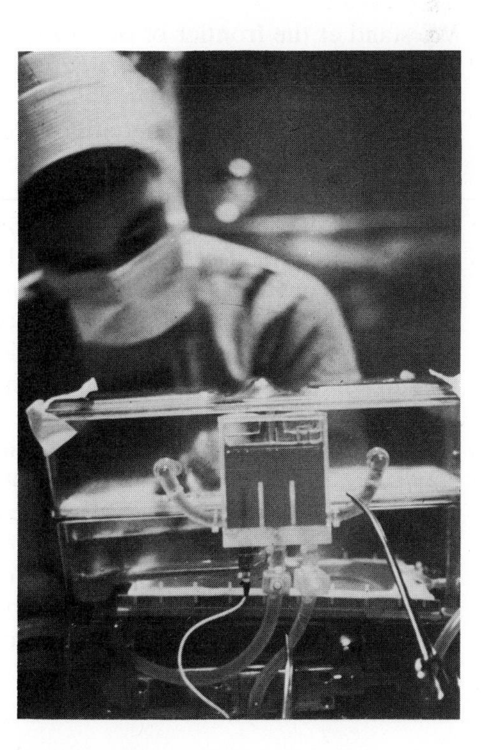

Figure 23–5. Perfused Kidney. A donor kidney is being perfused with a solution to keep it in "healthy" state of metabolism. The kidney will later be transplanted into a "selected" recipient. The selection of recipients, especially difficult today because of the shortage of donors, involves "matching up" the kidney donor and recipient as closely as possible in order to keep transplant rejection at a minimum. (Photo by Erich Hartman, reprinted from *Modern Medicine*, © The New York Times Media Company, Inc.)

could definitely stabilize the human gene pool with the "most desirable traits" based on the concept of morality prevailing at that time. Such engineering followed by cloning could alleviate the genetic load that every human carries. Just think, a society with no mental retardation, blindness, deafness, or need for kidney dialysis, etc.

We must also realize that by the time engineering and cloning are available in humans they will have been *exploited* to the *n*th degree in plants and lower animals. As mentioned, a number of species of plants and two of animals have already been cloned. The potential benefits of the use of genetic engineering and cloning in domesticated plants and animals are mind staggering; we cannot even imagine the new types of fruits, meats, and vegetables that can come from the use of these techniques.

Perhaps one possibility that has not been mentioned in the already enormous amount of material on the subject of genetic manipulation of humans is that, once we can determine the type of "people" we will foster, we will undoubtedly be able to design an environment for them.

Genetic engineering and cloning in general will alleviate inherited diseases, bypass sterility, provide the sex type desired, preserve family likeness, and provide humankind with the potential for the expression of tolerance and *peace*. Perhaps René Dubos was correct in claiming that the future is of our own creation and that humans have the *privilege* and the *responsibility* of shaping themselves and the future.

final reflections

We stand at the frontier of our own evolution. We can intervene in our genetic future. Oh, if only our *desire* to know were equalled by our *ability* to understand. For twenty-two chapters of this text we discussed *what* we are doing for humankind. Too often *what* we did was done simply because it worked and not because we fully understand *why* it worked or the implications if it did work. Whether it is artificial insemination, inspecting a chromosome karyotype, perscribing a diet for a genetic defect, arguing the decision over fetal research made in a court of law, giving advice on the "pitfalls" of consanguineous mating, being concerned over the amount of radiation in our environment, discussing the incidence of an inherited ethnic disorder, or referring to the widespread repercussions of politics on science, we can never be sure that the decisions we have made or will make are *the best*. We only know that not to decide an issue *is* to decide the issue. Decisions in a republic are made *for* the majority but not always *by* the majority. Now, as humanity stands at the brink of making the most important decision in the history of the human species, whether we will make *deliberate* changes in our DNA, there is much indecision and fear—but are we not prepared to make the decision? If we are not prepared, then who or *what* is? Man must determine his destiny—we have come too far to stop. If, as Thomas Wolfe said, we "can't go home again," research must be continued; if it is not, we may go the way of the dinosaur.

what we can anticipate from continued genetic research

One very recent and encouraging event that occurred, in keeping with the fact that research in science of genetics will continue, was the Conference on Recombinant DNA Molecules that took place at Asilomar, Pacific Grove, California, in February 1975. The conference was held because of a letter that appeared in *Science* calling for a voluntary moratorium on certain experiments because of the potential hazards involved in linking together DNA of different species. The underlying question was whether scientists involved in their search for fundamental knowledge have the right to *create new* molecules of DNA that are potentially a health hazard to humanity.

The conference was attended by 140 scientists from 16 nations. The discussion dealt principally with the scientific aspects of the remarkable technological advances in the fusion of genetic information, taking genes from separate species and combining them into a new molecule of DNA to be expressed in a recipient cell. Basically, two types of experiments were discussed, Type I and Type II.

Type I experiment involves the physical and genetic construction of "new" self-replicating bacterial plasmids (units of DNA that exist in the bacterial cytoplasm free of the bacterial chromosome) that code for the resistance to antibiotics or for the production of bacterial toxins in strains that were once toxin free (Figure 23–6). The danger in such experiments is that one or more of our common bacterial pathogens, which can now be controlled by antibiotics, might become resistant via the construction of such plasmids. The transfer of these plasmids could become widespread very quickly, putting presently controllable diseases *out of control.* Even more dangerous, gene recombination experiments during the construction of these bacterial plasmids, might give rise

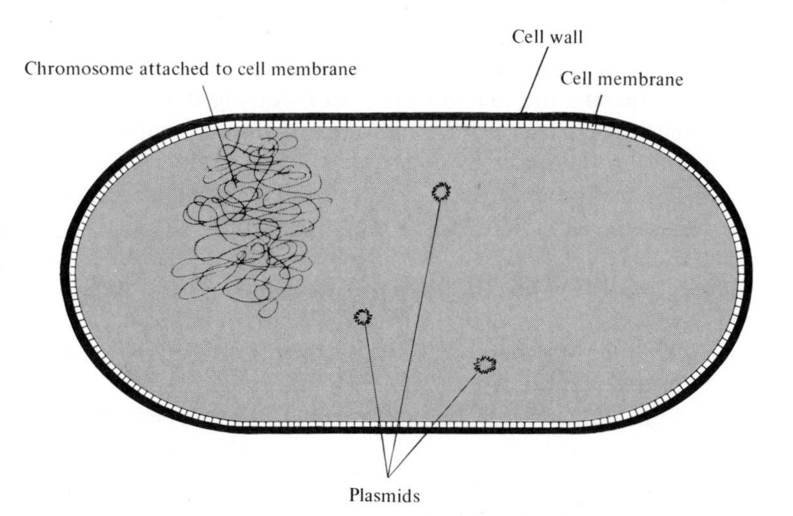

Figure 23–6. The Bacterial Plasmid. A general diagram of a bacterial cell showing the bacterial chromosome attached to the cell membrane. The plasmids are circular molecules of DNA that are self-replicating. A plasmid has the capability to join the bacterial chromosome, becoming an integrated unit of that chromosome, to leave the chromosome, taking extra genetic information with it, or to remain as a separate genetic unit (replicon) in the cytoplasm. The plasmid DNA is transmitted to daughter cells on cell division and is infectious in that the plasmid can be transferred to a second cell without chromosome transfer. In *E. coli* there are generally two plasmids per cell. The number of plasmids varies with bacterial species.

to new gene combinations that could cause diseases we have *never faced before.* We could have the beginning of a disaster worse than the bubonic plague. With respect to building molecules of DNA that code for a lethal toxin, what if a common human colon bacterium like *Escherichia coli* became genetically converted into a producer of a deadly toxin and through experimental error such organisms contaminate the environment, say, by getting into water supplies? A great loss of life could occur even before the problem was identified.

Neither of the two possibilities can be thought of as having a low probability of occurrence. To the contrary, naturally occurring resistance transfer factors, plasmid molecules of DNA that confer resistance to up to five antibiotics at one time, have been known to be transmitted between the enteric bacteria (bacteria that inhabit the human intestinal tract) for years. For example, there are strains of *Hemophilus influenzae* (causes acute respiratory infections) that carry a plasmid that offers *Hemophilus* resistance to the antibiotic ampicillin. But we have been able to come up with a substitute antibiotic to destroy them. What if the constructed plasmid contains a "new" antibiotic-insensitive gene, and a substitute cannot be found? Recent plasmid reconstruction experiments reveal that a variety of foreign DNA can be inserted into a plasmid. Stanley Cohen (1975) reported on the successful construction of bacterial plasmids that contain DNA of *Drosophila* and the toad *Xenopus laevis.* These plasmid chimeras (a hybrid of bacterial and animal DNA) were then used to transform bacterial cells. The transformed bacterial cells reproduced the animal genes generation after generation, and the toad DNA or nucleotide sequences of the toad were successfully transcribed by the bacterial system into an RNA product.

In addition, investigators at Stanford University and the University of California, Santa Barbara, have recombined DNA of yeast that codes for the synthesis of the amino acid leucine with bacterial *E. coli* DNA. The *E. coli* cell, which was unable to synthesize the leucine prior to receiving the yeast DNA, was then able to synthesize leucine. Further, the investigators from the Santa Barbara laboratory have preliminary evidence that genes or DNA from the fruit fly (*Drosophila*) can also be expressed in *E. coli* bacteria (Skinner, 1976).

The two California research teams hope that their ability to have yeast DNA expressed in *E. coli* will lead to learning about the distribution and function of genetic control regions in higher organisms. The researchers originally "thought it would be very difficult to find a piece of eukaryotic DNA that would be expressed in a prokaryote." "It turns out that it is relatively simple and we've isolated several examples. The success rate is high enough so that I would expect that a good deal of eukaryotic DNA probably will be expressed" (Skinner, 1976).

Type II experiments involve the joining of DNA from oncogenic virus (virus that cause tumors) to the plasmid DNA of bacterial cells. Because the plasmid is self-replicating, it is possible to make many copies of such tumor plasmid gene associations. Should these be transmitted between bacteria, once harmless bacteria may have the potential of "creating" tumors.

what we can anticipate from continued genetic research

The plasmids are not generally necessary for the bacterial cells' survival except in cases where the plasmid confers antibiotic or viral resistance, should they come in contact with these agents. What is dangerous is the fact that plasmids are very infectious, that is, are transmitted quickly between cells, and are also transducible, carried by virus. Recent studies show that bacterial virus invade both animal and human cells (Merril, 1974). With the possibility of virus-mediated plasmid transport, it becomes clear that once a "dangerous" molecule of DNA becomes transportable by virus it would be difficult to stop.

Lawyers also participated in the Asilomar Conference. Alexander Capron of the University of Pennsylvania Law School said that whether or not scientists are calling upon the public for support of research, if the research involves a potential public hazard, public discussion of such research is necessary. He also made the point that, as crucial as research seems to be to the achievement of progress, scientists should be prepared for the eventuality that the public may not agree to support such science. And Roger Dworkin of the University of Indiana Law School told the scientists of their liability to lawsuits. For example, if a laboratory worker becomes ill as a result of these "exotic" experiments and does not respond to conventional therapy, then the scientists would be open to suit (Merril, 1974).

The question we may ask is: If such "gene-grafting" experiments are so potentially dangerous, why do them? It's like asking the question of why use nuclear energy for a source of energy. There is a great potential danger, but society is using nuclear reactors for the production of energy. Obviously, then, it's a question of balance: does the *promise* of benefit outweigh the *peril.* But how do we decide when the balance is favorable and to what degree? Gene-grafting in bacterial and animal cells offers many benefits in pure and applied sciences. In pure science, for example, scientists hope to gain new insight into the mechanisms of cell division—to differentiate between cancerous and normal cells. In the applied sciences, scientists hope to construct micro-organisms to produce insulin for use by the diabetic (the production of insulin is now a very expensive process of extracting the insulin from the pancreas of dead cows); to handle major environmental problems, such as digestion of synthetic compounds, both toxic and nontoxic, from plastics to insecticides; to become a reserve energy source by taking in low grade silage and other refuse materials and converting them into high grade protein; and to introduce genes for the fixation of nitrogen in cereals and other grain plants such that the requirement for expensive fertilizers would be minimal. The limits to benefits are unknown and only through the manipulation of the DNA will the potential become known. What if we were able to graft the genes for chlorophyll production into an animal cell and the hybrid functioned?

David Baltimore said at the opening of the conference, "We're here because a new technology of molecular biology appears to have allowed us to abort the standard events of evolution by making combinations of genes that could be unique in natural history. These pose special hazards while they offer enormous benefits. We're here, in a sense, to balance the benefits and hazards right now, and to offer a strategy that will minimize the hazards of the future" (McBride, 1975).

However, the Asilomar meeting, at times, was very heated and it was difficult to reach agreement on the central issues; there were scientists from foreign countries, scientists from different disciplines, and scientists holding different points of view from within the same discipline.

The major problem in trying to set guidelines for research that *appears* dangerous and has *potential* benefits is that there are no precedents and there can be none for *predicting* the results of Type I and Type II experiments. Besides the debate based on biological grounds and on purpose and intent of the experiments, there were questions concerning the morality of doing such research.

Before the meeting adjourned, a consensus was reached that further experiments be classified into low, medium, and high risk categories according to the "risk" involved. Further, that each type experiment be considered for possible safeguards. For example, low and medium risk experiments could be carried out on mutant bacteria that could not live outside the test tube or strains of bacteria could be used that "self-destruct," that is, die on exposure to air. Similarly, bacteriophage and plant and animal viruses must be so constructed as to be inactive under normal ecological conditions. Equally important may be the construction of plasmids that cannot be transported by virus. It was also agreed that certain high risk experiments *were too hazardous* to be undertaken within existing university laboratories, for example, the construction of vaccina or variola (small pox) viral–lambda gene graft DNA molecules, that is, small pox DNA linked to bacterial virus DNA. We really have no way of knowing what would really happen short of doing the experiments. But it is obvious that it appears to be a dangerous experiment. Do we really need this type of investigation? Who knows? It may revolutionize medicine but it could mean the demise of a large portion of humanity or perhaps of plants and animals.

The moratorium on DNA recombinant molecular research lasted for seven months (July 1974 to February 1975). It was lifted with an agreement that the biohazard potential of any experiment should be matched by carefully planned methods and procedures for the biological and physical containment of the investigator's work. Sidney Brenner suggests that one of the biggest biohazards is our lack of knowledge (Berg, 1975). Scientists hope that personal commitment to safety and respect for human life will be sufficient motivation for scientists to control the type of research that will be done in the future. However, some scientists, not convinced that personal responsibility is a sufficient safeguard, have encouraged federal granting agencies to require that researchers provide a detailed explanation of their biological containment methods and facilities. Future meetings similar to the one held at Asilomar will be held annually to review the guidelines set up at the first meeting, and it has been recommended that a newsletter and central stock center be established for the investigators so that the scientific community can monitor itself. However, peer evaluation and self-monitoring are not easy. Currently, scientists involved in recombinant DNA research are divided into two groups: those who feel the guidelines written up at Asilomar are too lax and those who feel the guidelines are too restrictive. It is becoming apparent that if the scientists involved in this research continue to disagree on the use of guidelines, someone will have to explain the entire rationale of the guidelines to the United States Congress and the public. The result may be new suggestions on how to control the techniques for experiments with recombinant DNA on a national basis (Wade, 1976).

what we can anticipate from continued genetic research

The Asilomar conference *is* a milestone; scientists are seeking guidance and want to share the burden of responsibility. It is not easy for a scientist to set aside a thoroughly fascinating piece of research. It should also be pointed out that investigations covered in the moratorium on gene-grafting are necessary to the possibility of genetic engineering of humans. Frog, mouse, and fruit fly genes have already been transplanted into *E. coli* and expressed (*The Sciences*, 1974; Skinner, 1976).

In the discussion of diversifying natural selection in Chapter 3, it was concluded that the number of possible genotypes as a result of genetic recombination during sexual reproduction is astronomical. In fact, it has been calculated that the number of potential genotypes is greater than the total of the electrons and protons that make up everything in the universe (Dobzhansky, 1950). Thus we can see that of all the humans that ever lived or will live only a *very small percentage* of all possible genotypes that could occur have occurred or will occur.

If these calculations are correct, the number of possible genotypes is so large that it is impossible for every possible genotype to have been tested in nature for its "fitness." In fact, only *a very small* percentage of all possible genotypes can possibly be tested. So it is that of those genotypes that have been tested—yours and mine—have occurred by chance. Is it sensible to continue to rely on this chance phenomenon? The human species is surviving in the current environment, but are we the biologically best that we can be? We do not know because all possible genotypes will never be tested. So we can sit back and let someone else drive the bus or we can *try* to drive it ourselves. We can correct the defective genotype that nature has provided us with and provide ourselves with new and more abundant quality food, building materials, medicine, happiness, and, above all, perhaps peace of mind. These changes will not come easily. There is a price to be paid and that is a part of our fear. What will be the price? What is the form of payment?

But has nature been less demanding? Not really; nature has selected against millions of species. They are extinct and we may also become extinct if left to nature. We pay the price in watching those less fit being cared for because they lost in the roulette of gene combination according to the rules of natural selection. Are we certain we cannot do better than nature? Surely we can improve on our social behavior. *Can we?* We will have to be careful—domestic and wild animals bred in captivity quickly lose their former patterns of behavior. In several species of fish the behavioral change is so drastic that they fail to tend their young.

Herman Muller stated "If we fail to act now to eradicate genetic defects, the job of ministering to infirmities would come to consume all the energy that society could muster for it, leaving no surplus for general, cultural purposes." But there is a subtle flaw in this kind of thinking. First, there is a moral issue of whether *statistics* can be used to determine who should be born or in establishing the rights of society and second, Muller makes the assumption that only our genes are necessary for taking humankind into the future. It must also be recognized that our society, our cultural heritage, is also very important in establishing the future of humankind.

Aristotle said that "all men have a desire to know"; yet we do not know what nature (natural selection) has in store for us any more than we can predict the outcome of our *desire* to drive our own evolutionary bus. Nevertheless, we are going to *try* to drive that bus—it is inevitable! The problem most likely to remain after gene manipulation is a common occurrence and we are well in command of the bus is that humans will still not know the meaning and purpose of their existence. For that, it is said, we must turn to the scriptures. Dostoevsky's Grand Inquisitor noted that "the secret of man's being is not only to live but to have something to live for." What do *we* live for? Today we seek physical comfort and mental satisfaction, together with a moral philosophy that allows us to have both without reservations. We like to think of what man *ought to be*; yet we do not have a clear idea of what he is. The psychiatrist Viktor Frankl concludes, based on his internment in a World War II Nazi concentration camp, that

> Man is neither dominated by the will-to-pleasure nor by the will-to-power, but by what I should like to call man's *will-to-meaning*; that is to say, his deep-seated striving and struggling for a higher and ultimate meaning to his existence. Man is groping and longing for a meaning to be fulfilled by him and by him alone; in other words, for what we would call a mission.

If humankind has a "mission" it will survive. As Nietzsche said, "He who has a *why* to live can bear with almost any *how*." I know not what our mission is, but it is evident that our *unrest* lies in our capacity to envision *possibilities*, many of which were discussed in this text.

Dostoevsky's Grand Inquisitor referred to human hunger as a search for meaning, a desire so great that the search may drive men to sacrifice individual freedom in order to escape the frustration that the search may cause. If we could define man in terms of knowledge, values, and morals, we might have a better idea of what and who man is and *what* his search will find! Our *self*-concepts are based on self-understanding, ideas and assumptions of what we think we are. These assumptions form individual and social attitudes that manifest themselves in our actions, actions that can make possible a future of *promise* or *peril*. It is fitting that we conclude this text and the chapter on the future with St. Francis of Assisi's prayer: "God grant me the serenity to accept the things I cannot change, the courage to change the things I can, and the wisdom to know the difference." The *wisdom* to know the difference!

references

what we can anticipate from continued genetic research

Berg, Paul, et. al. 1975. Asilomar Conference on Recombinant DNA Molecules. *Science, 188*:991–94.

Cohen, Stanley N. 1975. The Manipulation of Genes. *Scientific American, 233*(1):25–33.

Dobzhansky, Theodosius G. 1950. The Genetic Basis of Evolution. In W. S. Laughlin and R. H. Osborne (eds.), *Human Variation and Origins.* Freeman, San Francisco.

Early, Kathleen. 1975. Disinheriting Disease. *The Sciences, 15*:19–23.

Fletcher, Joseph. 1974. *The Ethics of Genetic Control.* Anchor/Doubleday, New York.

Galston, Arthur W. 1974. Molding New Plants. *Bios, 83*:94–96.

Goran, Morris. 1974. *Science and Anti-Science.* Ann Arbor Science Publishers, Ann Arbor, Mich.

Gurdon, J. B. 1968. Transplanted Nuclei and Cell Differentiation. *Scientific American, 219*(6): 24–35.

Hadsell, Robert. 1971. Gene Defect Corrected. *Chemical Engineering News, 78*: 9–10.

Huxley, Aldous. 1932. *Brave New World.* Harper & Row, New York.

Kass, Leon R. 1971. The New Biology: What Price Relieving Man's Estate. *Science, 174*: 779–88.

Luria, Salvador A. 1972. Ethical Aspects of the New Perspectives in Biomedical Research. *The Challenge of Life,* Birkhauser Verlag, Stuttgart.

McBride, Gail. 1975. Gene Grafting Experiments Produce Both High Hopes and Grave Worries. *Journal of the American Medical Association, 232*: 337–41.

Merril, Carl R. 1974. Bacteriophage Interactions With Higher Organisms. *Transactions of the New York Academy of Sciences, 36*(3): 245–308.

Munyon, William, et al. 1971. Transfer of Thymidine Kinase to Thymidine Kinaseless L Cells by Infection with Ultraviolet-Irradiated Herpes Symplex Virus. *Journal of Virology, 7*: 812–20.

Osborn, Frederick. 1960. A Return to the Principles of Natural Selection. *Eugenics Quarterly, 7*: 204–11.

Qasba, P. R., and H. V. Aposhian. 1971. DNA and Gene Therapy: Transfer of Mouse DNA to Human and Mouse Embryonic Cells of Polyoma Pseudovirons. *Proceedings of the National Academy of Sciences USA, 68*: 2345–49.

The Sciences. 1974. Quanta. *14*(Sept.): 4.

Skinner, K. J. 1976. Expression of Yeast DNA in *E. coli* Achieved. *Chemical and Engineering News, 54*: 21–22.

Steward, F. C. 1970. From Cultured Cells to Whole Plants: The Introduction and Control of Their Growth and Morphogenesis. *Proceedings of the Royal Society (London), B175*: 1–30.

Time. 1976. The Making of a Gene. September 13, p. 60.

Wade, Nicholas. 1976. Recombinant DNA: Guidelines Debated at Public Hearing. *Science, 191*: 834–36.

glossary of selected terms

ABO blood group—major blood antigen system containing types A, B, AB, and O, depending on the presence or absence of the A and B antigens.

Anchondroplasia—one type of inherited dwarfism.

Acrocentric—centromere is very close to one end of the chromosome.

Adaptation—the development of a characteristic that results in improved chances of survival and reproduction.

Agglutination—clumping.

Albino—an individual that lacks melanin pigment.

Alkaptonuria—a heritable disease that results from a failure to break down alkapton (2,5-dihydroxyphenylacetic acid); the alkapton excreted in the urine turns black when exposed to the air.

Allele—one member of a gene pair that occupies a particular site on homologous chromosomes.

Amino acid—an organic compound composed of both carboxyl (COOH) and amine ($-NH_2$) groups. A series of amino acids constitute a protein.

Amniocentesis—the removal of amniotic fluid and fetal cells by means of a needle inserted into the amniotic sac.

Amniotic fluid—the liquid medium surrounding the fetus in the amniotic sac.

Anaphase—a stage of nuclear division where the movement of chromosomes from the equator begins with the division of the centromeres and ends with the poleward movement of chromosomes.

Androgen—the general term for any substance acting as a male hormone.

Anemia—blood that is low in number of red blood cells.

Anencephaly—an open cranium; absence of brain and spinal cord.

Aneuploidy—a variation in the number of chromosomes by whole numbers but not by an entire set (e.g. monosomy, $2n - 1$).

Ångstrom unit—1×10^{-8} centimeter.

Anthropoid—from the suborder Anthropoidea; includes monkeys, apes, and man.

Anthropologist—one who studies the evolution of man.

Antibody—a protein produced in a living organism that neutralizes a specific antigen.

Antigen—any substance that causes the production of an antibody when introduced into a living organism

Arterioles—the small branches of an artery that end in capillaries.

Asexual reproduction—the development of a new individual from a single cell or from a group of cells in the absence of a sexual process.

Ashkenazi—the eastern European, Yiddish-speaking division of the Hebrews.

Autoimmune—an organism that builds antibodies against its own tissues.

Autosome—a chromosome not associated with the sex of an individual, not a sex chromosome.

Autotroph—organism that manufactures organic nutrients from inorganic materials, self-feeder.

Bacterium—a unicellular prokaryotic organism in which the genetic material is dispersed through the cytoplasm (no nuclear membrane).

Barr body—sex chromatin of normal females that can be seen in the body cells (squamous epithelial cells). Females are sex-chromatin positive; normal males are sex-chromatin negative.

Biochemical genetics—the study of hereditary functions at the level of the molecules within a cell.

Biochemical pathway—conversion of one substrate to another through the use of enzymes.

Biochemistry—the study of chemical functions in living organisms.

Biogenesis—the axiom that life originates only from preexisting life.

Birth defect—congenital deformity or malfunction.

Blood group—blood type based on the type of antigen the individual produces.

Buccal mucosa—the lining of the mouth.

Capillaries—smallest blood vessels in the body, often the width of a single cell.

Carbohydrate—macromolecule consisting of carbon, hydrogen, and oxygen; the main component of our diets (sugars, starches).

Carcinogen—an agent that induces cancer.

Carrier—an apparently normal heterozygote who "carries" a recessive gene paired with a dominant gene.

Catalyst—a substance that speeds up a chemical reaction but does not take part in it.

Centromere—a small spherical zone of each chromosome with which the spindle fibers become associated during mitosis and meiosis.

Characteristic—an individual component of a phenotype (e.g., eye color).

Chiasma—crossover configuration between two chromatids that is seen during prophase I of meiosis.

Chloroplast—a cytoplasmic body in plants that contains several pigments, particularly chlorophyll.

Chromatid—one of the two visibly distinct longitudinal halves of a chromosome; shares a common centromere with a sister chromatid.

Chromatography—a means of separating closely related compounds by allowing a solution or mixture of them to seep through a differentially absorbent material.

Chromosomes—microscopic rod-shaped bodies that carry the genes.

Classical genetics—Mendelian genetics; a study of genetics on an organismal rather than cellular level.

Clone—a population of cells that is derived from a single cell by mitosis.

Coagulate—clot or clump together, especially of blood cells.

Codominants—members of a gene pair each of which produces its own phenotypic effect in heterozygotes.

Codon—a unit of three nucleotides (bases) in RNA that is specific for one amino acid.

Complementary base pairing—the process by which specific purine bases are matched with specific pyrimidine bases.

Conception—the penetration of the egg by a sperm cell and the formation of a zygote.

Conjugation—the uniting of two bodies (e.g., synapsed chromosomes in meiosis) or two organisms during sexual reproduction.

Consanguinity—marriage of two individuals related by descent from a relatively close common ancestor.

Cooley's anemia—thalassemia major.

Copulation—intercourse; the insertion of the penis into the vagina.

Cross—a mating or bringing together of genetic material from different genotypes to achieve recombination.

Crossing-over—exchange of segments of homologous chromosomes.

Cystic fibrosis—an inherited childhood disease involving the mucus-secreting glands.

Cytoplasm—the protoplasm of a cell other than the nucleus.

Dalton—a unit equivalent to 1/12 the mass of a carbon-12 atom (1.661×10^{-24} gram).

Diakinesis—stage of meiosis just before metaphase I (last prophase stage).

Differentiation—change in the structure and/or function of embryonic cells toward a more specialized state.

Dihybrid—an individual that is heterozygous in two pairs of genes.

Diploid—an individual or cell with two complete sets of chromosomes.

Diplotene—stage of meiosis in which chromosomes all visibly double prior to diakinesis.

Dizygotic twins—twins that are the products of two ova.

DNA (deoxyribonucleic acid)—a double-stranded, helically coiled nucleic acid molecule that is the genetic material of living organisms and many viruses.

Dominant—the member of a gene pair that expresses itself in heterozygotes.

Dosage compensation—any mechanism by which the effective dosages of sex-linked genes in organisms with an XX–XY or XX–XO mechanism of sex determination are made equal.

Down's syndrome—mongolism, a disease characterized by mental retardation, oriental eyes, short stature, abnormal palm prints, and malformations of the heart, ears, hands, and feet; generally caused by presence of an extra number 21 chromosome.

Duplication—a structural change of a chromosome that results in the doubling or repeating of a section of the genome.

Elan vital—the vital force or impulse of life, the spark of life.

Electron microscope—a machine that focuses a beam of electrons onto the object being examined and then onto a fluorescent screen for immediate viewing or onto a photographic plate; gives enormous magnification of object being viewed.

Electrophoresis—process that allows the migrations of proteins and enzymes in an electrical field.

Enzyme—a protein that regulates the rate of a specific biochemical reaction without being used up in the reaction.

Epistasis—the domination of one gene by a second, nonallelic, gene.

Erythroblastosis fetalis—destruction of fetal red blood cells due to *Rh* incompatibility with the mother; causes severe anemia.

Erythrocyte—red blood cell.

Escherichia coli—a species of human intestinal bacteria often used in biological research.

Estrogen—female hormone regulating secondary sexual characteristics.

Ethnic group—a population group united by manners and customs and an association (present or past) with a particular geographic region.

Eugenics—means and methods of social control to improve the hereditary qualities of future generations.

Evolution—the transformation of an organism in a way that descendents differ from their predecessors.

Expressivity—degree of expression of a genotype in the phenotype.

Fermentation—the breakdown and rearrangement of molecules to supply energy in the absence of oxygen.

Fetus—stage of human development from the ninth week until birth.

"Fixity of species"—the axiom that all species, once determined, remain constant (exact opposite of the theory of organic evolution).

Fossil—remains of an organism hardened and preserved in rocks.

Fundamentalist—one who believes in the literal interpretation of the Bible.

Galactosemia—autosomal recessive disease characterized by the accumulation of galactose in the blood.

β-Galactosidase—the enzyme that catalyzes hydrolysis of lactose to galactose and glucose.

Gamete—a mature reproductive cell that is capable of forming a zygote by fusing with a cell of similar origin but of opposite sex.

Gene—a unit of the genetic material (DNA) localized in the chromosome that determines a hereditary trait.

Gene frequency—the ratio of one type of allele to all alleles at that genetic locus in a breeding population.

Gene interaction—an interaction between alleles or nonallelic genes of the same genotype in the production of a particular phenotype.

Gene pool—the total of all genes in an interbreeding population at a given time.

Genetic mapping—the process of determining the location of genes and distances between genes on a chromosome.

Genetics—the science of heredity.

Genetic code—base triplets in DNA that specify the amino acids to go into the polypeptide chains.

Genetic collapse—when the concentration of lethal genes in a population becomes great enough to cause the population to die out.

Genetic disease—any malfunctioning of the body directly caused by malfunctioning genes.

Genetic load—average number of lethal genes in a population.

Genetic recombination—the ability of the genes to segregate and assort themselves into all possible pairs during meiosis.

Genetic screening—the testing of individuals for a particular genetic trait.

Genocide—the destruction of individuals having an undesirable genotype.

Genotype—the sum total of the genetic information of an individual.

Germ cells—generative cells in multicellular organisms; distinguished from the somatic cells.

Gonads—sexual glands; ovaries of the female, testes of the male.

Gonadotrophins—hormones produced by the pituitary that act only on the gonads.

Gynandromorph—an individual composed of both male and female parts.

Haploid—an individual or cell with a single complete set of chromosomes.

Hardy–Weinberg formula—a mathematical statement of the relationship between gene frequencies and genotype frequencies within populations that describes genetic equilibrium ($p^2 + 2pq + q^2$).

Hemizygous—possessing only one X chromosome; male.

Hemoglobin—iron-containing protein pigment in red blood cells that transports oxygen.

Hemolytic—causing the red blood cells to lyze or burst.

Hemophilia—an X-linked recessive disease characterized by the inability of the blood to clot.

Heredity—the passing of traits through successive generations.

Hermaphrodite—an individual having both male and female sex organs.

Heterotroph—an organism that must take in organic nutrients to survive.

Heterozygous—having alternate members of gene pairs of one or more genes in homologous chromosome segments.

Histones—a heterogeneous group of proteins rich in basic amino acids and variable in their amino acid compositions and sequences; they are found complexed with chromosomal DNA in eukaryotic cells.

Holandric—Y-linked; females never have the trait and males pass it to all of their sons.

Homologous chromosomes—a "matched pair" of chromosomes morphologically alike and bearing the same gene loci.

Homozygous—having identical alleles of one or more genes in homologous chromosome segments.

Hybridization—the crossing of two genetically different individuals to produce a hybrid offspring.

Hydrocephalus—"water head"; an excess of fluid in the head at birth.

Hydrogenated—with hydrogen added.

Hypoplasia—arrested development of an organ (smaller than normal); immaturity of an organ.

Hypothalamus—a center of the autonomic nervous system located near the bottom of the brain.

Immune response—production of an antibody to an antigen; phagocytic action of cells.

Immunogenetics—the study of gene-controlled synthesis of antibodies and their reaction with antigens.

Immunoglobulins—protein globulins in the blood that provide immune responses to diseases.

Inbreeding—the mating of closely related individuals.
Inorganic—matter other than that of living things (water, minerals, etc.).
Interphase—the stage of the cell cycle in which the cell is not actively dividing.
Intersex—an individual, often sterile, who shows secondary sexual characteristics of both sexes.
In vitro—in the test tube.
Karyotype—the size, shape, and general appearance of chromosomes in mitotic metaphase.
Lethal (semilethal) gene—a gene that, when expressed before (after) age 1 year, results in the eventual death of the organism.
Linkage—the association in inheritance of certain genes because they are located on the same chromosome.
Lipid—a fat-like substance.
Locus—the position of a gene on a chromosome.
Lyon hypothesis—that in any given cell of a female one X chromosome is active and the other is inactive.
Lysis—the disintegration of a cell by the destruction of the cell membrane (cell wall).
Malignant—threatening to produce death.
Mammal—member of the class Mammalia; mammals nurse their young.
Meiosis—nuclear reduction divisions in which the diploid chromosome number is reduced by half.
Melanin—a dark brown to yellow pigment found in cells of the skin, hair, and iris of the eye; also in many plant species.
Menstruation—bleeding caused by the sloughing off of the uterine lining.
Metabolism—the sum total of all chemical functions in the body.
Messenger RNA (mRNA)—ribonucleic acid that is transcribed along a given molecule of DNA and then serves as a template for protein synthesis.
Metabolic block—a block in a metabolic pathway that prevents the completion of a product.
Metabolite—a product of metabolism.
Metacentric—centromere is in the middle of a chromosome.
Metaphase—stage of nuclear division in which the chromosomes are located at the equatorial plane of the cell prior to centromere division.
Mitosis—nuclear division that produces identical daughter cells.
Molecule—a structure containing two or more atoms.
Mongolism—*see* Down's syndrome.
Monohybrid—an individual that is heterozygous for one gene pair.
Monosomic—an individual that lacks one chromosome of a set.
Morphology—the science of structure.
Mosaic—composed of cells of two or more types.
Multiple alleles—a series of alleles, any one of which may occupy the same locus in a chromosome.
Mutagen—any agent that causes a mutation.
Mutation—a change in genotype that has no relation to the individual's ancestry.
Natural selection—environmental selection of individuals best adapted to live and reproduce.
Neural—having to do with the nervous system.

Nondisjunction—the irregular distribution of homologous chromosomes or sister chromatids to the poles of the cells.

Nucleic acid—a chain of nitrogenous bases individually linked to sugar and phosphate.

Nucleotide—a molecule consisting of phosphate, a five-carbon sugar (either ribose or deoxyribose), and a purine or a pyrimidine.

Nucleus—an internal cell structure containing the chromosomes and nucleoli.

Obesity—being overweight.

Oogenesis—female egg production through meiosis.

Operator site—segment of DNA in the sperm; controls structural genes.

Operon—a system of genes (cistrons), operator, and promoter sites that regulates a given genetically controlled metabolic activity.

Organic—chemical molecules generally produced by living organisms; contain carbon.

Ovum (pl. **ova**)—an egg.

Pathogenic—disease causing.

Pedigree—a record of an individual's ancestral history showing inheritance patterns for a given trait or traits.

Penetrance—whether or not an individual of a given genotype expresses the corresponding phenotype (see Expressivity).

Phagocytized—antigen or foreign particle engulfed and destroyed by leukocytes.

Phenocopy—the expression of a phenotype due to environmental factors rather than genetic ones.

Phenotype—the characteristics (appearance) of an organism produced by the interaction between the organism's genotype and the environment.

Photoreactivation—a reversal of the mutagenic effects of ultraviolet light by exposure of irradiated cells to white light.

Photosynthesis—the process by which green plants convert carbon dioxide and water into sugar using chlorophyll.

Pituitary gland—a major gland in the body that produces several hormones, including the gonadotrophins.

Placenta—a wall of tissue formed during pregnancy that acts as a screen between the fetus and the uterus.

Plasma—fluid part of blood as opposed to suspended material.

Pleiotropy—the influence of a single gene on more than one trait (multiple expression).

Point mutation—a single base change in DNA.

Polygenic—involving more than one gene.

Polymer—a long molecular structure of repeating identical units.

Polydactyly—multiple fingers or toes.

Polymerase—an enzyme that catalyzes polymer formation.

Polymorphic—displaying several alternative forms.

Polypeptide—several amino acid residues joined by peptide bonds.

Polyploid—an individual or cell having more than two sets of chromosomes.

Population genetics—the story of genetic principles of heredity at the population level.

Prebiotic—before living things (cells).

Primary constriction—a constriction in the chromosome that is determined by and associated with the centromere region.
Primordial—primitive, fundamental, first formed.
Probability—the estimated chance of an event occurring.
Progeny—the results of a particular mating; offspring.
Progesterone—a female sex hormone regulating egg production.
Protein—made up of a series of amino acids, a polypeptide of a given size.
Prototroph—a wild-type organism that is able to grow on minimal medium.
Pseudohermaphrodite—an individual who is wrongly classified as a bisexual on the basis of physical abnormalities.
Puberty—the period in life when the secondary sexual characteristics develop and hormone balances change; reproductive potential becomes realizable.
Pure-breeding line—a strain of individuals that are homozygous for all genes being considered.
Race—family, tribe, or nation belonging to the same stock.
Radiation—process of emission or diffusion of radiant energy from a given source.
Random mating—an individual of one sex has the equal probability of mating with any individual of the opposite sex.
Recessive—the allele that is not expressed in the presence of its dominant allele (the individual is a heterozygote).
Recombination—the new association of genes in the progeny.
Recurrence risk—the probability of any trait showing up again.
Regulator gene—gene responsible for producing repressor protein.
Replication—reproduction by copying from a template; synthesis of new DNA from pre-existing DNA.
Repressor—a protein produced by a regulator gene that inhibits the action or expression of an operator gene.
Reverse mutation—an inheritable change in a mutant gene capable of restoring the original nucleotide sequence.
Rh positive (+)—a person who manufactures antigen D.
Ribose—the sugar found in ribonucleic acid (RNA).
Ribosomes—small organelles found in the cell cytoplasm that are composed of RNA and proteins and function in translating mRNA into protein.
RNA (ribonucleic acid)—single-stranded nucleic acid found mainly in the nucleolus and ribosomes; contains uracil (where DNA contains thymine) and ribose sugar.
- **mRNA** (messenger RNA)—carries the genetic message from the DNA in the nucleus to the ribosomes in the cytoplasm for the synthesis of protein.
- **rRNA** (ribosomal RNA)—in asociation with protein, forms the ribosomes.
- **tRNA** (transfer RNA)—transports specific amino acids to ribosomes for assembly into protein.

Segregation—the separation, usually during meiosis, of a gene pair and the distribution of the alleles to different cells.
Serum—portion of the blood that carries specific immunoglobulins.
Sex chromosomes—accessary chromosomes associated with the organism's sex.
Sex-influenced—an autosomal trait that is found in both sexes but expressed more often in one sex than the other.

Sex-limited—an autosomal trait that is expressed in one sex or the other, but not both.

Sex-linkage—linkage of genes that are located on the sex chromosomes of eukaryotic cells; Inheritance pattern differs from genes located on the autosomes.

Sephardim—one of the three major Hebrew groups.

Somatic cell—a body cell that normally has two sets of chromosomes, one set from each parent.

Speciation—the genetic divergence of a population into two or more separate groups that can no longer interbreed.

Spermatocyte—any of the sperm mother cells that undergo meiosis to produce four spermatids.

Spontaneous generation—the theory that life arises spontaneously from nonliving material.

Strata—layers of the earth's crust.

Structural gene—gene that transcribes a message for the synthesis of a protein.

Submetacentric—centromere is nearer to one end of the chromosome than the other (just off center).

Synapsis—the pairing of homologous chromosomes.

Syndrome—a group of symptoms that occur together and characterize a particular disease.

Testicular feminization—a sex-limited abnormality in which an XY male has a female appearance because the testes secrete estrogen rather than androgen.

Testosterone—a primary male sex hormone produced by the testes.

Thalassemia—an inherited anemia that shows codominance; an individual heterozygous for the trait has thalassemia minor, while the homozygote exhibits the severe form, thalassemia major.

Thalidomide—a tranquilizer taken by pregnant mothers during the 1960's that was found to induce phenocopies of phocomelia.

Transcription—a process in which the genetic code contained in DNA is used to order complementary sequences of nucleotides into mRNA molecules.

Transduction—a virus-mediated transfer of genetic information.

Transformation—absorption of naked extracellular DNA into a cell.

Translocation—shift of a chromosome segment to another area of the same chromosome or to a different chromosome.

Triple X—trisomy for the X chromosome.

Trisomic—a cell or individual with one extra chromosome in a set ($2n + 1$).

Tryptophan—an amino acid.

μm (micrometer)—1×10^{-6} meter.

Uracil—a nitrogenous base (pyrimidine) that is found in RNA, not in DNA.

Vaccine—preparation of killed pathogens introduced into an organism to raise the antibody level.

Vasectomy—the severing of the vas deferens to prevent sperm from reaching the urethra.

Vestigial—degenerate, not fully developed.

Virus—a noncellular parasitic organism on the submicroscopic level; composed of a nucleic acid and protein shell.

Wild type—the predominant phenotype encountered in a natural breeding population.
X chromosome—one of the sex chromosomes; found in both sexes.
X-linked—located on the X chromosome.
X0 syndrome—Turner's syndrome, 45 chromosomes, chromatin negative.
XXY syndrome—Klinefelter's syndrome, 47 chromosomes, chromatin positive.
Y chromosome—a sex chromosome found only in males.
Zygote—the cell that results from the fusion of two gametes.

Italicized page numbers indicate illustrations.

index